NANOTECHNOLOGY SCIENCE AND TECHNOLOGY

ADVANCED MATERIALS

STUDIES AND APPLICATIONS

Nanotechnology Science and Technology

Additional books in this series can be found on Nova's website under the Series tab.

Additional e-books in this series can be found on Nova's website under the e-book tab.

NANOTECHNOLOGY SCIENCE AND TECHNOLOGY

ADVANCED MATERIALS

STUDIES AND APPLICATIONS

IVAN A. PARINOV
SHUN-HSYUNG CHANG
AND
SOMNUK THEERAKULPISUT
EDITORS

nova
publishers
New York

Additional color graphics may be available in the e-book version of this book.

Library of Congress Cataloging-in-Publication Data

Advanced materials (Nova Science Publishers)
Advanced materials: studies and applications / editors, Ivan A. Parinov, Shun-Hsyung Chang, and Somnuk Theerakulpisut (Vorovich Mechanics and Applied Mathematics Research Institute, Southern Federal University, Russia, and others).
pages cm. -- (Nanotechnology science and technology)
ISBN 978-1-63463-749-7 (hardcover)
1. Materials. I. Parinov, Ivan A. II. Chang, Shun-Hsyung, III. Theerakulpisut, Somnuk. IV. Title.
TA403.A258 2014
620.1'1--dc23

2014047200

Published by Nova Science Publishers, Inc. † New York

CONTENTS

PREFACE

The advanced materials and their applications based on nanotechnology and piezoelectric approaches have found a tremendous interest in the modern science and techniques. The book presents processing techniques, physics, mechanics, and applications of novel materials. The book concentrates on some nanostructures, ferro- and magnetoelectric crystals, materials and composites, materials for solar cells, polymeric composites etc. There are present nanotechnology approaches, modern piezoelectric techniques, and also studies of the structure-sensitive properties of the materials. Great attention is devoted to novel devices with high accuracy, longevity and extended possibilities to work into wide temperature and pressure ranges, etc., which show characteristics defined by used materials and composites with improved properties opening new possibilities in study of various physical processes, in particular transmission and receipt of signals under water.

This collection of 30 papers presents selected reports of the 2014 International Symposium on "Physics and Mechanics of New Materials and Underwater Applications" (PHENMA-2014), which has been taken place in Khon-Kaen, Thailand, 27-29 March, 2014 (http://phenma2014.math.sfedu.ru), devoted to the 50th anniversary of Faculty of Engineering, Khon Kaen University and sponsored by the Russian Foundation for Basic Research, National Science Council of Taiwan, South Scientific Center of Russian Academy of Sciences, New Century Education Foundation (Taiwan), Ocean & Underwater Technology Association, Unity Opto Technology Co., EPOCH Energy Technology Corp., Fair Well Fishery Co., Formosa Plastics Co., Woen Jinn Harbor Engineering Co., Lorom Group, Longwell Co., Taiwan International Ports Co., Ltd. and South Russian Regional Centre for Preparation and Implementation of International Projects.

The thematic of the PHENMA-2014 continues ideas of previous symposia PMNM-2012 (http://pmnm.math.rsu.ru) and PHENMA-2013 (http://phenma.math.sfedu.ru), whose results have been published in the edited books "Physics and Mechanics of New Materials and Their Applications", Ivan A. Parinov, Shun Hsyung-Chang (Eds.), Nova Science Publishers, New York, 2013, 444 p. ISBN: 978-1-62618-535-7 and "Advanced Materials - Physics, Mechanics and Applications". Springer Proceedings in Physics. V. 152. Shun-Hsyung Chang, Ivan Parinov, Vitaly Topolov (Eds.), Springer Heidelberg, New York, Dordrecht, London, 2014, 380 p. ISBN: 978-3319037486, respectively.

The presented papers are divided into four scientific directions: (i) processing techniques of advanced materials, (ii) physics of advanced materials, (iii) mechanics of advanced materials, and (iv) applications of advanced materials.

Into framework of the first theme are considered, in particular the search and development of smart materials, investigation of $NaTaO_3$ photocatalyst materials, studies of the doped ZnO thin films grown by dual-plasma-enhanced metal-organic chemical vapor deposition and pulsed laser deposition, characterization of ZnO nanocrystals, researches of sapphire laser processing and optimization of cure processing of polymeric composite.

The second direction covers structure and properties of spinel-perovskite lead-free magnetoelectric composites, finite-element modeling piezoelectric and magnetoelectric composites, porous piezoceramics and ceramic matrix composites, piezoelectric elements with inhomogeneous polarization, acoustic emission diagnostics of the kinetics of physicochemical processes in liquid and solid media, MDSM structures and thermoelectric capacitor with the semiconductor plate.

From viewpoint of mechanics in the third section are studied the problem on flat punch sliding on boundary of elastic half-plane, the coupled boundary-value problems of anisotropic elasticity, voscoelasticity, elastodynamics, poroelastodynamics and numerical methods (BIE and BEM) used for their study. Moreover, there are discussed the semi-inverted method in the problem of plastic deformation of cylindrical shell, and stress-strain state of layered anisotropic elliptic construction under action of impulse load.

The fourth direction demonstrates novel results in modeling and experimental studies of ultrasound transducers for medical applications, energy efficiency of piezoelectric generators, optic devices for measurement of displacements, based on method of control object highlighting by using laser interferometer, LEDs for marine applications, transmission signals under water, protection of equipment from excessive pressure and study of rigidity of the Novikov gearing teeth.

The book is addressed to students, post-graduate students, scientists and engineers taking part in R&D of nano-materials, ferro-piezoelectrics and other advanced materials and composites, presented in the book, and also participating in manufacture of different devices based on them having broad applications in different areas of modern science and technique, in particular in underwater communication.

October, 2014

Part I. Processing Techniques of Advanced Materials

In: Advanced Materials - Studies and Applications
Editors: I.A. Parinov, S.-H. Chang, S.Theerakulpisut
ISBN: 978-1-63463-749-7

Chapter 1

SEARCH AND DEVELOPMENT OF SMART MATERIALS BASED ON SOLID SOLUTIONS OF SYSTEMS WITH DIFFERENT CHEMICAL COMPOUNDS

L. A. Reznichenko*, L. A. Verbenko, O. N. Razumovskaya, L. A. Shilkina, S. I. Dudkina, A. G. Abubakarov, A. V. Pavlenko, I. M. Aliev, S. Kh. Alikhadzhiyev, G. N. Tolmachev and H. A. Sadykov

Research Institute of Physics, Rostov State University, Rostov-on-Don, Russia

ABSTRACT

The structural and physical properties of $Pb(Ti, Zr)O_3 - \sum^4_{n=1}(PbB'_{1-\alpha}B''_\alpha O_3)_n$ solid solutions markedly different in electric hardness were correlated with the average electronegativity of the metal atoms on the *B* site (EN_B). The electric hardness of the solid solutions was shown to be proportional to EN_B (*B*-O bond covalence). The observed correlations can be used to predict properties of new piezoelectric ceramics.

Dielectric permittivity measurements in $Pb(Ti, Zr)O_3 - \Sigma^4_{n=1}(PbB'_{1-\alpha}B''_\alpha O_3)_n$ solid-solution systems demonstrate that the permittivity $\varepsilon^T_{33}/\varepsilon_0$ of the solid solutions correlates with the molar ratio from the more soft-electric to the less soft-electric component of $PbB'_{1-\alpha}B''_\alpha O_3(\gamma)$. With increasing γ, $\varepsilon^T_{33}/\varepsilon_0$ rises because of the decrease in the uniform cell distortion parameter. The permittivity is also sensitive to structural disordering at $\gamma > 1$ and reduction in the density of the materials at $\gamma >> 1$.

The structural and dielectric ($\varepsilon^T_{33}/\varepsilon_0$) properties of $Pb(Ti, ZrO_3) - PbB'_{1-\alpha}B''_\alpha O_3 - PbB'''_{1-\beta}B''''_\beta O_3$ solid solutions were studied along the morphotropic region. Depending on the nature of the substituent cations, the properties of the solid solutions either vary monotonically with the content of the soft-electric component or exhibit extrema at a certain average electronegativity of the B-site cations.

Conditions for fabricating $(Ba_{0.5}Sr_{0.5})Nb_2O_6$ ceramics are optimized. It is shown that materials synthesized at $T = 1250$ °C and sintered at $T = 1375 - 1400$ °C exhibit the best ceramic characteristics. The $(Ba_{0.5}Sr_{0.5})Nb_2O_6$ solid solution is found to be a ferroelectric

---

* Email: Ilich001@yandex.ru

with a diffuse phase transition. The possibility of fabricating $(Ba_{0.5}Sr_{0.5})Nb_2O_6$ thin films from ceramic targets prepared in the resulting optimum regimes is shown.

Nonstoichiometric bismuth ferrite Bi_xFeO_3 ($0.88 \leq x \leq 1.04$) fabricated via solid-state synthesis is investigated by means of X-ray diffraction and electrical methods. The periodicity of the formation of impurity phases and variations in the structural parameters and densities of ceramic samples are established. The correlation between the number of impurity phases and characteristics of the dielectric spectra in the cation/anion-deficient ($0.88 \leq x < 1.00$) and cation/anion-excess ($1.00 < x \leq 1.04$) nonstoichiometry regions of Bi_xFeO_3 is confirmed. The observed effects are explained for the actual (defect) structure of bismuth ferrite.

1. Effects of B-Cation Electronegativity and B-O Bond Covalence on the Properties of Multicomponent Solid Solutions Based on Lead Zirconate Titanate

Earlier [1,2], we investigated the effect of the electronegativity of *B*-cations, EN_B, on the structural and electric properties of multicomponent solid solutions based on lead zirconate titanate (PZT),

$$Pb(Ti, Zr)O_3 - \sum^{4}_{n=1}(PbB'_{1-\alpha}B''_{\alpha}O_3)_n, \quad (1)$$

containing roughly equal amounts of non-PZT components and close in Ti/Zr ratio. The results of those studies demonstrate that the electric hardness, characterizing the stability of the domain structure to external influences, is proportional to $EN_{B''}$- (at fixed B') and EN_B (B = B', B", Ti, Zr) [1] and varies non-monotonically (passes through a maximum) with $EN_{B'B''}$.

As a measure of electric hardness [1,2], we took the uniform distortion parameter δ_t [3], since it determines the strain resulting from domain reorientations. With increasing $EN_{B''}$- and EN_B, δ_t rises (linearly in the case of EN_B), as do the mechanical quality factor (Q) and sound velocity (v_R), whereas tan δ_t, the relative permittivity of poled samples ($\varepsilon^T_{33}/\varepsilon_0$), the electromechanical coupling coefficient K_p, and the piezoelectric charge coefficient decrease, which also points to an increase in the electric hardness of the solid solutions.

The B" cations could be classified according to their electric hardness: Mn was found to be the most hard-electric, Co exhibited medium electric hardness; and Mg, Zn, and Ni were the most soft-electric. As shown recently [4], Ni is the most soft-electric cation in this series. The effect of B' and B" cations on the electric hardness of solid solutions illustrates in Table 1 [2].

Table 1. Effect of B' and B" cations on the electric hardness of solid solutions

Electric Hardness	B"	B'
Soft-electric Medium	≠ Mn, Co	Any
	Mn, Co	≠ W (except for $W_{1/2}Mn_{1/2}$)
Hard-electric	≠ Mn	Any
	Mn	≠ W (except for $W_{1/2}Mn_{1/2}$)

Experimental studies of solid solutions in more than 150 quaternary, quinary, and senary systems over wide ranges of non-PZT contents (up to 30 mol %, without predominance of the soft or hard-electric component) and Ti/Zr ratios (1.1 – 1.4) allowed the earlier derived relations to be extended to a broad spectrum of solid solutions markedly different in properties, without compositional limitations.

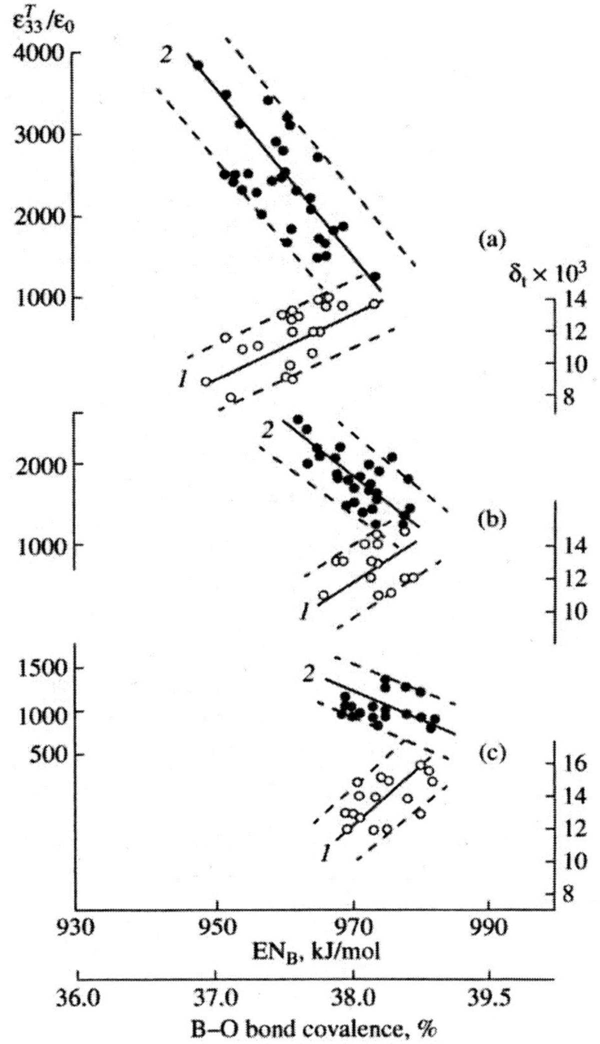

Figure 1. Plots of (*1*) δ_t and (*2*) $\varepsilon^T_{33}/\varepsilon_0$ *vs* EN_B and *B*-O bond covalence for (a) soft-electric, (b) medium hardness, and (c) hard-electric solid solutions (1).

Figure 1 displays the plots of δ_t and $\varepsilon^T_{33}/\varepsilon_0$ *vs* EN_B (solid lines) for solid solutions differing in electric hardness. The dashed lines in Figure 1 show the spread in parameters. The plots are linear and demonstrate that the electric hardness of the multicomponent solid solutions increases with increasing EN_B. A few features of these data are worth noting:

1) With increasing electric hardness, the plots shift to higher EN_B values, which provide additional evidence that the electric hardness increases with EN_B.

2) The slope of the plot of $\varepsilon^T_{33}/\varepsilon_0$ *vs* EN_B decreases and the slope of the plot of δ_t *vs* EN_B increases with increasing electric hardness, which reflects the decrease in the maximum values of $\varepsilon^T_{33}/\varepsilon_0$ and the increase in the maximum values of δ_t. Moreover, the observed variation in the slope of the plot of $\varepsilon^T_{33}/\varepsilon_0$ *vs* EN_B indicates that the sensitivity of the solid solutions to EN_B depends on their electric hardness: with increasing EN_B, the $\varepsilon^T_{33}/\varepsilon_0$ of the soft-electric solid solutions decreases more rapidly than that of the hard-electric solid solutions.
3) The spread in $\varepsilon^T_{33}/\varepsilon_0$ and δ_t decreases slightly with decreasing electric hardness (see Table 2). This is probably due to the wide diversity of $PbB'_{1-\alpha}B''_{\alpha}O_3$ constituent oxides in the soft-electric solid solutions, including both compositionally ordered antiferroelectrics (tungstates) and disordered ferroelectrics (niobates), drastically different in properties (e.g., t_c ranging from 130 °C in $PbNb_{2/3}Ni_{1/3}O_3$ to 400 °C in $PbW_{1/2}Cd_{1/2}O_3$).

Table 2. Spread in $\varepsilon^T_{33}/\varepsilon_0$ and δ_t

Electric Hardness	$\pm\Delta\ \varepsilon^T_{33}/\varepsilon_0$, %	$\pm\Delta\ \delta_t$, %
Hard-electric	29.0	16.5
Medium	29.5	17.0
Soft-electric	32.5	18.0

In hard-electric hosts, always containing manganese, the effect of some constituent oxides is reduced by the presence of Mn^{3+}, a Jahn-Teller active cation. Its contribution seems to play a key role in determining the ionic displacements and resulting dipole moment, which increases the electric hardness of such systems.

Figure 2 shows the plots of K_p and d_{31} *vs* EN_B and *B*-O bond covalence for solid solutions differing in electric hardness. These plots have a number of features in common with the plots in Figure 1: (i) they are also linear and shift to higher values of EN_B with increasing electric hardness; (ii) the largest spread in K_p and d_{31} is observed for soft-electric solid solutions; (iii) the slope of the plot d_{31} *vs* EN_B decreases and the slope of the plot of K_p *vs* EN_B increases with increasing electric hardness. At the same time, $\varepsilon^T_{33}/\varepsilon_0$ and d_{31} vary similarly, whereas δ_t and K_p vary in opposite ways. This can be accounted for using the relations [1,2,5]:

$$d_{31} \sim \varepsilon^T_{33} P_r;\ K_p \sim (\varepsilon^T_{33})^{1/2} P_r, \quad (2)$$

where P_r is the residual reorientational polarization in the piezoelectric ceramic.

These relations follow from thermodynamic relationships [6] and experimental data [7]. The similarity in the variations of $\varepsilon^T_{33}/\varepsilon_0$ and d_{31} is due to the strong dependence of d_{31} on $\varepsilon^T_{33}/\varepsilon_0$, whereas K_p is a stronger function of P_r. Since the P_r of tetragonal solid solutions decreases with increasing 5 [1,2,5] (on account of the associated decrease in the degree of domain reorientation [1]), so does K_p, as demonstrated in Figures 1 and 2.

In contrast to the data in Figures 1 and 2, the tan δ_t *vs* EN_B data for solid solutions differing in electric hardness can be represented by a single plot (Figure 3); the same refers to the v_R *vs* EN_B data. Both tan δ and v_R vary linearly with EN_B. With increasing EN_B, tan δ decreases and v_R increases, indicating an increase in electric hardness. In both plots, the data points corresponding to hard-electric solid solutions lie at higher EN_B, which provides additional evidence that the electric hardness rises with EN_B.

The Q vs. EN_B data can be represented by three distinct lines corresponding to different electric hardnesses. With increasing EN_B, Q increases only gradually for the soft-electric solid solutions and very rapidly for the hard-electric ones.

It is of interest to turn to bond covalence, which determines the bond strength, structure type, and chemical composition of materials and, in conjunction with other factors, plays a crucial role in determining their physical and physicochemical properties [8].

As is well known [9], the character of chemical bonding can be judged from the difference in electronegativity, ΔEN, between the elements involved in bonding: with decreasing ΔEN, the bond covalence increases. The plot of the A(*B*)-O bond covalence against ΔEN (data from [9]) is shown in Figure 4 (curve 1). It can be seen that the B"-O bond covalence can be increased by raising $EN_{B''}$, since the electronegativity of oxygen (2219 kJ/mol) is always higher than that of B" in the appropriate valence state (Figure 4, curve 2). In other words, $EN_{B''}$ can be taken as a measure of the B"-O bond covalence.

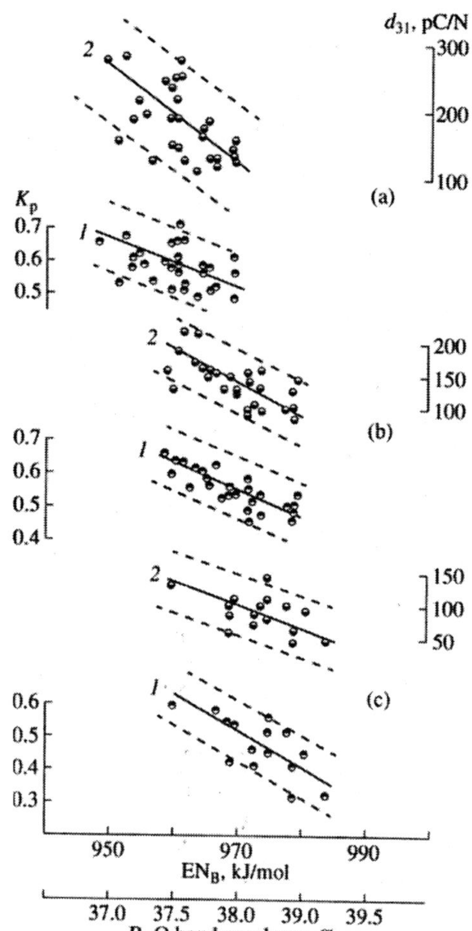

Figure 2. Plots of (1) K_p and (2) d_{31} *vs B* – O bond covalence for (a) soft-electric, (b) medium hardness, and (c) hard-electric solid solutions (1).

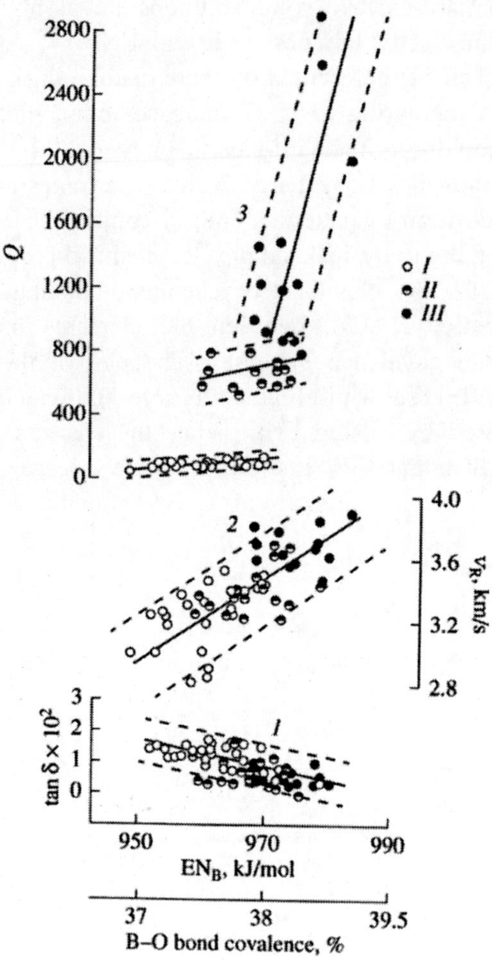

Figure 3. Plots of (1) tan δ, (2) v_R, and (3) Q *vs* EN_B and B – O bond covalence for (I) soft-electric, (II) medium – hardness, and (III) hard – electric solid solutions (1).

Consequently, the above correlations between the properties of solid solutions and EN_B hold as well in the case of the B – O bond covalence. Clearly, the use of EN_B for assessing the bond covalence in a given anion environment has the advantage that there is no need to calculate ΔEN.

The data in Figures 1 – 3 demonstrate that the bonding configuration on the B-site plays a key role in determining the piezoelectric and dielectric properties of the solid solutions studied: a slight increase in the B – O bond covalence, from 36.8 to 39.2%, leads to a change from soft-electric to hard-electric materials.

The correlations found in this work make it possible to predict the physical properties of multicomponent PZT-based solid solutions (ferroelectric ceramics) if their compositions are known and will be helpful in choosing solid solutions with desired properties.

So, the $\varepsilon^T_{33}/\varepsilon_0$, d_{31}, K_p, tan δ, Q, and δ_t of tetragonal Pb(Ti, Zr)O_3 – $\sum^4_{n=1}(PbB'_{1-\alpha}B''_\alpha O_3)_n$ solid solutions markedly different in electric hardness and non-PZT content were correlated with the average electronegativity of the metal atoms on the B-site.

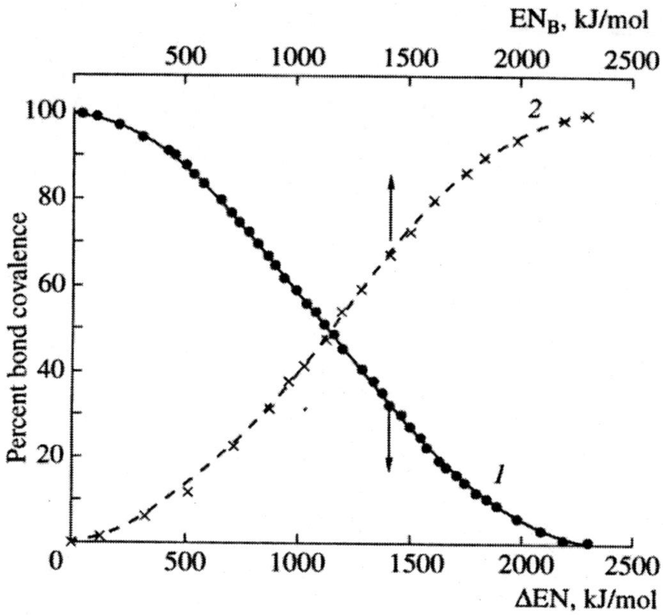

Figure 4. $B - O$ bond covalence as a function of (1) ΔEN and (2) EN_B.

The $PbB'_{1-\alpha}B''_{\alpha}O_3$ oxides were classified according to their effect on the electric hardness of the solid solutions. The properties of the solid solutions were found to vary linearly with EN_B, confirming the conclusion that their electric hardness increases with EN_B.

The $B - O$ bond covalence was evaluated from the electronegativities of the B metals and oxygen and was shown to vary little for the solid solutions with a wide range of piezoelectric properties.

The correlations found in this work can be used to predict properties of ferroelectric solid solutions and design new piezoelectric ceramics.

2. Dielectric Properties of $Pb(Ti, Zr)O_3 - \Sigma^4_{N=1}(PbB'_{1-A}B''_AO_3)_N$ Solid Solutions

Earlier, we described correlations between the dielectric, crystal-chemical, and structural properties of $Pb(Ti,Zr)O_3 - \Sigma^4_{n=1}(PbB'_{1-\alpha}B''_{\alpha}O_3)_n$ solid solutions containing roughly equal amounts of non-PZT components [1]. Here, we consider materials in which the ratio of such components, γ, varies over a wide range.

The objective of this research was to study the dielectric properties of quaternary and quinary solid solutions with $0 < \gamma \leq 11$ as a function of γ.

The solid solutions studied are close in Ti/Zr ratio and total content of non-PZT components. All of them have tetragonal structures and compositions near the morphotropic boundary. By way of illustration, we present data for the systems $PbTiO_3 - PbZrO_3 - PbNb_{2/3}Zn_{1/3}O_3 - PbW_{1/2}Mg_{1/2}O_3$ (I), $PbTiO_3 - PbZrO_3 - PbNb_{2/3}Zn_{1/3}O_3 - PbNb_{1/2}Mg_{1/3}O_3$ (II), and $PbTiO_3 - PbZrO_3 - PbNb_{2/3}Zn_{1/3}O_3 - PbW_{1/2}Mg_{1/2}O_3$ - $PbNb_{2/3}Mn_{1/3}O_3$ (III). We studied solid solutions with Ti/Zr = 1.11 – 1.16 (I), 1.23 – 1.35 (II), and 1.21 – 1.23 (III), and

Σ PbB'$_{1-\alpha}$B"$_{\alpha}$O$_3$ = 19.2 – 21.04 (I), 18.5 – 21.6 (II), and 14.2 – 17.4 (III) mol %. The values of γ were 0.12 – 10.1 (I), 0.135 – 11 (II), 0.15 – 3.05 (III). We determined the uniform distortion parameter δ [3] and measured the dielectric permittivity $\varepsilon^T_{33}/\varepsilon_0$ of poled ceramics fabricated by hot pressing and conventional solid-state reactions. As γ, we took the molar ratio from the more soft-electric (smaller δ and, accordingly, lower T_c [10]) to the less soft-electric materials.

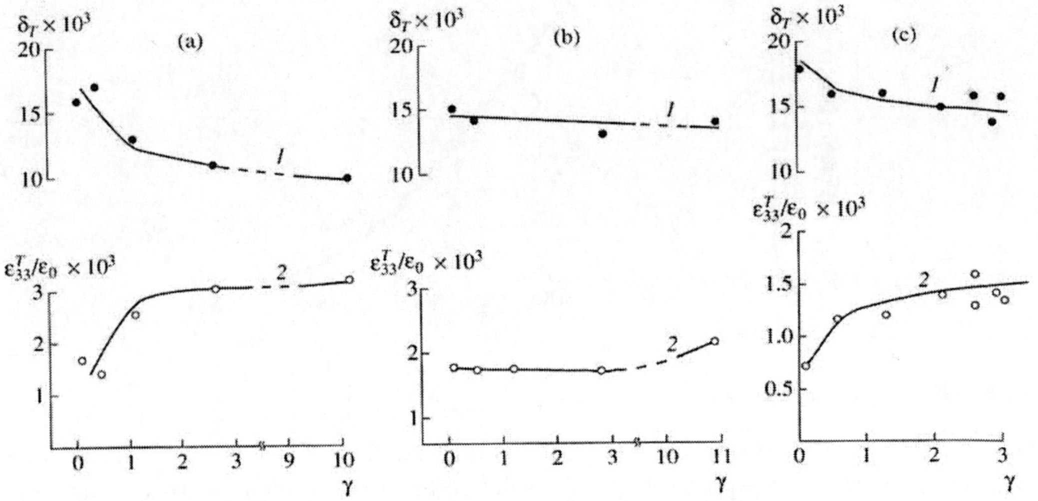

Figure 5. Plots of (1) δ_T and (2) $\varepsilon^T_{33}/\varepsilon_0$ *vs* γ for systems (a) I, (b) II, and *(c)* III.

Figure 5 shows the plots of δ and $\varepsilon^T_{33}/\varepsilon_0$ *vs* γ for sections equidistant from the morphotropic regions in systems I – III. It can be seen that, as the content of soft-electric components increases, δ decreases systematically, in accordance with the continuity and consistency principles, which are fulfilled in solid-solution systems. In systems I and III, δ decreases most rapidly for γ < 1 and shows only a gradual decrease for γ > 1. Since $\varepsilon^T_{33}/\varepsilon_0 \sim 1/\delta$ [11], $\varepsilon^T_{33}/\varepsilon_0$ varies in the opposite way, increasing rapidly for γ < 1.

A few features of the data are worth noting. In system I, δ continues to decrease at γ > 1, while $\varepsilon^T_{33}/\varepsilon_0$ varies insignificantly. This behavior seems to be due to the reduced density of the ceramics with large γ, which reduces $\varepsilon^T_{33}/\varepsilon_0$ [12]. By contrast, in system III, a slow variation in δ is accompanied by a marked rise in $\varepsilon^T_{33}/\varepsilon_0$. It is reasonable to assume that the presence of different amounts of non-PZT components in a solid solution favors structural disordering, thereby raising its dielectric permittivity [13]. To check this assumption, we measured the integral widths of the 111 and 222 X-ray peaks (b_{111} and b_{222}), the size of coherently scattering domains in the [111] direction (D_{111}), dislocation density (ρ_{D111}), and microstrain $\Delta d_{111}/d_{111}$ (Figure 6).

The nonmonotonic variations of these parameters, with extrema at γ ~ 1, point to a reduced structural perfection of the materials with γ < 1 and γ > 1. The ceramics with γ = 1 (having roughly equal contents of non- PZT components) have the most perfect microstructure (*D* is maximal, while $\Delta d/d$ and ρ_D are minimal). Deviations of γ from unity are likely to cause structural disordering, as evidenced by the increased scatter in electrical properties ($\overline{\Delta x}, \overline{\delta x}$) and broadening of the phase transition Δ in solid solutions of systems I and III with γ ≠ 1 (Figure 7) [8].

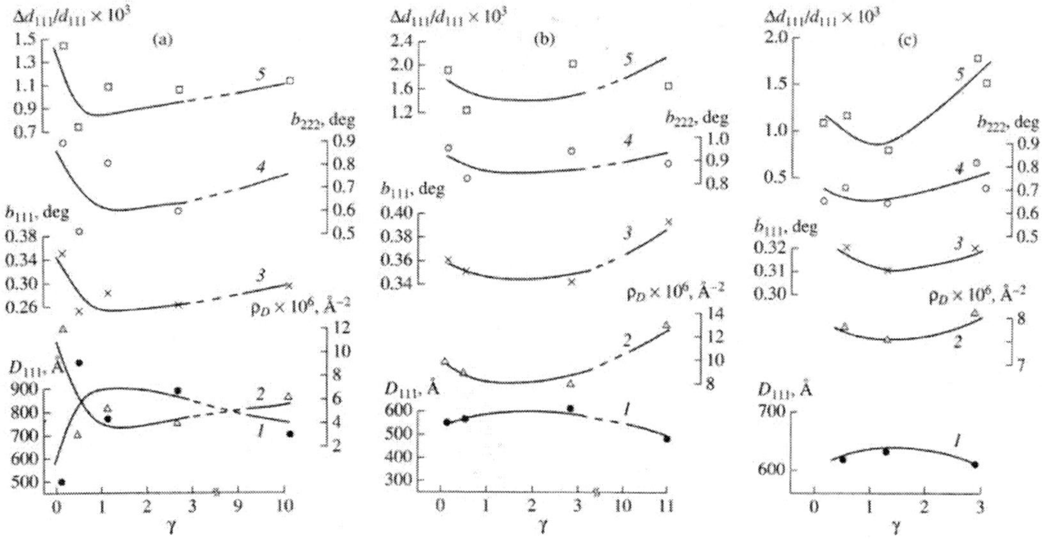

Figure 6. Plots of (*1*) D_{111}, (*2*) ρ_D, (*3*) b_{111}, (*4*) b_{222}, and (*5*) $\Delta d_{111}/d_{111}$ *vs* γ for systems (a) I, (b) II, and (c) III.

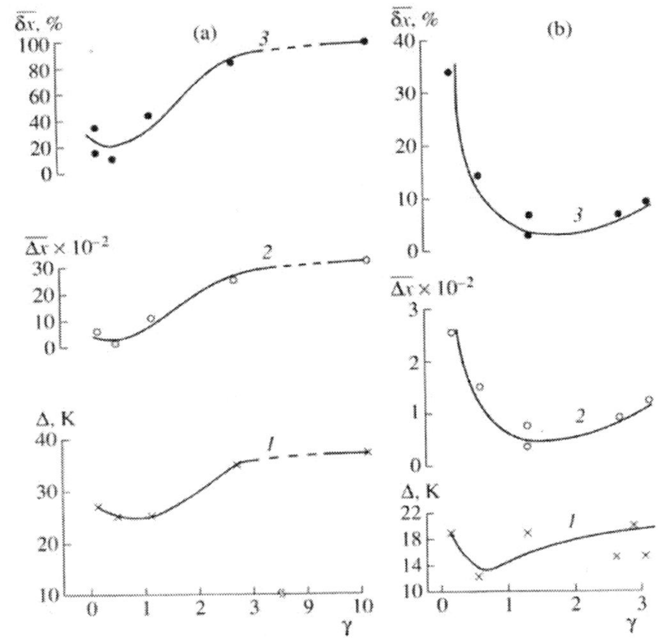

Figure 7. Structural disordering at deviations of γ from 1, evidenced by the increased scatter in electrical properties ($\overline{\Delta x}, \overline{\delta x}$) and broadening of the phase transition Δ in solid solutions of systems I and III with γ ≠ 1.

The fact that, in system I, $\varepsilon^T_{33}/\varepsilon_0$ rises only slightly at $\gamma > 1$, despite the structural disordering, suggests that the density of the ceramics has a strong effect on their dielectric permittivity. Comparison of the data in Figures 5 – 7 indicates that, in systems I and III, all the measured parameters vary more rapidly than those in system II. One possible reason is that heterovalent substitution of Nb for W produces vacancies, which have a significant effect on the properties of the solid solutions. Thus, the variation of $\varepsilon^T_{33}/\varepsilon_0$ with γ is sensitive to the structural perfection (disordering), density, and vacancies. If the contents of the $PbB'_{1-\alpha}B''_{\alpha}O_3$ compounds present differ only slightly, the influence of δ prevails, and a rapid decrease in δ is accompanied by a sharp rise in $\varepsilon^T_{33}/\varepsilon_0$. Otherwise, the effects of structural disordering and density may be stronger.

So, our results for $Pb(Ti, Zr)O_3 - \Sigma^4_{n=1} (PbB'_{1-\alpha}B''_{\alpha}O_3)_n$ solid solutions demonstrate that their dielectric permittivity correlates with γ (molar ratio of the more soft-electric to the less soft-electric component of $PbB'_{1-\alpha}B''_{\alpha}O_3$). With increasing γ, $\varepsilon^T_{33}/\varepsilon_0$ rises because of the decrease in the uniform cell distortion parameter.

At $\gamma > 1$, $\varepsilon^T_{33}/\varepsilon_0$ is also sensitive to the structural perfection of the solid solutions (degree of disordering), their density, and vacancies produced by heterovalent substitutions.

3. Properties of Multicomponent Solid Solutions Based on Lead Zirconate Titanate near Morphotropic Regions

The physical properties of ferroelectric solid solutions are known to show extrema near morphotropic phase transitions [1,16,17]. This effect is very pronounced when morphotropic regions are investigated along secant rays [1]. At the same time, the variations in the properties of solid solutions along morphotropic boundaries (within the morphotropic region or outside it, ~ 1 – 2 and ~ 10 – 15 mol % away from the boundary) have not yet been studied.

A physicochemical method for investigation of morphotropic regions in multinary systems was proposed by Fesenko et al. [1]. In this method, multinary systems are represent by multidimensional simplexes with a dimensionality smaller by unity than the number of components (tetrahedron at $k = 4$, pentatop for $k = 5$, and hexatop for $k = 6$). To find compositions with extremal properties in a $Pb(Ti, Zr)O_3 - \Sigma^4_{n=1} (PbB'_{1-\alpha} B''_{\alpha}O_3)_n$ system, the morphotropic region is investigated along rays emerging from the $PbTiO_3$ corner. The shape, boundary, and extent of the morphotropic region are refined by analyzing the structural and physical properties of the solid solutions with compositions along these rays. By studying a large number of systems with $k = 4 - 6$, we accumulated a great number of experimental data on the properties of solid solutions both in the tetragonal (*T*) + rhombohedral (*R*) phase field and near the morphotropic boundary in the single-phase region. This allowed us to analyze the properties of solid solutions along morphotropic boundaries.

We examined solid solutions based on lead zirconate titanate (PZT) in the systems:

1) PZT – $PbNb_{2/3}Zn_{1/3}O_3$ – $PbNb_{2/3}Mg_{1/3}O_3$,
2) PZT – $PbNb_{2/3}Co_{1/3}O_3$ – $PbNb_{2/3}Mg_{1/3}O_3$,
3) PZT – $PbNb_{2/3}Mg_{1/3}O_3$ – $PbW_{1/2}Ni_{1/2}O_3$,
4) PZT – $PbNb_{2/3}Zn_{1/3}O_3$ – $PbW_{1/2}Mg_{1/2}O_3$.

These solid solutions were investigated earlier in great detail over wide composition ranges [1]. The systems in question offer the possibility to compare the properties of solid solutions containing isovalent (Nb – Nb) and heterovalent (Nb – W) B' –B''' combinations. Heterovalent substitutions may produce more complex changes in structure and, hence, in properties, e.g., owing to an increased vacancy concentration [1].

Figure 8a (taken from [1]) shows a composition tetrahedron in which the *ABC* plane represents a morphotropic region and sections *I – IV* pass through the morphotropic region in different directions. It can be seen that the regions common to the morphotropic region and one of the sections extend from the system Pb(Ti, Zr)O_3 – $PbB'_{1-\alpha}B''_{\alpha}O_3$ (component K_3) to Pb(Ti, Zr)O_3 – $PbB'''_{1-\beta}B''''_{\beta}O_3$ (component K_4) (sections *I – III*) or from the system Pb(Ti, Zr)O_3 to $PbB'_{1-\alpha}B''_{\alpha}O_3$ – $PbB'''_{1-\beta}B''''_{\beta}O_3$ (section *IV*). The dashed lines indicate the *T – R* phase boundary. Figures 8b and lc show secant rays in sections *I* and *IV* and the compositions (crosses) along the rays analyzed earlier [1]. In this work, we consider compositions lying along the morphotropic region a fixed distance from it.

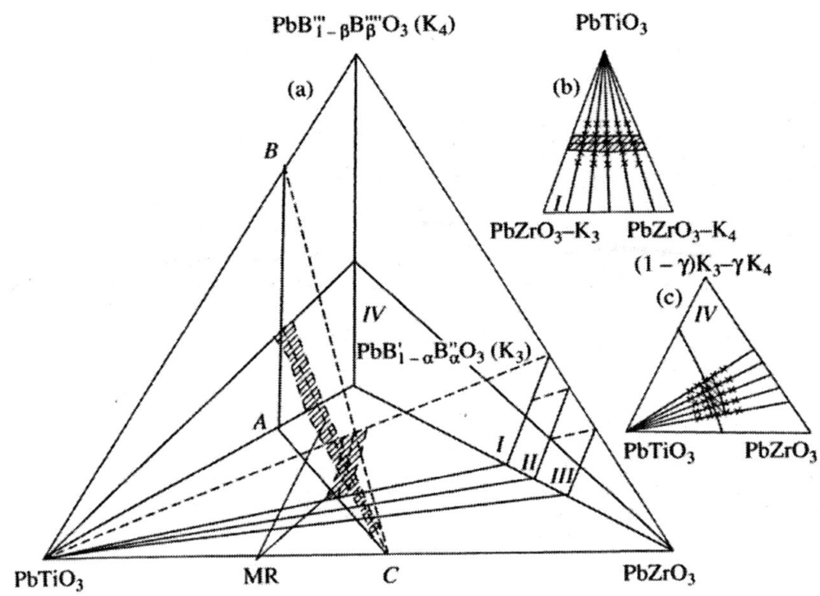

Figure 8. (a) Composition tetrahedron and morphotropic region (MR) schematically represented by *ABC*-plane; (b, c) sections *I* and *IV*.

Figure 9 displays the dependences of the uniform distortion parameter (δ) [10] and relative permittivity of poled samples ($\varepsilon^T_{33}/\varepsilon_0$) on content of the more soft-electric component along the morphotropic region in section *I* for different $PbTiO_3$ contents (specified at the curves in mol%).

These data demonstrate that, in going from the system Pb(Ti, Zr)O_3 – K_3 to Pb(Ti, Zr)O_3 – K_4 (replacement of $PbB'_{1-\alpha}B''_{\alpha}O_3$ by $PbB'''_{1-\beta}B''''_{\beta}O_3$), there are two types of correlations, depending on the nature of the B' and B''' cations in the components K_3 and K_4.

In the systems with isovalent B' – B''' combinations (Nb – Nb), $\varepsilon_{33}/\varepsilon_0$ rises monotonically with the content of the more soft-electric component, while δ decreases (Figure 9a, system *1*, solid curves). Farther away from the morphotropic region, the parameters vary more gradually. This behavior is attributable to the slight decrease in the B" – O bond covalence associated with the decrease in the average electronegativity (EN) of the metals in the *B*-site upon partial substitution of Mg (EN = 730 kJ/mol) for Zn (EN = 870 kJ/mol). As shown by Fesenko et al. [1], a decrease in EN and bond covalence is accompanied by a decrease in the electric hardness of the solid solutions, which leads to a rise in $\varepsilon^T{}_{33}/\varepsilon_0$ and a reduction in δ. As the difference in EN between B" and B"" increases (at B' = B'''), $\varepsilon^T{}_{33}/\varepsilon_0$ and δ vary more rapidly. The dashed lines in Figure 9a show the variation of $\varepsilon^T{}_{33}/\varepsilon_0$ with the content of the more soft-electric component in system 2, where B" = Co (EN = 835 – 1090 kJ/mol, depending on the oxidation state) and B"" = Mg (EN = 730 kJ/mol).

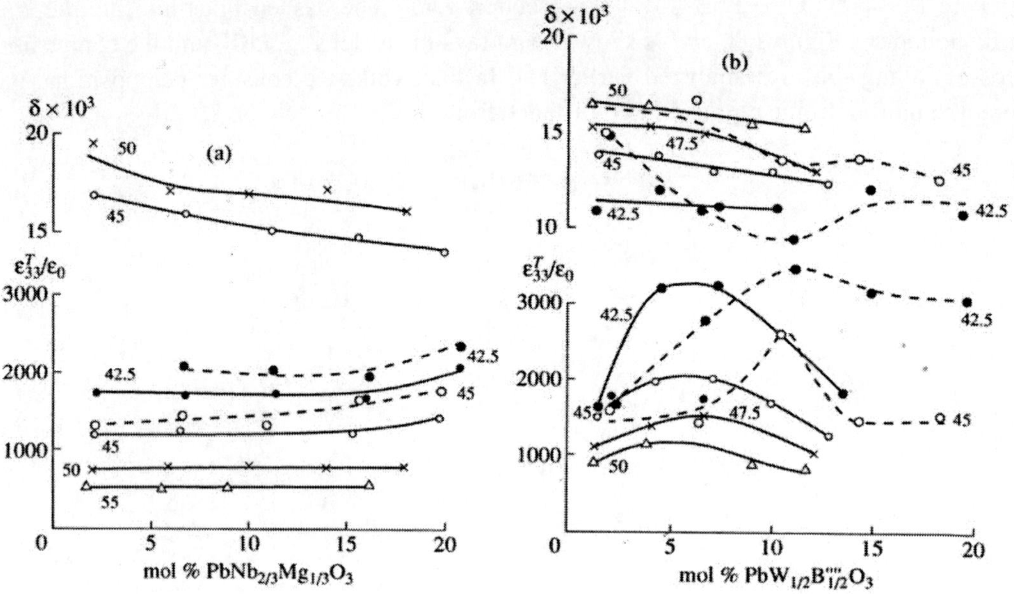

Figure 9. Plots of δ and $\varepsilon_{33}/\varepsilon_0$ (a) *vs* $PbNb_{2/3}Mg_{1/3}O_3$ content for systems *1* (solid curves) and *2* (dashed curves) and (b) *vs* $PbW_{1/2}B''''_{1/2}O_3$ content for systems *3* (solid curves) and *4* (dashed curves); section *I*; numbers at the curves specify mol % $PbTiO_3$.

In the systems with heterovalent B' – B''' combinations (Nb – W), $\varepsilon^T{}_{33}/\varepsilon_0$ and, occasionally, δ vary non-monotonically (Figure 9b).

These results can hardly be understood within the approximation of noninteracting ions distributed at random. This refers primarily to the W^{6+}-containing ternary perovskites $PbW_{1/2}{}^{6+}B_{1/2}{}^{2+}O_3$. As is well known [10], owing to the large difference in charge state between the *B*-cations ($\Delta z = 6 - 2 = 4$), these phases are always ordered, since the energy gain (driving force of ordering) is proportional to Δz. The formation of $PbW_{1/2}{}^{6+}B_{1/2}{}^{2+}O_3$-based solid solutions is accompanied by some frustration of the long-range order in the arrangement of B^{2+} and W^{6+}, which typically leads to a sharp decrease in Curie temperature T_C. At the same time, the relaxor behavior of the $PbW_{1/2}{}^{6+}B_{1/2}{}^{2+}O_3$-based solid solutions [18–20] indicates that some ordered nanoclusters persist, but the interaction between such clusters is reduced by electric-field and strain fluctuations. A similar picture is observed in the classical relaxor ferroelectric $PbNb_{2/3}Mg_{1/3}0_3$ [20,21].

The presence of compositionally ordered, W^{6+}-containing clusters in the materials in question is evidenced by the $T_C(x)$ data for $(1 - y)Pb(Ti, Zr)O_3 \cdot y[(1 - x)PbB'_{1-\alpha}B''_{\alpha}O_3 \cdot xPbB'''_{1-\beta}B''''_{\beta}O_3]$ solid solutions in systems *1– 4*. Since all the constituent perovskite oxides in these solid solutions contain the same A cation (only *B*-site substitutions), their T_C must vary linearly with x [22], In Figure 10, the $T_C(x)$ dependences calculated in a linear approximation are compared with experimental data for several solid-solution systems. It can be seen that, in systems *1* and *2* (B' = B" = Nb), the measured T_C varies roughly linearly with x and there is good agreement between calculation and experiment. At the same time, the measured T_C of the W^{6+}-containing solid solutions differs drastically from the calculation results. As in other perovskite solid-solution systems [18–20,23,24], the minimum in $T_C(x)$ in systems *3* and *4* is likely due to the formation of compositionally ordered W^{6+}-containing clusters. The fact that no super-lattice reflections corresponding to B' (B''') or W^{6+} ordering were detected in the X-ray patterns of these solid solutions seems to be evidence that the clusters are very small. The maximum in $\varepsilon^T_{33}/\varepsilon_0$ is probably due to changes in the amount and/or size of the clusters.

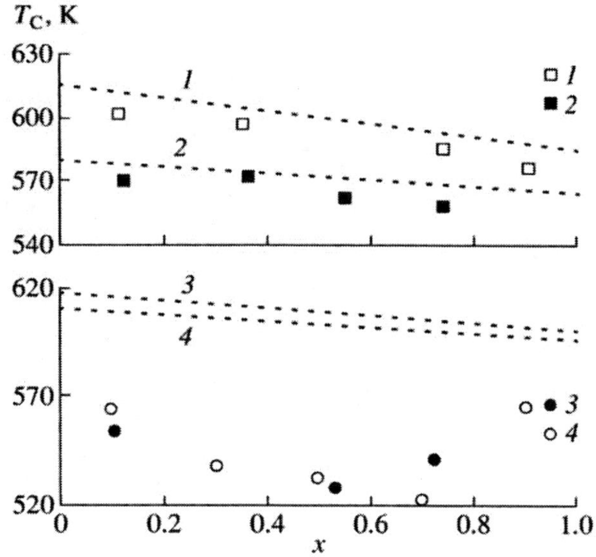

Figure 10. Composition dependences of T_C calculated in a linear approximation (dashed lines and experimental data (points) for (1) $0.8PbTi_{0.55}Zr_{0.45}O_3 \bullet 0.2[(1-x)PbNb_{2/3}Zn_{1/3}O_3 \bullet xPbNb_{2/3}Mg_{1/3}O_3]$, (2) $0.75PbTi_{0.6}Zr_{0.4}O_3 \bullet 0.25[(1-x)PbNb_{2/3}Mg_{1/3}O_3 \bullet xPbNb_{2/3}Co_{1/3}O_3]$, (3) $0.8PbTi_{0.55}Zr_{0.45}O_3 \bullet 0.2[(1-x)PbNb_{2/3}Zn_{1/3}O_3 \bullet xPbW_{1/2}Mg_{1/2}O_3]$, and (4) $0.85PbTi_{0.53}Zr_{0.47}O_3 \bullet 0.15[(1-x)PbNb_{2/3}Mg_{1/3}O_3 \bullet xPbW_{1/2}Ni_{1/2}O_3]$ solid solutions.

The extrema in the composition dependences of structural parameters provide further evidence for ordering of the solid solutions: the maximum in $\varepsilon^T_{33}/\varepsilon_0$ is accompanied by a maximum in the size of coherently scattering domains (D_{111}), a minimum in the half width of the 222 diffraction peak (b_{222}), and a minimum in lattice strain ($\Delta d/d_{111}$). The width of the phase transition (Δ) and the scatter in dielectric properties ($\Delta\lambda^1$, $\delta\lambda^1$) also pass through a minimum (Figure 11, solid lines). The higher structural perfection of the solid solutions near the maximum in $\varepsilon^T_{33}/\varepsilon_0$ is likely associated with compositional ordering. The presence of vacancies may also have a certain effect on the properties of the solid solutions under consideration [1].

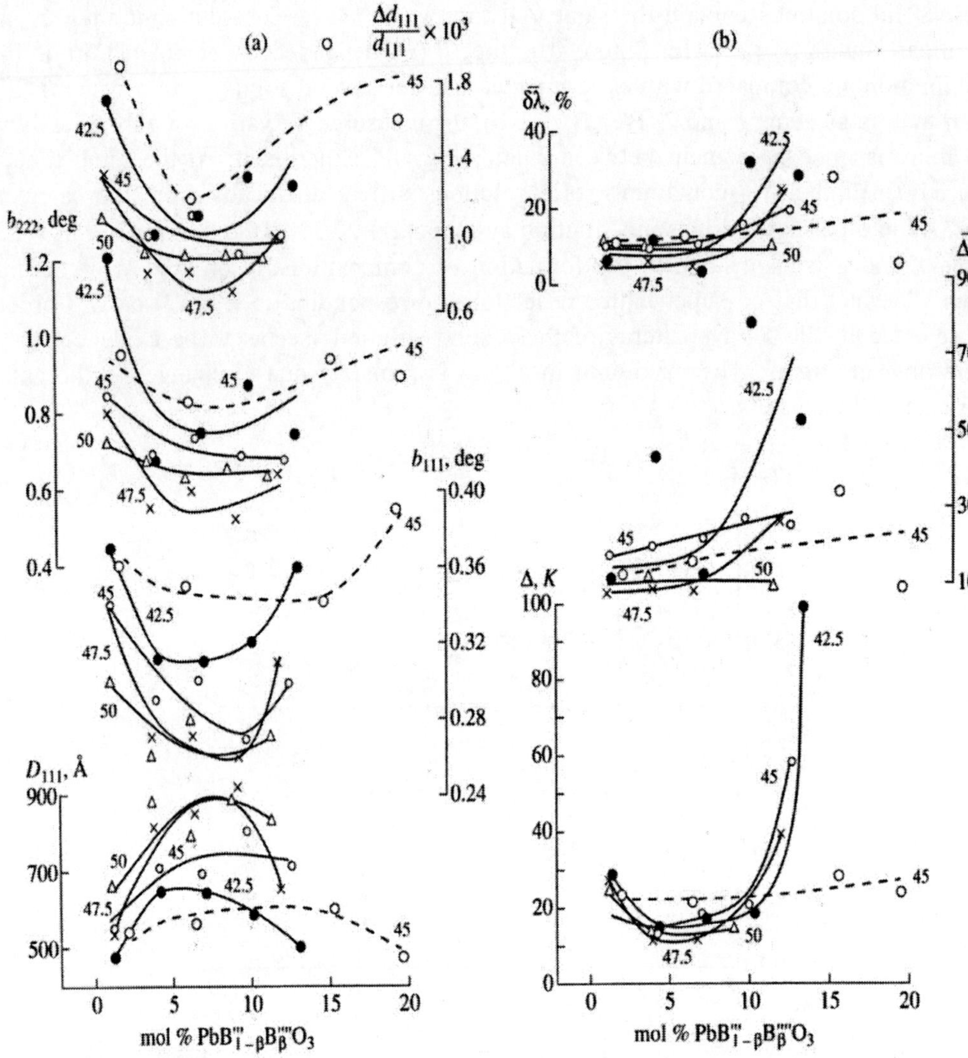

Figure 11. Plots of (a) D_{111}, b_{111}, b_{222}, $\Delta d_{111}/d_{111}$, (b) Δ, Δλ, and δλ *vs* $PbW_{1/2}Ni_{1/2}O_3$ content for system *3* (solid curves) and *vs* $PbNb_{2/3}Mg_{1/3}O_3$ content for system *1* (dashed curves); section *I*; numbers at the curves specify mol % $PbTiO_3$.

It is interesting to note that, in the systems with isovalent substitutions (Nb – Nb), there is also a composition range of enhanced structural perfection (Figure 11a, dashed lines), but the dependences of the microstructural parameters on the non-PZT content are, in this case, weaker, and Δ, Δλ, and δλ are insensitive to structural perfection (Figure 11b, dashed lines).

The more gradual variation in δ and $\varepsilon^T_{33}/\varepsilon_0$ farther away from the morphotropic region (Figure 9b) and the decrease in $\varepsilon^T_{33}/\varepsilon_0$ (Figure 9a) demonstrate that the effect of the morphotropic region, which is known to be characterized by extremal properties [1], becomes weaker.

Figure 12 displays the variations in δ and $\varepsilon^T_{33}/\varepsilon_0$ along the morphotropic region in the direction from Pb(Ti, Zr)O_3 to $(1-\gamma)K_3+\gamma K_4$ over a wide range of K_3+K_4 concentrations (section *IV*, systems *1*, *3*, *4*). The plot shows δ and $\varepsilon^T_{33}/\varepsilon_0$ as functions of the total of two components, which are in certain ratios in the corners of the secant triangles. The general trends of these data are similar to those described in [25], where we examined the effect of $\sum PbB'_{1-\alpha}B''_{\alpha}O_3$ content on the properties of PZT-based solid solutions differing in electric hardness.

The composition dependences of $\varepsilon^T_{33}/\varepsilon_0$ and δ in systems *1* and *4* show maxima and minima, respectively, at 21 mol % $\sum PbB'_{1-\alpha}B''_{\alpha}O_3$, in agreement with earlier data (= 21 mol %) for soft-electric systems [25].

The plots for the most soft-electric solid solutions (system *3*) show no extrema up to 30 mol % of $\sum PbB'_{1-\alpha}B''_{\alpha}O_3$ also in agreement with earlier results [25].

So, the structural and dielectric properties of Pb(Ti, Zr)O_3 – $PbB'_{1-\alpha}B''_{\alpha}O_3$ (K_3) – $PbB'''_{1-\beta}B''''_{\beta}O_3$ (K_4) solid solutions were studied along the morphotropic region in different directions.

In going from the system Pb(Ti, Zr)O_3 – K_3 to Pb(Ti, Zr)O_3 – K_4 (replacement of K_3 by K_4), two types of correlations were found, depending on the nature of the B' and B''' cations in components K_3 and K_4. At B' = B''' = Nb, $\varepsilon^T_{33}/\varepsilon_0$ rises steadily and δ decreases as the content of the more soft-electric component increases, which is attributable to the lower electronegativity of this component. At B' = Nb and B''' = W, there are a maximum in $\varepsilon^T_{33}/\varepsilon_0$ and a minimum in δ, which can be accounted for by the presence of ordered clusters in the W^{6+}-containing solid solutions. The amount and size of the clusters depend on their composition.

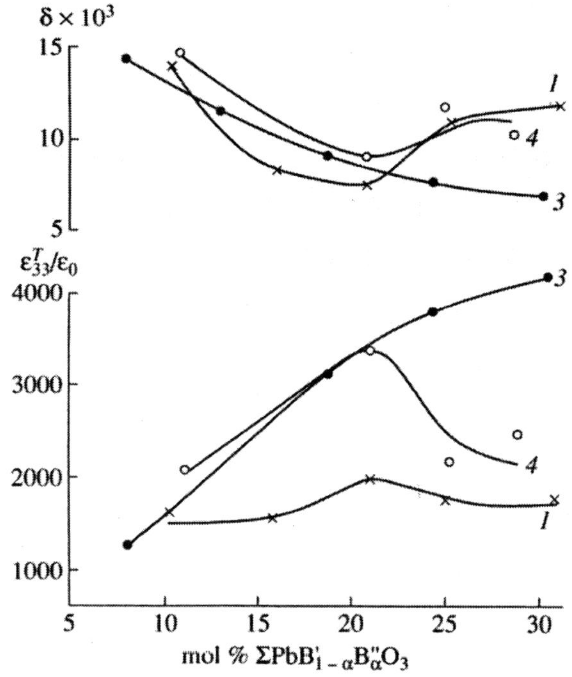

Figure 12. Plots of δ and $\varepsilon^T_{33}/\varepsilon_0$ *vs* $\sum PbB'_{1-\alpha}B''_{\alpha}O_3$ content for section *IV* in systems *1*, *3*, and *4*.

Farther away from the morphotropic region and deeper into the tetragonal phase region, the extrema broaden, since the effect of the morphotropic region becomes weaker.

The present data on the variation of the ferroelectric properties of solid solutions in different directions along the morphotropic region make it possible to predict properties of new materials.

4. Optimizing Conditions of Fabrication and the Properties of $BaNb_2O_6 - SrNb_2O_6$ Binary Ceramics

With their extraordinary electro-optical, dielectric, piezoelectric, and pyroelectric properties [26], barium and strontium niobates $BaNb_2O_6$ and $SrNb_2O_6$ are promising for use in pyroelectric detectors, phase inverters, electro-optical modulators, lasers, and so on [27]. However, the complex structure of tungsten bronzes (as these media are classified), the polymorphism of $BaNb_2O_6$, and its non-isomorphism to the second end member of the system ($SrNb_2O_6$) restrict the number of solid solutions based on them [28].

The above circumstance, along with the multistage formation of $(Ba_{1-x}Sr_x)Nb_2O_6$ solid solutions, makes it difficult to fabricate the latter without impurities of foreign phases that seriously complicate the use of these solid solutions as functional materials in the form of single crystals, ceramics, and thin films. It is therefore important to optimize the conditions of their fabrication and establish regularities in the forming of the structure and dielectric properties of $BaNb_2O_6$, $SrNb_2O_6$, and solid solutions based on them, and this is the aim of this section.

4.1. Experimental

Our objects of study were $BaNb_2O_6$ and $SrNb_2O_6$ compounds and $(Ba_{0.5}Sr_{0.5})Nb_2O_6$ solid solutions used to fabricate thin ferroelectric films. All objects were fabricated via double solid-state synthesis. Synthesis temperatures and times were chosen within the ranges T_{syn1} = (1200 – 1350) °C, τ_{syn1} = (6 – 10) h, T_{syn2} = (1350 – 1375) °C, and τ_{syn2} = 2 h. In dependence on their composition, the ceramics were sintered at T_{sin} = 1350 – 1400 °C for τ_{sin} = 2 h. Experimental density ρ_{exp} of the samples was determined by hydrostatic weighing in octane. X-ray diffraction patterns were recorded on DRON-2 and DRON-3 diffractometers in the Bragg – Bretano geometry with filtered CoK_a radiation using the 0 – 20 method in the step-by-step scanning mode of 20 through 0.020; the pulse collection time at each point was 2 s. Temperature dependences of the real part of complex permittivity e'/e_0, where s_0 is the electric constant, and loss tangent tan δ at T = 30 – 300 °C in the frequency range f = 25 – 10^6 Hz were investigated using a measuring bench based on an E7-20 LCR meter. Hysteresis loops were recorded on a Sawyer – Tower oscillographic setup at room temperature and a frequency of 50 Hz. This allowed us to estimate residual polarization P_R and coercive field E_c of the investigated samples.

4.2. Results and Discussion

Figure 13 shows the dependences of the relative intensities of the impurity phase lines on T_{syn1} for the investigated objects. Analysis of these curves revealed that as the annealing temperature rose the impurity content in $BaNb_2O_6$ fell considerably and then remained almost invariable. At T_{syn1} < 1250 °C and > 1350 °C, double barium niobate crystallizes mainly in the hexagonal phase; the rhombic phase becomes at T = 1300 °C the basic one. In the synthesis of double strontium niobate, the maximum purity of the products is reached at $T \sim$ 1250 °C. Pure $(Sr_{0.5}Ba_{0.5})Nb_2O_6$ is obtained at $T \sim$ 1250 °C; in this case, an increase in annealing multiplicity does not noticeably reduce the amount of the impurity phase, and the obtained solid solution forms in the tetragonal syngony with the unit cell parameters of a = 12.355 Å and c = 3.896 Å. The choice of the optimum sintering temperatures (T_{sin}) for a number of samples fired at 1300 – 1400 °C showed that the ceramics fabricated at T_{sin} = 1375 – 1400 °C were almost pure and had the highest density (see Figure 14).

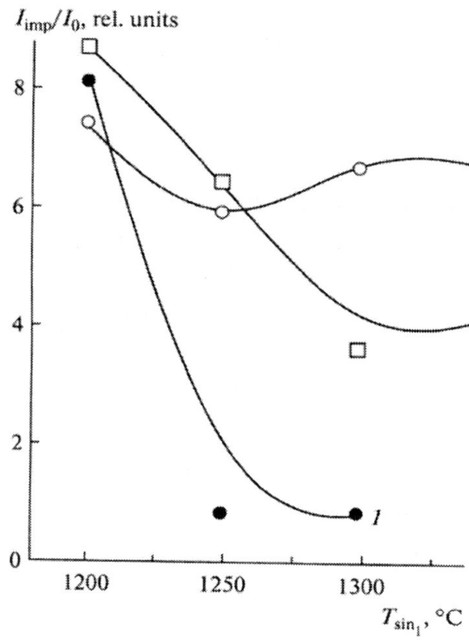

Figure 13. Dependences of the relative intensities of the impurity phase on the temperature of the first firing of (1) $Ba_{0.5}Sr_{0.5}Nb_2O_6$, (2) $BaNb_2O_6$, and (3) $SrNb_2O_6$.

Figures 15 and 16 present the results of investigations of the dielectric characteristics of $Ba_{0,5}Sr_{0,5}Nb_2O_6$ ceramics sintered at T_{sin} = 1400 °C. It can be seen that the hysteresis loops (Figure 15) are extended. The residual polarization and coercive fields at $E = 1.5 \times 10^6$ V/m are equals to ~0.024 C/m^2 and ~7.2 × 10^5 V/m, respectively. At room temperature and frequency $f = 10^4$ Hz, we have $e'/s_0 \sim 900$ and tan $\delta \sim 0.05$. As the temperature rises (Figure 16), maxima appear in the curves $\varepsilon'/\varepsilon_0$ (7) and tan δ (T) at T = 90 – 110 °C, which correspond to the transition from the ferroelectric (FE) to the paraelectric (RE) phase. As the frequency rises, these maxima shift toward higher temperatures. Above the Curie temperature (T_C) at T > 125 °C, frequency stratification and an increase in dispersion in proportion to the rise in T

are observed in dependences $\varepsilon'/\varepsilon_0$ (7), and a sharp rise in tan δ (T) is observed in the tan δ (T) curves. The observed behavior of $\varepsilon'/\varepsilon_0$ (T, f) and tan δ (T, f) near the FE → PE phase transition allows us to classify the investigated object as a ferroelectric with a diffuse phase transition, which is consistent with the results reported in [29]. This is due to fluctuations in the chemical composition of the material as a result of the disordering of Ba and Sr atoms in the structure. Because of these fluctuations, the parameters that determine the properties of the material change from one micro-part to another, broadening the region of the phase transition. The behavior of the dependences $\varepsilon'/\varepsilon_0$ (T, f) and tan δ (T, f) observed at $T > T_C$ is most likely produced by the rise in the electric conductivity of BSN ceramics at high temperatures.

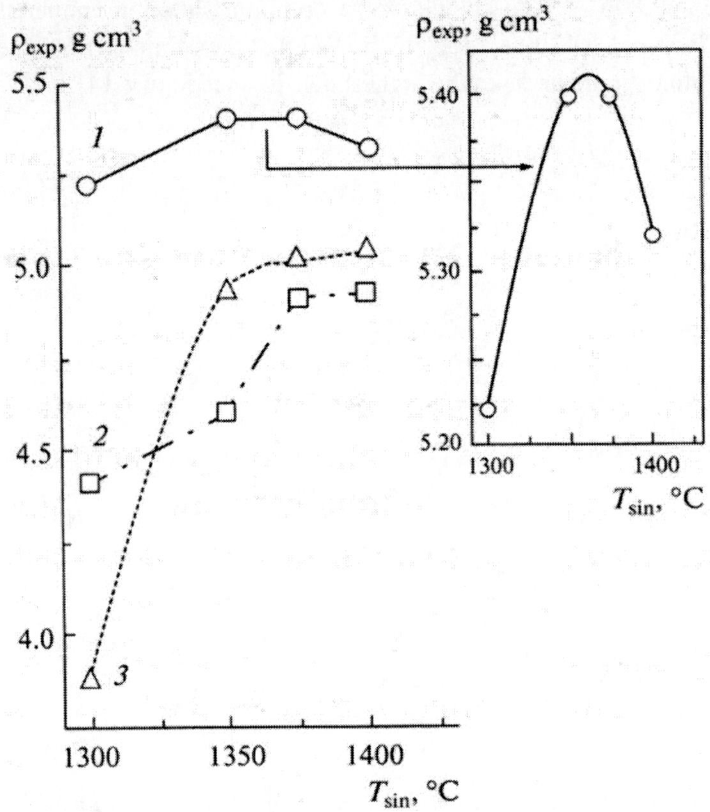

Figure 14. Dependences of experimental density ρ_{exp} on sintering temperature T_{sin} for (1) $Ba_{0.5}Sr_{0.5}Nb_2O_6$, (2) $BaNb_2O_6$, and (3) $SrNb_2O_6$.

In this study, we attempted to fabricate $Ba_{05}Sr_{05}Nb_2O_6$ thin films on MgO 100 and Si 100 substrates using the targets described above. The films were sputtered in a pure oxygen atmosphere at a pressure of 0.5 torr in one stage (without annealing) using the technique described in [30]. We obtained single-phase films with 001 orientation in sputtering times of 1 – 15 min on MgO and 30 – 60 min on the Si-substrate. The rate of film growth, determined using laser radiation reflected specularly from a substrate in [31], was 10 – 12 nm/min. Stable X-ray reflections were observed at film thicknesses of no more than 10 nm, testifying the high structural quality of the films.

Discharge plasma optical spectra were investigated during film sputtering using the technique described in [30]. The lines of atomic and ionized oxygen and strontium were observed in a discharge in the oxygen pressure range of 0.3 – 0.5, and there were no lines of Nb and Ba. Since metal component stoichiometry was retained in the films, we may conclude that the target material was transferred to the substrate mainly in the cluster form, as in the BST targets [32]. The observed dust component of the target material near the target surface upon He-Ne laser irradiation of this region [33] confirmed this assumption.

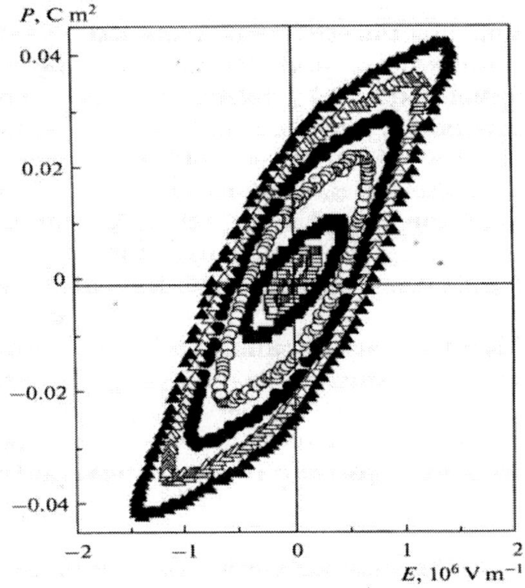

Figure 15. Dielectric hysteresis loops for $Ba_{0.5}Sr_{0.5}Nb_2O_6$ ceramics at different electric field strengths and room temperature.

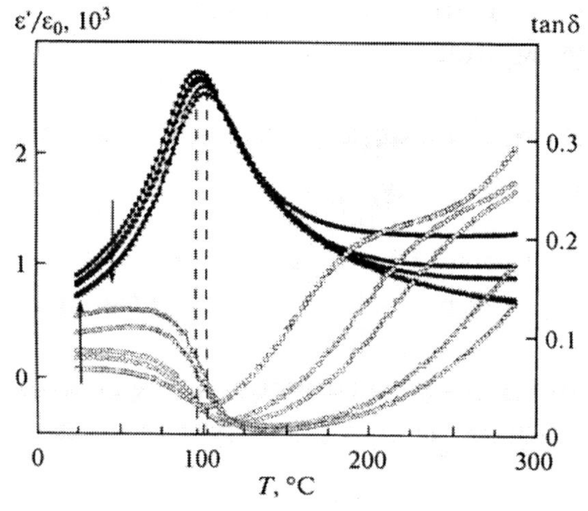

Figure 16. Temperature dependences of $\varepsilon'/\varepsilon_0$ (black symbols) and tan δ (white symbols) for $Ba_{0.5}Sr_{0.5}Nb_2O_6$ ceramics in the range of 30 – 300 °C at frequencies of $f = 10^4$, 5×10^4, 10^5, 5×10^5, and 10^6 Hz (the direction of f growth is shown by arrows).

5. Effect of Nonstoichiometry on the Structure and Dielectric Properties of Bismuth Ferrite

Bismuth ferrite is the best-known multiferroic with ferroelectric order up to 830 °C and antiferromatic order up to 370 °C. The solid-state synthesis of pure $BiFeO_3$ is complicated by the formation of uncontrolled amounts of the stable accompanying compounds $Bi_{25}FeO_{39}$ and $Bi_2Fe_4O_9$ [34] and, as a consequence, by low reproducibility of the properties of ceramic samples.

As was established for $AgNbO_3$ and NaNbO in [35–37], the formation of impurities in ceramics fabricated via solid-state synthesis is due to a deficiency of *A* positions in the ABO_3 structure (structural nonstoichiometry), which occurs upon a crystallographic shift (CS), i.e., a plane defect characteristic of oxygen-octahedral compounds of the ReO_3 type that contain elements with variable valence in the *B*-position [38,39]. After a CS, the stiochiometric mixture of initial components contains excess *A* cations, which, depending on their properties, are either localized in intergrain spacers or partially displace *B* cations from their positions with the formation of impurity compounds. Such phenomena can accompany the synthesis of $BiFeO_3$. As was demonstrated in [35–37], one way of eliminating the impurity phases is to create artificial nonstoichiometry. The aim of this study was to establish regularities in the formation of the structure and dielectric properties of nonstoichiometric bismuth ferrite, and to determine the optimum composition for ensuring the purity of the final product.

5.1. Experimental

Our objects of study were bismuth ferrite Bi_xFeO_3 ceramics, where x changes within $0.88 < x < 1.04$. Cation/anion-excess ceramics of $Bi_{1+y}FeO_{3+3/2y}$, (y = 0.00 – 0.04) composition obviously form in the range of $1.00 << x \leq 1.04$, and cation/anion-deficient vacancy-saturated ceramics of $Bi_{1-y}FeO_{3-3/2y}D_{3/2y}$ composition (y = 0.00 – 0.12; D is the vacancy) form in the range of $0.88 < x < 1.00$. The samples were obtained using conventional ceramic technology via two-stage solid-state synthesis at $T_1 = T_2$ = (750 – 850) °C and exposure times τ_1 = 10 h and τ_2 = 2 h. The synthesized granular powders were sintered at T_{sint} = (800—950) °C for T_{sint} = (1 – 2) h, depending on their composition. The initial reagents were high-purity Bi_2O_3 and Fe_2O_3. The experimental density of the samples was determined by weighing in octane; the relative density, using the formula $\rho_{rel} = (\rho_{exp}/\rho_{X\text{-}ray}) \times 100\%$.

X-ray powder diffraction patterns were recorded on a DRON-3 diffractometer in the Bragg - Bretano geometry with filtered CoK_α radiation. The objects of study were ground to eliminate the surface effects, stresses, and textures that occur during synthesis of ceramics. The parameters of the reduced rhombohedral (R) cell were calculated using the standard technique in [7]. The relative measurement errors were $\delta a = \delta b = \delta c = \pm$ 0.05% for the linear parameters, $\Delta\infty = \pm$ 5% for the angular parameters, and $\Delta V = \pm$ 0.07% for the volume. During the X-ray diffraction analysis of Bi_xFeO_3, the $Bi_{25}FeO_{40}$ and $Bi_2Fe_4O_9$ impurity phase contents were estimated from the relative intensities of the corresponding strong lines I/I_1, where I is the impurity phase line intensity and I_1 is the Bi_xFeO_3 strong line intensity.

The dielectric spectra were investigated using an Aglient E4980A LCR meter in the temperature and frequency ranges of 25 – 600 °C and 20 Hz – 2.0 MHz, respectively.

5.2. Results and Discussion

We determined the parameter *a*, experimental and theoretical volumes V_{exp} and V_{theor} of the reduced R-cell of Bi_xFeO_3. The theoretical volume was calculated using the formula [10], in which *L* is the unstrained cation – oxygen bond length with regard to the coordination number of a cation by oxygen, calculated using Belov – Bokii ionic radii, and *n* is the cation valence. Along with relative experimental density ρ_{exp}, which characterizes the ceramic quality, we calculated the relative theoretical density of the samples, $\rho_{rel.theor} = (\rho_{X\text{-}ray}/\rho_{theor}) \times 100\%$, which characterizes the structural quality (imperfection), using V_{theor}.

Figure 1 shows the above-mentioned structural parameters as functions of *x*. It can be seen that the experimental values periodically change as *x* grows. The dependence $V_{exp}(x)$ does not correlate with the behavior of $V_{theor}(x)$ in the range of $0.88 < x < 1.02$ and only at $x > 1.02$: $V_{exp}(x)$ decreases together with $V_{theor}(x)$. The behavior of $\rho_{rel\cdot theor}$ is surprising, since it should rise with growth of *x* from 0.88 up to 1.00 due to the decrease in the number of defects in the $BiFeO_3$ structure with falling the degree of stoichiometry. Instead, $\rho_{rel.theor}$ decreases to $x = 0.02$ and then rises slightly. In addition, we can see that the $Bi_{25}FeO_{40}$ content grows sharply in the range of $0.97 < x < 1.00$ and then drops to zero at $x = 1.04$, even though the Bi^{3+} content exceeds that of the stoichiometric composition by 4%. The $Bi_2Fe_4O_9$ content diminishes with growth of *x*, and reaches minimal value at $x = 1.04$.

As we showed in [35,40,41], the behavior of the dependences of the structural characteristics in Figure 1 is typical of other perovskite compounds and their solid solutions with deficiencies of the *A* positions. In ABO_3 compounds with perovskite structure that contain elements with variable valence in the *B*-position, the CS results to drop in V_{exp} relative to V_{theor} calculated for a perfect stoichiometric structure [7], except with $PbTiO_3$, for which $V_{exp} > V_{theor}$ ($V_{exp} = 63.28$ A^3, $V_{theor} = 62.71$ A^3 and $V_A = 0.57$ A^3). The situation is even more pronounced for $BiFeO_3$: $V_{exp} = 62.37$ A^3, $V_{theor} = 61.26$ A^3 and $V_A = 1.11$ A^3. The authors of [43] validated the assumption that in $PbTiO_3$, a certain number (~2%) of $Pb+^2$ ions are found in oxygen octahedra extended along the CS planes, leading to significant excess of V_{exp} over $V_{theor\ r}$. Since Bi and Pb atoms are similar [44], we may assume that Bi can also occupy such positions.

The structural nonstoichiometry of $BiFeO_3$ is apparently higher than that of $PbTiO_3$; therefore, replacement of $Fe+^3$ ions from the oxygen octahedra with $Bi+^3$ ions leads to a much larger excess of V_{exp} over V_{theor} than in $PbTiO_3$. One of the reasons for the occurrence of impurities in nonstoichiometric $BiFeO_3$ ceramics could thus be structural nonstiochiometry of no lesser than 3%, according to our estimate. For this reason, the impurity content of $Bi_{25}FeO_{39}$ grows sharply when $x > 0.97$. Since the formation of CS planes is related to the periodic accumulation, ordering, and exclusion of oxygen vacancies, the behavior of all the experimental quantities in Figure 17 is also periodic.

We experimentally observed two pronounced maxima of the impurity phase contents at Bi excesses and deficiencies of ~ 4 – 6%. Since $BiFeO_3$ is located at the stability boundary of the perovskite structure, suggesting excessive strength of chemical bonds [10], we may assume that its reduction is in the first case related to the formation of a number of vacancies in the *A* sublattice that are allowed by the $BiFeO_3$ structure, and to both the compensation for Bi loss during sublimation and possible incorporation of Bi in the *B*-sublattice [10] (internal modification [43]) in the second case.

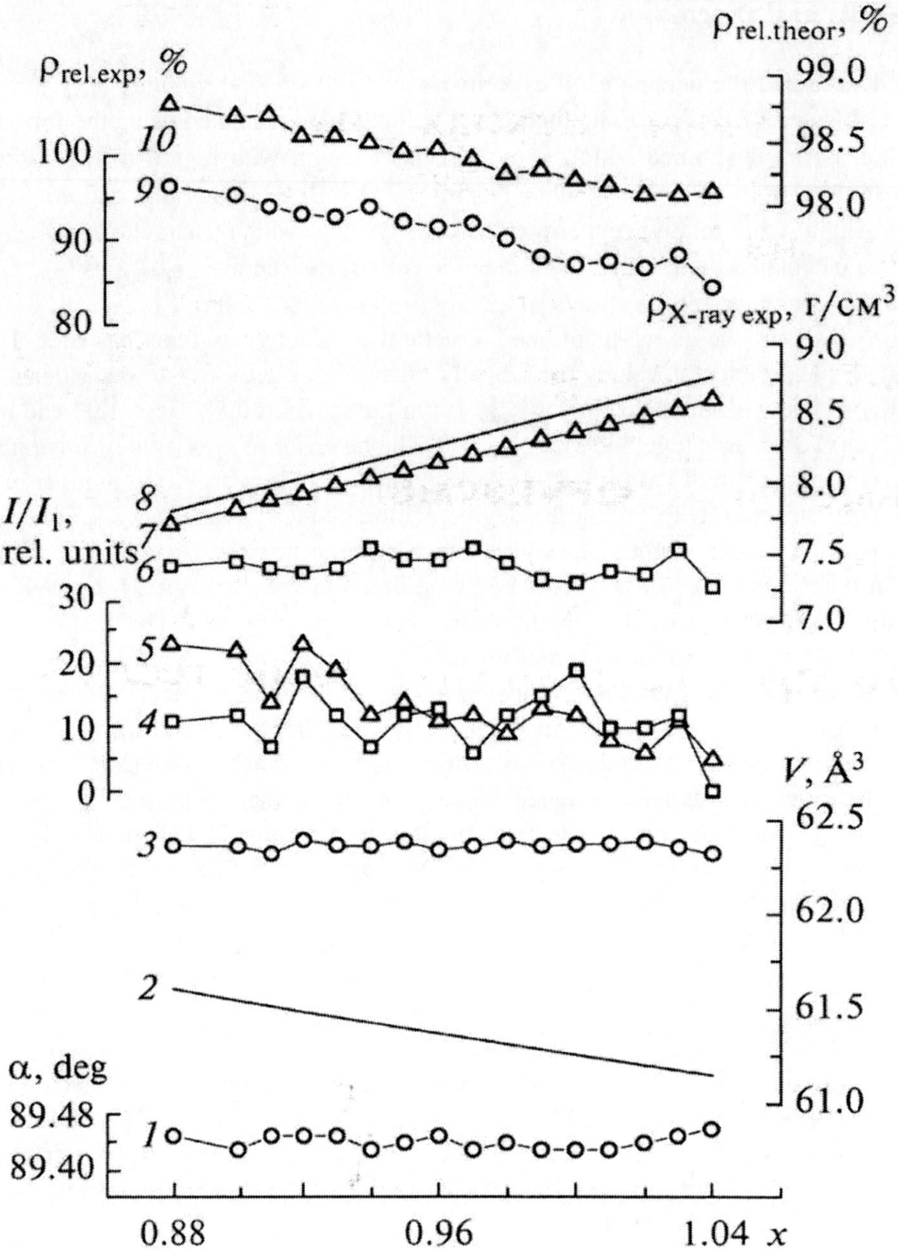

Figure 17. (*1*) α, (*2*) V_{theor}, (*3*) V_{exp}, (*4*) line I/I_1 of $Bi_{25}FeO_{40}$ impurity phase content, (*5*) line I/I_1 of $Bi_2Fe_4O_9$ impurity phase content, (*6*) ρ_{exp}, (*7*) $\rho_{X\text{-ray}}$, (*8*) ρ_{theor}, (*9*) $\rho_{rel.\ exp}$ and (*10*) $\rho_{rel.\ theor}$, for Bi_xFeO_3 as functions of *x*.

Figure 18 shows the dielectric spectra for stoichiometric and nonstoichiometric bismuth ferrite. It can be seen clearly that the compositions with minimum impurity content exhibit the most stable dielectric properties. Moreover, the number of impurities correlates with the value of $\varepsilon/\varepsilon_0$ at room temperature, confirming the impurities have strong effect on penetrating conduction in the investigated objects.

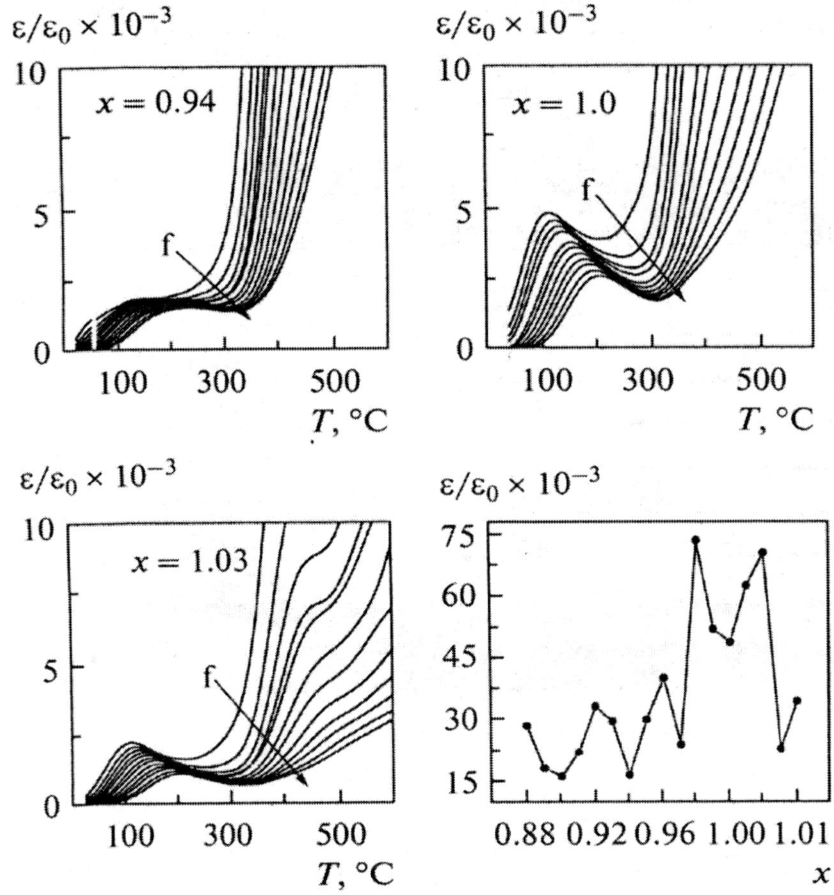

Figure 18. Dielectric spectra for stoichiometric and nonstoichiometric bismuth ferrite and dependence of relative permittivity $\varepsilon/\varepsilon_0$ on degree of nonstoichiometry x at frequency of 100 Hz and temperature of 25 °C.

6. Effect of the Conditions of Structure Formation on the Properties of BST Ceramics

Today, test and measurement equipment of the new generation is developing rapidly. Its basis is electrically controlled devices of millimeter or submillimeter ranges. There are phase shifters, delay lines, resonators and filters. These items include thin (down to nanoscale) ferroelectric films. Solid solutions of the BST $(Ba_{0.8}Sr_{0.2})TiO_3$ and $(Ba_{0.7}Sr_{0.3})TiO_3$ have low voltage polarization switching and maximum quality factor (the ratio of the differential phase shift to the level of insertion loss) and are traditionally used as targets for the deposition of ferroelectric thin films.

By creating these films a sensitivity of the properties of complex oxides with a perovskite structure to their thermodynamic previous history and thermal nanosegregation of their chemical composition must be considered. This requires a detailed selection of optimal conditions for the establishment BST targets. The purpose of this study is selection of optimum conditions of their synthesis and sintering.

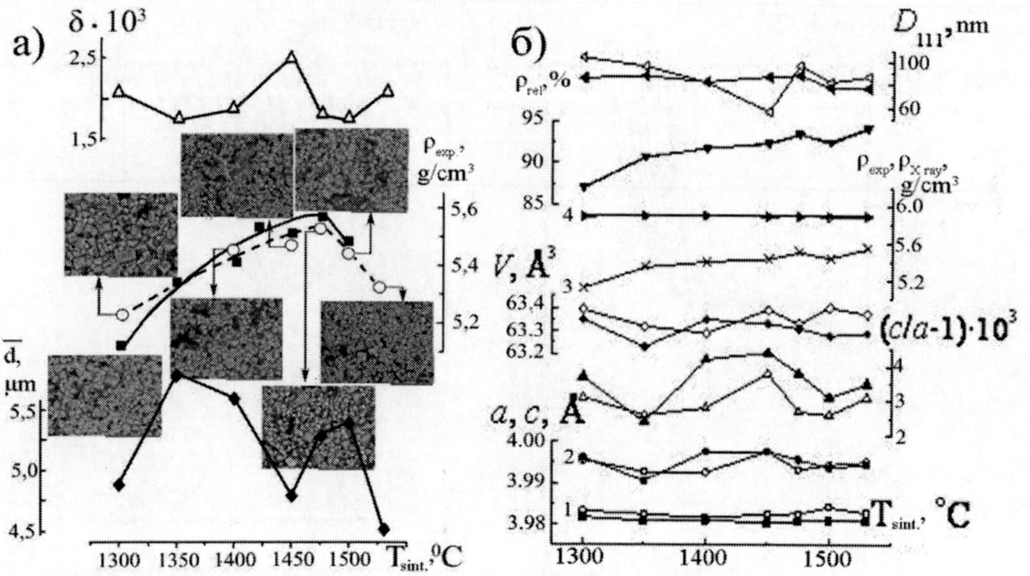

Figure 19. Dependences of (a) ceramic characteristics ($\bar{d}$ is the average grain size; $\rho_{exp.}$ is the experimental density (the results obtained in two independent experiments), δ is the homogeneous deformation parameter) and (b) structural parameters on sintering temperature of the samples.

Analysis of dependences of the structural parameters and grain structure of the composition of the ceramics and features of phase formation in the manufacture of polycrystalline BST-containing components was conducted. It was found that the critical dependence of the properties of BST-ceramics on the conditions of their production was depended primarily on the influence of liquid phase on the features of recrystallization at the sintering ceramics. The correlations between structural features of the ceramic pieces (targets) and thermal-frequency dependences of dielectric characteristics of the studied objects were identified and present in Figure 19.

Acknowledgments

This work was supported in part by the Russian Foundation for Basic Research, grant No. 99-02-17575, Russian Department of Education and Science, grants Nos. 213.01-11/2014-21, 213.01-2014/012-VG, 3.1246.2014/K and Russian Federal Aim Program (State Contract No. 14.575.21.0007).

References

[1] Fesenko, E. G., Dantsiger, A. Ya.; Razumovskaya, O. N., *New Piezoelectric Ceramics*, Rostov State University Press, Rostov-on-Don, 1983 (In Russian).

[2] Dantsiger, A. Ya., *Multicomponent Ferroelectric Oxide Solid Solutions and Related High-Performance Piezoelectric Ceramics, DrSc Dissertation (Physics and Mathematics),* Rostov State University Press, Rostov-on-Don, 1986 (In Russian).

[3] Fesenko, E. G.; Filip'ev, V. S.; Kupriyanov, M. F., *Fiz. Tverd. Tela (Leningrad)*, 1969, *vol. 11(2)*, 466 – 471.

[4] Reznichenko, L. A.; Razumovskaya, O. N.; Dudkina, S. I., *Proc. Int. Conf. on the Fundamental Problems in Piezoelectric Device Technology, P'ezotekhnika-99*), Rostov State University Press, Rostov-on-Don, 1999, *vol. 1*, pp.109-123 (In Russian).

[5] Dantsiger, A. Ya.; Razumovskaya, O. N.; Reznichenko, L. A.; Dudkina, S. I., *High-Performance Piezoelectric Ceramics: Optimized Approaches*, Paik, Rostov-on-Don, 1995 (In Russian).

[6] Devonshire, A. F., *Philos. Mag.*, 1951, *vol. 42(333)*, 1065 – 1079.

[7] Berlincourt, D.; Kerran, D.; Jaffe, H., In: *Physical Acoustics: Principles and Methods,* W.P. Mason (Ed.), Academic, New York, 1964, *vol. 1, part A*. (Translated into Russian *Fizicheskaya akustika,* Mir, Moscow, 1966, *vol. 1, part A*, pp. 204 – 326.

[8] Karapet'yants, M. Kh.; Drakin, S. I., *Constitution of Matter*, High School, Moscow, 1967 (In Russian).

[9] Povarennykh, A.S., In: *Zapiski LJki: Otdelenie Vsesoyuznogj Mineralogicheskogo Obschestva.,* Moscow, 1962, *part 1*, pp. 3-27 (In Russian).

[10] Fesenko, E. G., *Perovskite Family and Ferroelectricity*, Atomizdat, Moscow, 1972 (In Russian).

[11] Venevtsev, Yu. N.; Politova, E. D.; Ivanov, S. A., *Ferro- and Antiferroelectrics of the Barium Titanate Family*, Khimiya, Moscow, 1985 (In Russian).

[12] Haertling, G. H., *US Atomic Energy Committee Report, No. 3/1*, 1 – 9.

[13] Nomura, S.; Uchino, K., *Ferroelectrics,* 1982, *vol. 41(l)*, 117-132.

[14] Kasandrova, O. N.; Lebedev, V. V. *Data Processing*, Nauka, Moscow, 1970 (In Russian).

[15] Bokov, A. A., *Zh. Eksp. Tear. Fiz.,* 1997, *vol. 111(5)*, 1817 – 1832.

[16] Jaffe, B.; Roth, R. S.; Marzullo, S., *J. Appl. Phys.,* 1954, *vol. 25(6)*, 809 – 810.

[17] Berlincourt, D.; Cmolik, C.; Jaffe, H., *Proc IRE*, 1960, *vol. 48(2)*, 220 – 229.

[18] Smolenskii, G. A.; Krainik, N. N.; Agranovskaya, A. I., *Fiz. Tverd. Tela (Leningrad)*, 1961, *vol. 3(3)*, 981 – 990.

[19] Shvartsman, V. V.; Politova, E. D.; Stefanovich, S. Yu., *Ferroelectrics,* 1999, *vol. 235(3/4)*, 267 – 268.

[20] Chen, I. W.; Li, P.; Wang, Y., *J. Phys. Chem. Solids,* 1996, *vol. 57(10)*, 1525 – 1536.

[21] Cross, L. E., *Ferroelectrics,* 1987, *vol. 76(1 – 4)*, 241 – 267.

[22] Raevskii, I. P.; Reznichenko, L. A.; Prokopalo, O. I.; Fesenko, E. G., *Zh. Tekh, Fiz,,* 1978, *vol. 48(12)*, 2601 – 2603.

[23] Uchino, K.; Nomura, S., *Ferroelectrics,* 1978, *vol. 17(1 – 4)*, 505 – 510.

[24] Isupov, V. A.; Isupova, E. N.; Sokolova, N. Yu.; Smolenskii, G. A., *Izv. Akad. Nauk SSSR, Neorg. Mater.,* 1984, *vol. 20(5)*, 825 – 827.

[25] Reznichenko, L. A.; Dantsiger, A. Ya.; Razumovskaya, O. N.; et al., Effect of $\sum PbB'_{1-\alpha}B''_{\alpha}O_3$ Content on the Properties of $Pb(Ti, Zr)O_3$-Containing Multicomponent Solid Solutions, *XV All-Russia Conf. on the Physics of Ferroelectrics, VKS-XV*, Rostov-on-Don, 1999, p. 282 (In Russian).

[26] Lines, M. E.; Glass, A. M., *Principles and Applications of Ferroelectrics and Related Materials*, Oxford University Press, London, 1977.

[27] Koo, J.; Hyeok, J. J.; Bae, B.-S., *J. Amer. Ceram,* 2001, *vol. 84(1)*, 193.

[28] Lee, W.-J.; Fang, T.-T., *J. Amer. Ceram,* 1998, *vol. 81(1)*, 193.

[29] Kuz'minov, Yu. S., *Ferroelectric Crystals for Laser Radiation Control*, Nauka, Moscow, 1982 (In Russian).

[30] Mukhortov, V. M.; Golovko, Yu. I.; Tolmachev, G. N.; Klevtsov, A. N., *Zh. Tekh. Fiz.*, 1992, *vol. 72(5)*, 22.

[31] Zinchenko, S. P.; Kovtun, A. P.; Tolmachev, G. N., *Tech. Phys*., 2009, *vol. 54(11)*, 1689.

[32] Mukhortov, V. M.; Golovko, Yu. I.; Mamatov, A. A., *Collection of Southern Scientific Center of Russian Academy of Sciences Scientific Works, vol. 2: Physics, Mechanics, Engineering*, Southern Scientific Center of Russian Academy of Sciences Press, Moscow, 2007 (In Russian).

[33] Zinchenko, S. P.; Tolmachev, G. N., *Plasma Phys. Rep,* 2013, *vol. 39(13)*, 107.

[34] Carvalho, T. T.; Tavares, P. B., *Mater. Lett*., 2008, *vol. 62*, 3984.

[35] Reznichenko, L. A.; Shilkina, L. A.; Gagarina, E. S., et al., *Kristallografiya*, 2004, *vol. 49(5)*, 909.

[36] Kravchenko, O. Yu.; Gadzhiev, G. G.; Omarov, Z. M.; et al., *Inorg. Mater.,* 2011, *vol. 47(6)*, 679.

[37] Kravchenko, O. Yu.; Gadzhiev, G. G.; Omarov, Z. M.; et al., *Inorg. Mater.*, 2011, *vol. 47(8)*, 919.

[38] Petrenko, A. G.; Prisedskii, V. V., *Structure Defects in Ferroelectrics*, Uchebno-metodicheskii kabinet po vysshemu obrazovaniyu pri Minvuze USSR, Kiev, 1989.

[39] Rao, C. N. R.; Gopalakrishnan, J., *New Directions in Solid State Chemistry*, Cambridge University Press, Cambridge, 1997.

[40] Reznichenko, L. A.; Shilkina, L. A.; Razumovskaya, O. N.; et al., *Inorg. Mater.,* 2003, *vol. 39(2)*, 139.

[41] Andryushina, I. N.; Reznichenko, L.A.; Shilkina, L. A.; et al., *Ceram. Int.,* 2013, *vol. 39(2)*, 1285.

[42] Reznichenko, L. A.; Shilkina, L. A.; Titov, S. V.; et al.; *Inorg. Mater.*, 2005, *vol. 41(5)*, 492.

[43] Titov, S. V.; Shilkina, L. A.; Razumovskaya, O. N.; et al., *Inorg. Mater.,* 2001, *vol. 37(7)*, 718.

[44] Ugai, Ya. A., General and Inorganic Chemistry, High School, Moscow, 1997 (In Russian).

In: Advanced Materials - Studies and Applications
Editors: I.A. Parinov, S.-H. Chang, S.Theerakulpisut
ISBN: 978-1-63463-749-7

Chapter 2

HYDROTHERMAL PREPARATION OF $NaTaO_3$ PHOTOCATALYST MATERIALS

Min Yen Yeh[1,•], Yi Cheng Lee[1], Kun Fu Hsu[1], Chyi-Da Yang[1], Cheng-Liang Huang[1], Po-Hsun Lei[2] and Shun-Hsyung Chang[1]
[1]Department of Microelectronics Engineering, National Kaohsiung Marine University, Kaohsiung, Taiwan
[2]Department of Electro-Optical Engineering and Graduate Institute of Electro-Optical and Materials Science National Formosa University, Huwei, Yunlin, Taiwan

ABSTRACT

In this study, $NaTaO_3$ powders were prepared by hydrothermal synthesization at the temperatures of 140 – 200°C. The $NaTaO_3$ close to stoichiometry composition were obtained by hydrothermal synthesis at the temperature of 180°C for 12 hr. The as-hydrothermal-synthesized $NaTaO_3$ grains reveal uniform-cubic shape with the particle size of ~200 nm. The $NaTaO_3$ samples were further calcined at the temperatures of 800 – 900°C for 12 hr and the grain sizes increased to 200 – 400 nm. The $NaTaO_3$ powders were perovskite structures which are thought to have better photocatalytic characteristics. The crystallinities of the $NaTaO_3$ samples were well improved as they were treated with calcinations. The photo-energies of the $NaTaO_3$ were lower as they were treated with calcinations. Photo-energies of the as-hydrothermal synthesized $NaTaO_3$ and that treated with post-calcination at 800°C were 3.9 and 3 eV, respectively.

1. INTRODUCTION

Dye pollution coming from textile printing and dyeing processes in the outflow water has become a major source of environmental pollution. Even now people can't find an efficient and useful method to reduce the dye residuals to an acceptable level [1–3].

• E-mail: minyen@webmail.nkmu.edu.tw.

Over the last decade, a lot of efforts have been devoted to the development of photocatalysts used for degrading organic water pollutants [4–6]. The photocatalytic process was an important technology for destructing organic pollutants resulting in eco-friendly products. The pollutant conversion was conducted directly in the contaminant source. Moreover the photocatalysts are recyclable materials. The photocatalytic reaction occurs as photo-induced molecular reaction takes place at the catalyst surface. Electron/hole pairs with free electrons (e^-) escaped in the empty conduction band leaving positive holes (h^+) in the valence band will be generated when the photocatalyst is under excitation using a line with a higher energy than the photocatalyst's band gap. As the electron-hole pairs diffuse out to the surface of photocatalyst a series of chemical reactions occurs and water (H_2O) is split into OH^{-1} and oxygen. The OH^{-1} is so highly active to react with organic materials (R-H) into CO_2 and H_2O. Through such a reaction toxic organic and inorganic materials will be dissociated in air or solution.

Semiconductor oxide photo-catalysis for solving environmental pollution has many clear advantages as compared to other oxidation processes such as low cost, no addition of chemicals, and room temperature operation. Recently semiconductor oxides of titanium (Ti), bismuth (Bi), zinc (Zn) and tin (Sn) have received much attention for use as photocatalyst materials [2, 7–9]. TiO_2 with high photosensitivity, photochemical stability and large band gap (~388 nm) is widely used as semiconductor photocatalyst. As TiO_2 is used in the photocatalytic purification, a UV lamp with an exciting line of energy higher than that of TiO_2 should be applied. A lot of efforts have been devoted to enhance photochemical response from UV to visible/solar radiation. Recently $NaTaO_3$ with perovskite-type alkali tantalates structure were found to have better photocatalytic activities for splitting water under UV illumination than those of TiO_2 photocatalysts [10, 11].

Up to now $NaTaO_3$ photocatalysts were prepared by solid state reaction [12], sol–gel synthesis [13], molten-salt flux method [14], spray pyrolysis [15], hydrothermal synthesis [16, 17], solvothermal method [18], alkalide reduction [19], and microwave synthesis [20]. Most of the commercialized semiconductor photocatalysts are produced by high temperature solid state reaction at over 1000°C and the resulting powder requires milling. The resulting particle size is lack of control.

It is a long-time and high-temperature process and its grain size is larger with a lower photocatalytic activity [21]. An alternative synthesis is needed for obtaining good photocatalyst with uniform and fine size particle. Hydrothermal technique is thought to be a good choice for synthesizing high performance photocatalyst. It has the advantages of homogeneity and high purity and is widely applied to prepare nano-scale photocatalysts. As compared to other semiconductor fabrication technique, hydrothermal is a simple wet-chemical synthesis method with advantages of low cost of preparing nano-materials at a lower temperature.

In this work, hydrothermal method was used to synthesize $NaTaO_3$ photocatalysts. Sodium hydroxide (NaOH) and tantalum pentaoxide (Ta_2O_5) are used for the raw materials for synthesizing tantalum sodium ($NaTaO_3$). The hydrothermal synthesization of tantalum sodium is conducted at a temperature of 140 – 200°C, and then treated with calcinations at 800 – 900°C for 12 hr. The structural properties of $NaTaO_3$ are analyzed using X-ray diffraction (XRD), scanning electron microscopy (SEM), and energy dispersive spectroscopy (EDX).

2. Experimental Methods

Commercial Ta_2O_5 (99.99%) and semiconductor grade NaOH were used as the start materials. The start materials of 2.4 g NaOH and 0.882 g Ta_2O_5 with Ta:OH in a mole ratio of 1:15 were mixed in 40 ml deionized water and magnetic stirred for 30 min. The solution was then poured into a Teflon container. The container was sealed in an autoclave shielded with stainless steel outside. Figure 1 shows the process of the hydrothermal synthesis of $NaTaO_3$. The autoclave was put into a circular fan oven for synthesis at 140 – 200 °C for 12 hr. After the synthesis a centrifugal process was used to separate residual solvent and solid products in two layers. The upper layer of the residual solvent was poured out and then fresh deionized water was filled to mix with the solid products again waiting for next centrifugal process. The repeat centrifugal and filling fresh deionized water processes were called washing processes for eliminating the residual solvent at ambient temperature. As washing process was conducted several times purified products were collected and sent into an oven for drying the powder. The obtained powders were cooled to room temperature naturally. In order to investigate the influence of high temperature annealing on the characteristics of the powders, some of the as-hydrothermal-synthesized powders were treated with calcination processes. The calcinations process was carried out at 800 – 900 °C for 12 hr in ambient air with a heating rate of 15 °C /min and then the as-calcined samples were cooled down naturally.

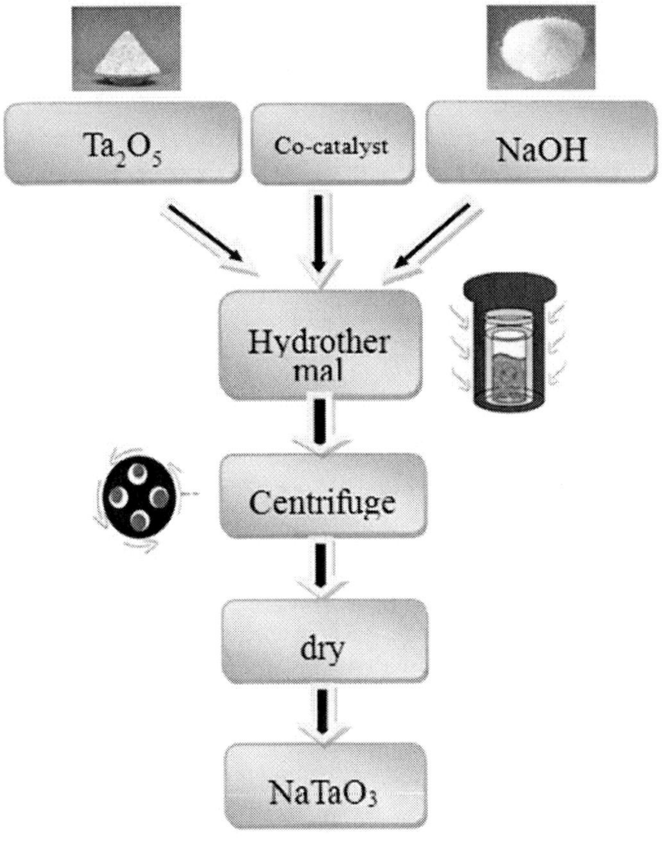

Figure 1. Scheme of hydrothermal synthesis of NaTaO3.

Structural properties of all samples were examined by X-ray diffraction (XRD) using PANalytical X'Pert Pro with Cu Kα radiation (θ = 1.54065 Å). The operation voltage and current were set at 45 kV and 40 mA, respectively. The 2θ scan range is 20-80° with a scan rate of 1°/min. Scanning electronic microscope (JEOL ISM-6700F) was used to characterize the morphology and microstructure. EDX analysis was measured using an OXFORD INCA ENERGY 400. The analyzer 1280 from n&k Technology, scanning from 190 to 1000 nm, was used to measure the UV-visible spectra. The composition ratios of the $NaTaO_3$ powders were evaluated by using energy dispersive spectrometry (EDS). The average absorption coefficient was derived from absorption coefficients (α) of the CZTS in the range of 400 – 700 nm. The optical energy gap of the $NaTaO_3$ sample was determined by extrapolating the linear part of the square of the absorption coefficient of the CZTS as a function of photon energy [22]. Raman scattering measurements were conducted under excitation by a 532-nm LASER line. The LASER beam was focused to a spot of ~1 μm in diameter through a 100× magnification objective lens.

3. Results and Discussions

Figure 2 shows XRD patterns for the samples prepared by hydrothermal synthesis at 180°C for 12 hr and then treated with calcinations at different temperatures for 12 hr. The as-hydrothermal synthesized sample reveals perovskite structures of $NaTaO_3$ as recognized from the XRD reflection planes of (020), (200), (220), (202), (301), (321) and (400) according to JCPD card 25-0863. Besides the $NaTaO_3$ structures NaO_3, Na_2O_2, $Na_xTa_xO_x$ and residual starting Ta_2O_5 were found in the samples. The existence of residual starting material of Ta_2O_5 means that it does not thoroughly react with NaOH. The concentration of NaOH has been reported to be a critical factor for synthesizing NaOH [23]. Lower concentration (< 0.1 mol/L) of NaOH for synthesizing with Ta_2O_5 would obtain un-reacted Ta_2O_5 in products. They used 0.5 mol/L in the synthesis with 0.442 g Ta_2O_5 and the result reveals no residual Ta_2O_5 in the products. Our NaOH concentration of 1.5 mol/L (3 times to the pointed concentration) was used to synthesize 0.882 g Ta_2O_5 (two times the report's Ta_2O_5) for the same synthesis time of 12 hr but the un-reacted Ta_2O_5 was still found in our samples. This means that the least concentration of NaOH needed for complete reaction with Ta_2O_5 is not proportional to the volume of Ta_2O_5 used for the same reaction time. For complete reaction of Ta_2O_5 in synthesizing $NaTaO_3$ the required synthesis temperature and time are not proportional to the Ta_2O_5 volume and NaOH concentration. The complete-hydrothermal reaction of Ta_2O_5 into $NaTaO_3$ needs enough concentration of NaOH and sufficient reaction time at a suitable reaction temperature. In our synthesis system the Ta_2O_5 complete reaction time of more than 12 hr is required. The XRD patterns in Figure 2 show that the $NaTaO_3$ powders have perovskite structures which are thought to have better photocatalytic characteristics. Figure 2 (b)–(d) revealed the XRD crystallinities of $NaTaO_3$ samples as-calcined at different temperatures for 12 hr. The crystallinities of the $NaTaO_3$ samples were well improved as the samples were treated with calcinations as compared to the non-calcined sample.

The crystallinities were readily improved as the $NaTaO_3$ samples were treated with calcinations. That means that calcination temperature plays an important role for obtaining a better crystalline $NaTaO_3$. In report [24] they used NaOH and Ta_2O_5 as starting materials and directly calcine them into $NaTaO_3$. Their $NaTaO_3$ XRD-crystal-diffraction peaks appeared when the calcination temperature was higher than 700°C. As compared to the report by XRD characterization our crystalline $NaTaO_3$ was obtained at a lower temperature (180°C) using the method of hydrothermal synthesis.

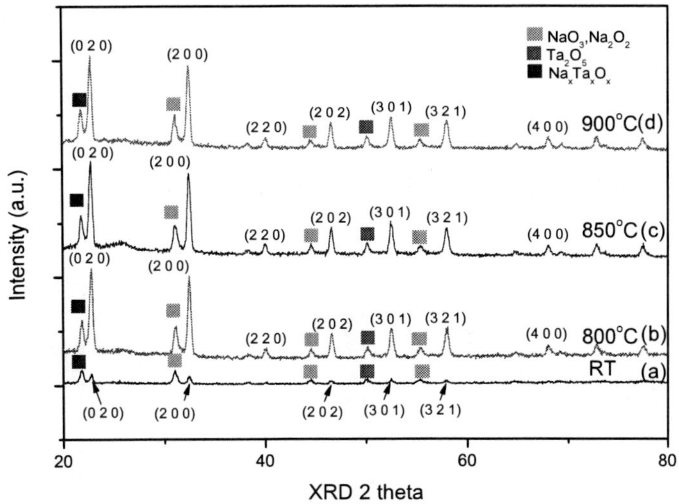

Figure 2. XRD patterns of NaTaO3 as-hydrothermal-synthesized at (a) 180 °C, and then treated with calcination at (b) 800°C, (c) 850°C, and (d) 900°C.

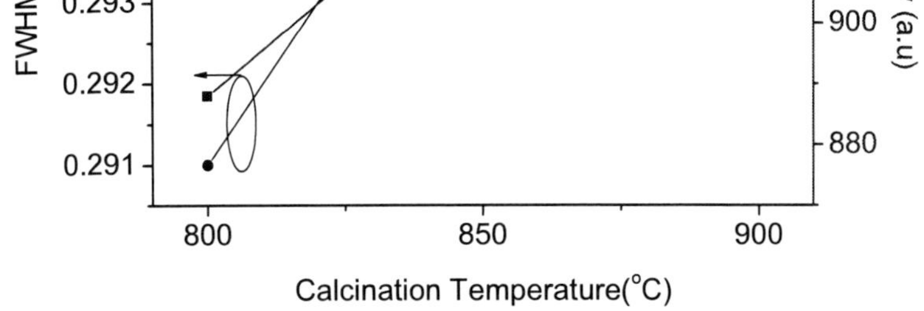

Figure 3. XRD (020) FWHM and intensity of as-hydrothermal synthesized $NaTaO_3$ treated with calination at different temperatures.

Figure 3 shows the crystalline properties of $NaTaO_3$ as a function of calcination temperature. The crystallinity of the as-calcined $NaTaO_3$ was characterized by full width at half maximum (FWHM) and intensity of the XRD (020) peak. A better crystallinity of

$NaTaO_3$ calcined at a temperature of 800°C was obtained as evaluated from the FWHM values.

The morphologies of $NaTaO_3$ photocatalysts were determined by scanning electron microscopy (SEM). As shown in Figure 4, all of the photocatalyst grains reveal cubic structures. The grain sizes are much smaller than that obtained by traditional solid state reaction method (> 1 μm) [25]. It is also obvious that grain size as shown in Figure 4(a) is much uniform of ~200 nm. This means that the reaction is homogenous using hydrothermal synthesis in this work. As the samples were treated with calcinations part of the $NaTaO_3$ side by side were aggregated to form larger cube structures. Since the aggregation of $NaTaO_3$ during calcinations is not a homogenous process, the resulting $NaTaO_3$ cubic grains are not uniform in the range of 200 – 400 nm as shown in Figure 4(b)-(d). The cubic-grain sizes of the $NaTaO_3$ are summarized in Table 1.

Figure 4. SEM morphologies of $NaTaO_3$ hydrothermal synthesized at (a) 180 °C and then treated with post-calcination at (b) 800°C, (c) 850°C and (d) 900°C.

Table 1. Grain size of as-hydrothermal synthesized $NaTaO_3$ and that treated with calcinations

	Grain size (nm)			
Calcination / Hydrothermal	RT	800 °C	850 °C	900 °C
180 °C	200 – 220	200 – 400	200 – 400	200 – 400

The analyzer 1280 from n&k Technology, scanning from 190 to 1000 nm, was used to measure the UV-visible spectra. The UV–visible diffuse transmittance spectra of the $NaTaO_3$

powder samples were applied to obtain their absorption coefficients. From [26], the absorption coefficient of material has a dependence on material's energy gap,

$$\alpha = A(h\nu - E_g)^{1/2}, \tag{1}$$

where A is a constant, $h\nu$ is the photon energy and E_g is the energy band gap. A direct transition can then be found by the linear dependence of $(\alpha h\nu)^2$ on $h\nu$. Thus optical-energy gap can be determined by the cross-point when we extrapolate the linear part of the square absorption coefficients as a function of photon energy to cross the axis corresponding to factor of zero absorption. Figure 5 shows the photo-energies of $NaTaO_3$. The as-hydrothermal synthesized $NaTaO_3$ had a photo-energy of ~ 3.9 eV (317.9 nm) and it is sensitive to UV-light irradiation. As the $NaTaO_3$ was treated with further calcinations its photo-energies were lower to a value of ~ 3 eV (413 nm). Nowadays most efforts have been devoted to develop high sensitive photo-catalysts by doping itself using rare-earth metal elements [27, 28] or non-metal elements [24]. Another effort was devoted to develop high sensitive photo-catalysts for use in a near visible band or directly in solar irradiation. Figure 5 reveals that the photo-energies of the $NaTaO_3$ are lower as they are treated with calcinations. A further degrade experiments of the samples related to different light sources are necessary to obtain the photocatalytic efficient data.

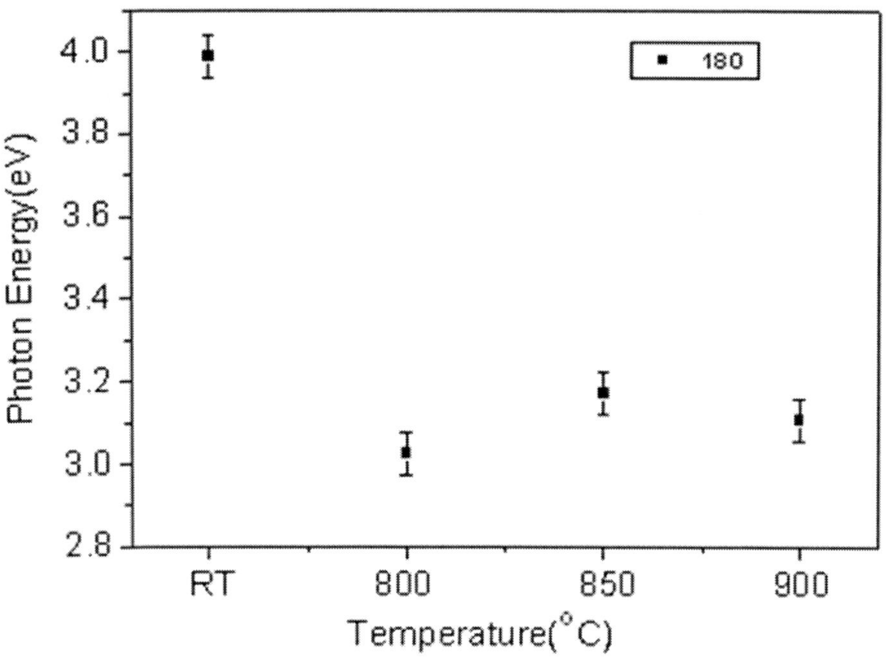

Figure 5. Photo-energies of as-hydrothermal synthesized (RT) $NaTaO_3$ and samples treated with post-calcination.

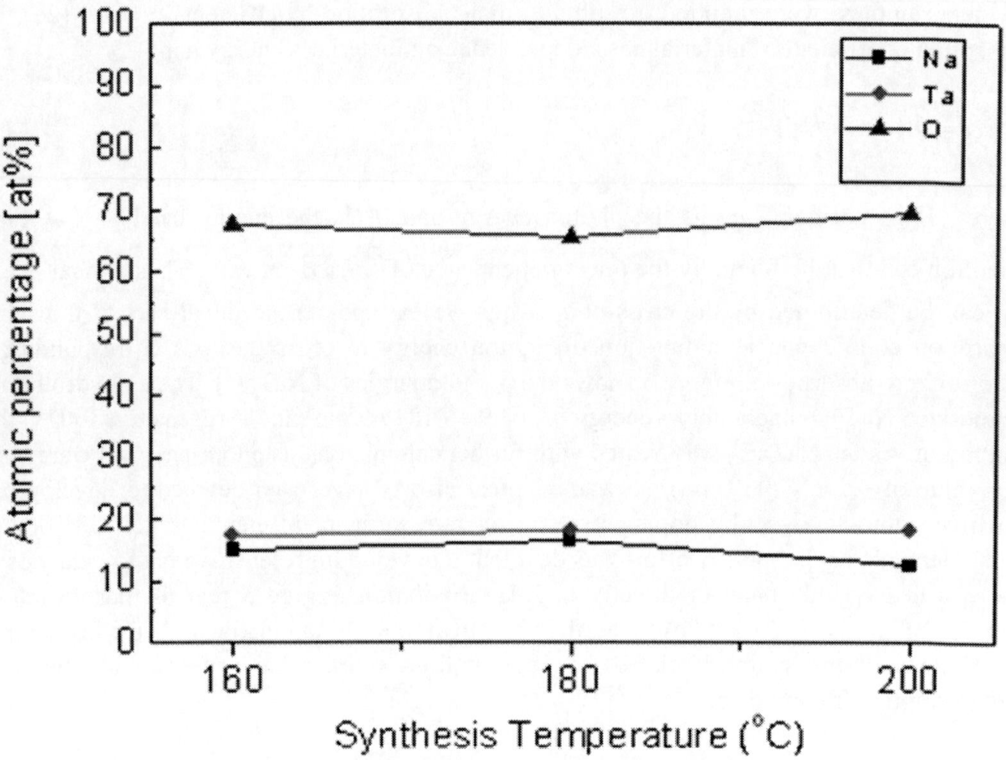

Figure 6. Composition ratio of $NaTaO_3$ as a function of synthesis temperature.

Figure 6 reveals the compositions of $NaTaO_3$ synthesized at different temperatures using hydrothermal methods. The obtained oxygen compositions in the samples were higher than the stoichiometry composition of $NaTaO_3$. The Na compositions were a little lower than Ta. From Figure 6, the $NaTaO_3$ close to stoichiometry composition were obtained by hydrothermal synthesis at a temperature of 180°C for 12 hr.

CONCLUSION

In this work, hydrothermal method was used to synthesize $NaTaO_3$ photocatalysts. Sodium hydroxide (NaOH) and tantalum pentaoxide (Ta_2O_5) are used for the raw materials for synthesizing tantalum sodium ($NaTaO_3$). The hydrothermal synthesization of tantalum sodium is conducted at a temperature of 140 – 200°C. The $NaTaO_3$ close to stoichiometry composition were obtained by hydrothermal synthesis at a temperature of 180°C for 12 hr. The $NaTaO_3$ powders were perovskite structures which are thought to have better photocatalytic characteristics. The calcinations process was carried out at 800 – 900°C for 12 hr. The crystallinities of the $NaTaO_3$ samples were well improved as they were treated with calcinations as compared to the non-calcined samples. The photo-energies of the $NaTaO_3$ were lower as they were treated with calcinations. Photo-energies of the as-hydrothermal synthesized $NaTaO_3$ and that treated with post-calcination at 800°C were 3.9 and 3 eV, respectively.

REFERENCES

[1] Özer, D.; Dursun, G.; Özer, A., *J. Hazard. Mater.*, 2007, vol. 144, 17.
[2] Zhang, W. W.; Zhang, J. Y.; Chen, Z. Y.; Wang, T. M., *Catal. Commun.*, 2009, vol. 10, 1781.
[3] Franco, A.; Neves, M. C.; Ribeiro Carrott, M. M. L.; Mendonc, M. H.; Pereiraa, M. I.; Monteir, O.C. *J. Hazard. Mater.*, 2009, vol. 161, 545.
[4] Hoffmann, M. R.; Martin, S. T.; Choi, W.; Bahnemann, D. W., *Chem. Rev.*, 1995, vol. 95, 69.
[5] Huang, H. H.; Tseng, D. H.; Juang, L. C., *J. Hazard. Mater.*, 2008, vol. 156, 186.
[6] Pouretedal, H. R.; Norozi, A.; Keshavarz, M. H.; Semnan, Abolfazl, *J. Hazard. Mater.*, 2009, vol. 162, 674.
[7] Hua, X. S.; Zhang, Y. J.; Ma, N. H.; Li, X. F.; Wang, H. W., *J. Hazard. Mater.*, 2009, vol. 172, 256.
[8] Li, L. Z.; Yan, B., *J. Alloy Compd.*, 2009, vol. 476, 624.
[9] Hou, L. R.; Yuan, C. Z.; Peng, Y., *J. Hazard. Mater.*, 2007, vol. B139, 310.
[10] Kato, H.; Asakura, K.; Kudo, A., *J. Am. Chem. Soc.*, 2003, vol. 125, 3082.
[11] Liua, J. W.; Chen, G.; Lia, Z. H.; Zhang, Z. G., *Int. J. Hydrogen Energy*, 2007, vol. 32, 2269.
[12] Kato, H.; Asakura, K.; Kudo, A., *J. Am. Chem. Soc.*, 2003, vol. 125, 3082.
[13] Hu, C. C.; Teng, H., *Appl. Catal. A: General*, 2007, vol. 331, 44.
[14] Porob, D. G.; Maggard, P. A., *J. Solid State Chem.*, 2006, vol. 179, 1727.
[15] Kang, H. W.; Park, S. B., *Advanced Powder Technology*, 2010, vol. 21, 106.
[16] Liu, J. W.; Chen, G.; Li, Z. H.; Zhang, Z. G., *Int. J. Hydrogen Energy*, 2007, vol. 32, 2269.
[17] He, Y.; Zhu, Y. F.; Wu, N. Z., *J. Solid State Chem.*, 2004, vol. 177, 3868.
[18] He, Y.; Zhu, Y., *Chem. Lett.*, 2004, vol. 33, 900.
[19] Nelson, J. A.; Wagner, M. J., *J. Am. Chem. Soc.*, 2003, vol. 125, 332.
[20] Yan, S. C.; Wang, Z. Q.; Li, Z. S.; Zou, Z. G., *Solid State Ionics*, 2009, vol. 180, 1539.
[21] Xu, J. S.; Xue, D. F.; Yan, C. L., *Mater. Lett.*, 2005, vol. 59, 2920.
[22] Kamoun, N.; Bouzouita, H.; Rezig, B., *Thin Solid Films*, 2007, vol. 515, 5949.
[23] He, Y.; Zhu, Y.; Wu, N., *Journal of Solid State Chemistry*, 2004, vol. 177, 3868.
[24] Liu, D. R.; Wei, C. D.; Xue, B.; Zhang, X. G.; Jiang, Y. S., *Journal of Hazard. Mater.*, 2010, vol. 182, 50.
[25] Torres-Martínez, L. M.; Cruz-López, A.; Juárez-Ramírez, I.; Meza-de la Rosa, M. E., *J. Hazard. Mater.*, 2009, vol. 165, 774.
[26] Pankove, J. I.; Optical Processes in Semiconductors; Dover Publication: New York, 1975; pp 34-95.
[27] Iwase, A.; Kato, H.; Okutomi, H.; Kudo, A., *Chem. Lett.*, 2004, vol. 33, 1260.
[28] Porob, D. G.; Maggard, P. A., *Journal of Solid State Chemistry*, 2006, vol. 179, 1727.

In: Advanced Materials - Studies and Applications
Editors: I.A. Parinov, S.-H. Chang, S.Theerakulpisut
ISBN: 978-1-63463-749-7

Chapter 3

CHARACTERIZATION OF ANTIMONY DOPED ZNO FILMS DESIGNED BY PULSED LASER DEPOSITION

A. S. Puzikov*[1]*, N. V. Lyanguzov*[2]*, D. G. Nesvetaev*[1]*,
***I. N. Zakharchenko*[1] *and E. M. Kaidashev*[1,2,*]**
[1]Vorovich Mechanics and Applied Mathematics Research Institute, Southern Federal University, Rostov-on-Don, Russia
[2]Physics Department, Southern Federal University, Rostov-on-Don, Russia

ABSTRACT

High quality epitaxial films of ZnO doped by 0.7 and 2.3 at. % of Sb on sapphire substrates were grown by PLD and *in situ* annealed in oxygen. The variation of electrical, structural and lattice vibration properties were systematically studied in a dependence of oxygen pressure used for the films growth. Both ZnO lattice parameters *a* and *c* were expanded upon Sb doping. Hall measurements showed that the films have conductivity of n-type. The films with 0.7 at. % of Sb grown at higher oxygen pressure show lower electon density. The conductivity of Sb doped films is decreased for films with higher Sb content. The experimantal conditions to produce ZnO:Sb (0.7 at. %) films with high optical quality were found. The intensity of the A_1(LO) phonon mode is found to be higher for ZnO:Sb films with higher electron concentration.

Keywords: Pulsed laser deposition, antimony doped ZnO films, X-ray diffraction, energy-dispersive X-ray microanalysis, Hall measurements, Raman spectroscopy

* Corresponding author: E. M. Kaidashev. Vorovich Mechanics and Applied Mathematics Research Institute, Southern Federal University, Stachki 200/1, 344090, Rostov-on-Don, Russia. Physics Department, Southern Federal University, Zorge 5, 344090, Rostov-on-Don, Russia. E-mail: kaidashev@mail.ru.

1. Introduction

Being a wide band gap semiconductor with a large exciton binding energy of 60 meV, ZnO is an ideal host material for applications in photonics, spintronics and photocatalysis. p-type doping of ZnO without degradation of the optical quality is the major challenge limiting the design of many devices of photonics such as laser light emitting diodes based only on ZnO semiconductor. Antimony doped p-type ZnO thin films were grown by pulsed laser deposition (PLD), molecular beam epitaxy (MBE), and thermal oxidation recently [1,2,3]. However, the experimental conditions reported by different groups are sometimes in controversy and the effects responsible for the p-type conductivity in ZnO are poorly studied yet. Below we report the systematic study of the influence of the oxygen pressure on the resistivity, carriers concentration and carrier mobility of ZnO films doped by 0.7 and 2.3 at% of Sb. The structural changes, film composition and the crystal structure dynamics change upon doping are also characterized.

2. Experimental

Antimony doped ZnO films were deposited by PLD on c-Al_2O_3 substrates at the temperature of 550 °C. The radiation of KrF laser ($\lambda = 248$ nm) was focused on the surface of ZnO:Sb_2O_3 ceramic target with Sb concentration of 1 or 2 at % to have a fluence of ~2 J/cm^2. The distance between target and substrate was 50 mm. The films were deposited at different oxygen ambient pressures in the range of 2×10^{-3} – 5×10^{-2} mbar. To reduce the structural strains a thin ZnO buffer layers were deposited prior the ZnO:Sb film growth. The films were *in situ* annealed at 650 °C after the growth in oxygen (2×10^{-3} mbar). The XRD patterns of the samples were recorded by X-ray diffraction apparatus DRON-7 with Co Kα, Kα_1, Kβ incident radiation. The elemental composition were investigated by energy-dispersive X-ray microanalysis (EDX) which is a part of the scanning electron microscope Zeiss SUPRA 25. RT Raman spectra were recorded in a backscattering geometry using a Renishaw inVia Reflex Raman spectrometer. The 514.5 nm line from a cw Ar+ ion laser was used for excitation. The samples were irradiated in the direction normal to a substrate, and Raman spectra were recorded in the reverse direction without an analyzer, which corresponds to scattering geometry of *z(xx)z'* and *z(xy)z'*. The electrical properties were studied at room temperature by Hall measurements using Van der Pauw configuration. Photoluminescence (PL) measurements were carried out under excitation by 325 nm He–Cd laser line at room temperature. The samples were excited under the angle of 15° to the surface normal with the PL emission being collected along the surface normal. The emission was dispersed by a Solar TII MSDD1000 monochromator (1800 mm^{-1}) with spectral resolution better than 0.1 nm.

3. Results and Discussion

EDX spectra and corresponding SEM images of antimony doped ZnO films are showen in Figure 1. The films deposited using ZnO targets doped by 1 and 2 at % of Sb showed the antimony concentration of 0.7 and 2.3 at %, correspondingly. Antimony concentrations in resulting films deposited at oxygen pressure of 5×10^{-3} and 3×10^{-2} are shown in Table 1. The oxygen pressure variation does not affect the films stoichiometry in this pressure range.

The XRD spectra of ZnO: Sb films show intense lines corresponding to the family of ZnO {001} planes (Figure 2). The films thickness was measured to be ~400 nm, which is not sufficient to fully absorb the incident X-ray beam.

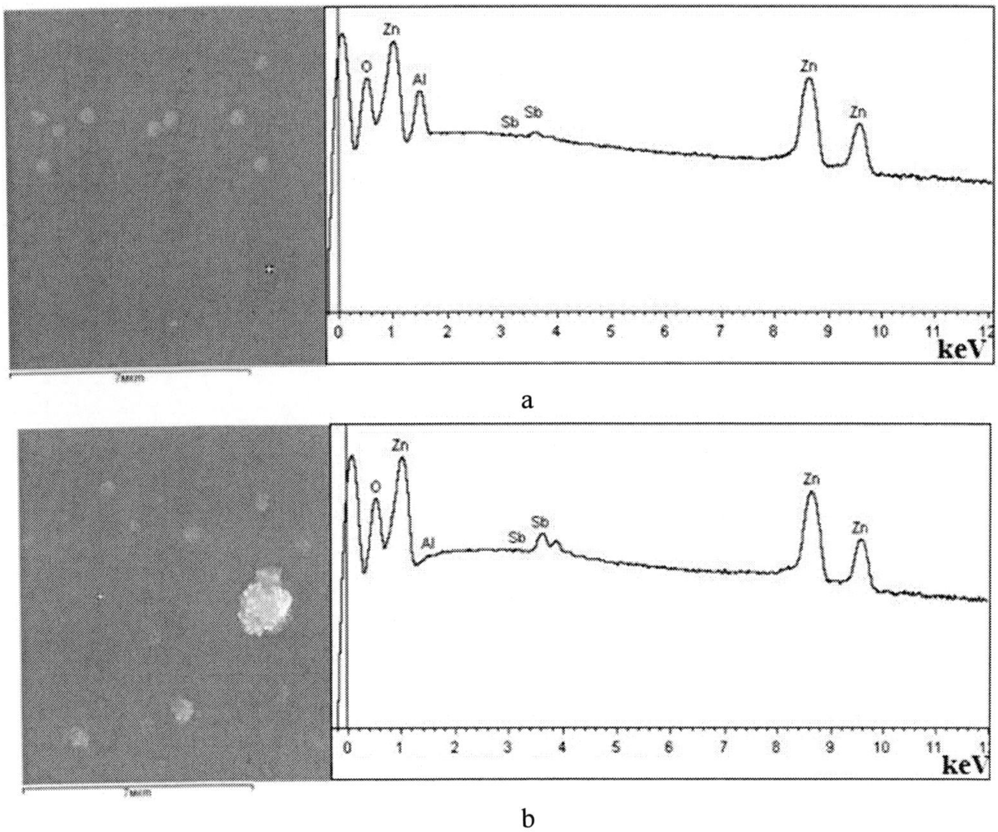

Figure 1. EDX spectra (right) and SEM images (left) of ZnO: Sb films designed using targets with antimony concentration of 1 at % (a) and 2 at % (b).

Table 1. Sb concentration of ZnO:Sb films deposited at various oxygen pressures

Concentration of Sb in ceramic targets, at. %	O_2 pressure, mbar	Sb concentration in film, at. %
1	5×10^{-3}	0.7
	3×10^{-2}	0.71
2	5×10^{-3}	2.3
	3×10^{-2}	2.4

Thus, Al_2O_3 (110) lines were present in XRD spectra. The presence of such set of peaks claims that crystallographic orientations are $(001)_{ZnO} \parallel (110)_{Al2O3}$. Thus, the film is epitaxial, or at least an axial texture is present. However, a wide band around 22° related to Sb secondary phase was detected. The evidence of epitaxial growth was supported by φ-scan of {114} plane (Figure 3).

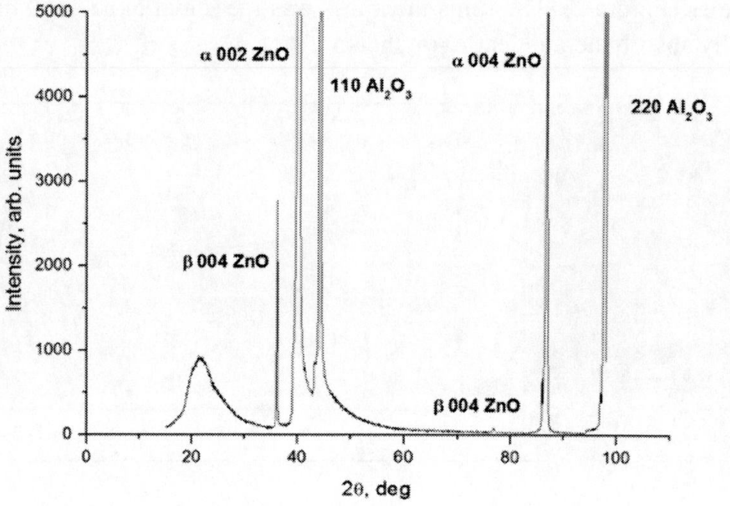

Figure 2. XRD pattern of as grown and annealed ZnO:Sb thin film deposited at 2×10^{-3} mbar and 550 °C.

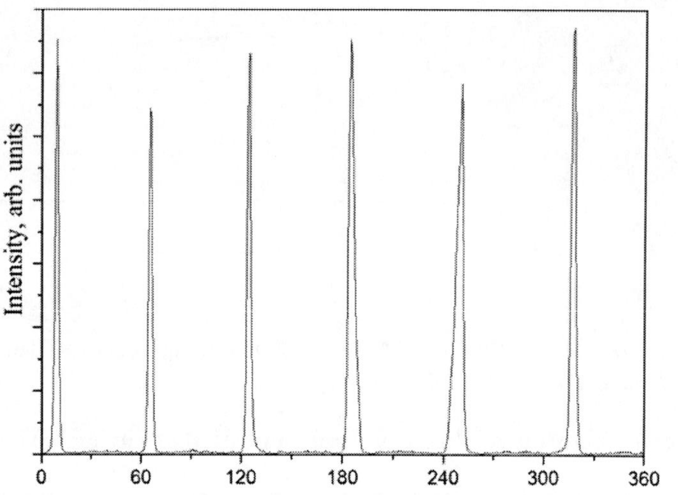

Figure 3. φ-scan of {114} plane of ZnO:Sb thin film deposited at the temperature 550 °C at oxygen pressure of 2×10^{-3} mbar. The film was *in situ* annealed at 650 °C in oxygen ambient pressure of 2×10^{-3} mbar after the growth.

Crystal lattice parameters *c* and *a* were also calculated from XRD data for all the samples. Both parameters were found to be greater than those known for a single crystalline ZnO. The presence of mechanical stresses in the ZnO films upon Sb doping can be explained by slightly larger ionic radius of Sb^{3+} (0.076 nm) than that of Zn^{2+} (0.074 nm). Indeed, Sb atoms likely occupy Zn sites rather those of O because the ionic radius of Sb^{3+} is comparable with the one of Zn^{2+} and much larger than that of O^{2-}. The phenomenon of the p-type conductivity in Sb doped ZnO is normally explained in terms of a Sb_{Zn}-$2V_{Zn}$ complex formation [4]. In this case two Zn atoms should be missing and a lattice shrinkage is expected [5]. However the lattice expansion was observed in our case instead of shrinkage reported for ZnO:Sb films of a p-type [5]. Thus, it seems that the substitution of a Zn ion by Sb ion occurred without the loose of the second Zn atom. Hall measurements showed a conductivity of n-type for all the samples studied. The electron concentration n, mobility μ and the sample resistivity ρ are given in Table 2. The electron concentration and resistivity of the films versus oxygen pressure used during deposition are shown in Figure 4.

It is remarkable that for the series of films deposited by ablating ZnO targets with 1 at. % and 2 at. % of Sb the carrier concentration was varied by change the oxygen content. The films deposited using ZnO:Sb (1 at. %) show the decrease of the carriers (electrons) concentration for the increased oxygen pressure. It seems, that the amount of an oxygen vacancies decreases for higher ambient pressure. However, all the films show reliable n-type of conductivity even under high pressure regime.

The electron concentration was decreased more than for an order of magnitude by the increased twice Sb concentration in the target and varied between $\sim 5\times 10^{18}$ and 4×10^{15} cm^{-3}. A local electron concentration maximum of 5×10^{18} cm^{-3} and the corresponding resistivity minimum of 3.017×10^{-2} Ω • cm were observed for the films grown at the oxygen pressure of 7×10^{-3} mbar (Figure 4b). Raman spectra of Sb doped ZnO films (1 and 2 at. %) deposited at the same oxygen pressure of 5×10^{-3} mbar are shown in Figure 5. All Raman spectra reveal a broad band in the range of 500-600 cm^{-1} (Figure 5). The band has 3 pronounced maxima at 511, 531 and 575 cm^{-1}. The last one is well known and corresponds to A_1(LO) phonon of ZnO. Remaining peaks were attributed to Sb-related in-band modes which are coupled to the LO phonon modes of the host lattice [6].

Table 2. Electrical properties of ZnO: Sb films measured at room temperature

Concentration of Sb in target, at. %	Oxygen pressure, mbar	Resistivity, Ω·cm	Mobility, cm^2/V·s	Electron concentration, cm^{-3}
1	2×10^{-3}	3.187×10^{-3}	45.41	4.313×10^{19}
	$5\cdot 10^{-3}$	4.187×10^{-3}	38.08	3.915×10^{19}
	$7\cdot 10^{-3}$	4.728×10^{-3}	38.98	3.387×10^{19}
	$3\cdot 10^{-2}$	6.298×10^{-3}	35.97	2.756×10^{19}
	$5\cdot 10^{-2}$	2.114×10^{-2}	20.94	1.409×10^{19}
2	2×10^{-3}	1.231	37.01	1.37×10^{17}
	5×10^{-3}	1.517×10^{-1}	12.48	3.294×10^{18}
	7×10^{-3}	3.017×10^{-2}	42.30	4.893×10^{18}
	10^{-2}	5.226×10^{-1}	13.08	9.128×10^{17}
	2×10^{-2}	40.32	38.58	4.013×10^{15}

All the samples show conductivity of n-type.

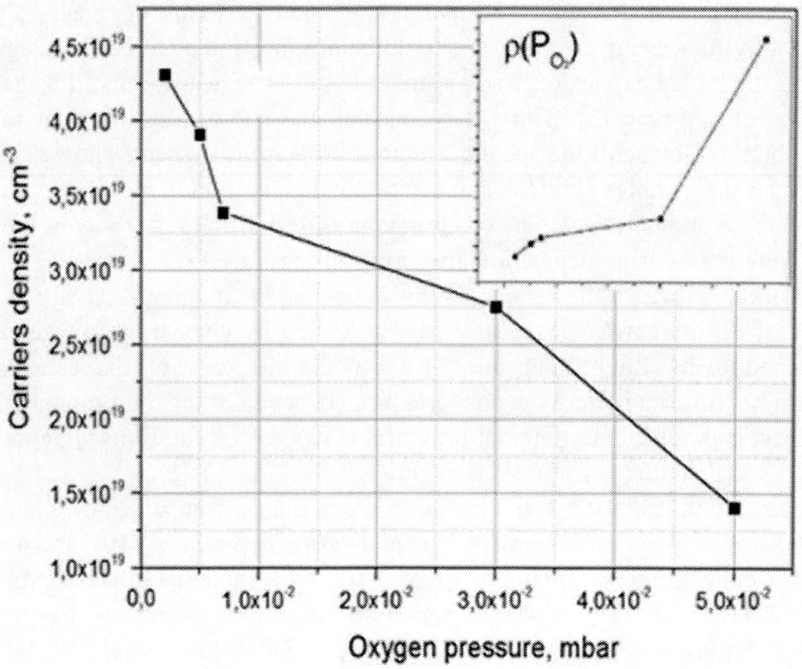

a

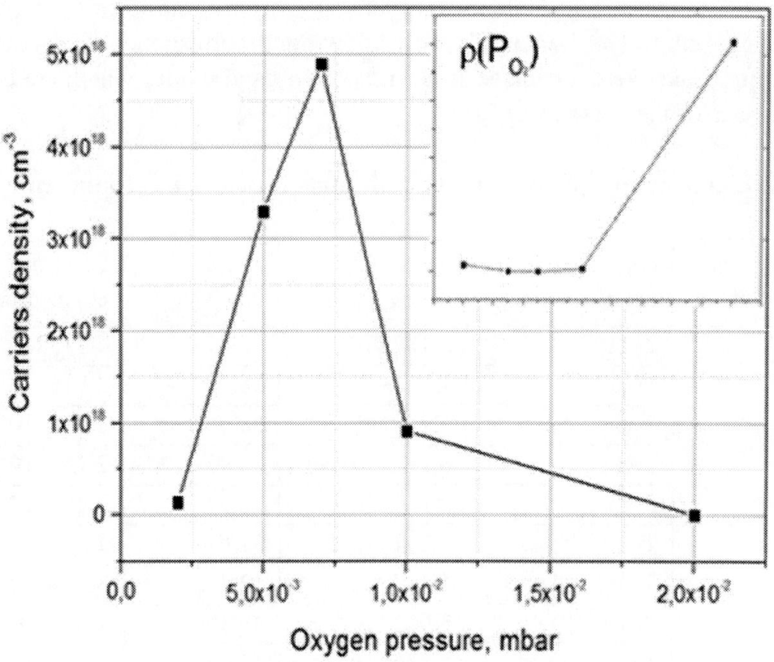

b

The insets show the resistivity of the samples versus oxygen pressure.

Figure 4. The electron concentration of the ZnO:Sb films versus oxygen pressure used during the film deposition by ablating ZnO targets with 1 at. % (a) and 2 at. % of Sb (b) respectively.

Another Sb doping induced vibration near 277 cm^{-1} was detected similar to the cases when ZnO was doped by gallium, aluminum and iron [7].

The phonon at 365 cm^{-1}, was not observed in antimony doped ZnO films previously, but it was detected in the phosphorus doped zinc oxide [8]. Two wide phonon modes which appear at 641 and 683 cm^{-1} upon doping were attributed to Sb–O–Sb vibrations [9-11].

ZnO films doped by 0.7 at. % of Sb deposited at higher oxygen pressure show lower carriers density. A strong correlation between A_1(LO) phonon intensity, oxygen pressure used for deposition and, as a result, the electron concentration can be followed in Raman spectra (Figure 6). The intensitiy ratio $A_1(LO)/E_2^{high}$ increases monotonically for films with higher oxygen deficiency and higher electron concentration (Figure 7).

ZnO films doped by Sb 2 at. % of Sb show similar effect (Figure 8). The maximum of the $A_1(LO)/E_2^{high}$ intensities ratio was observed for films grown at the oxygen pressure of 7×10^{-3} mbar, which corresponds to the highest concentration of electrons as in seen in Figure 4b. The position of non-symmetrical phonon E_2^{high} observed at 437 for bulk ZnO has shifted from 1.5 cm^{-1} to higher wave numbers, indicating an induced strain in the ZnO lattice upon Sb doping.

Photoluminescence of Sb doped ZnO films heavily depends on the oxygen pressure used during the growth.

Further annealing in oxygen was used for the further improvement of the optical properties. It was experimentally found, that the ZnO films doped by ~1 at. % of Sb grown at the temperature of 550 °C in ambient oxygen pressure of $2\cdot10^{-3}$ mbar and *in situ* annealed at 650 °C have very high optical quality (Figure 9).

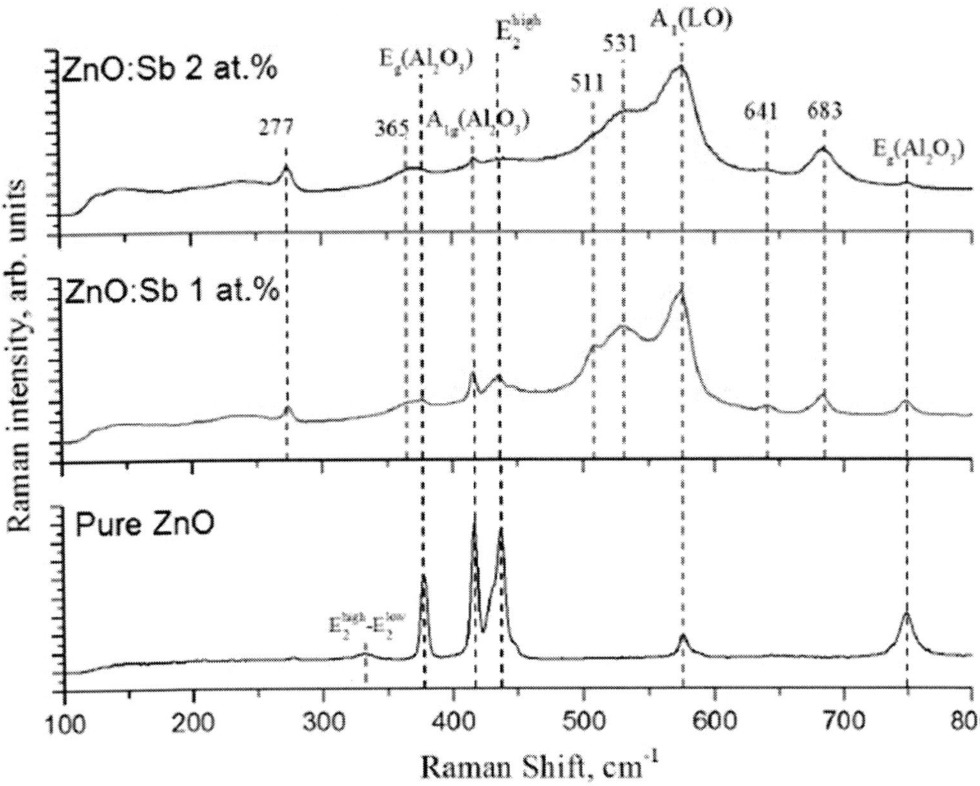

Figure 5. Raman spectra of Sb doped ZnO films (1 and 2 at. %) and spectrum of the pure ZnO film deposited at the same oxygen pressure of 5×10^{-3} mbar.

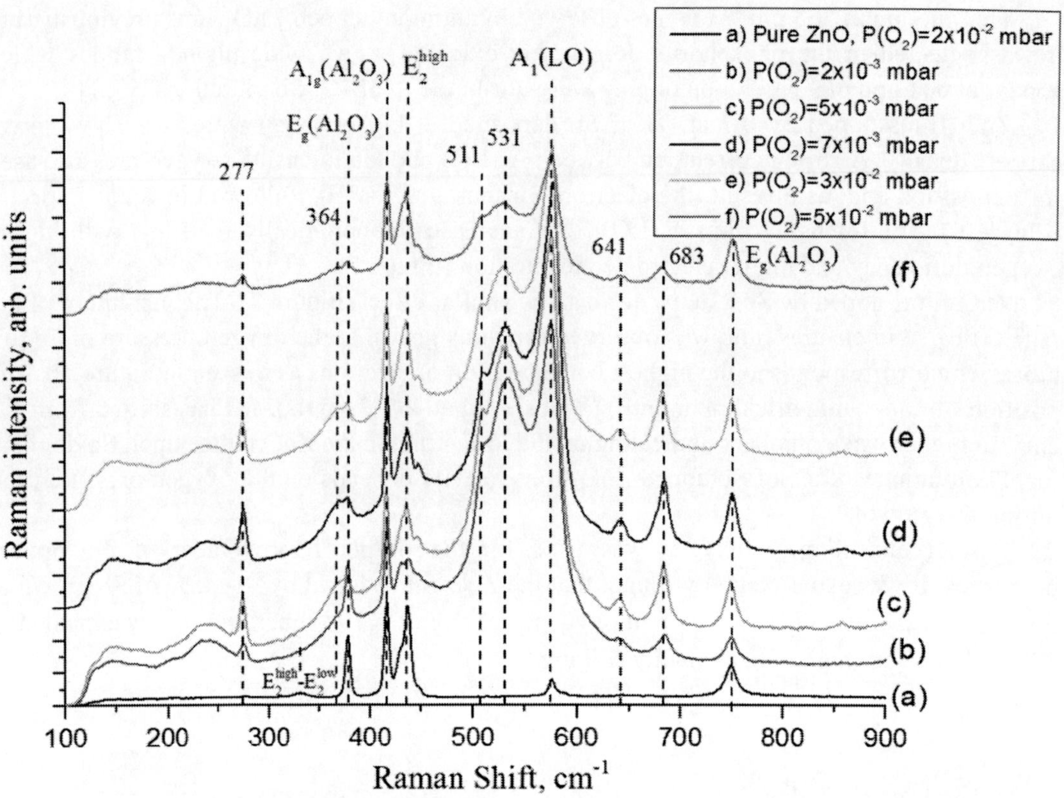

Figure 6. Raman spectra of ZnO films doped by 1 at. % of Sb.

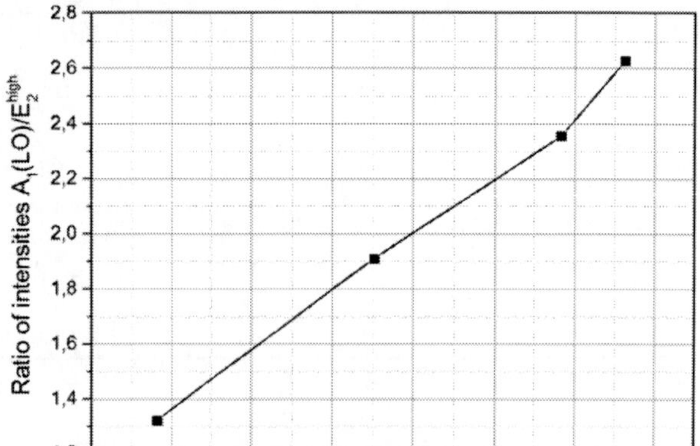

Figure 7. A_1(LO) to E_2^{high} mode intensity ratio versus electron concentration in ZnO:Sb film.

No green band in the range of 500-600 nm is detected which indicates a low concentrations of point defects related to oxygen vacancies. Meanwhile the ZnO exciton peak was red shifted upon Sb doping which is in line with correlates with narrowing of ZnO band gap upon Sb doping observed earlier [12,13].

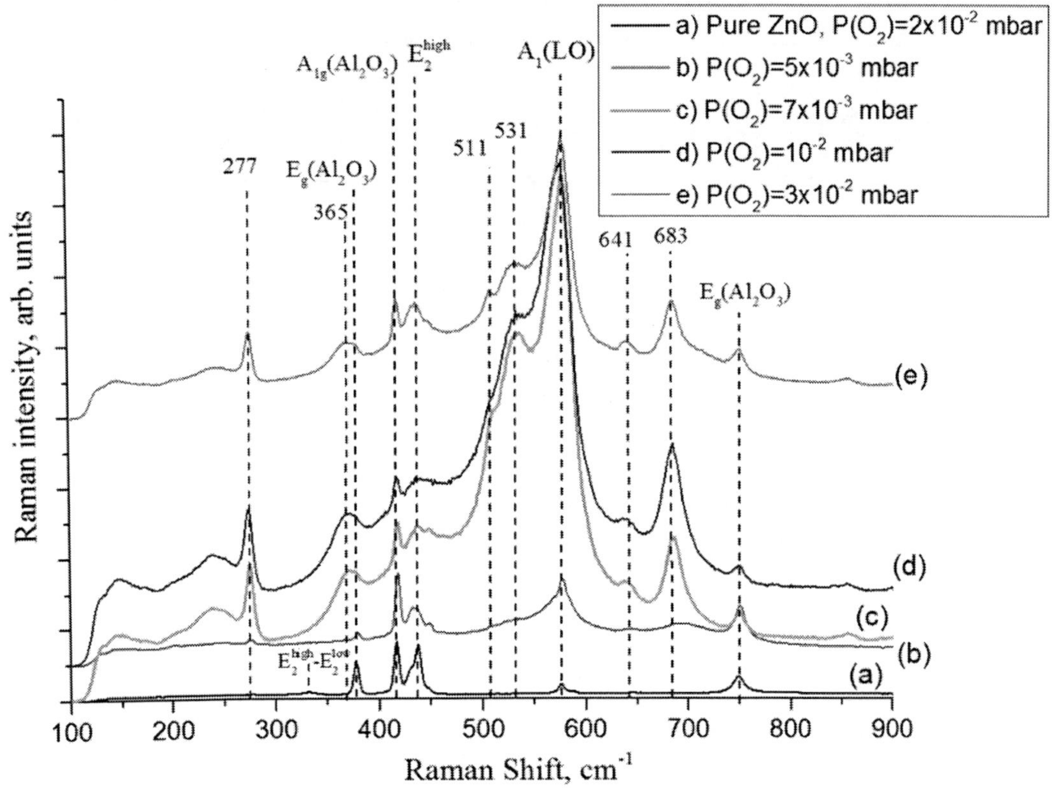

Figure 8. Raman spectra of ZnO:Sb films with Sb concentration 2 at. %.

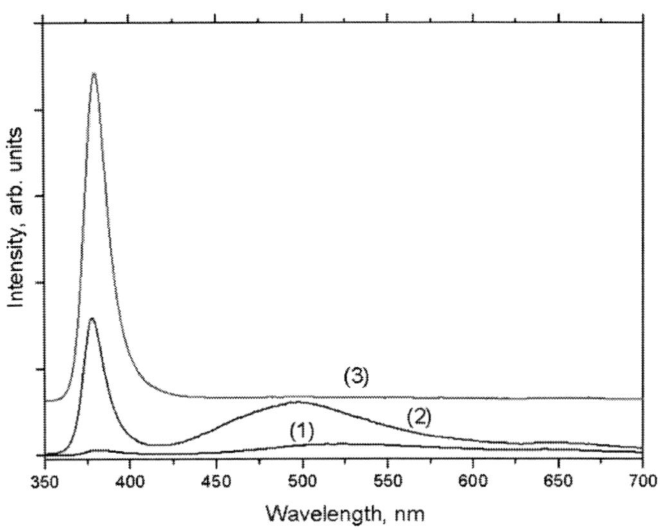

Figure 9. Room-temperature PL spectra of ZnO:Sb films (1 at. %): grown at temperature 550 °C in oxygen ($2 \cdot 10^{-3}$ mbar), annealed at 750 °C in oxygen ($2 \cdot 10^{-3}$ mbar) (1), not-annealed (2) and annealed at 650 °C in oxygen ($2 \cdot 10^{-3}$ mbar) (3).

CONCLUSION

High quality epitaxial films of ZnO doped by 0.7 and 2.3 at. % of Sb were grown on sapphire substrates by PLD. The variation of electrical, structural and lattice vibration properties were systematically studied in a dependence of oxygen pressure used for the films growth. Hall measurements showed that the films have conductivity of n-type. Both ZnO lattice parameters *a* and *c* were expanded upon Sb doping. ZnO films doped by 0.7 at. % of Sb grown at higher oxygen pressure show lower carriers density. The conductivity of Sb doped films is decreased for films with higher Sb content. ZnO:Sb (0.7 at. %) films grown at the temperature of 550 °C in ambient oxygen pressure of 2×10^{-3} mbar and annealed in-situ at 650 °C show low concentrations of defects related to oxygen vacancies and have very high optical quality. Meanwhile the ZnO exciton peak show red shift upon Sb doping. The intensity of the A_1(LO) phonon mode is found to be higher for ZnO:Sb films with higher electron concentration. A local electron concentration maximum of 5×10^{18} cm^{-3} was found for ZnO:Sb (2.3 at. %) films grown at the oxygen pressure of 7×10^{-3} mbar.

ACKNOWLEDGMENTS

This work is supported by the Russian Education and Science Ministry, the project №16.219.2014/K "Development of synthesis methods, theoretical and experimental study of zinc oxide based nanostructures to create elements of photodetectors, optical nanoantennes, piezosensors and chemosensors".

REFERENCES

[1] Aoki, T., Shimizu, Y., Miyake, A., Nakamura, A., Nakanishi, Y., Hatanaka, Y., *Phys. Status Solidi B,* 229, 911, (2002).

[2] Chu, S., Lim, J. H., Mandalapu, L. J., Yang, Z., Li, L., Liu, J. L., *Appl. Phys. Lett.,* 92, 152103 (2008).

[3] Pan, X. H., Guo, W., Ye, Z. Z., Liu, B., Che, Y., He, H. P., Pan, X. Q., *J. Appl. Phys.*, 105, 113516 (2009).

[4] S. Limpijumnong, S. B. Zhang, S. H. Wei, and C. H. Park, *Phys. Rev. Lett.* 92, 155504 (2004).

[5] S. Tian, D. Zeng, C. Xie, X. Zhao, *Mat. Lett.* 116, 363 (2014).

[6] Friedrich, F., Sieber, I., Klimm, C., Klaus, M., Genzel, Ch., *Appl. Phys. Lett.*, 98, 131902 (2011).

[7] Bundesmann, C., Ashkenov, N., Schubert, M., Spemann, D., Butz, T., Kaidashev, E. M., Lorenz, M., Grundmann, M., *Appl. Phys. Lett.*, 81, 10 (2003).

[8] Ye, J. D., Gu, S. L., Zhu, S. M., Liu, S. M., *Appl. Phys. Lett.*, 88, 101905 (2006).

[9] Cody, C. A., DiCarlo, L., Darlington, R. K., *Inorg. Chem.*, 18, 1572 (1979).

[10] Naidu, B., Pande, M., Sudarsan, V., Vatsa, R., Tewari, R., *Chem. Phys. Lett.*, 474, 180 (2009).

[11] Gilliam, S. J., Jensen, J. O., Banerjee, A., Zeroka, D., Kirkby, S. J., Merrow, C. N., *Spectrochimica Acta Part A*, 60, 425 (2004).

[12] Guo, W., Allenic, A., Chen, Y. B., Pan, X. Q., Che, Y., Hu, Z. D., Liu, B., *Appl. Phys. Lett.*, 90, 242108 (2007).

[13] Samanta, K., Bhattacharya, P., Katiyar, R. S., *J. Appl. Phys*. 108, 113501 (2010).

In: Advanced Materials - Studies and Applications
Editors: I.A. Parinov, S.-H. Chang, S.Theerakulpisut
ISBN: 978-1-63463-749-7

Chapter 4

Influence of Zn Vapor Supersaturation on Morphology and Optical Properties of ZnO Nanorods Grown by CVD Technique

V. I. Pushkariov[1,2], A. L. Nikolaev[1,2] and E. M. Kaidashev[1,2,*]

[1]Vorovich Mechanics and Applied Mathematics Research Institute, Southern Federal University, Rostov-on-Don, Russia
[2]Physics Department, Southern Federal University, Rostov-on-Don, Russia

Abstract

ZnO nanorod and needle-like nanorod arrays were grown by catalyst free chemical vapor deposition of thermally evaporated metallic Zn in Ar/O_2 flow in various synthesis conditions. Morphology and optical properties of the nanorods were modified by changing a supersaturation near the substrate surface varying Zn vapor and oxygen concentrations. Array of ZnO nano-needles with excellent optical quality were designed at 550 °C.

Keywords: Chemical vapor deposition (CVD), pulsed laser deposition (PLD), supersaturation, nanowires, nano-needles, photoluminescence

1. Introduction

Zinc oxide is a promising semiconducting and piezoelectric material with direct band gap (~3.37 eV) and large exciton binding energy (~60 meV). During the last decade one-dimensional nanocrystalline ZnO structures such as nanowire, nanorods, nanobelts and nanocables are emerging as advanced materials for nanoscale devices.

* Corresponding author: E. M. Kaidashev. Vorovich Mechanics and Applied Mathematics Research Institute, Southern Federal University, Stachki 200/1, 344090, Rostov-on-Don, Russia. Physics Department, Southern Federal University, Zorge 5, 344090, Rostov-on-Don, Russia. E-mail: kaidashev@mail.ru.

ZnO nanorod arrays are used as active elements of such nanodevices as LEDs [1], nanosensors [2], UV lasers [3] photodetectors [4], etc.

The development of new technologies of design the highly oriented ZnO nanorod arrays with needed electronic and optical properties are of particular interest.

ZnO nanorod arrays are produced by many techniques such as hydrothermal method, pulsed laser deposition (PLD), chemical vapor deposition (CVD) and metal-organic chemical vapor deposition (MOCVD), thermal evaporation, etc.

Chemical vapor deposition of thermally evaporated metallic Zn in Ar/O_2 flow is relatively cheap and effective method to produce nanorod arrays of high optical quality. The method allows reduce synthesis temperature by 550–600 °C. The structures produced at such low temperature have much less structural defects and demonstrate higher optical quality.

Below the influence of vapor supersaturation and oxygen concentration on ZnO nanorods morphology and optical properties is studied. The possibilities of a catalyst free thermal deposition technique at various synthesis conditions are examined.

2. Experimental

Nanowire arrays were grown onto Si (100) substrates without catalyst. ZnO buffer layers 100 nm thick were preliminary deposited by PLD. Irradiation of pulsed KrF eximer laser (248 nm, 10 Hz) was focused onto ZnO target to have a fluence of ~2 J/cm^2. The rotating target was positioned at 5 cm from the substrate. ZnO film was deposited by 2000 laser shots in oxygen ambient pressure of 2×10^{-2} mbar at the substrate temperature of 500 °C.

The nanorods were grown in a horizontal 32 mm quartz tube positioned inside a resistive furnace. Metallic zinc granule (99.996 %, 500 mg) was positioned in a small quartz tube (1 cm in diameter) as shown in Figure 1. A substrate to Zn granule distance was chosen to be 1 cm. The nanorods growth process was studied upstream to Zn granule where Zn vapor can be transported only by diffusion. The substrate was positioned at various distances from the right open side of a small tube.

In contrast to previous studies [5-7] we use fully opened small tube. Thus, the concentrations of species in the vapor phase reaching the substrate are different from those used in previous studies. An argon and oxygen flows were controlled by two MKS mass flow controllers. The quartz chamber was pumped down to 2 mbar and Ar flow (240 sccm) was introduced. A pressure was maintained to be 11 mbar.

After the furnace was heated up to 600 °C (during 35 min) an oxygen flow (8 sccm) was also introduced. Highly oriented ZnO nanorods array was grown during 15 minutes.

The surface morphology of the grown structures was analyzed by scanning electron microscopy (SEM Zeiss SUPRA 25). Photoluminescence (PL) measurements were carried out under excitation by 325 nm He–Cd laser line at room temperature.

The samples were excited under an angle of 15° to the surface normal with the PL emission being collected along the surface normal. The emission was dispersed by a Solar TII MSDD1000 monochromator (1800 mm^{-1}) with spectral resolution better than 0.1 nm.

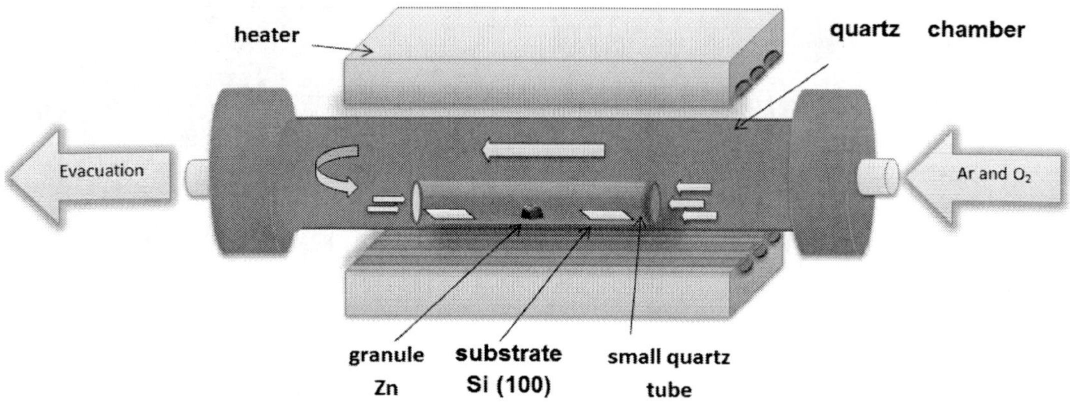

Figure 1. First configuration of the experimental setup used for nanostructures growth.

3. Results and Discussion

SEM images and PL spectra of nanorod arrays are shown in Figure 2. The length and the optical quality of ZnO rods were found to be dependent on the substrate position in the inner tube of a reactor. An average diameter of the rods was ~50 nm, whereas the height of rods grown on the substrate is increased for longer substrate-to-source distances from 200 nm on the Zn source side to 800 nm on the oxygen rich side (Figure 3). Nanorods grown on the substrate positioned closer to the small tube opening, shown in Figure 2a, have less defects related to an oxygen vacancies.

The morphology of nanorods can be controlled by varying the concentrations of Zn and O atoms in the gas phase and the vapor supersaturation. The needle-like nanorods were designed by change the Zn/O ratio during the growth (Figure 4).

A substrate was positioned in a small tube at the distance of 4.5 cm from the Zn granule downstream the gas mixture flow as it is shown in Figure 1. After the chamber was pumped down to 2 mbar an Ar flow (240 sccm) was introduced. A pressure was maintained to be 100 mbar. The Zn granule was heated up to the temperature of 560 °C during 25 minutes. When the necessary temperature of 560 °C was reached the oxygen flow (5 sccm) was also introduced and the gas mixture pressure was reduced by 11 mbar. An array of highly oriented ZnO needle-like nanorods was grown during 30 minutes. Afterwards the pressure was rapidly increased by 100 mbar and the furnace heating was stoped.

The needle-like nanorods have similar shape to those designed by Pan et al. [8]. A rapid increase of pressure changes the supersaturation in gas phase. Photoluminescence spectra show that the designed structures have an intense exciton emission peak in the ultraviolet spectrum range near 380 nm. The samples show very weak green photoluminescence, which claims a low concentration of the defects related to oxygen vacancies.

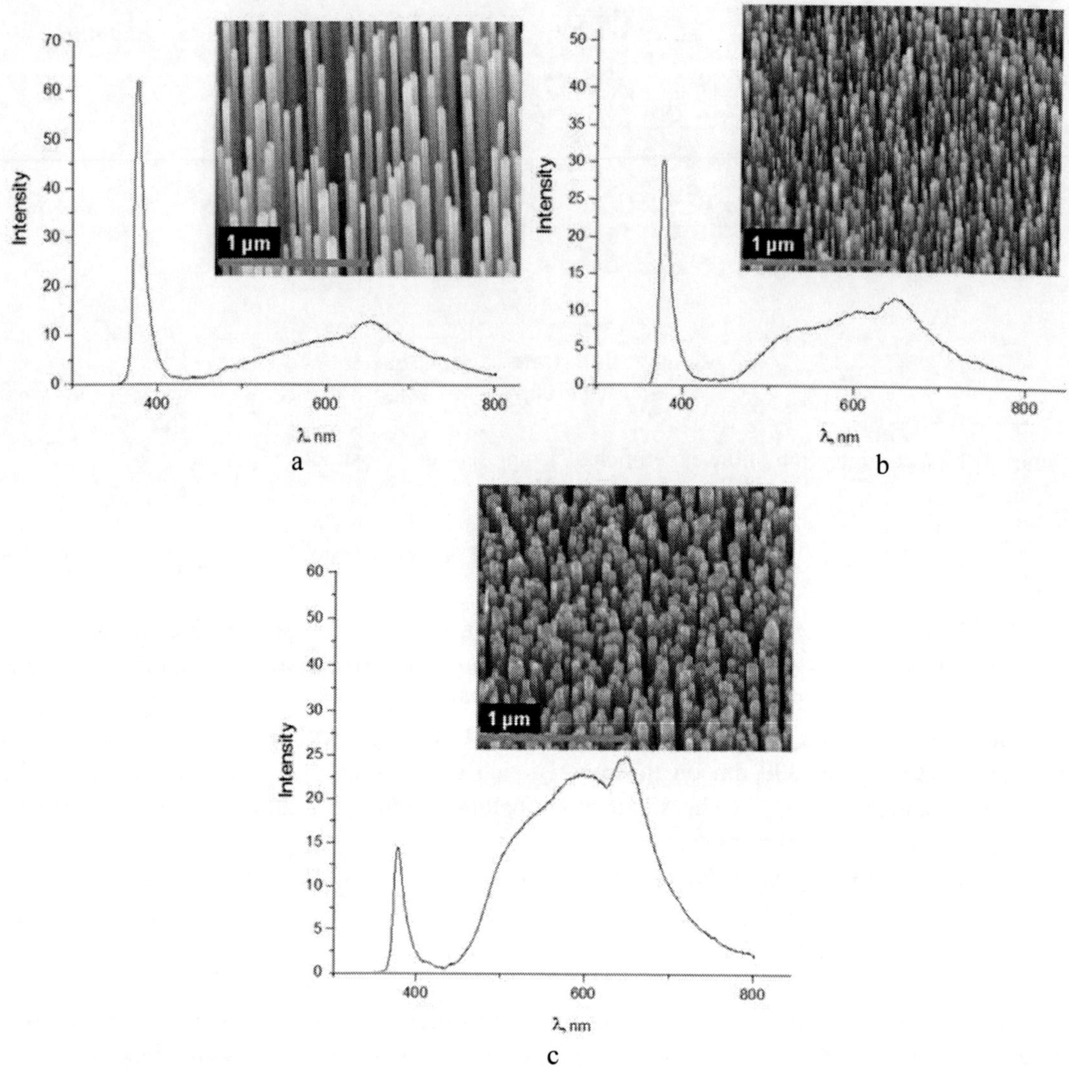

Figure 2. SEM images and PL spectra of nanorods grown at 35 (a), 40 (b) and 45mm (c) from the right side of the small tube.

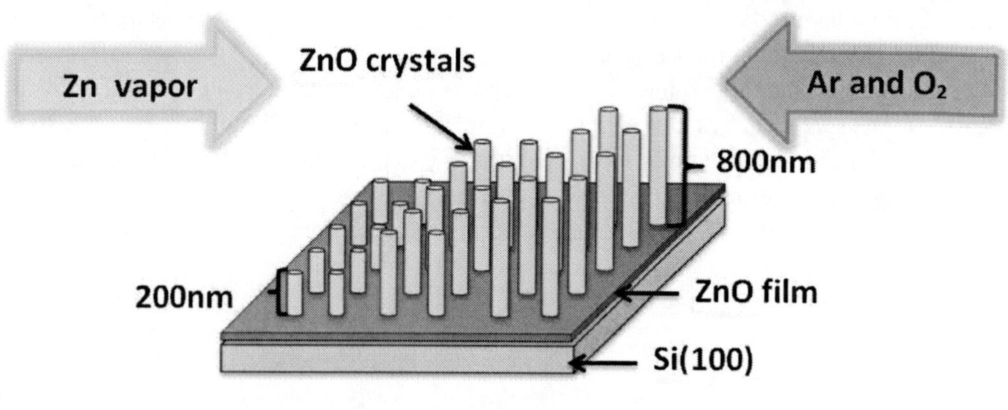

Figure 3. Sketch of nanorods distributed on the substrate.

Conclusion

ZnO nanorod and needle-like nanorod arrays were grown by chemical vapor deposition of thermally evaporated metallic Zn in Ar/O_2 flow.

The method allows growing nanostructures at quite low temperature (550-600 °C).

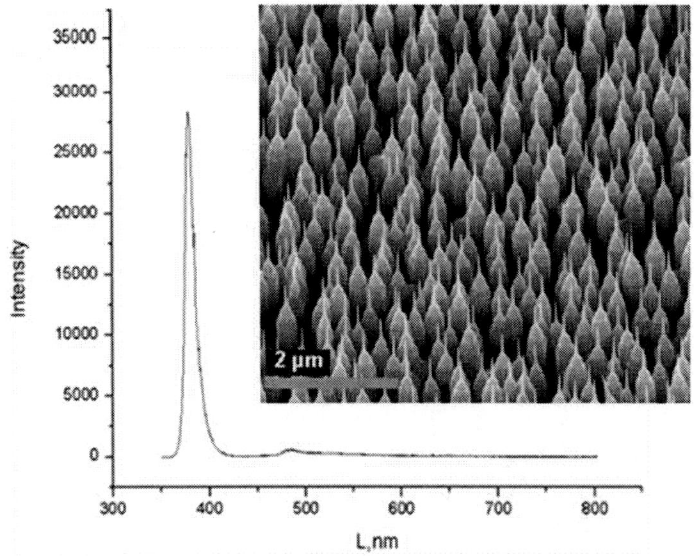

a

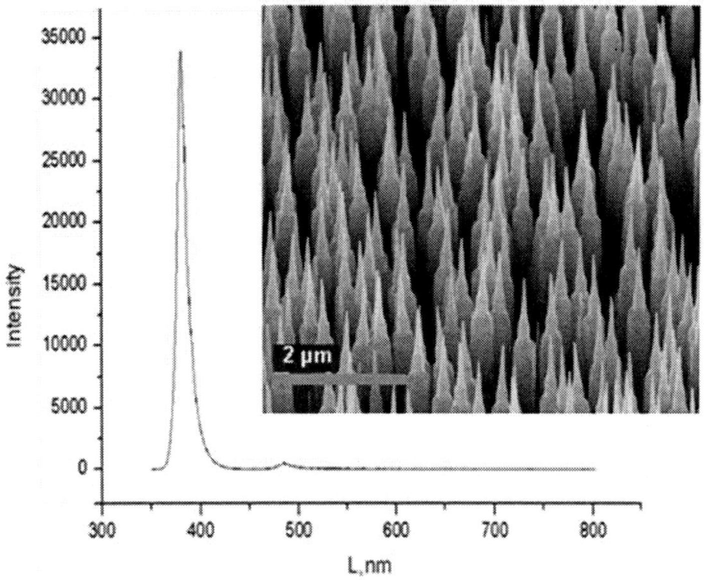

b

Figure 4. SEM images and PL spectra of nanorods grown by changing pressure during the synthesis. The images show the nanorods morphology grown at various Zn concentrations in the gas phase (distances from the granule Zn to substrate were 40 (a) and 50 mm (b), respectively.

Varying the Zn vapor and oxygen concentrations and Ar/O_2 mixture pressure near the substrate surface one can control the supersaturation in the gas phase and, thus, change the morphology and optical properties of the nanorods.

Array of ZnO nano-needles with excellent optical quality were designed at 550 °C. Such rods are of great interest for applications in quantum optoelectronics, in particular, for ultraviolet photodetectors and light-emitting diodes.

ACKNOWLEDGMENTS

The work was carried out with the financial support of the Russian Education and Science Ministry, the project №16.219.2014/K "Development of synthesis methods, theoretical and experimental study of zinc oxide based nanostructures to create elements of photodetectors, optical nanoantennes, piezosensors and chemosensors."

REFERENCES

[1] Sun, X. W., Huang, J. Z., Wang, J. X., Xu, Z., *Nano Lett.* 8, 1219 (2008).

[2] Kim, D. Y., Son, J. Y., *Electrochemical and Solid-State Letters* 12, 109 (2009).

[3] Huang, M. H., Mao, S., Feick, H., Yan, H., Wu, Y., Kind, H., Weber, E., Russo, R., Yang, P., *Science*, 1897 (2001).

[4] Chien, Y. L., Shoou,-J. C., Sheng, P. C., Ching, T. L., Che, F. K., Hong, M. C., *Applied Physics Letters*, 89, 153101 (2006).

[5] Park, J., Choi, H., Siebein, K., Singh, R. *Journal of Crystal Growth*, 258, 342 (2003).

[6] Chang, P., Fan, Z., Wang, D., Tseng, W., Chiou, W., Hong, J., Lu, J., *Chem. Mater.*, 16, 5133 (2004).

[7] Red'kin, A. N., Makovei, Z. I., Gruzintsev, A. N., Dubonos, S. V., Yakimov, E. E., *Inorganic Materials*, 43, 253 (2007).

[8] Pan, N., Wang, X., Zhang, K., Hu, H., Xu, B., Li, F., Hou, J., *Nanotechnology*, 16, 1069 (2005).

In: Advanced Materials - Studies and Applications
Editors: I.A. Parinov, S.-H. Chang, S.Theerakulpisut
ISBN: 978-1-63463-749-7

Chapter 5

PHOTOELECTRIC PROPERTIES OF METAL-SEMICONDUCTOR-METALL STRUCTURE BASED ON ZnO NANORODS DESIGNED THERMAL EVAPORATION AND CARBOTHERMAL METHODS

***D. A. Zhilin*[1,2], *N. V. Lyanguzov*[2], *V. I. Pushkariov*[1,2], *L. A. Nikolaev*[1,2], *V. E. Kaydashev*[2] and *E. M. Kaidashev*[1,2,*]**

[1]Vorovich Mathematics, Mechanics and Computer Sciences Institute, Southern Federal University, Rostov-on-Don, Russia

[2]Physics Department, Southern Federal University, Rostov-on-Don, Russia

ABSTRACT

Photoelectric characteristics of two types of metal-semiconductor-metal (MSM) structures based on ZnO nanorod arrays and two Schottky barriers were studied. The nanorods with low and high amount of deep levels, related to oxygen defects were designed by catalyst-free CVD from Zn vapor (T = 600 °C) and by catalyst-free carbothermal synthesis (T = 950 °C). The nanorods grown at lower temperature in oxygen rich gas flow show much less oxygen defect than those grown at elevated temperature.

Detectors with high amount of deep defect levels show beter performance under green illumination, while the deffect-free structures show higher sensitivity to UV light. Deffect free structures demonstrate fast photocurrent rise and recovery times of ~1 μs and ~10 μs, respectively.

* Corresponding author: E. M. Kaidashev. Vorovich Mathematics, Mechanics and Computer Sciences Institute, Southern Federal University, Stachki 200/1, 344090, Rostov-on-Don, Russia. Physics Department, Southern Federal University, Zorge 5, 344090, Rostov-on-Don, Russia. E-mail: kaidashev@mail.ru.

1. Introduction

Zinc oxide, being a semiconductor with direct wide band gap of 3.37 eV and large exciton binding energy of 60 meV, has been extensively studied as promising material for UV/green photodetectors and solar cells [1], light emitting diods [2] and elements of thin film electronics and optoelectronics [3-5]. Recent advances in design of crystalline ZnO nanostructures, such as nanorods, nanowires, nanotubes, tetrapods, nanocrystals and nanoagregates with large surface area provide new possibilities for light harvesting [6-10] and photocatalytic applications [11-14].

Metal-semiconductor-metal (MSM) photodetectors based on ZnO and Schottky barriers ZnO/Pt or ZnO/Au show faster photoresponse over conventional *p-n* junction [15,20]. Zink oxide with low/high concentration of oxygen vacancies allows design light selective photodiods efficient in UV/green range of spectrum [17-19].

The photodetectors based on array of nanorods have enhanced light scattering and trapping properties because of their developed surface morphology and large surface area. The efficiency of photo- generated electrons harvesting heavily depends on the defect states, working as traps for the carriers on the nanowire surface [16].

Below, photoelectric characteristics of two types of metal-semiconductor-metal (MSM) structures based on ZnO nanorod arrays and two Schottky barriers were studied. Two types of photodetectors based on nanorods with high/low amount of oxygen defect are characterized.

2. Experimental

To design MSM structures based on ZnO nanorods the following procedure was used. Thin buffer ZnO films were deposited onto oxidized Si(100) wafers by PLD in oxygen ambient pressure of 2×10^{-2} mbar. ZnO target was positioned at a distance of 5 cm from a substrate. The irradiation of KrF (248 nm, 10 Hz) laser was focused on the surface of rotating target to give a fluence of 2.3 J/cm^2. The temperature of the substrate was maintained at 500 °C. The preliminary deposited buffer layer reduced the structural strains during the nanorods growth. Then, ZnO nanorods with high/low oxygen deficiency were designed by two methods: (i) catalyst-free chemical vapor deposition by thermal evaporation of metallic Zn in Ar/O_2 mixture [21], and (ii) catalyst-free carbothermal synthesis [22,23]. The setups used for synthesis are illustrated in Figure 1a and 1b, respectively. Catalyst-free chemical vapor deposition was carried out in an argon (240 sccm)/oxygen (8 sccm) mixture flow at a synthesis pressure of 15 mbar and at a substrate temperature of 600 °C. Carbothermal growth was carried out using ZnO:C tablet with molar ratio of 1:1. An argon flow of 200 sccm, a pressure of 20 mbar and a substrate temperature of 950 °C were used for synthesis.

Two Au contact pads were deposited by PLD onto array of nanorods and formed two ZnO/Au Schottky contacts as shown in Figure 2. A substrate with perpendicular standing nanorods was positioned at the angle of 45° to the axis of the plasma plume. A continuous film was grown onto a dense standing nanorods array. A designed Au/ZnO(nanorods)/ZnO (film)/ZnO(nanorods)/Au structure with two Schottky barriers is schematically shown in Figure 2.

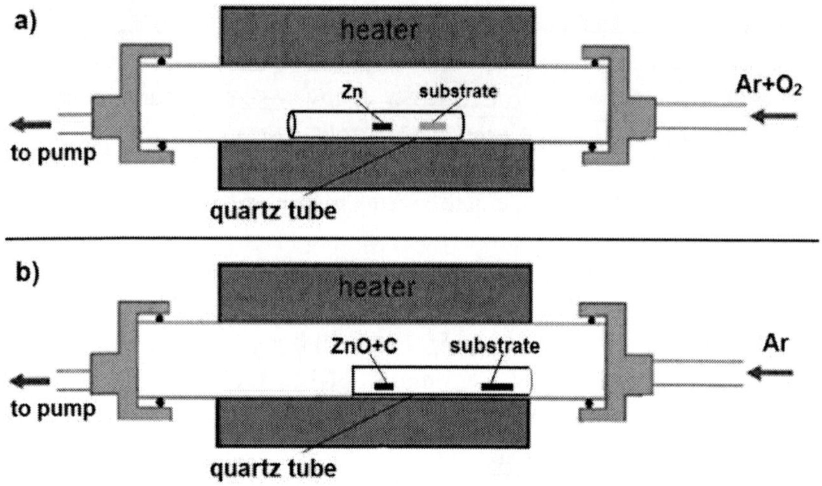

Figure 1. Setups used for nanorod arrays growth. For catalyst-free chemical vapor deposition the Zn precursor is located in a small quartz tube, which is open on both sides (a). For catalyst-free carbothermal synthesis the left side of a small quartz tube is closed (b).

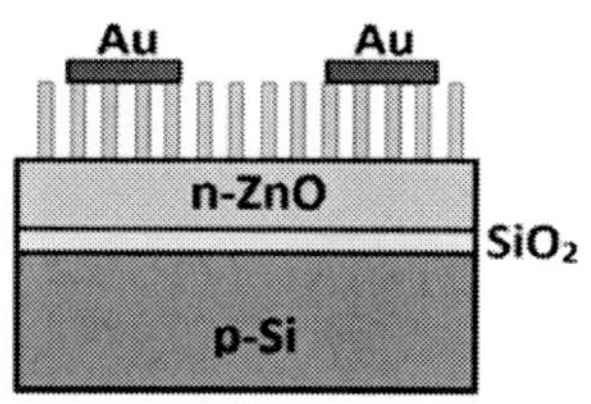

Figure 2. Metal-semicondcuctor-metal structure with double Schottky barriers and efficient light trapping.

Photoluminescence of the designed rods was studied under using a 325 nm He–Cd laser line at room temperature. The samples were excited under ~~an~~ the angle of 15° to the surface normal with the PL emission being collected along the surface normal. The emission was dispersed by a Solar TII MSDD1000 monochromator (1800 mm^{-1}) with spectral resolution better than 0.1 nm.

To study the photoelectrical properties the photodetectors the *I-V* characteristics were recorded under illumination by UV radiation of continuous He-Cd laser (325 nm, 0.7 mW) and by radiation of blue (405 nm), green (518 nm) and red (631 nm) photodiods with the power of 0.7 mW. The used diodes had emission band width of ~ 10 nm. For time-resolved phototesponse measurements the 355 nm Nd-YAG laser line and a bias of 5V were used.

3. Results and Discussion

The photoluminescence spectra of the samples designed by CVD of thermally evaporated metallic Zn (at 600°C in Ar/O_2 flow) and by carbothermal method (at 950°C, in argon flow) show that the rods have different amount of oxygen vacancies. Indeed, the rods grown at lower temperature and in more rich oxygen ambient show intense ZnO exciton related luminescence in the UV region (near 380 nm) and a low broad green band centered at ~525 nm and related to oxygen defects (Figure 3a). The nanorods grown in pure Ar, flow at elevated temperatures in opposite show intense green band and a weak UV luminescence (Figure 3b). The fact that the nanorods grown at lower temperature in more rich oxygen ambient have less structural defects related to oxygen vacancies well established in many studies [24].

The *I-U* curves of MSM photodetectors Au/ZnO(nanorods)/ZnO(film)/ZnO(nanorods)/ Au were measured under HeCd illumination at 325 and laser photodiods at 405, 518 and 631 nm. The structures designed by two methods demonstrated better selective sensitivity in the visible or ultraviolet regions of the spectrum, respectively (Figure 4).

The MSM structure with ZnO NRs prepared by the method of thermal evaporation, has high photosensitivity in the UV region (325 nm) of 563 mA/W, and lower photosensitivity of 67.6 mA/W under green illumination (~518 nm) (Figure 4b).

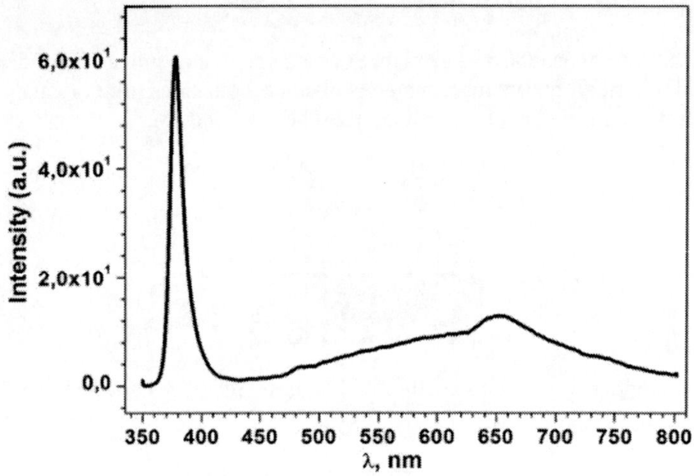

a

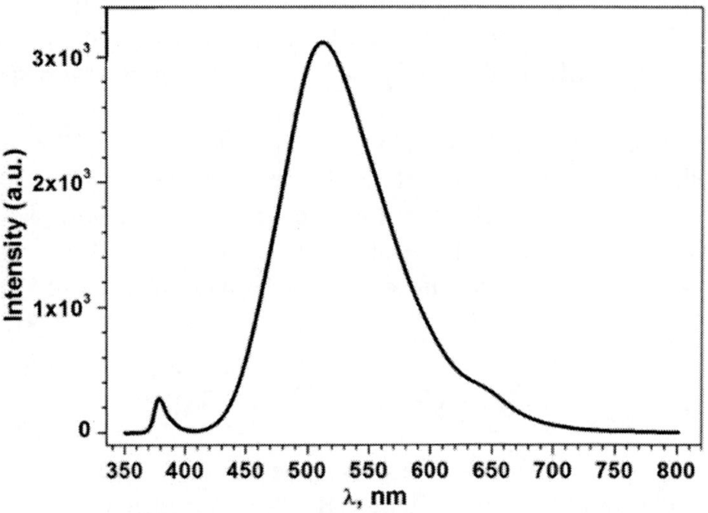

b

Figure 3. PL spectra of ZnO NRs, obtained by using catalyst-free CVD of thermally evaporated metallic Zn (a) and catalyst-free carbothermal method (b).

For structures, prepared by carbothermal synthesis, the photosensitivity values were 21.9 mA/W and 154 mA/W under UV and green illumination, respectively, at bias voltage of 5V (see Figure 4a). For the detectors with low amount of deep defect levels the electrons photo-generated under UV light illumination (325 nm) near the Schottky barrier can easily transfer to Au, be fast isolated from the parasitic recombination in ZnO and work enhance the photocurrent.

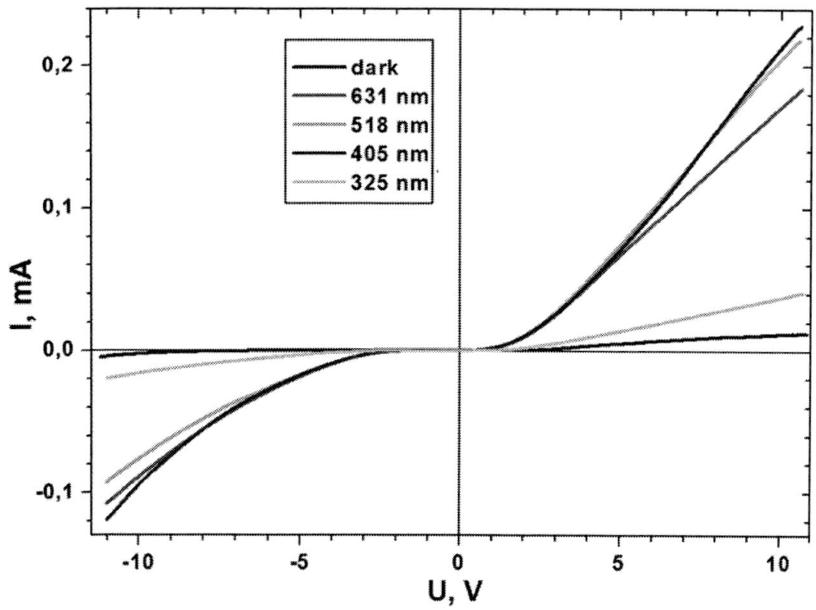

a

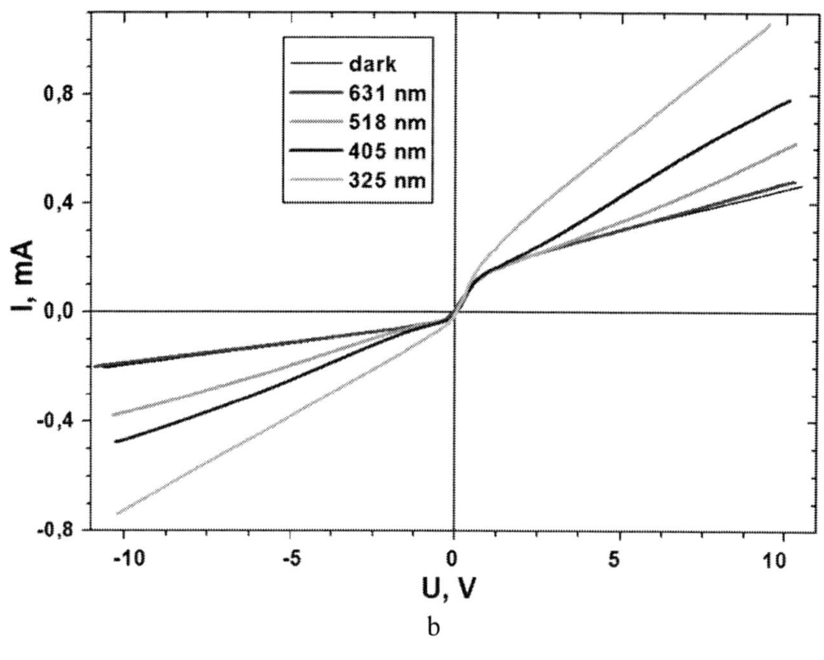

b

Figure 4. *I-U* curves of MSM structure: (a) ZnO NRs processed by carbothermal method, (b) ZnO NRs processed by thermal evaporation.

In defect rich ZnO (with pronounced green photoluminescence) the electrons photo-generated by UV light are fast trapped by deep defect levels and the energy is thermally dissipated or the green recombination happens. The wires grown at poor oxygen ambient normally suffer from high oxygen deficiency of the surface. These defects work like traps for photogenerated electrons killing the carrier diffusion in nanowires. The electrons excited by green light and populating the levels higher than Fermi level of Au are likely tunnel to Au via ZnO surface defect states on the Au/ZnO boundary and, thus, enhance photocurrent.

The time-resolved photo-response of MSM structure with ZnO nanorods, designed by the thermal evaporation method is shown in Figure 5. Under illumination by 10 ns laser pulse at 355 nm and applied bias voltage of 5V the photocurrent rise time was ~1 μs. The fall time measured at full-width at half-maximum was ~10 μs.

CONCLUSION

Two types of the MSM structures with low and high amount deep levels, related to oxygen defects were designed by catalyst-free CVD from Zn vapor and by catalyst-free carbothermal synthesis. Detectors with high amount of deep defect levels show beter performance under green illumination, while the deffect-free structures show higher sensitivity to UV light. Deffect free structures demonstrate fast photocurrent rise and recovery times of ~1 μs and ~10 μs, respectively. The studied photodetectors are of great interest in UV/green radiation harvesting and photocatalysis.

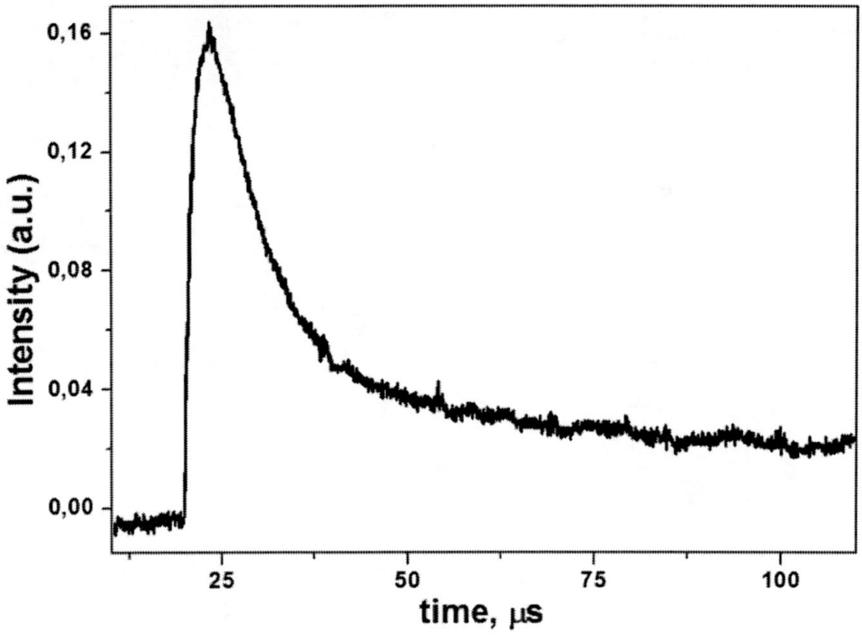

Figure 5. Time-resoled photoresponse of MSM detector with nanorods showing low level of deffects related to oxygen vacancies.

ACKNOWLEDGMENTS

The work was carried out with the financial support of the Russian Education and Science Ministry, the project No. 16.219.2014/K "Development of synthesis methods, theoretical and experimental study of zinc oxide based nanostructures to create elements of photodetectors, optical nanoantennas, piezosensors and chemosensors".

REFERENCES

[1] Mridha, S., Basak, D., *J. Applied Physics*, 101, 08102 (2007).

[2] Dae-Kue Hwang, Min-Suk Oh, Jae-Hong Lim, Seong-Ju Park, *Journal of Physics D: Applied Physics,* 40, 387-412 (2007).

[3] Cheng-Pin Chen, Pei-Hsuan Lin, Liang-Yi Chen, Min-Yung Ke, Yun-Wei Cheng, Jian-Jang Huang, *Nanotechnology*, 20, 245204 (2009).

[4] Pau, J. L., Piqueras, J., Rogers, D. J., Hosseini Teherani, F., Minder, K., McClintock, R., Razeghi, M., *J. Applied Physics*, 107, 033719 (2010).

[5] Lee, J. H., Lee, J. Y., Kim, J. J., Kim, H. S., *Journal of the Korean Physical Society*, 56, № 1, 429-433 (2010).

[6] Law, M., Greene, L. E., Johnson, J. C., Saykally, R., Yang, P. D., *Nature Mater.* 4, 455 (2005).

[7] Han, J. B., Fan, F. R., Xu, C., Lin, S. S., Wei, M., Duan, X., Wang, Z. L., *Nanotechnology*, 21, 405203 (2010).

[8] Chen, W., Zhang, H. F., Hsing, I. M., Yang, S. H., *Electrochem. Commun.* 11, 1057 (2009).

[9] Zhang, Q. F., Chou, T. R., Russo, B., Jenekhe, S. A., Cao, G. Z., *Angew. Chem.: Int. Edit.* 47, 2402 (2008).

[10] Zhang, Q., Cao, G., *Nanotechnology*, 6, 91 (2011).

[11] Liu, Z., Hou, W., Pavaskar, P., Aykol, M., and Cronin, S. B., *Nano Lett.* 11, 1111 (2011).

[12] Udawatte, N., Lee, M., Kim, J., Lee, D., *ACS Appl. Mater. Interfacces* 3, 4531 (2011).

[13] Chen, H. M., Chen, C. K., Chen, C. J., Cheng, L. C., Wu, P. C., Cheng, B. H., Ho, Y. Z., Tseng, M. L., Hsu, Y. Y., Chan, T. S., Lee, J. F., Liu, R. S., and Tsai, D. P., *ASC Nano*. 6, 7362 (2012).

[14] Liu, M., Nam, C. Y., Black, C. T., Kamcev, J., and Zhang, L., *J. Phys. Chem. C.* 117, 13396 (2013).

[15] Liu, K. W., Ma, J. G., Zhang, J. Y., Lu, Y. M., Jiang, D. Y., Li, B. H., Zhao, D. X., Zhang, Z. Z., Yao, B., Shen, D. Z., *Sol. State Electron* 51, 757 (2007).

[16] Tsai, D. S., Lin, C. A., Lien, W. C., Chang, H. C., Wang, Y. L., and He, J. H., *ACS Nano,* 5, 7748 (2011).

[17] Peng, S. M., Su, Y. K., Ji, L. W., Wu, C. Z., Cheng, W. B., and Chao, W. C., *J. Phys. Chem. C*, 114, 3204 (2010).

[18] Liu, X., Gu, L., Zhang, Q., Wu, J., Long, Y., Fan, Z., *Nature Comm*., 5 (2014) 4007.

[19] Bai, Z., Chen, X., Yan, X., Zheng, X., Kang, Z., and Zhang, Y., *Phys. Chem. Chem. Phys*., 16 (2014) 9525.

[20] Leonard J. Brillson, Yicheng Lu, *Journal of Applied Physics*, 109, 121301 (2011).
[21] Pushkariov, V. I., Lyanguzov, N. V., Kaidashev, E. M., *Inzhenernyj Vestnik Dona,* 4 (2013).
[22] Lyanguzov, N. V., Kaidashev, E. M., Zakharchenko, I. N., Bunina, O. A., *Technical Physics Letters*, 9, 767-770 (2013).
[23] Lyanguzov, N. V., Zhilin, D. A., Kaidashev, E. M., In: *Advanced Materials Physics, Mechanics and Applications, Series: Springer Proceedings in Physics,* 152, № 4, 37-44 (2014).
[24] Alexey, A. S., Samuel, A. F., Stefan, T. B., Richard, A. C., Huub J. J. van Dam, Paul, S., *Faraday Discussions*, 134, 267-282 (2007).

In: Advanced Materials - Studies and Applications
Editors: I.A. Parinov, S.-H. Chang, S.Theerakulpisut
ISBN: 978-1-63463-749-7

Chapter 6

STUDY OF LASER PROCESSING SAPPHIRE

S. P. Malyukov* and Yu. V. Klunnikova
Engineering Technological Academy,
Southern Federal University, Taganrog, Russia

ABSTRACT

The study of interaction of laser radiation with matter is one of the important directions of science and technology. The laser equipment (LIMO, Lissotschenko Mikrooptik GmbH, Germany) with a wave length of 532 nm and 1064 nm allows one to carry out wide studies of laser radiation influence on objects. The investigation of processing a sapphire by using laser radiation was conducted. The problems of choice of the correct laser power and radiation time are discussed. These investigations were performed in the Scientific and Educational Center «Laser Technologies» of the Nanotechnology, Electronics and Equipment Construction Institute of the Southern Federal University and in the Zelinsky Institute of Organic Chemistry of the Russian Academy of Science (Moscow).

1. INTRODUCTION

Sapphire crystals demonstrate certain physical properties (high temperature of melting, chemical and radiation stability, high hardness and transparency); so they can be widely used in microelectronics, quantum electronics, optics of high resolution, and nanotechnology [1, 2].

Today two main areas of sapphire applications exist, namely (i) the constructional materials with unique mechanical characteristics, and (ii) the optical materials with unique properties. The equipment or its elements operate in the aircraft-building, space equipment, chemical technologies and so on at simultaneous influence of the medium, radiation, high temperatures, pressure and mechanical impact. Any materials are subject to intensive corrosion and erosion in such extreme conditions. The structure of polycrystalline materials

* Email: malyukov@fep.tti.sfedu.ru

and, therefore, their mechanical properties change significantly in extreme conditions due to the processes of recrystallization, grain boundary corrosion etc. The diffusion speed on boundaries of blocks grows with temperature, radiation dose and operation time, causing the material damage. These phenomena in sapphire occur in a much lesser degree [3].

The substrate fabrication is one of the most important areas of sapphire application. The substrates are used for epitaxy of semiconductor (Si, GaN, AlGaN and many others) films and processing the integrated circuits. The advantages of the sapphire substrate include the possibility to operate at high temperatures and mechanical impacts, and also the possibility to fabricate big substrates. They are used even when their lattice parameters do not correspond to parameters of the heteroepitaxial structures. The sensors of pressure, weight, humidity, IR-radiation detectors (HgCdTe films on sapphire) and many other goods are produced on the sapphire substrates [4].

The bright blue and green light-emitting diodes (LEDs) are developing quickly. The areas of their applications also expand from traffic lights to treatment of Alzheimer illness [4, 5]. The GaN/sapphire structure is used for blue LEDs. The concentration of carriers (holes) in GaN layer reaches of 10^{16} – 10^{18} cm^{-3}. Sapphire substrates have diameters of 2, 3 and 4 inches and their permissible deviation from the plane surface (0001) is ± 0.1°. The quality of finishing processing changes in the range from 0.2 to 2 nanometers.

The expansion of laser applications requires development of new technological processes which could not be performed without laser use. These processes are only on the early stage of development; they will be introduced in industry in the near future. Fast prototyping, creation of images in the transparent materials, the combined processing, etc. belong to such processes. The use of laser radiation in fabrication of material allows one to provide high quality of the final goods, processing productivity, ecological purity, and also economy of human and material resources.

The problem of the laser processing consists in providing an arrangement of ions in the near-surface region of material and reducing a quantity of structure defects.

The variety of equipment, used for the fabrication of material by laser radiation, is extremely wide. The majority of producers submit on the market the laser technological set-ups instead of separate lasers. These set-ups include devices of external optics, operational tables, manipulators, processing robots, and software, required for implementation of particular technology. Experimental studies of the sapphire fabrication and film formation on sapphire surface were performed with the equipment (LIMO 100-532/1064-U model), which consisted of the laser with a wave length of 532 nm (green line) and infrared Nd:YAG laser with the fixed wave length of 1064 nm, the pulse continuance of 84 ns and power of a pulse which can be in the range from 0.1 to 100 W. The measurement device of power was used for the power testing of the laser pulse. The laser radiation demonstrates the following unique properties: high intensity (power) of electromagnetic energy flux, high monochromaticity and high degree of coherence in time and space. So it differs from other sources of electromagnetic energy by a very narrow orientation of light distribution.

The laser set-up LIMO 100-532/1064-U is characterized by the controllable diameter of laser spot and high degree of laser beam homogeneity, because an error of size definition is no more than 1.5 % (during operation of the device in recommended conditions). The main units of the system are the following: (1) diode laser, (2) attenuator, (3) optical table, (4) elements of the optical system, (5) small table for substrates, (6) movement system of the table, and (7) video camera (Figure 1).

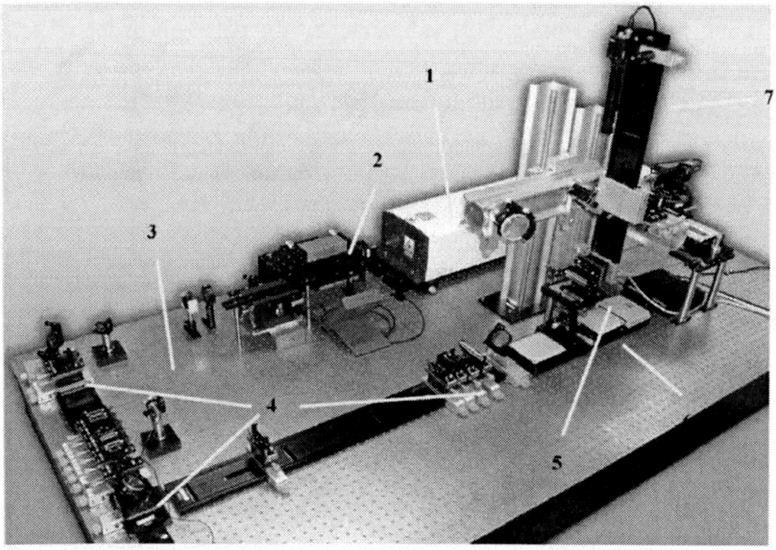

Figure 1. Main units of system (description of the blocks is present into text).

The following units are the parts of the laser device LIMO [6, 7]:

i) optical system for transformation of round section of the laser beam into linear one;
ii) attenuator, operated by the step engine and intended to adjust the laser output power (the output power changes from 96 % down to 1% with a step of 1%);
iii) small mobile table for substrates (intended for their scanning), operated by the step engine;
iv) testing system of the laser beam focus with the video camera for precise testing of the beam at scanning;
v) measurement device of power, which locates along the beam direction;
vi) two personal computers for program testing the laser and scanning system for green and red lines.

The main characteristics of the laser set-up LIMO 100-532/1064-U are present in Table 1 [7].

Table 1. Main characteristics of laser set-up LIMO 100-532/1064-U

Wave length, nm	532 – 1064
Length of laser line, mm	60
Width of laser line, μm	10
Maximum laser power, W	140
Operational frequency of pulses, kHz	100
Pulse duration, ns	45
Laser radiation homogeneity	No less than 98.5 %

The required laser power, regulated by the attenuator, depends on the sample material and treatment of technological process. The measurement device of power is necessary for testing the laser power relaxation (Figure 2). During optical adjustment of the power relaxation by using mobile scanner C, it is necessary mirror B in order to redirect the laser beam to the working head of the measurement device of power. After the laser power adjustment, it is required to remove the mirror from optical way of the beam. Block of mask positioning A locates between the lens of field positioning B and the scanner C. The exact sample position can be regulated in vertical and horizontal directions with an accuracy of 10 μm.

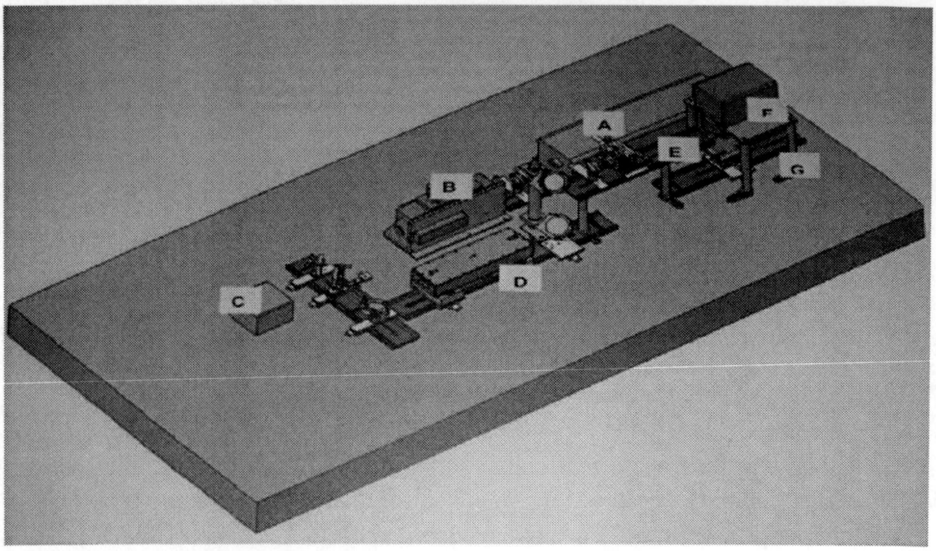

Figure 2. Scheme of laser work.

2. Investigation Results

The experimental studies of laser radiation impact (wave length of 1064 nm) on the considered interface "solid body/liquid" were conducted. The absorbing solutions $KMnO_4$, $FeCl_3$, KCl were chosen as liquids for carrying out the experiments. In order to study the laser beam influence on the interface "solid body/liquid", it was fabricated the experimental set-up (aluminum case with anode covering), shown in Figure 3.

The laser radiation is subject to refraction and considerable absorption in liquid in transition from atmosphere. The absorption of the laser radiation in liquid medium occurs due to the dispersion from heterogeneities in the liquid and molecular absorption. Thus the long-wave range is absorbed more than the short-wave one. The radiation absorption is combined with phenomena of the liquid dissociation, connected with formation of bubbles into liquid. Due to the intense radiation, the bubbles can grow to significant sizes and cause considerable dispersion of radiation.

High intensity of laser radiation leads to coalescence of bubbles and formation of steam-gas channel along the laser beam.

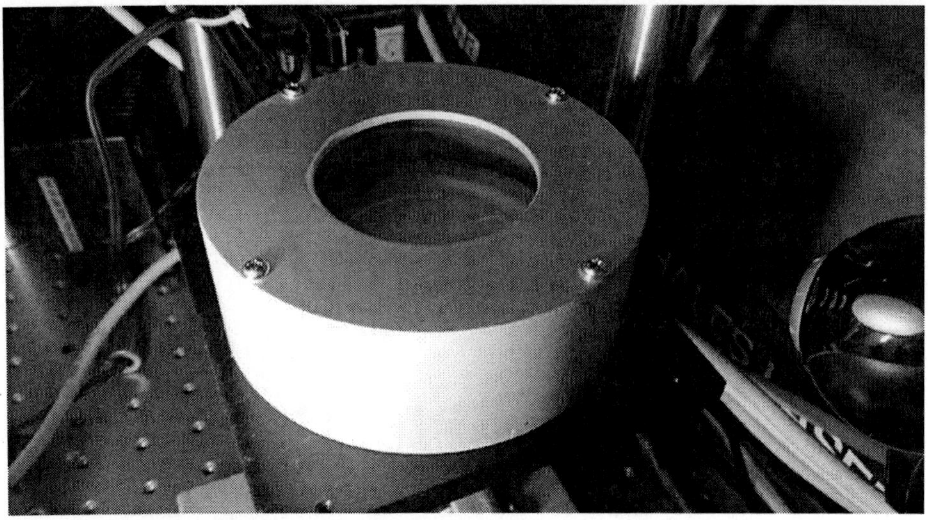

Figure 3. Experimental set-up.

Heterogeneities in liquid as well as impurities with high coefficient of absorption lead to additional considerable dispersion of laser radiation. Thus the heterogeneities heat strongly, causing a liquid heating and boiling. The interaction of laser radiation with liquid is accompanied by light and hydraulic effects, leading to the transformation of laser radiation energy into mechanical energy of shock waves.

The high-temperature zone forms at the initial moment, when laser radiation interacts with particles in liquid. During this process, the gas cavity forms which pulls out liquid with great speed from the particles. Thus the impact load arises, which extends in all directions. The temperature growth in the influence zone increases amount of vapors in the gas cavity and causes pressure jump in it. The liquid pressure on the cavity leads to sudden increase of gas pressure. As a result, the process becomes similar to an explosion [8].

The gas cavity extends inside the liquid with increasing speed. The cavity becomes spherical and extends slightly. High temperature of gases in the cavity causes intense evaporation of liquid from surface of the formed torch. The dense steam envelops the torch cavity and increases with the gas cavity growth.

The material damage is generally defined by features of laser radiation influence on the material in liquid medium, thickness of liquid layer along the laser beam, light and hydraulic effects, and also thermal properties of liquid.

The thickness of liquid layer depends on features of dispersion and absorption of laser radiation, namely the wider liquid layer, separating from material surface at the beam entrance in liquid, causes the greater power losses. The depth of the formed crater decreases and its diameter increases slightly, when the liquid layer thickness increases [8–12].

The pulse laser radiation of the considered interface "transparent solid body/liquid" leads to film sedimentation on the solid surface. The effect of film sedimentation on surface of the transparent solid body at laser radiation of liquid is a new effect; it has not been investigated earlier. For the first time, a similar effect was described by Voronov V. V., Dolgayev S. I. and Shafeev G. A. [13–21], who used the laser on copper vapors in their investigations. The film sedimentation is caused by thermal decomposition (pyrolysis) in focus of laser beam of the substances dissolved in water. The sapphire was used as the substrate. Since radiation was focused on interface of sapphire with solution through absorbing film, its optical thickness was inversely proportional to the intensity of the laser radiation.

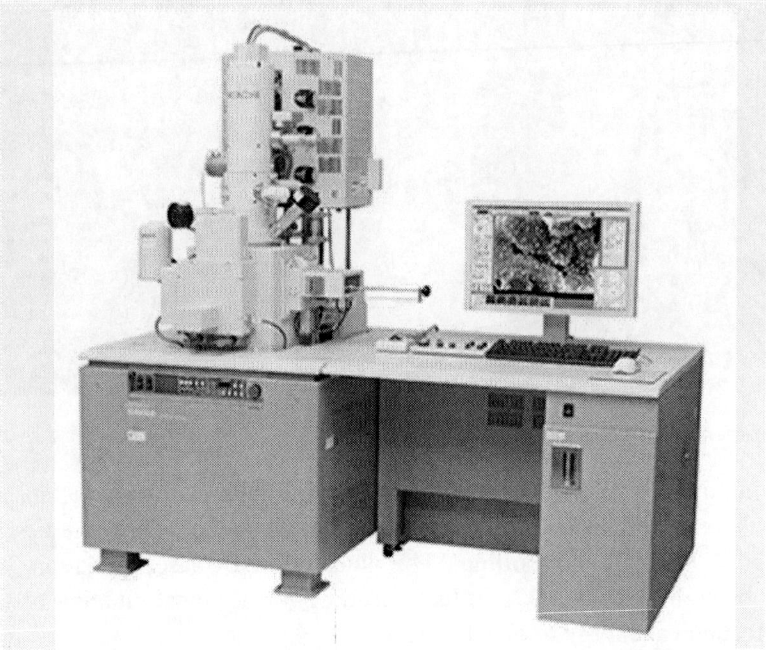

Figure 4. Scanning electron microscope Hitachi SU8000.

The structure of the sample surface was analyzed by using Hitachi SU8000 field-emission scanning electron microscope (FE-SEM) (Figure 4) of the Zelinsky Institute of Organic Chemistry of the Russian Academy of Sciences (Moscow). The features of this microscope are the following ones: (i) high resolution even at the low accelerating voltage, (ii) multiple detecting systems, (iii) wide set of objects and measurement ranges of various parameters in a whole imaging, (iv) the possibility to optimize the set-up for scientific and educational activities.

The target-oriented approach was utilized for optimization of measurements [22]. Before the measurements, the samples were mounted on a 25 mm aluminum specimen stub and fixed by conductive silver paint. Metallic coating with a thin film (6 nm) of platinum/palladium alloy (80/20) was deposited by using the magnetron sputtering method as described elsewhere [23]. The observations were carried out by using the scanning electronic microscope Hitachi SU8000. Images were obtained in secondary electron mode at the accelerating voltage of 10 kV and operation distance of 8 – 18 mm. Morphology of the samples was studied taking into account a possible influence of metallic coating on the surface [23].

EDX-SEM studies and mapping were performed by using Oxford Instruments X-max EDX system. Preliminary, the samples were coated with a thin film (6 nm) of palladium by utilizing Cressington 208 carbon coater.

The image of sapphire surface morphology is present in Figure 5, a. The mass-spectra of the sample are given in Figure 5, b.

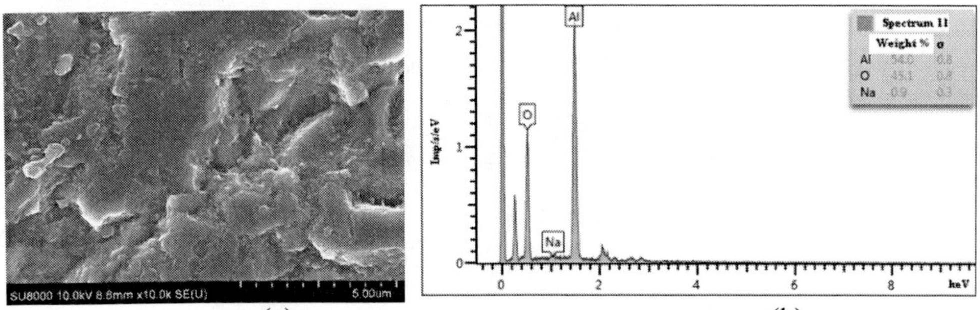

(a) (b)

Figure 5. Image of sapphire surface morphology (a) and mass-spectra of the sample (b).

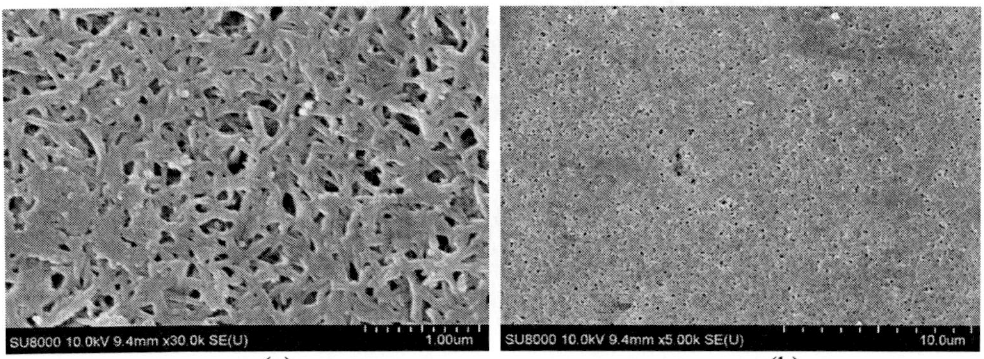

(a) (b)

Figure 6. Images of insoluble in water film of Fe_2O_3 on the sapphire surface.

The images with yellow color of water insoluble film of Fe_2O_3 on the sapphire surface are present in Figure 6. This film was obtained in 3% $FeCl_3$ solution in presence of water at the laser processing the sapphire after 4 passes of the 96 W-Nd:YAG laser beam. Mass-spectra of the fabricated sample are given in Figure 7.

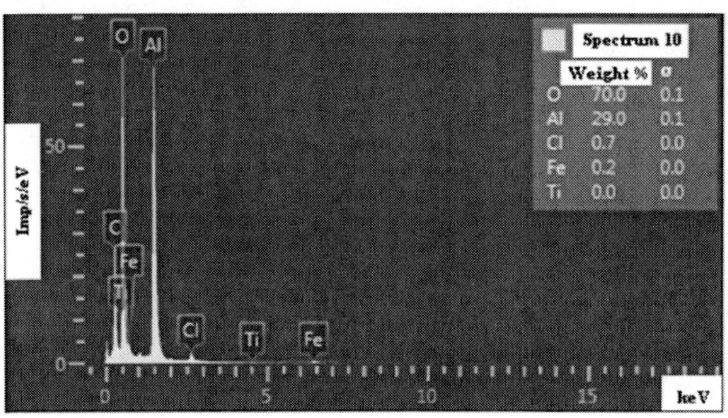

Figure 7. Mass-spectra of the fabricated sample with Fe_2O_3 film.

In order to analyze the spectral characteristics and define the film structure, a sample with irregular film was fabricated (see Figure 8). The direction in the sample for mass-spectra measurements is shown in Figure 9, a. The sample's mass-spectra that characterize distributions of aluminum, iron, chlorine and oxygen are given in Figure 9, b-e. The obtained results demonstrate presence of Fe_2O_3 film on the sapphire surface.

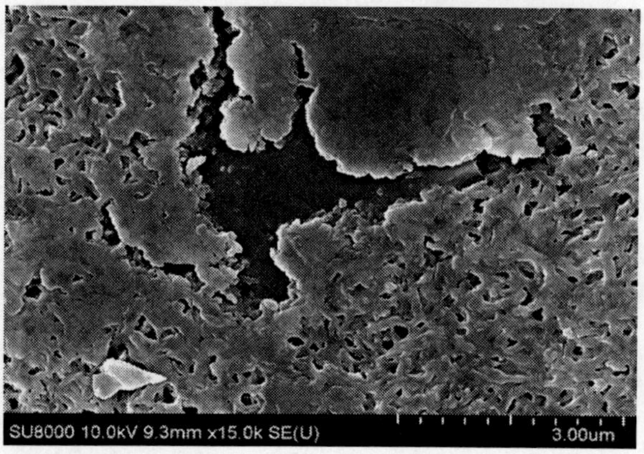

Figure 8. Sample with irregular film.

The image of MnO_2 film with pink color on the sapphire surface is present in Figure 10, a. The film was obtained in $KMnO_4$ solution in presence of water at the laser processing the sapphire after 10 passes of the 96 W-laser beam.

The mass-spectra of the obtained sample are shown in Figure 10, b.

The laser processing the sapphire in KCl solution in presence of water after 8 passes of the laser beam with a power of 96 W led to formation of a white oxide film of ~ 1 μm thickness. The experimental studies were also conducted in presence of $KMnO_4$ solution in water for borosilicate glass which was used as a substrate. The pink film of MnO_2 was obtained on the surface of borosilicate glass of $B_2O_3 - SiO_2 - Na_2O$ system. We observed an increase of the film thickness at growth of the substrate temperature.

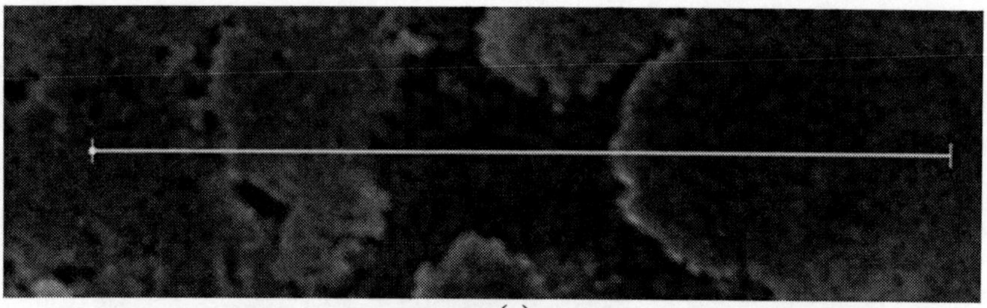

(a)

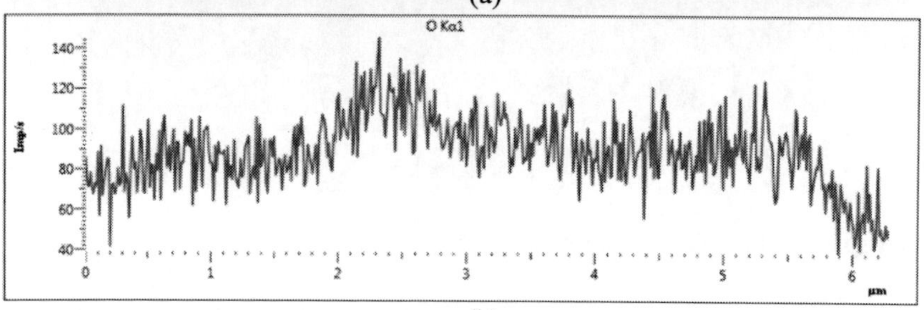

(b)

Figure 9. (Continued).

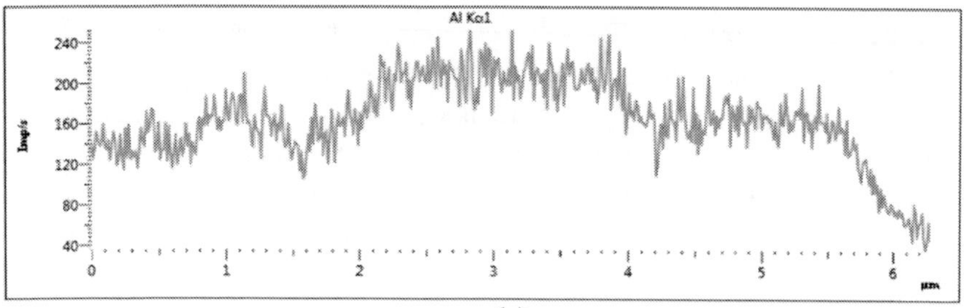

(c)

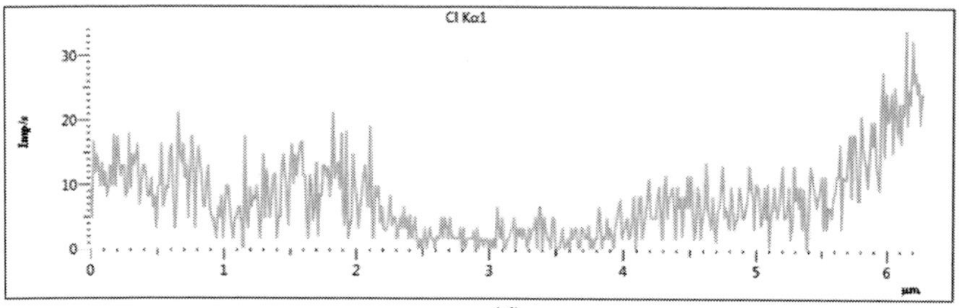

(d)

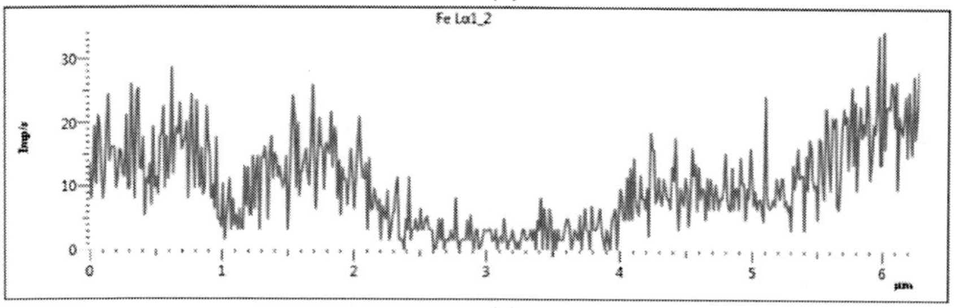

(e)

Figure 9. Mass-spectra measurement direction in the irregular film sample (a), and distributions of aluminum (b), iron (c), chlorine (d) and oxygen (e) along this direction.

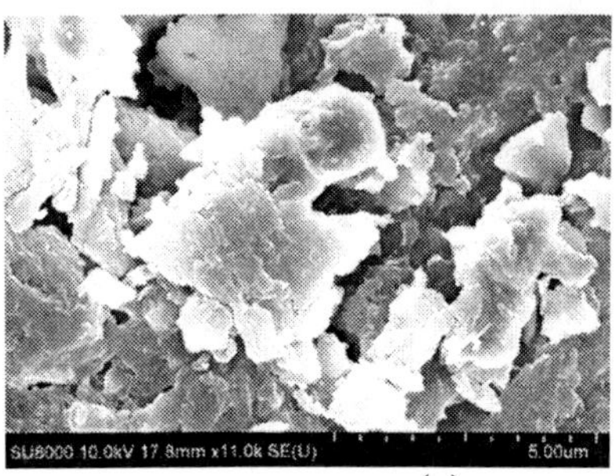

(a)

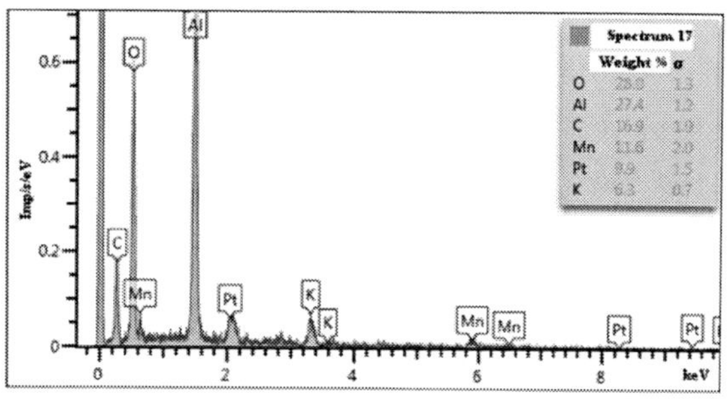

(b)

Figure 10. Film sample photo (a) and mass-spectra of the studied sample (b).

Conclusion

The existence of an absorbing film in the conditions of multipulse laser influence can lead to a substrate ablation. Not only the film forming, but also the water under supercritical conditions influences this process. These conditions can be characterized with high pressure and temperature at which sapphire becomes well soluble in water [15].

The area of high pressure and temperature is formed by influence of laser radiation on the considered interface "solid body/water solution" in beam focus during the laser pulse in liquid. This area adjoins to the sapphire surface and can lead to crystallite structure defects and formation of microcracks on its surface. The sapphire surface is affected by the liquid under supercritical condition at the multipulse mode of radiation of the interface. This condition contributes to the sapphire ablation and demands further investigation of the processing influence on the formation mechanisms of residual stresses and microcracks in sapphire. The experiments on influence of laser radiation on the sapphire surface without medium, which absorbs water, were conducted. The sapphire ablation was observed without water medium at the same parameters of laser radiation. Influence of several laser pulses on sapphire leads to formation of crater on its surface. The crater diameter is close to diameter of the focused laser beam. The crater depth grows with increasing the laser pulses.

The advantage of this approach is a possibility to realize it in open air. The morphologies of the processed sapphire surface and films depend on the scanning beam speed, number of laser beam passes and some other conditions.

The investigation of ablation and film sedimentation processes by laser radiation is very interesting for thin-film optics and microelectronics. The properties of oxide materials such as their deoxidizing and oxidizing catalytic activity, as well as combination of high general conductivity with ionic conductivity, caused by diffusion of loosely coupled oxygen, can be applied in development of new thin-film high-temperature ionic devices like solid oxide fuel elements and gas-dividing membranes. Therefore, the investigation results of formation of the films and film structures can be very useful for development of new gas-sensitive sensors and specific thin-film coatings of fuel elements, as long as they combine the properties of catalysts and conducting elements. Finally, the method of experimental study of sapphire and borosilicate glass, fabricated by using laser processing, was developed and implemented in practice under various conditions with the help of LIMO laser set-up. Chemical analysis of the obtained samples was also carried out. The laser radiation ablation of solid materials has certain practical interest, because this method is characterized by formation of microstructures that demand further detailed studies of their properties and applications.

ACKNOWLEDGMENTS

Electron microscopy research was carried out at the Department of Structural Studies of the Zelinsky Institute of Organic Chemistry (Moscow). The authors would like to thank the State Unitary Enterprise "Rostovoblfarmatsiya" (Rostov-on-Don) for assistance in conduction of the experiments.

REFERENCES

[1] Malyukov, S. P.; Klunnikova, Yu. V., In: *Advanced Materials Physics, Mechanics and Applications, Series: Springer Proceedings in Physics,* Shun Hsiung-Chang, Ivan A.

Parinov, Vitaly Yu. Topolov (Eds.). Springer Cham, Heidelberg, New York, Dordrecht, London, 2014, Chapter 6, *vol. 152*, pp. 55 – 69.

[2] Malyukov, S. P.; Klunnikova, Yu. V., In: *Nano- and Piezoelectric Technologies, Materials and Devices*, Ivan A. Parinov (Ed.). Nova Science Publishers, New York, 2013, Chapter 5, pp. 133-150.

[3] Dobrovinskaya, E. R.; Lytvynov, L. A.; Pishchik, V. V. *Sapphire. Material, Manufacturing, Applications*. Springer, New York, 2009.

[4] Dobrovinskaya, E. R.; Lytvynov, L. A.; Pishchik, V. V. *Sapphire & Other Corundum Crystals*, Institute of Monocrystals Press, Kharkov, 2002 (In Russian).

[5] Wallace, J. *Laser Focus World,* 2003, *vol. 19*.

[6] Kireev, P. S. *Physics of Semiconductors*. Higher School, Moscow, 1969 (In Russian).

[7] Borisenko, V. E. *Solid-phase Processes in Semiconductors at Pulse Heating*. Science and Technique, Minsk, 1992 (In Russian).

[8] Cherviakov, G. G. *Interaction of Laser Radiation with Substance*. SFedU Press, Taganrog, 2005 (In Russian).

[9] Rybalkin, N. N., et al. *Laser Processing Metals*. Mechanical Engineering, Moscow, 1975 (In Russian).

[10] Chun, M. K.; Rose, K., *Appl. Phys.*, 1970, *vol. 41*, 614.

[11] Grigoryants, A. G.; Shiganov, I. N.; Misyurov, A. I. *Technological Processes of Laser Processing*. Bauman MGTU Press, Moscow, 2008 (In Russian).

[12] Altudov, Yu. K.; Garitsyn, A. G. *Laser Microtechnologies and Their Applications in Electronics*. Radio and Communication, Moscow, 2001 (In Russian).

[13] Dolgaev, S. I.; Lyalin, A. A., Simakin, A. V.; Shafeev, G. A. *Quantum Electronics*, 1996, *vol. 1*, 67.

[14] Voronov, V. V.; Dolgaev, S. I.; Shafeev, G. A. *Reports of the Russian Academy of Sciences*, 1998, *vol. 4*, 465.

[15] Voronov, V. V.; Shafeev, G. A.; Prokhorov, A. M. In: *Sapphire: Structure, Technology and Applications*, Isaias Tartaglia (Ed.), Nova Science Publishers, New York, 2013, pp. 75 – 100.

[16] Dolgaev, S. I.; Karasev, M. E.; Kulevskiy, L. A.; Simakin, A. V.; Shafeev, G. A. *Quantum Electronics*, 2001, *vol. 7*, 593.

[17] Simakin, A. V.; Shafeev, G. A. *Laser Physics*, 1994, *vol. 3*, 610.

[18] Shafeev, G. A.; Obraztsova, E. D.; Pimenov, S. M. *Appl. Phys. A*. 1997, *vol. 65*, 29.

[19] Dolgaev, S. I.; Lyalin, A. A.; Simakin, A. V.; Shafeev, G. A. *Appl. Surf. Sci.*, 1996, *vol. 96*, 491.

[20] Dolgaev, S. I.; Lyalin, A. A.; Simakin, A. V.; Shafeev, G. A.; Voronov, V. V. *Appl. Surf. Sci.,* 1997, *vol. 109*, 201.

[21] Dolgaev, S. I.; Shafeev, G. A.; Voronov, V. V. *Appl. Phys. A*. 1997, *vol. 66*, 201.

[22] Kachala, V. V.; Khemchyan, L. L.; Kashin, A. S.; Orlov, N. V.; Grachev, A. A.; Zalesskiy, S. S.; Ananikov, V. P. *Russ. Chem. Rev.,* 2013, *vol. 82*, 648.

[23] Kashin, A. S.; Ananikov, V. P. *Russ. Chem. Bull.*, 2011, *vol. 60*, 2602.

In: Advanced Materials - Studies and Applications
Editors: I.A. Parinov, S.-H. Chang, S.Theerakulpisut ISBN: 978-1-63463-749-7

Chapter 7

MULTIOBJECTIVE OPTIMIZATION OF THE CURE CYCLE AT THE CLOSED-MOULD FORMING OF THICK WALLED POLYMERIC COMPOSITE STRUCTURES

***S. Shevtsov*[1,*]*, I. Zhilyaev*[1]*, A. Evlanov*[2]*, I. Tarasov*[3]*, M. Flek*[4] *and L. Chinchan*[4]**

[1]Department of Aircraft Systems Mechanics and Technologies, South Center of Russian Academy of Science, Rostov-on-Don, Russia
[2]Department of Composite Manufacturing, Beriev Aircraft Co., Taganrog, Russia
[3]APGroup LLC, Saint-Petersburg, Russia
[4]Rostvertol Helicopters Co., Rostov-on-Don, Russia

ABSTRACT

High-loaded aircraft composite structures are produced by autoclave or closed moulding processes when consolidation of fibers and matrix is completed at the same time as the part is shaped. Full curing schedule includes a pre-heating stage lasting until resin viscosity reaches its minimum level, then pressure supply to remove gas bubbles and excess resin, and finally consolidation of resin at higher temperature to its full polymerization. Composite states should be changed as uniformly as possible through thick-walled pieces. Main processing problems, encountered during the curing of parts with large variations in thickness, include porosity, resin-rich areas, resin-dry areas, insufficient consolidation, degraded mechanical properties, and distortion which can result from non-uniform cure. As a solution of these problems we suggest a model-based control approach, which uses a mathematical model for epoxy-based thick-walled composite structures curing. This model involves coupling kinetic equation of resin cure with heat transfer equation and takes into account an exothermal heat, change of the heat capacity and heat transfer coefficient at the resin phase transition from liquid to gel and further to solid state, where external temperature field is held by distributed heaters. As a way of generating multiobjective optimal control, we have conducted a multiple finite

[*] Corresponding author: aeroengdstu@list.ru.

element transient analysis of the developed model, which produced the Pareto points in the space of the design variables.

1. Introduction

Closed moulding is a popular process of manufacturing composite parts due to its high-efficiency in manufacturing complex-shaped parts for aircraft and ship structures. This process begins with placing premixed compound in a mould. As the mould is closed and heated, resin matrix flows and fills mould cavity and fine crevasses between fibers in a composite part. Pressure supply and higher temperature cause the raw prepreg material to take the desired shape, defined by the mould geometry. At the first stage of curing process resin viscosity reaches its minimum level, caused by external heat and the heat of reaction, and then increases rapidly due to molecular chain extension. The next step of this reaction is gelation, when the chains start to cross-link, the resin no longer flows and most individual reaction rates significantly decrease. The third step at high cross-linking is vitrification [1–3].

The mould is heated inside the autoclave or by the set of electric elements and their temperature is monitored and controlled by the automatic control loop [1,2]. Heating and cooling rates as well as supplied pressure are controlled together to ensure uniform curing throughout the part and to reduce the possibility of residual stresses causing structural deficiencies or distortions [1,2,4,5]. Complex heating/pressure cycles are necessary to meet the requirements of chemical reactions and to ensure that resin viscosity is optimum when pressure increases. So, since resin viscosity falls with temperature increasing until the resin begins to cross-link chemically (gel), it is important to supply pressure before gelation process beginning to remove entrapped gases and excess resin. This will help to avoid severe porosity and resin-rich areas.

It is sometimes possible to increase the cure temperature or preheat resin providing faster cross-linking reactions to reduce cure time [6]. But this can lead to significant loss of the composite product quality, which appears as distortions and delaminations. In some cases this problem can be avoided by gelling resin at low temperature and slowly increasing the temperature by a linear ramp up to its maximal value [1,2,7]. The temperature-time control law usually incorporates the dwell section (isothermal hold) to prolong the time of consolidation and volatiles removal. This hold also pre-reacts resin and reduces the danger of intensive exothermic reactions that can occur in thick laminates [4], and also allows temperature to become more uniform; this is very important for structures with large variations in thickness.

The main difficulty in large composite parts manufacturing is unobservability of the material state during cure process. Well-known methods including Differential Scanning Calorimetry (DSC), Dielectric Analysis (DEA) and others can be effectively used only in laboratory conditions [7]. Therefore, a model-based control strategy used here assumes the formulation of a mathematical model of the controlled cure process that allows an access to both observed and hidden parameters of the process, and building an appropriate control law [8–10].

The presented model of epoxy based composites curing incorporates the kinetics of thermoset resin reactions changing its phase state, thermal capacitance during cure, heat transfer within the whole technological system and multi-input control of heaters. Basing on this model we formulate and solve the problem of synthesis optimal temperature schedule using the Pareto frontier approach, which is highly efficient in cases when process quality is characterized by two or more objectives. In this study we use 8 design variables and select such objectives as conversion and conversion's deviation, averaged over the cured composite body. Due to large dimensions of the design space there is a difficulty in reconstructing objective's topology [12]. To overcome the difficulty in searching the best set of design variables, we use a slice in one of the objectives' subspace which possesses its admissible values and we also perform the fitting of Pareto subset for another objective that allows us to visualize the set of feasible points in the design space.

2. Equation of Cure Kinetics and Identification of its Parameters

Regardless of dimensionality and geometry of the modeled cured polymeric composite system, each model requires identification of all thermal and rheological resin parameters at all stages of resin cure cycle. For that reason, the first step before statement of an optimization problem is to describe thermo-physical properties of the materials. Most of these properties are temperature dependent. For example, specific heat capacity C_r of thermosetting resin depends on both reaction temperature T and conversion α during the curing process. The degree of epoxy cure α is defined as a ratio of energy released during partial reaction to total enthalpy of the reaction. This definition has been widely used in determining epoxy cure state and can be applied to evaluate the exothermic energy rate in a heat transfer model. DSC is widely used to determine reaction and specific heat while line source or point source methods are applied for thermal conductivity determination [7,8,10,11].

An exhaustive mathematical description of cure process was first given in kinetic equation applied by Kamal [13] for epoxy resin and used with some extensions in many studies [3,14–16]. Most common requirements to kinetic equations include correlation between experimental and modeled dependencies of differential heat flow on temperature and time, the model ability to fit both one and two staged cure reactions, and minimum amount of adjustable model parameters determined by using experimental data. Two staged cure reactions are the most difficult to simulate [10,11]. In this investigation we use already developed kinetic model [10] but with modified technique of model parameters identification.

DSC experiments have been performed using NETZSCH DSC 204 F1 equipment. The studied resin was subjected to heating at such rates as 1, 2, 5, 10 °C/min from 30 °C up to 250 °C. As for the studied resin, an increasing heating rate leads to increasing of gelation and vitrification temperature. After reheating DSC curves don't have any peaks. That confirms full completion of polymerization process at the first heating step, i.e., resin reaching the degree of cure whose value is 1.

After excluding contribution of material specific heat capacity in total heat flux using this definition

$$dQ/dt = CvT + dQ_s/dt\,, \tag{1}$$

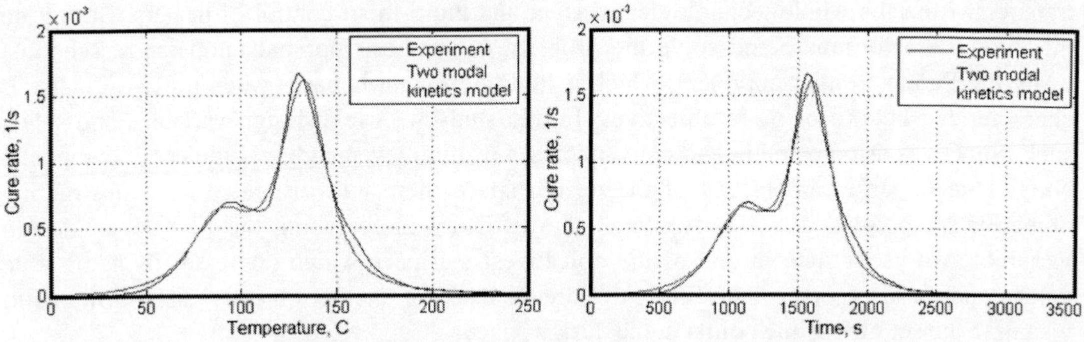

Figure 1. Cure rate dependencies on temperature (left) and time (right), obtained in DSC experiments at the heating rate of 5 K/min.

where Q_s is the heat evolved due to exothermal effects, the dependencies of net exothermal heat flux on temperature and time were obtained. The following relationships were used to calculate the time dependencies for cure rate $d\alpha/dt(t)$:

$$\frac{d\alpha}{dt}(t) \equiv \frac{1}{Q_s}\frac{dQ_s}{dt}\ ;\ Q_s = \int_0^{Tp}\left(\frac{dQ_s}{dt}\right)dt\ , \tag{2}$$

where time dependencies for differential heat flow were approximated by the 3rd order splines and integration spread over time interval between start and stop of evolving heat. The resulting dependencies are shown in Figure 1. These dependencies together with total evolved heat values Q_s and heating rate vT were further used for identification of kinetic equation parameters.

The main feature of obtained kinetic curves is their two-mode character, which confirms two staged cure reaction in the studied resin.

As all kinetic models, which have been considered by Koorevaar [3] cannot be adjusted to describe the behavior of resin at two stages of the cure reaction and do not give a good correlation with experimental data, we used an earlier suggested kinetic model [10] which consists of 8 parameters:

$$\frac{d\alpha(T,\alpha)}{dt} = \left[k_1 e^{-((\alpha-\alpha_t)/\beta_t)^2} + k_2\cdot\alpha^m\right]\cdot(1-\alpha)^n, \tag{3}$$

where exponential terms are

$$k_i = A_i\exp(-E_i/RT)\quad i=1,2\ . \tag{4}$$

In last two equations all designations mean the following: k_i - rate constants, A_i - pre-exponential factors, E_i - activation energies, R - universal gas constant, m,n - reaction orders, α_t,β_t - location and width of the first peak, and t - curing time. The first summand in brackets ensures start and fast suppression of the first reaction when the degree of cure is comparable to α_t, and the second one describes the second reaction, which gives a contribution at relatively high conversion.

Equation (3) has been realized programmatically as MATLAB's *m*-function with 2 input parameters (time and degree of cure α) whereas 8 parameters (A_i, E_i; $i = 1, 2$; m, n, α_t, β_t) that should be determined and temperature are defined as global variables. The numerical integration of equation (3) by Runge-Kutta solver used zero initial condition for α. As the objective function to be minimized, we used curing time average squared discrepancy between experimental and modeled values of the reaction rate. At each of iterations, genetic algorithm (GA) requested the objective function that depended on the eight identifiable parameters. In turn, the objective function calling the solution of the kinetic differential equation with identifiable parameters values that were appointed by GA. All identified parameters have been transformed to the logarithmic scale to facilitate a monitoring of the running GA.

Using above-described technique of numerical processing of DSC results, empirical dependence for heat capacity $C_r(\alpha)$ has been suggested:

$$C_r(\alpha) = [C_f - (C_f - 1.15C_s) \cdot H(\alpha - \alpha_t, \delta\alpha)](0.8 + 0.2e^{-\alpha}), \tag{5}$$

where C_f and C_s are specific heat of uncured (liquid) and fully polymerized (solid) resin respectively; conversion α_t corresponds to the jump of heat capacity at the phase transition; and the jump width $\delta\alpha$ in the smoothed Heaviside function *H* for the studied resin. These values are defined by the following empirical relationships:

$$\alpha_t = 0.05 + 0.45\tanh^2(T_t/12); \; \delta\alpha = 0.1 + 0.4[1 - \exp(T_t/15)] \tag{6}$$

Equation (5) describes the jump of heat capacity at the phase transition, and broadening of this jump with increasing heating rate. Relationship for the dependence of resin thermal conductivity similar to equation (5) can be obtained after a replacement $k_r \to C_r$; $k_f \to C_f$; $k_s \to C_s$. Both empirical dependencies for specific heat and thermal conductivity on the degree of cure are shown in Figure 2.

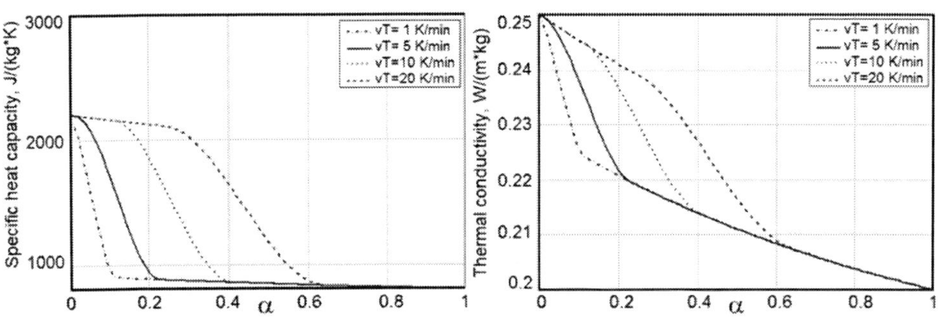

Figure 2. Dependencies of specific heat capacitance (left) and thermal conductivity (right) on conversion α at the different heating rate.

Found values of parameters in Equations (3) – (6) have then been used for modeling and optimization of curing process for large composite parts with varied thickness.

3. Modeling of Cure Kinetics and Heat Transfer for Closed Mould Cure

When a cured part occupy the volume in a heated mould, temperature and rheology state of composite material will change in its various points. Hence, temperature and conversion fields change over time, influencing on each other and heat transfer is strongly coupled with spatially-temporal evolution of α and T. The similar problem has been considered in many studies [9, 10, 17, 18] with somewhat simplified statement. In this chapter we consider this coupled problem using the example of a composite spar for a helicopter tail rotor blade. The problem statement for a cured composite part includes kinetics equation (3) for conversion coupled with heat transfer equation

$$\rho_c C_c \, \partial T/\partial t + \nabla(-k_c \nabla T) = Q_{exo}, \tag{7}$$

where composite material density ρ_c is defined through resin density ρ_r, reinforcing fibers density ρ_{fb}, and relative quantity of fibers ν_{fb} according to the mixing rule [11, 19]

$$\rho_c = \rho_{fb}\nu_{fb} + \rho_r(1-\nu_{fb}), \tag{8}$$

the heat transfer coefficient of filled composite [17]

$$k_c = k_r \cdot \frac{(1+\nu_{fb})k_{fb} + (1-\nu_{fb})k_r}{(1-\nu_{fb})k_{fb} + (1+\nu_{fb})k_r}. \tag{9}$$

In the last equation an actual value of heat transfer coefficient k_r for resin matrix is defined by the function of conversion

$$k_r = \left[k_f - (k_f - 1.145k_s)\cdot H(\alpha-\alpha_t, \delta\alpha)\right]\cdot(0.8 + 0.02e^{-\alpha}) \tag{10}$$

Specific thermal capacity C_c of a composite in equation (7) is also defined by the mixing rule

$$C_c = C_{fb}\nu_{fb} + C_r(1-\nu_{fb}), \tag{11}$$

and intensity of internal heat sources is defined according to the following relationship

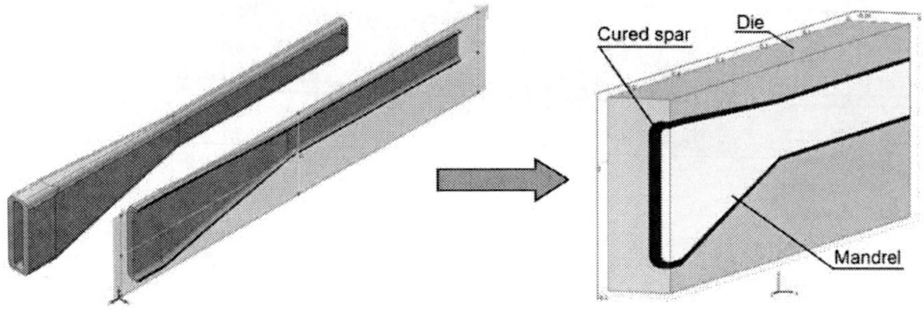

Figure 3. A half of the CAD model of a cured spar (left) and the finite-element model of the whole curing system (right).

$$Q_{exo} = Q_{tot}(\partial T/\partial t)(\partial \alpha/\partial t). \tag{12}$$

All these equations are active in the spar body only, but the heat transfer equation (7) with appropriate coefficients is active everywhere. All boundary conditions depend on the cured piece and mould geometry (see Figure 3) and also on the scheme of controlled heating. On the end surfaces insulation boundary conditions are taken as

$$-\mathbf{n}\cdot(-k\nabla T) = 0; \quad -\mathbf{n}\cdot(-\nabla \alpha) = 0. \tag{13}$$

At the interfaces of metal and cured composite, continuity boundary conditions are used as follows

$$-\mathbf{n}_s\cdot(-k_c\nabla T) - \mathbf{n}_m\cdot(-k_m\nabla T) = 0, \tag{14}$$

where lower indices indicate: s – composite spar, m – metallic mould and mandrel, and $\mathbf{n}$ - unit normal vector to the boundary.

Due to changes of the cured spar cross-section (see Figure 3) heat is supplied to the external surface of a mould by a set of independently controlled electric heaters. As far as the cured spar has geometry with symmetry, we study in the finite element implementation only a half of the curing system that is heated by two spatially separated heaters highlighted in red in Figure 4. Boundary conditions for the incoming heat flux are

$$-\mathbf{n}\cdot(-k\nabla T) = q(y,z,t). \tag{15}$$

All other external mould boundaries irradiate heat under the Stefan–Boltzmann law

$$-\mathbf{n}\cdot(-k\nabla T) = \varepsilon\sigma(T_{amb}^4 - T^4), \tag{16}$$

where T_{amb} is an ambient temperature, σ is a Stefan–Boltzmann constant, and ε is surface emissivity equal to 0.5.

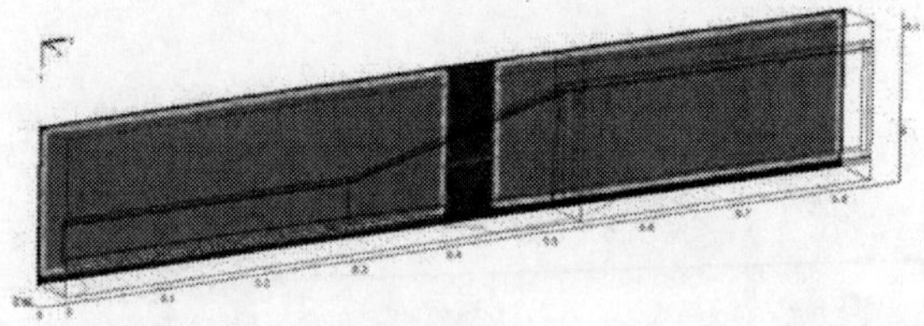

Figure 4. Spatial distribution of heated area on the mould boundary.

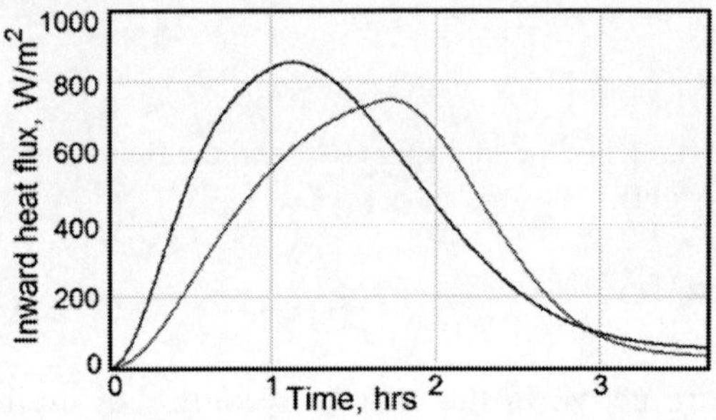

Figure 5. Two examples of time-dependence for the heaters performance.

Conversion from CAD format of previously developed model to the finite-element (FE) format was performed by Comsol Multiphysics built-in converter. FE model was formulated by coupled Heat Transfer and Partial Differential Equation (PDE) modes where the kinetics equation (3) was present in a standard PDE form. A ready model can be simulated for performing of transient analysis in real time mode.

For achieving good flexibility of control providing uniform distribution of conversion inside the cured composite part immediately before supplying compressive pressure as well as at the end of full curing cycle, both electric heaters (see Figure 4) should be controlled by their own time dependencies of performance. These dependencies (see Figure 5) must be written as functions of time, depending on parameters, whose amount should provide necessary flexibility. On the other hand, this amount should be as small as possible to diminish the value of freedom degree in an optimization problem. Moreover, the objective functions (which are to be set) should be sensitive enough to the variation of these parameters.

The following parameterization has been suggested for one step heating (to gelation or full curing states)

$$Q(t) = W \begin{cases} Q_1(t) = Q_1^0 \tanh^2(t/\tau_1) & \text{at} \quad t \le 2\tau_1; \\ Q_1(t) - Q_2^0[1 - e^{-[(t-2\tau_1)/\tau_2]2}] & \text{at} \quad t > 2\tau_1, \end{cases} \tag{17}$$

where Q_1^0, Q_2^0 are parameters determining peak and stable values of the generated heat flux, τ_1, τ_2 are time constants, and multiplier W provides the right value and dimensionality of the fluxes. This law provides the gradually increasing heat flux, which then reaches its maximum value, and afterwards decreases and moves to the stationary value, which depends only on heat energy which radiates from mould surfaces.

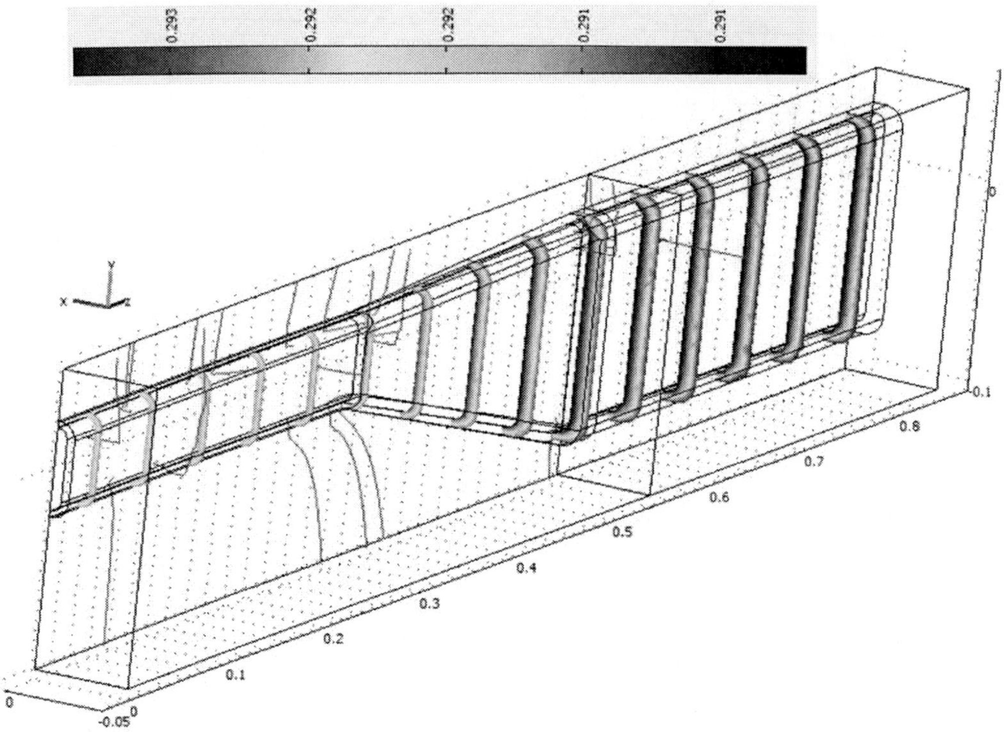

Figure 6. Example of postprocessing at the end of integration time; the color map corresponds to the degree of cure.

When numerically solving equations (3), (7) – (17) for the system in Figure 4, the final integration time was assumed to be equal to 3 hours. The integration has been carried out with the distributed solver, which integrates the kinetic equation and heat transfer equation separately with the time step of 120 sec. The example of postprocessing result for spatial distribution of cure degree (colored segments) and for temperature gradient (streamlines) is shown in Figure 6. As one can see the conversion deviation along the cured body is within the range of $0.29 \le \alpha(\mathbf{r}) \le 0.31; \quad \forall \mathbf{r} \in \Omega$.

4. Multiobjective Optimization of the Cure Cycle

Two different stages of curing process should be optimized: the stage of resin gelation with the following pressure supply, and the further stage of final polymerization, which starts from an optimal state based on the previous step. We present the developed approach using the most important first stage of gelation as an example. There are 8 parameters $\{Q_i^0\}_{i=1,..4}$ and $\{\tau_i\}_{i=1,..4}$ for two electric heat sources, which fully determine control signals. We assumed the average value of conversion α as the curing quality criteria which should be as close as possible to the desired degree of cure $\overline{\alpha}$ as well as standard deviation of the degree of cure $\langle \Delta\alpha \rangle$ at the end of monitored curing stage, by minimizing the expressions

$$\min_{\{Q,\tau\}} \langle \alpha - \bar{\alpha} \rangle \equiv \sqrt{\int_{\Omega} (\alpha - \bar{\alpha})^2 d\omega \Big/ V}\ ;\ \min_{\{Q,\tau\}} \langle \Delta\alpha \rangle \equiv \sqrt{\int_{\Omega} [\alpha - \sqrt{\langle \alpha - \bar{\alpha} \rangle}]^2 d\omega \Big/ V}\ , \tag{18}$$

that were calculated by integration over the whole cured composite volume V. These integrations were carried out at the end of simulated real time T_F. Variation of T_F allows slowing down or speeding up the overall cure cycle. All varied parameters in equation (18) for each heater are constrained by the set of inequalities

$$Q_2^0 - Q_1^0 \leq 0;\ \ \tau_1 + \tau_2 \leq \tilde{\tau},, \tag{19}$$

where constant $\tilde{\tau}$ is a predefined part of the full simulation time T_F.

According to the used approach we can focus on the set of choices that are Pareto-efficient. A number of numerical experiments at the curing time $T_F = 3$ hours was carried out to reproduce topology of the Pareto set. Our results demonstrate that the desired value of average conversion $\bar{\alpha}$ is reached easily for the given value of T_F, while providing variation $\langle \Delta\alpha \rangle$ within a restricted interval was the most difficult part. For effective filling of the design space eight-dimensional volume, the ranges of each design variable variation have been divided into 10 equal segments, where the designed variables take random values. However, it will take 10^8 computations to fill out all the cells in the selected volume of at least one point. Therefore, after first simulations the expected ranges of parameters have been narrowed, and then a special fitting procedure performed for the reconstruction of level lines. This fitting procedure, which was especially intended to visualize the topology of Pareto set, includes dividing each interval of every parameter into three equal sub-intervals and filling the resulting sub-cells with the weighted sum of the best objective values in all surrounding nearest cells. These weights were adopted as exponents that completely attenuate at a distance of two lengths of the cell. The "ant swarm" strategy was used to fill the design subspace faster.

Due to high dimensionality of the objective points' set, it is difficult to imagine this topology. Even in case of two objectives and eight design variables, visual representation of the Pareto set is not well descriptive. For eliminating this disadvantage, we circumscribe the range of the calculated $\langle \alpha \rangle$ by the range of $0.29 \leq \langle \alpha \rangle \leq 0.31$, which corresponds to its feasible values. Then two-dimensional projections of the points in the design space should be built.

Numerically calculated points in the objective space, which are projected on two-dimensional design subspace and corresponding level lines of the objective (standard deviation of conversion in the cured body, which should be multiplied by 0.001) (see Figure 7) were built after about 10^5 simulations. But a substantial part of the obtained results corresponded to the case when input heat fluxes take negative values, i.e., require local cooling (see Figure 8). Since such control laws with final cooling are difficult to provide at manufacturing conditions, these regimes (points in eight-dimensional space) were excluded. Figure 7a shows all feasible design points whereas Figure 7b depicts design points separately (designs that provide cooling are depicted in green diamonds, and designs that do not provide cooling are depicted in red hexagrams). Due to elimination of such points (i.e., designs) as those highlighted in green in Figure 9, the area of acceptable objective values (filled by red points) became smaller, and the best objective values were somewhat worse than they generally are (see Figure 9).

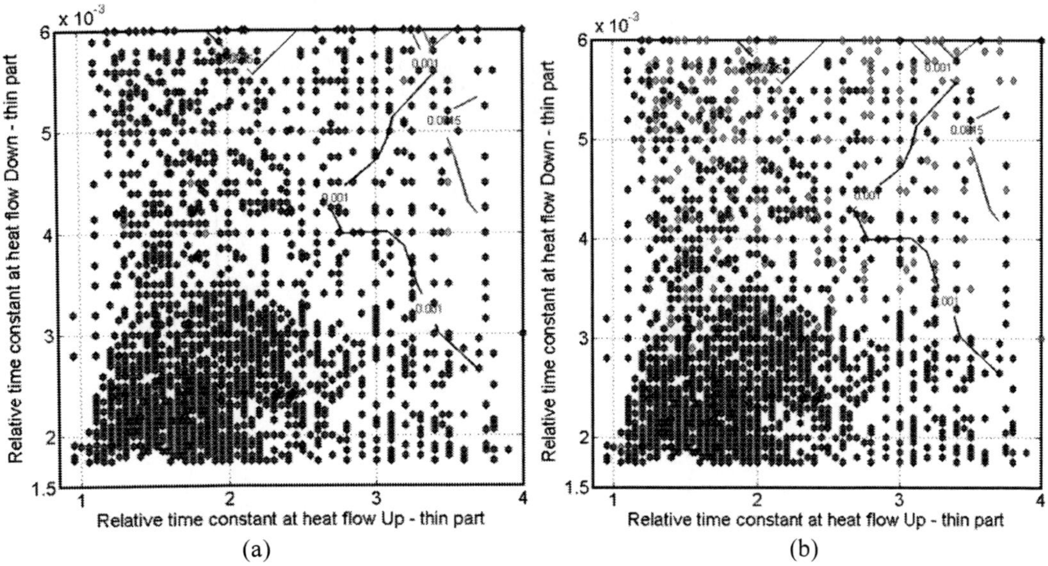

Figure 7. Scatter plots on the objective points' projection in subspace (τ_1, τ_2) for the heater, which is installed opposite the thin part of the cured spar; explanation can be found in the text.

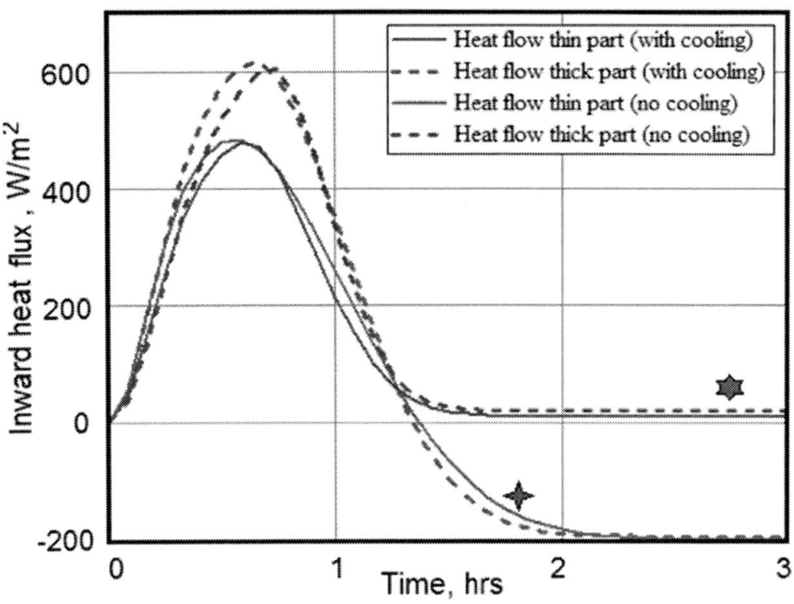

Figure 8. Two time histories of the heat fluxes that were applied to the thin part (solid lines) and to the thick part (dashed lines) of the cured spar.

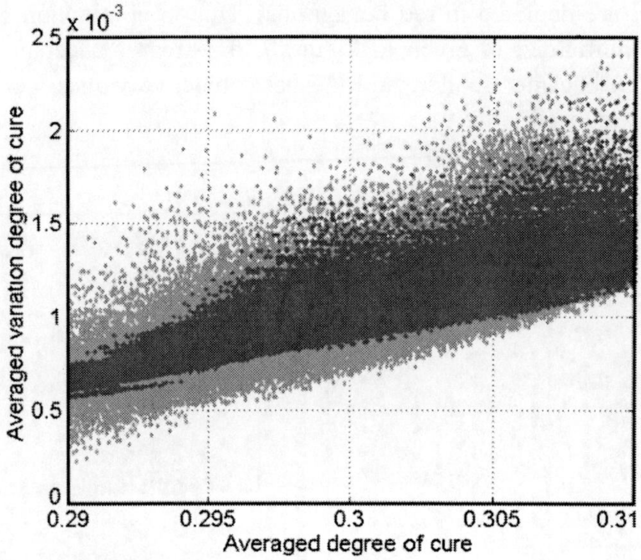

Figure 9. Scatter points in the objective space $(\langle\alpha\rangle,\langle\Delta\alpha\rangle)\frac{1}{2}$.

5. Results and Discussion

For better visibility of the obtained optimization results we present only subset of objective $\langle\alpha\rangle\in[0.29;\ 0.31]$, where all points provide schedules without cooling. Time constants τ in diagrams below (see Figure 10) are present in dimensionless units. The real value of each time constant is a result of multiplying the dimensionless value by 18 minutes. Similarly, the value of the internal heat flux acting upon the heated surface is a result of multiplying the dimensionless value (presented in diagrams) by 10 W/m^2.

The values of $\langle\Delta\alpha\rangle$ on the level lines should be multiplied by 10^{-3}. Markers on the diagrams correspond to the control laws, whose results are shown in Figure 11.

The diagrams in Figure 10 demonstrate a very complex topology of the feasible laws subset. Hence the objectives' dependencies on designed variables cannot be expressed by a simple analytical formula. This fact requires keeping all simulated data for the following application at the specific cure design.

All designed variables can vary greatly, but last two diagrams in Figure 10 show a very strong coupling between values of internal heat flows at phases of initial heating and its termination. It proves the necessity of accurate setting of heating parameters.

The time histories for the conversion, temperature and their deviations show that feasible curing results can be reached by using very different control parameters. The behavior of objectives in Figure 11 is similar enough whereas the control laws differ significantly. This result is very useful in practice because it provides some degrees of freedom for the technologists, allowing for changes in the curing schedule according to available equipment, and also find some ways to speed up the curing cycle.

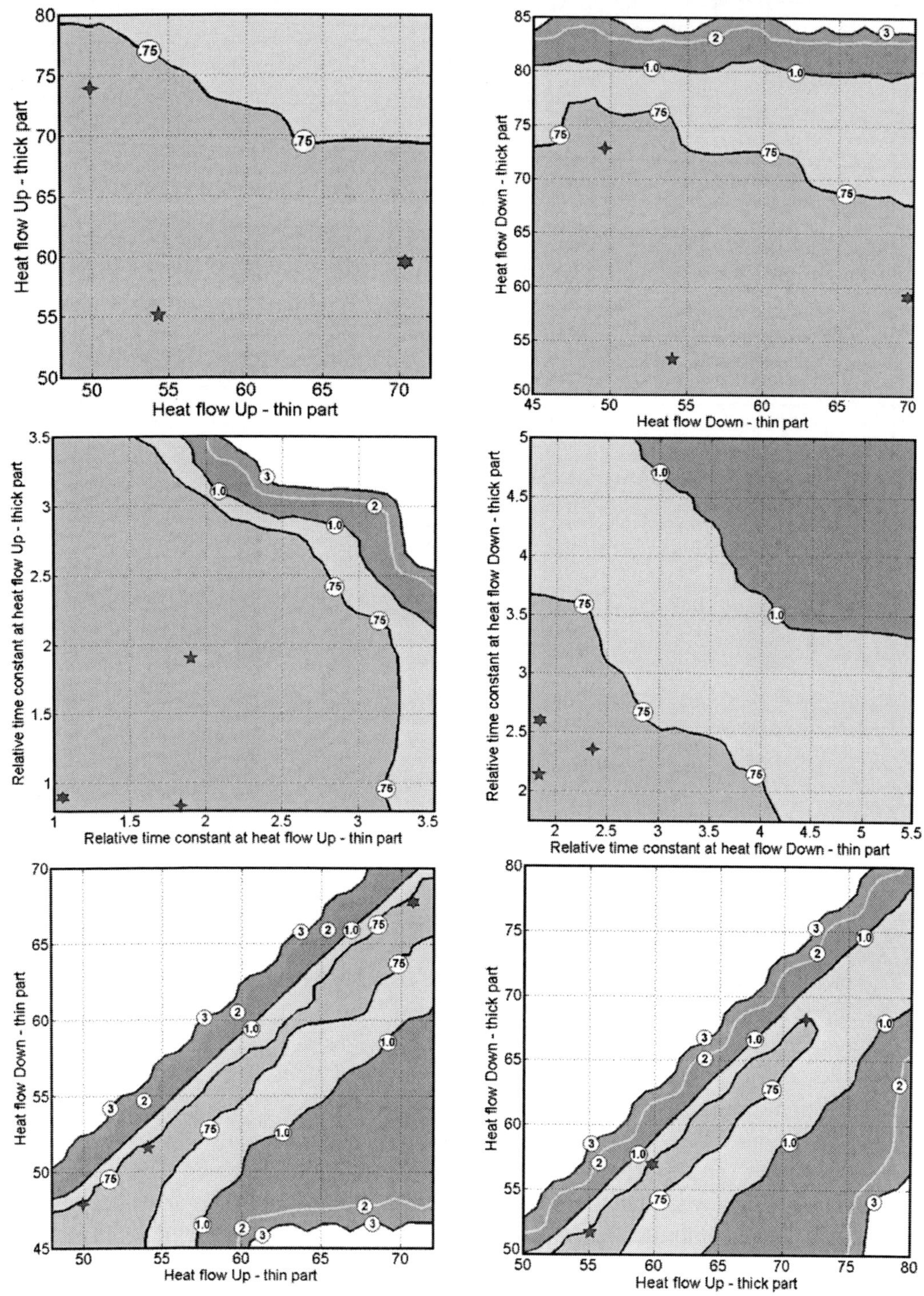

Figure 10. Some level lines diagrams for the objective showing average deviation of degree of cure at the value of $\langle\alpha\rangle \in [0.29; 0.31]$.

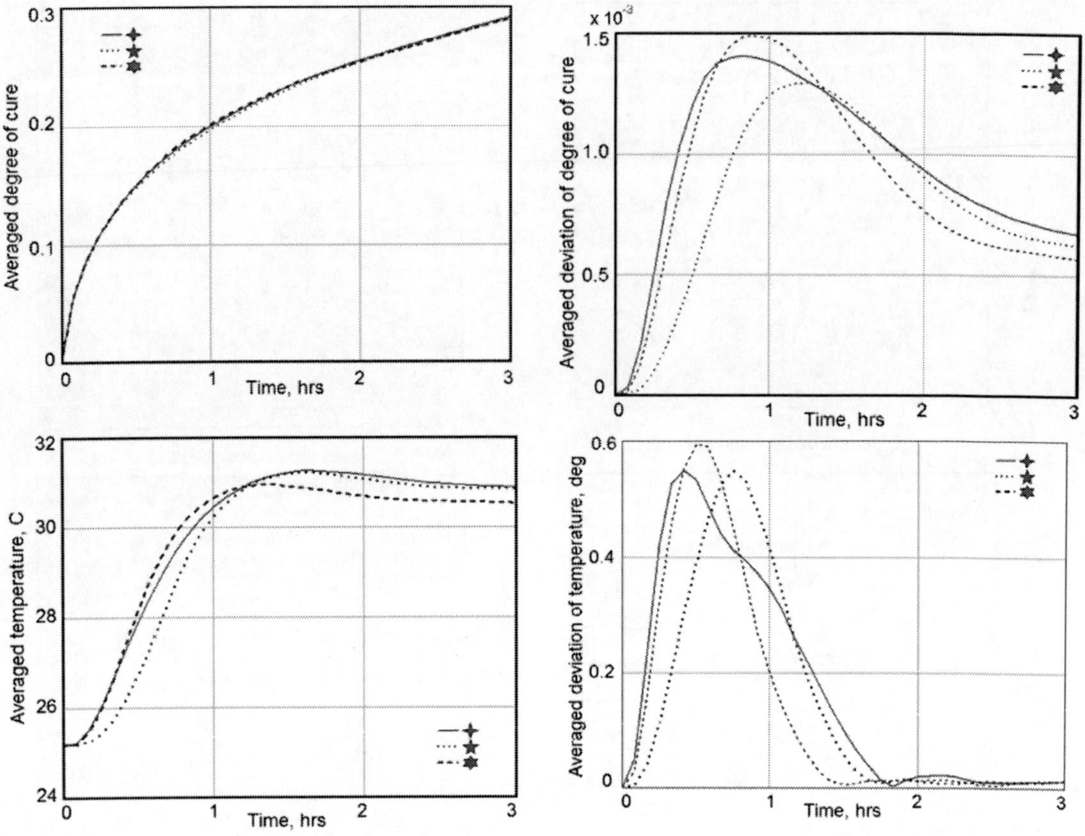

Figure 11. Time histories for the objectives simulated with the control laws depicted in Figure 10.

CONCLUSION

The model of spatially distributed curing process of epoxy matrix polymeric composite was developed and implemented using finite element method.

For avoiding formation of the internal peel and cavities due to excessive heating rate, a problem of optimal heating control was formulated and solved using suggested parameterization of heater's temporal performance, the developed finite element model, and the Pareto set reconstruction technique.

In an industrial process, it is cost-efficient to minimize cure time by increasing cure temperature, but this can lead to significant loss of the composite product quality. The developed model and optimization technique allow us to determine the optimal time for each stage of cure and control law parameters. The best choice of parameters for the used epoxy resin thermal schedule for a 3 hour 1st stage of cure was shown.

Only 30% of points in the design space are feasible (i.e., satisfy all technological restrictions).

All reported results were obtained assuming the open thermal control loop. However, all numerical and full-scale experiments show the good controllability of the heating with PD controller (which compensates small disturbances of ambient temperature) when using the measured temperature values in 8 spatially distributed points inside the steel body of the dies.

The suggested approach and simulation results allow choosing the best acceptable schedule in actual manufacturing conditions.

Acknowledgments

The authors wish to acknowledge the partial financial support from the Russian Foundation for Basic Research (Grant 14-08-31612) and the valuable technical assistance from Rostvertol Helicopters.

References

[1] Akovali, G. *Handbook of Composite Fabrication*. Shawbury, UK, RAPRA Technology LTD, 2002, 1–196.

[2] Baker, A; Dutton, S; Kelly, D. *Composite Materials for Aircraft Structures*. Reston, Virginia, AIAA ed., 2004, 1 – 597.

[3] Koorevaar, A. *Proceedings of the ISCM 2002 Conference*, Vollenhove, Nedherland. 2002, 12 p.

[4] Heinrich, C; et al. *Int. J. Engineering Science*., 2012, vol. 53, 85 – 111.

[5] Sorrentino, L; et al. *Proceedings of the* International Conference on Composite Materials ICCM17, Edinburgh, Scotland, 2009, 8 p.

[6] Garschke, C; Weimer, C; Parlevliet, PP; Fox, BL. *Composites: Part A.*, 2012, vol. 43, 935 – 944.

[7] Hardis, R; et al. *Composites: Part A*., 2013, vol. 49, 100 – 108.

[8] Melnik, RVN. *Model. Simul. Mater. Sci. Eng*., 2002. No. 10, 341 – 357.

[9] Saad, A; et al. *J. Reinforced Plastics and Composites*., 2012. vol. 31, 1388 – 1399.

[10] Shevtsov, S; Zhilyaev, I; Soloviev, A; et al. *Advanced Material Research*., 2012. *vol. 569*, 185 – 192.

[11] Shevtsov, SN; Flek, MB; Wu, JK. et al. In: *Advanced Materials. Physics, Mechanics and Applications*, S.-H. Chang, I. A. Parinov, V. Yu. Topolov (Eds.), Springer Proceedings in Physics, vol. 152, Chapter 7. Springer, Heidelberg, New York, Dordrecht, London, 2014, 71 – 88.

[12] Agrawal, G; et al. *Proceedings of the 10th AIAA/ISSMO Multidisciplinary Analysis and Optimization Conference*, Albany, New York, 2004, 12 p.

[13] Kamal, MR; Sourour, S. *Polymer Engineering and Science*., 1973. vol. 13, 59 – 64.

[14] Aboulaich, R; Boujena, S; Pousin, J. *J. Condensed Matter*., 2004. vol. 5(1), 4 – 9.

[15] Van Assche, G; Swier, S; Van Mele, B. *Thermochimica Acta*., 2002. *vol. 388*, 327 – 341.

[16] Wu, JK; Huang, JP; Shevtsov S; Chinchan, L. *Proceedings of the 2014 International Conference on Information Science, Electronics and Electrical Engineering ISEEE 2014*, Sapporo, Japan, 2014, 6 p.

[17] Garschke, C; Parlevliet, PP; Weimer, C; Fox, BL. *Polymer Testing*., 2013, vol. 32. 150 – 157.

[18] Liang, G; Chandrashekhara, K. *J. Applied Polymer Science*., 2004. vol. 916, 3513 – 3538.

[19] Harsch, M; Karger-Kocsis, J; Holst, M. *Eur. Polymer J*., 2007, vol. 43, 1168 – 1178.

PART II. PHYSICS OF ADVANCED MATERIALS

In: Advanced Materials - Studies and Applications
Editors: I.A. Parinov, S.-H. Chang, S.Theerakulpisut
ISBN: 978-1-63463-749-7

Chapter 8

STRUCTURE, DIELECTRIC AND MAGNETODIELECTRIC PROPERTIES OF SPINEL-PEROVSKITE LEAD-FREE MAGNETOELECTRIC COMPOSITES $(1-x)(Ni_{0.7}Zn_{0.3}Fe_2O_4)/x(Na, Li, Sr)NbO_3$

***M. V. Talanov**[*1], **L. A. Shilkina**[1], **V. M. Talanov**[2], **L. A. Reznichenko**[1] and **N. P. Shabelskaya**[2]*

[1]Research Institute of Physics, Southern Federal University, Rostov-on-Don, Russia
[2]South- Russian State Polytechnic University, Novocherkassk, Russia

ABSTRACT

Magnetoelectric composites $0.6(Ni_{0.7}Zn_{0.3}Fe_2O_4)/0.4(Na, Li, Sr)NbO_3$ and $0.2(Ni_{0.7}Zn_{0.3}Fe_2O_4)/0.8(Na, Li, Sr)NbO_3$ are prepared by conventional ceramic technology. Effect of temperature of composite sintering on their phase composition and structure is studied. It is stated that the composites obtained are two phase systems (perovskite and spinel, respectively). It is established that modulation of spinel structure is revealed more accurately at low temperatures of sintering (1180 °C), while modulation of perovskite structure is revealed at higher temperature (1220 °C). It is assumed that the discovered modulations of perovskite and spinel structures can be connected with extended defects such as crystallographic shear planes. It is shown that increasing temperature of composite sintering ($T_{\text{sint.}} \geq 1200$ °C) leads to disappearance of an impurity phase and the change of perovskite phase composition. The hysteresis behavior of curves $\varepsilon(H)$ and $MD(H)$ for composite $0.2(Ni_{0.7}Zn_{0.3}Fe_2O_4)/0.8(Na, Li, Sr)NbO_3$ has been revealed (the value of MD is equal to 0.3 %).

[*] Email: mvtalanov@sfedu.ru

1. INTRODUCTION

Coexistence and interrelation of magnetic and electric subsystems in magnetoelectric materials allow us to use them in fabrication of different transducers. By this, principal possibility of controlling magnetic characteristics by variations of electric field appears and *vice versa*. However at present there are almost no materials that could have magnitudes of magnetoelectric coefficients providing the practical supplements. As a result, in physics and chemistry of solids a new scientific-research direction is R&D of composites with magnetoelectric properties based on the heterophase structures having ferroelectric and magnetic components (phases). The advantage of the given approach is the possibility to select the magnetic and ferroelectric phases separately and choose the method of their connecting in one material (for example, layers, mixtures, etc.). To fabricate a composite with high magnetoelectric coefficients it is necessary to satisfy some conditions, namely: magnetic and ferroelectric components of material should have high value of magnetostrictive and piezoelectric parameters, low conductivity values of the components, and developed mechanical contact between these phases. A composite should have optimal content of the magnetic and ferroelectric phases, too. However, it is difficult for a single material to have this combination of the ferromagnetic and ferroelectric properties. Thus, ferromagnetic–ferroelectric/dielectric composite materials become an attractive subject. Although it is expected that they have the ability to show the productive and summary properties, but some physical properties (permeability and permittivity) are not so easy to be understood and modified. Electro-magnetic coupling also occurs in magnetoelectric effects which can be exhibited in composite materials. Therefore, the composite materials have been intensively studied both through fundamental studies to understand their physical properties and through synthesis of composites with various chemical formula components to find new functional properties. Ferromagnetic iron oxides (ferrites) and ferroelectric/dielectric complex oxides of niobium are good candidates for the components of composite materials since they have been widely used in microwave and millimeter wave devices. Composites with magnetic components with spinel structures and ferroelectric components with perovskite structures satisfy the above-mentioned conditions. Therefore they have become widespread [1, 2]. However there are poorly studied problems: mutual influence of phases which are different on structure and properties, the target optimization of compositions and methods for producing composites with high magnetoelectric coefficients. In the given chapter the results of research aimed to finding and processing new spinel - perovskite composites with high magnetoelectric coefficients have been presented. A magnetic component has been selected on the basis of a solid solution of spinels $NiFe_2O_4$ and $ZnFe_2O_4$, characterized by high values of resistance and magnetostrictive coefficients. In the study of composite structures $0.9BaTiO_3–0.1Ni_xZn_{1-x}Fe_2O_4$ and $Pb(Ti, Zr)O_3–Ni_xZn_{1-x}Fe_2O_4$ maximum magnetoelectric coefficient has been observed at $x = 0.2 – 0.3$ [3, 4]

Thus, to fabricate a composite structure the spinel $Ni_{0.7}Zn_{0.3}Fe_2O_4$ has been selected. The ferroelectric component was lead-free perovskite solid solutions on the basis of (Na, Li)NbO_3, characterized by relatively high piezoelectric activity and Curie temperature. The aim of this work is to establish the laws of formation of the structure of lead-free composites $0.6(Ni_{0.7}Zn_{0.3}Fe_2O_4)/0.4(Na, Li, Sr)NbO_3$ and $0.2(Ni_{0.7}Zn_{0.3}Fe_2O_4)/0.8(Na, Li, Sr)NbO_3$ obtained in the different processing conditions.

2. EXPERIMENTAL

In the most studies of composite materials, the ferromagnetic and ferroelectric/dielectric components are synthesized separately, and then composite materials are formed from the synthesized components. This is not an easy method since the final good is the result of two syntheses for ferromagnetic and ferroelectric/dielectric materials.

The composites studied in the work were prepared from mixtures of pre-formed spinel and perovskite. Synthesis of spinel structure $Ni_{0.7}Zn_{0.3}Fe_2O_4$ was carried out on ceramic technology. To intensify the process of synthesis of the samples the potassium chloride of the mark "chemically pure" was injected. Synthesis was performed according to [5] at 900 °C for 5 hours. Upon completion of the synthesis the samples were milled to grains with sizes lesser than 0.3 mm and washed from potassium chloride with distilled water. Synthesis of perovskite based on (Na, Li)NbO_3 was carried out at $T_{sint.}$ = (850 – 870) °C for 6 hours. Basic components of the composite have the following structural characteristics: $Ni_{0.7}Zn_{0.3}Fe_2O_4$ demonstrates cubic symmetry (*C*), the cell parameter a = 8.372 Å, $1/8V$ = 73.35 Å^3, (Na, Li, Sr)NbO_3 demonstrates phase mixture: from orthorhombic to monoclinic *O*(*M*) reduced cell and rhombohedral (*Rh*) in approximately equal amounts. Parameters of the *M* cell are: a = 3.912 Å, b = 3.876 Å, β = 90.78°, V = 59.32 Å^3, the *Rh* cell parameters: a = 3.907 Å, α = 89.17°, V = 59.58 Å^3 (at $T_{sint.}$ = 1200 °C). The starting materials were weighed with accuracy of 0.1 mg, homogenized in an agate mortar, placed in an alundum crucible and subjected to heat treatment at 900 °C for 2 hours. In order to determine the optimal conditions, sintering temperatures of 1110, 1150, 1180, 1200 and 1220 °C were selected and the samples were sintered with exposure for 2 hours. X-ray diffraction studies were carried out by using the device DRON 3.0 (CoK_α - radiation). The density measured (ρ_{meas}) was defined by the method of hydrostatic weighing in octane. The calculation of X-ray density ($\rho_{X\text{-}ray}$) was conducted by using the formula: $\rho_{X\text{-}ray} = 1.66M/V$, where M is the weight of formula unit in gram, V is the volume of the perovskite cell in Å^3. Relative density (ρ_{rel}) was calculated by using the formula: $(\rho_{meas}/\rho_{X\text{-}ray})\times 100\%$. The $Ni_xZn_{1-x}Fe_2O_4$ ferrite system has a cubic spinel structure with space group Fd3m. An elementary cell consists of eight formula units. One formula unit contains one cation located in tetrahedral site 8(a) and two cations located in octahedral site 16(d). Cations distribution on structurally nonequivalent positions of spinel lattice can be present by the following way:

$$(Zn_{\lambda 1}Ni_{\lambda 2}Fe_{1-\lambda 1-\lambda 2})[Zn_{1-x-\lambda 1}Ni_{x-\lambda 2}Fe_{1+\lambda 1+-\lambda 2}]O_4,$$

where λ_1 and λ_2 are the degrees of spinel inversibility [6 – 11]; in parentheses there are tetrahedral cations (they occupy Wyckoff position 8(a) of space group Fd3m), and in square brackets are shown octahedral cations (they occupy Wyckoff position 16(d) of space group Fd3m) [12].

The Ni^{2+} ions show a pointed preference for octahedral sites because of the favorable parameters of these ions to the charge distribution in the octahedral crystal field formed by oxygen ions [1, 7]. This means that λ_2 should be equal to zero ($\lambda_2 = 0$). Besides, Zn^{2+} ions prefer tetrahedral sites due to their readiness to form covalent bonds involving sp^3-hybrid orbitals [6, 12]. This means that $\lambda_1 = 1$. The linear increase in lattice constant with the Zn concentration may be attributed to the larger ionic radius of Zn^{2+} (0.74 Å) compared with that of Ni^{2+}(0.69 Å) [13]. Such a linear concentration dependence of lattice constant of cubic elementary cell of the $Ni_xZn_{1-x}Fe_2O_4$ solid solution manifests that our propositions $\lambda_1 = 1$ and $\lambda_2 = 0$ are quite proved [14]. Thus, cation distribution on structurally nonequivalent positions of the $Ni_xZn_{1-}xFe_2O_4$ spinel solid solution is expected to be like this: $(Zn_{0.3}Fe_{0.7})[Ni_{0.7}Fe_{1.3}]O_4$. Since nonmagnetic zinc divalent cations tend to sit at the 8(a) sites and the strongest exchange interaction is negative between ions in the tetrahedral 8(a) and octahedral 16(d) sites, the magnetization of the tetrahedral lattice in the zinc-mixed ferrite will be smaller than in the case of the simple ferrite, and then the saturation magnetization of the mixed ferrite at 0K is expected to increase with rising zinc content. For small concentrations of zinc, the linear rise has been found experimentally [15]. For larger concentrations, however, deviations are found, because the magnetic moments of the few remaining iron ions in tetrahedral sites are no longer able to align all the moments of the octahedral ions antiparallel to tetrahedral ones. So, the maximum saturation magnetization is obtained around Zn:M = 1:1, where the chemical formula becomes $Zn_{0.5}M_{0.5}Fe_2O_4$ [16].

The dependence of relative dielectric permeability (ε) on intensity of a constant magnetic field (H) was studied by using set-up including *LCR*-meter Agilent E4980A and electromagnet ($H = 0 - 0.7$ T). Measurements were carried out on the polarized samples (E = 5kV/cm at hold time of 40 minutes in room temperature). The magnetodielectric coefficient (MD) was determined by the formula: MD = $\{[\varepsilon(H) - \varepsilon(0)]/\varepsilon(0)\}\times 100\%$.

3. Results and Discussion

In Figure 1 the dependence of ρ_{meas} on T_{sint}, testifying strong influence of T_{sint} on ρ_{meas} $(\rho_{max} - \rho_{min})/\rho_{min} \approx 12\%$, where ρ_{max} is the maximum, and ρ_{min} is the minimum of ρ_{meas}, achieved by changing the given value of T_{sint}.

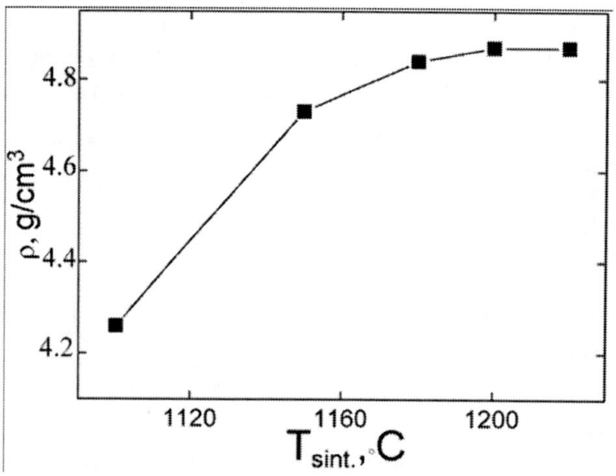

Figure 1. Dependence of the experimental density of the composite $0.6(Ni_{0.7}Zn_{0.3}Fe_2O_4)/0.4(Na, Li, Sr)NbO_3$ on T_{sint}.

The slowing growth and the steady-state of the dependence $\rho_{meas}(T_{sint.})$ at $T_{sint.} = 1180°C$, may be connected with achieving the melting temperature of the sample ($T_{sint.} = 1250$ °C). The maximum value of ρ_{rel} (95.6 %) is reached at $T_{sint.} = 1200$ °C.

Figure 2 shows the diffraction pattern of the studied composite $0.6(Ni_{0.7}Zn_{0.3}Fe_2O_4)$/ $0.4(Na, Li, Sr)NbO_3$ at $T_{sint.} = 1180$, 1200 and 1220°C.

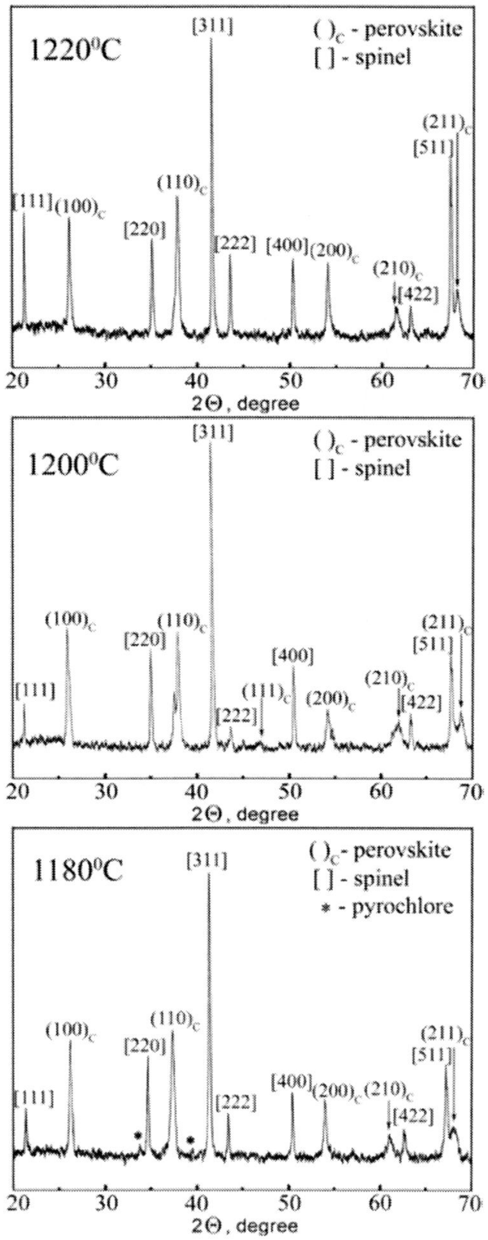

Figure 2. XRD pattern of the composite $0.6(Ni_{0.7}Zn_{0.3}Fe_2O_4)/0.4(Na, Li, Sr)NbO_3$ ($T_{sint.} = 1200$ °C) in the angular range $2\theta = 20 - 70°$. In square brackets are pointed spinel lines, in parentheses shown perovskite lines.

At $T_{sint.}$ = 1180 °C, the X-ray patterns, besides reflections of $Ni_{0.7}Zn_{0.3}Fe_2O_4$ and solid solutions on the basis of (Na, Li, Sr)NbO_3 system contains very little line (I/I_1 = 4, where I is the relative intensity of a strong reflex of an impurity phase, I_1 is the relative intensity of the strongest line on X-ray patterns) of outside phase (presumably pyrochlore). By increasing the $T_{sint.}$ up to 1200 °C and 1220 °C it leads to disappearance of lines of the impurity phase and increase of reflex intensity of the spinel phase. This is connected with attaining the optimal $T_{sint.}$ of spinel phase, at which more perfect crystal structure is formed. Thus the optimum value of $T_{sint.}$ = 1200 °C is chosen based on the X-ray analysis and the results of measurement of $\rho_{meas.}$

In Figure 3 X-ray lines $(200)_C$ and $(220)_C$ of perovskite phase of the composite 0.6($Ni_{0.7}Zn_{0.3}Fe_2O_4$)/0.4(Na, Li, Sr)NbO_3 are shown. In Figure 4 X-ray lines (400) and (440) of spinel phase are given.

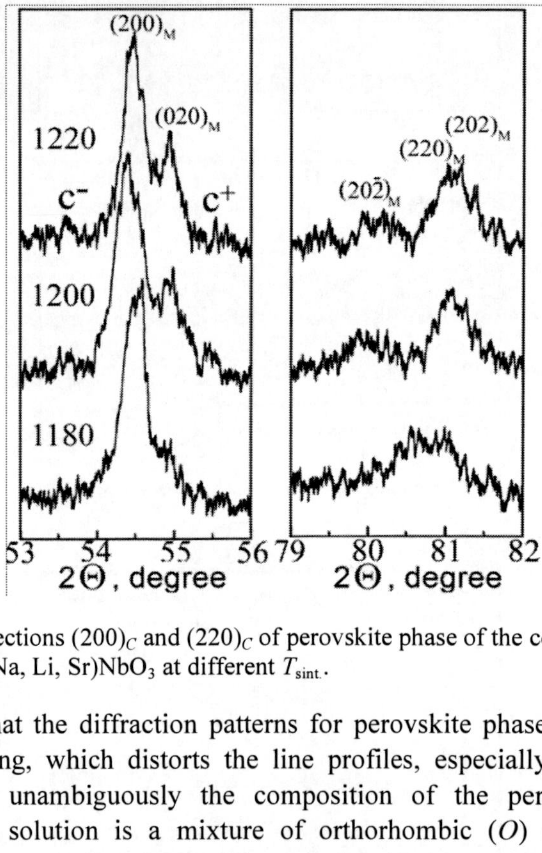

Figure 3. Diffraction reflections $(200)_C$ and $(220)_C$ of perovskite phase of the composite 0.6($Ni_{0.7}Zn_{0.3}Fe_2O_4$)/0.4(Na, Li, Sr)NbO_3 at different $T_{sint.}$.

Figure 3 shows that the diffraction patterns for perovskite phase is characterized by a strong diffuse scattering, which distorts the line profiles, especially the $(220)_C$. It is not possible to establish unambiguously the composition of the perovskite phase in the composite. The solid solution is a mixture of orthorhombic (O) and pseudocubic (not rhombohedral) phases at $T_{sint.}$ = 1180 °C. The pseudocubic phases disappear at $T_{sint.}$ = 1200 and 1220 °C and the presence of a small amount of Rh phase is not excluded.

Table 1 shows unit cell parameters of the phases of the composite, and density values of the ceramic samples (measured, X-ray and relative). It is seen that the volume of cells in both cases is reduced by increasing T_{sint} up to 1220 °C and the volume reaches values characteristic of initial components.

In Figure 3 it is seen that near to lines $(200)_C$ weak satellites are observed. They testify of structure modulation (in Figure 3 they are denoted as c^- and c^+ from smaller and larger angles θ, respectively). Modulation wavelength, Λ, calculated on the position of the satellite c^- with respect to the main peak (200), is equal to 127 Å. This value equals 32.5 times of the cell parameter. The calculation was performed using the formula: $\Lambda = (1/d_{200} - 1/d_{c^-})^{-1}$ [17].

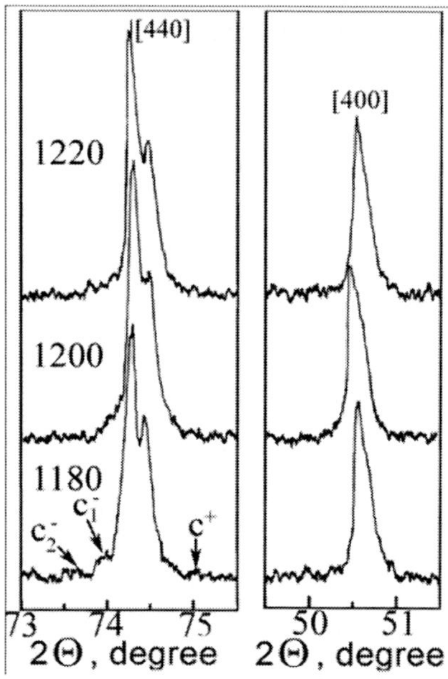

Figure 4. Diffraction reflexes [400] and [440] of spinel phase of the composite $0.6(Ni_{0.7}Zn_{0.3}Fe_2O_4)/0.4(Na, Li, Sr)NbO_3$ at different $T_{sint.}$ At $T_{sint.}$= 1180 °C diffusion maxima (satellites c_1^- and c_2^-) are clearly seen at the basis of the line (440) from the side of smaller angles.

Table 1. Structural characterization of phases of the composite $0.6(Ni_{0.7}Zn_{0.3}Fe_2O_4)/0.4(Na, Li, Sr)NbO_3$ and density values of the ceramic samples (measured, X-ray and relative)

T_{sint}, °C	Composition, I/I_1 relative units	Symmetry	Monoclinic cell parameters of perovskite				$\rho_{X\text{-ray}}$ of M phase, g/cm^3
			a, Å	b, Å	β, angle degree	V, Å^3	
1180	100-spinel 38-perovskite 4-pyrochlore	Cubic Monoclinic Pseudocubic	3.913 3.915	3.883	90.67	59.45 60.01	4.521
1200	38-perovskite	Monoclinic	3.914	3.880	90.76	59.44	4.543
1220	47-perovskite	Monoclinic	3.912	3.878	90.70	59.34	4.551

T_{sint}, °C	Composition, I/I_1 relative units	Symmetry	Cubic cell parameters of spinel		$\rho_{X\text{-ray}}$ cubic phase, g/cm^3	$\rho_{X\text{-ray}}$ composite, g/cm^3	ρ_{meas}, composite, g/cm^3	ρ_{rel}, %
			a, Å	$1/8V$, Å^3				
1180	100-spinel	Cubic	8.38	73.55	5.367	5.085	4.84	95.18
1200	100-spinel	Cubic	8.379	73.53	5.369	5.094	4.87	95.60
1220	100-spinel	Cubic	8.375	73.44	5.376	5.101	4.87	95.48

Half of the cell in value of wavelength modulation of the perovskite structure means that the cause of the modulation can be ordered extended defects such as crystallographic shear planes. In [18, 19] on the example of Ti- and Nb-containing compounds it is shown that the presence of ions of variable valence in the material structure may contribute to point defects – anionic vacancies. The latter are eliminated by shifting of the one part of the structure relative to the other along a particular crystallographic direction, forming extended defects – plane crystallographic shear. They can be regarded as a translational modulation of the initial structure [20].

Figure 4 shows that the diffraction of the spinel, unlike perovskite, is characterized by sharp, narrow lines, which indicates a well-formed structure. On the pattern diffraction, obtained at T_{sint} = 1180 °C, side satellites are clearly visible. They are located close to the line (440) from the smaller angles θ. They are indicated in Figure 4 as c_1^- and c_2^-, corresponding to the wavelength modulation Λ_1 = 421 Å (35.5 cells), Λ_2 = 210 Å, $\Lambda_2 = 1/2\ \Lambda_1$. Note that the modulation of the spinel structure is more apparent at low sintering temperatures, while the modulation of the perovskite structure occurs at higher T_{sint}. It can be explained by the following way. The ions with more stable valence state (Ni^{2+} and Zn^{2+}) are in the spinel composition. The iron ion is supposed to be in three-valence state. Disappearance of modulation in the structure of spinel phase at increasing $T_{sint.}$may be connected with formation of more perfect crystal structure at attainment of the optimal $T_{sint.}$

A perovskite phase contains an ion with a variable valence, Nb^{5+}. This ion restores at increase of temperature with formation of oxygen vacancies. These oxygen vacancies in process of their accumulation are ordered and excluded by crystallographic shift. In [21] it is shown that in stable high-temperature modification of α-Nb_2O_5, which is the main structure-forming reagent of a perovskite phase, the density of extended defects (the planes of the crystallographic shift) increases with increase of temperature.

At all temperatures of sintering the composite $0.2(Ni_{0.7}Zn_{0.3}Fe_2O_4)/0.8(Na, Li, Sr)NbO_3$, the $Ni_{0.7}Zn_{0.3}Fe_2O_4$ spinel lines and the lines of solid solution on the basis of the binary (Na, Li)NbO_3 system with perovskite structure are contained in X-ray pattern (see Figure 5). In Figure 5 the XRD patterns of composite are given for three values of T_{sint} (1180 °C, 1200 °C, 1220 °C). The spinel indices are shown without brackets, the perovskite indices are given in parantheses.

The perovskite component of the composite is a mixture of two phases, namely *Rh* and orthorhombic *O*(*M*) with the monoclinic reduced cell. The quantity of phases changes with growth of T_{sint}: *Rh*-phase prevails at T_{sint} = 1180 °C, these phases are contained approximately in equal quantities at T_{sint} = 1200 °C, and the quantity of *Rh*-phase increases a little at T_{sint} = 1220 °C.

In Figure 6, X-ray lines $(111)_c$, $(200)_c$ $(220)_c$ are shown which represent superposition of lines of *Rh*- and *O*(*M*)-phases. In the angle range θ = (44.5 – 45) there is a superstructure line. The elementary cell of *Rh*-phase corresponds to the doubled parameter *a* of reduced elementary cell of *O*(*M*)-phase which, in its turn, corresponds to the doubled parameter *b* of a monoclinic cell. In Figure 6 it is seen that profiles of lines are distorted by strong diffusion dispersion, especially in the field of big angles θ, and on wings of lines it is revealed in the form of diffusion maxima (in Figure 6 these maxima are pointed by arrows). Such diffraction picture can be consequence of periodic distribution of phases with various symmetries when dispersion occurs on phase boundaries.

X-ray patterns of spinel in difference from X-ray patterns of perovskite are characterized by accurate, narrow lines that testify about well formed structure. In Figure 7, X-ray lines 400 and 440 are shown at three temperatures of sintering. From Figure 7 it is seen that there are no fundamental differences of X-ray patterns, obtained at different temperatures. Only at T_{sint} = 1180 °C symmetric satellites c_1 and c_2 reflections (400) are visible, which show 2-wave modulation of spinel structure with wavelengths of 238 Å (28.5 cells) and 141 Å (~17 cells), respectively.

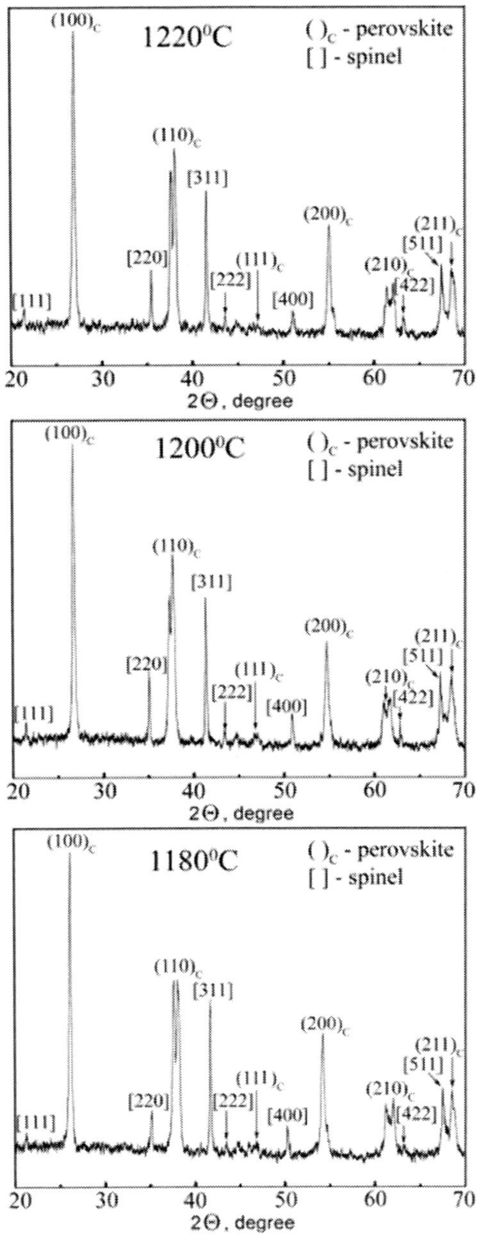

Figure 5. X-ray patterns of composite $0.2(Ni_{0.7}Zn_{0.3}Fe_2O_4)/0.8(Na, Li, Sr)NbO_3$ at all temperatures of sintering. The angle range 2θ is equal to 20 – 70°.

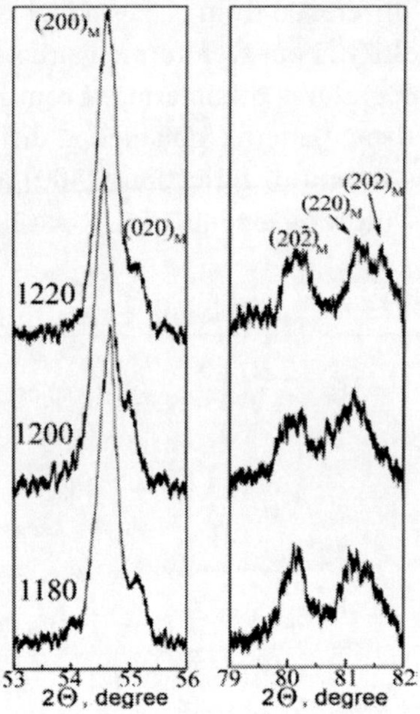

Figure 6. Diffraction reflections $(111)_c$, $(200)_c$ and $(220)_c$ of perovskite phase of composite $0.2(Ni_{0.7}Zn_{0.3}Fe_2O_4)/0.8(Na, Li, Sr)NbO_3$ at different temperatures of sintering.

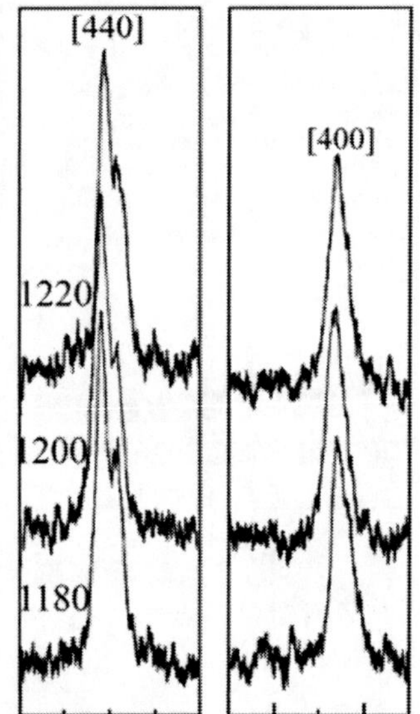

Figure 7. Diffraction reflections (400) and (440) of spinel at different sintering temperatures of composite $0.2(Ni_{0.7}Zn_{0.3}Fe_2O_4)$ / $0.8(Na, Li, Sr)NbO_3$. At T_{sint} = 1180 °C symmetric satellites c_1^-, c_2^- and c_1^+, c_2^+ are seen near (400) reflection.

Some increase in vibrations of a background in the field of wings of (440)-line is observed at all T_{sint}. This is a consequence of a disorder appearance in the direction [110].

Table 2. Relative intensity of lines of components of the composite $0.2(Ni_{0.7}Zn_{0.3}Fe_2O_4)/0.8(Na, Li, Sr)NbO_3$, structural parameters of cells of phases which are part of composite, density values of ceramic samples (measured, X-ray and relative)

T_{sint}, °C	Composition, I/I_1 relative units	Symmetry	Cell parameters of the reduced perovskite				$\rho_{X\text{-ray}}$ of perovskite phase, g/cm³
			a, Å	*b*, Å	*β*, angle degree	*V*, Å³	
1180	100-perovskite	*O(M)* *Rh*	3.911 3.904	3.874	90.68 89.16	59.24 59.49	4.55
1200	100-perovskite	*O(M)* Rh	3.910 3.906	3.877	90.76 89.18	59.27 59.58	4.54
1220	100-perovskite	*O(M)* *Rh*	3.910 3.904	3.874	90.63 89.18	59.22 59.51	4.55

T_{sint}, °C	Composition, I/I_1 relative units	Symmetry	Cubic cell parameters of spinel		$\rho_{X\text{-ray}}$ of spinel, g/cm³	$\rho_{X\text{-ray}}$ of compozite, g/cm³	ρ_{meas} of compozite, g/cm³	ρ_{rel} of compozite, %
			a, Å	1/8*V*, Å³				
1180	52-spinel	*C*	8.368	73.25	5.39	4.72	4.58	97.12
1200	45-spinel	*C*	8.370	73.30	5.39	4.71	4.64	98.46
1220	42-spinel	*C*	8.367	73.23	5.39	4.72	4.64	98.36

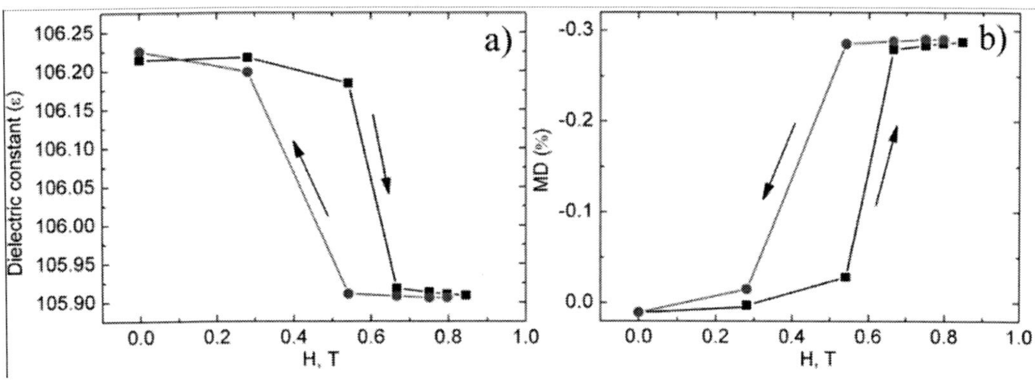

Figure 8. (a) $\varepsilon(H)$ and (b) MD(H)-dependences of $0.2(Ni_{0.7}Zn_{0.3}Fe_2O_4)$ / $0.8(Na, Li, Sr)NbO_3$ composite. The direct motion is shown by black color, the reverse motion is shown by red color.

Relative intensity of lines of the composite components, structural parameters and density of ceramic samples are specified in Table 2 for all temperatures of sintering. It is visible that cell volumes both perovskite and spinel are lesser than the volumes of initial composite components. Obviously, there is a partial mutual dissolution of the cations which are part of the perovskite and spinel. One of possible schemes of solid solutions interaction on the basis of system $(Na, Li)NbO_3$ and spinel $Ni_{0.7}Zn_{0.3}Fe_2O_4$ can be represented as follows. Partial substitution of Na^{+1} by any of the cations entering into spinel and having smaller ionic radius, than that of Na^{+1}, leads to reduction of the perovskite cell volume. Simultaneously, substitution of Nb^{+5} ions by the same size of Fe^{+3} ions (R = 0.66 Å and 0.67 Å at coordinational numbers 6) with formation of oxygen vacancies for conservation of electroneutrality may take place. The oxygen vacancies intensify formation process of the planes of crystallographic shift and, as a result, compacting of the crystal structure [6]. In Table 2 we can see that ceramics of the composite has very high relative density.

In Figure 8 are present $\varepsilon(H)$ and MD(H)-dependences of the composite 0.2 ($Ni_{0.7}Zn_{0.3}Fe_2O_4$)/0.8(Na, Li, Sr)NbO_3 fabricated at the room temperature. The considerable hysteresis between direct and reverse motion for both curves is observed. It is possible that partial grains squeezing of perovskite occur at increase of H due to the effect of magnetostriction of the spinel phase.

Mechanical pressure produced by spinel grains can lead to switching of the part of ferroelectric domains in the perovskite phase due to the direct piezoelectric effect. This explains both the existence of hysteresis in the $\varepsilon(H)$-dependence and decrease of ε values with increasing H. It is found that the maximum values of MD ~ 0.3%, which exceeds the similar parameter of lead-free composites based on $NiFe_2O_4$ and $Na_{1/2}Bi_{1/2}TiO_3$ [22]. Thus, the studied composites can be used in environmental-friendly ME-devices.

CONCLUSION

In the present work the technology for fabrication of uncontaminated composites 0.6($Ni_{0.7}Zn_{0.3}Fe_2O_4$)/0.4(Na, Li, Sr)NbO_3 and 0.2($Ni_{0.7}Zn_{0.3}Fe_2O_4$)/0.8(Na, Li, Sr)NbO_3 is developed.

Comparison of the composites with $x = 0.4$ and $x = 0.8$ shows that between them there are substantial differences not only in the correlation of the perovskite and the spinel phases, but even in the single phase.

(1) Perovskite phase in the composite with $x = 0.4$ is different from the composition of the starting component, in the composite with $x = 0.8$ perovskite composition corresponds to the initial solid solution.
(2) Volumes of reduced perovskite cell and spinel cell in the composite with $x = 0.4$ are significantly larger than their volumes in the original components of the composite, but in the composite with $x = 0.8$ they are smaller than in the starting components. This suggests that the interaction between solid solutions based on the system (Na, Li)NbO_3 and spinel composition $Ni_{0.7}Zn_{0.3}Fe_2O_4$ depends on the content of the composite components.
(3) Phase composition of the perovskite component in the composite with $x = 0.4$, depends on the sintering temperature, but in the composite with $x = 0.8$, the sintering temperature does not play a critical role.
(4) Composite with $x = 0.4$ without impurity forms only at high T_{sint}, but the composite with $x = 0.8$ forms at all T_{sint}.
(5) Ceramics density of the composite with $x = 0.4$ is lesser than the ceramics density of the composite with $x = 0.8$.
(6) Hysteresis behavior of curves $\varepsilon(H)$ and MD(H) for the composite 0.2($Ni_{0.7}Zn_{0.3}Fe_2O_4$)/0.8(Na, Li, Sr)NbO_3 is discovered (with value of MD = 0.3%).

REFERENCES

[1] Zvezdin Zheng, H.; Wang, J.; Lofland, S. E.; et al., *Science*, 2004, *vol. 303*, 661.
[2] Eerenstein, W.; Mathur, N. D.; Scott, J. F., *Nature*, 2006, *vol. 442*, 759.
[3] Upadhyay, S. K.; Reddy, V. R., *J. Appl. Phys.*, 2013, *vol. 113*, 114107.
[4] Srinivasan, G.; Rasmussen, E. T.; Hayes, R., *Physical Review B*, 2003, *vol. 67*, 014418.
[5] Talanov, V. M.; Shabelskaya, N. P., *Paten*t C1 2257953 Russia: B 01 J 23/86, 37/04, statement 29.01.2004; Publ. 10.08.2005, Bull. No.22 (In Russian).
[6] Talanov, V. M., *Energetic Crystallochemistry of Multisublattice Crystals,* Rostov State University Press, Rostov-on-Don, 1986 (In Russian).
[7] Talanov, V. M.; Bezrukov, G. V., *Solid State Commun*, 1985, *vol. 56*, 905.
[8] Talanov, V. M.; Bezrukov, G. V., *Phys. Status Solidi A*, 1986, *vol. 96*, 475.
[9] Talanov, V. M.; Bezrukov, G. V., *Phys. Status Solidi A*, 1986, *vol. 97*, 111.
[10] Talanov, V. M.; Bezrukov, G. V., *Solid State Commun*, 1990, *vol. 75*, 601.
[11] Talanov, V. M.; Bezrukov, G. V.; Men, A. N., *Phys. Status Solidi A*, 1989, *vol. 116*, 603.
[12] Krupichka, S., *Physics of Ferrites and Related Magnetic Oxides,* Mir, Moscow, 1976 (In Russian).
[13] Anwar, M. S.; Ahmed, F.; Koo, B. H., *Acta Materialia*, 2014, *vol. 71*, 100.
[14] Talanov, V. M., *Journal of Solid State Chemistry*, 1983, *vol. 48*, 86.
[15] Gorter, E. W., *Philips Res. Rep.*, 1954, *vol. 9*, 295 ; Gorter, E. W., *Philips Res. Rep.*, 1954, *vol. 9*, 321 (); Gorter E. W., *Philips Res. Rep.*, 1954, *vol. 9*, 403.
[16] Kamishima, K.; Nagashima, Y.; Kakizaki, K.; et al., *J. Phys. Soc. Jap.*, 2009, *vol. 78*, 124801.
[17] Guinier, A., *Radiocristallographie,* Moscow, 1961 (In Russian).
[18] Reznichenko, L. A.; Shilkina, L. A.; Titov, S. V.; et al., *Crystallography Reports*, 2003, *vol. 48*, 377.
[19] Reznichenko, L. A.; Shilkina, L. A.; Gagarina, E. S.; et al. *Crystallography Reports*, 2004, *vol. 49*, 820.
[20] Rao, C. N. R.; Gopalakrishnan, J., *New Directions in Solid State Chemistry,* Cambridge University Press, Cambridge, 1997.
[21] Reznichenko, L. A.; Akhnazarova, V. V.; Shilkina, L. A.; et al. *Crystallography Reports*, 2009, *vol. 54*, 483.
[22] Narendra, B. S.; Hsu, J.-H.; Chen, Y. S.; Lin, J. G., *J. Appl. Phys.*, 2011, *vol. 109*, 07D904.

In: Advanced Materials - Studies and Applications
Editors: I.A. Parinov, S.-H. Chang, S.Theerakulpisut
ISBN: 978-1-63463-749-7

Chapter 9

Finite Element Design of Piezoelectric and Magnetoelectric Composites with Use of Symmetric Quasidefinite Matrices

A. V. Nasedkin[*]
Department of Mathematics, Mechanics & Computer Science,
Southern Federal University, Rostov-on-Don, Russia

Abstract

The paper discusses the approaches to the determination of the effective material properties of magnetoelectric and piezoelectric composite materials based on the effective moduli method and the finite element techniques for solving the problems for the representative volumes. The paper suggests a set of algorithms with use of symmetric saddle point quasidefinite matrices for solving finite element equations. The developed algorithms can be used for computer design of magnetoelectric and piezoelectric composites.

1. Introduction

Recently there has been observed an increased interest to the investigations of active composite materials of complex structure that exhibit effective properties for many practical applications. For example, porous piezoelectric ceramic has extraordinary high volumetric piezosensitivity, low acoustic impedance and extended frequencies bandwidth. In this regard, the use of porous piezocomposite materials can significantly improve the main parameters of ultrasonic piezoelectric devices. Two-phase magnetoelectric composites consisting of piezoelectric and piezomagnetic phases demonstrate the ability of mutual transformation of electric and magnetic energies, whereas separately these phases do not have such property. Modern magnetoelectric composites have high efficiency of the magnetoelectric

[*] Corresponding author: nasedkin@math.sfedu.ru.

transformation, relatively high temperatures of phase transitions and significant technological resources. Mathematical and computer investigations of composite materials of complex structure enable to explain and simulate some of their important characteristics and predict the effectiveness of different relations and structure couplings of constitutive phases.

In the last years many models have been proposed to determine the effective moduli of magnetoelectric composites ([1–5], etc.) and piezoelectric composites ([6–12], etc.) In the present chapter we develop the effective moduli method and the finite element technique for magnetoelectric (piezomagnetoelectric or electromagnetoelastic) and piezoelectric composites. The finite element approximations of coupled problems for computer design of composite magnetoelectric and piezoelectric material are given in the form that uses the symmetric saddle point quasidefinite matrices. We also describe an algorithm to solve the resulting finite element equations. As an example, we demonstrate some results of finite element calculations of the most important effective piezomoduli for porous piezoceramic material.

2. Effective Moduli Method for Magnetoelectric Composites

Let Ω be a volume of a two-phase composite heterogeneous body composed of two materials Ω_e and Ω_m ($\Omega = \Omega_e \cup \Omega_m$), where the phase Ω_e has the piezoelectric properties and the phase Ω_m has the piezomagnetic properties; $\Gamma = \partial\Omega$ is the boundary of the region; $\mathbf{n}(\mathbf{x})$ is the vector of external unit normal to Γ; $\mathbf{u} = \mathbf{u}(\mathbf{x})$ is the displacement vector-function; $\varphi = \varphi(\mathbf{x})$ is the electric potential function; $\phi = \phi(\mathbf{x})$ is the magnetic potential function.

Both phases Ω_e and Ω_m can consist of separate, generally speaking, disjointed sub-regions $\Omega_e = \cup_k \Omega_{ek}$, $\Omega_m = \cup_l \Omega_{ml}$, that in the aggregate have common boundaries and do not overlap each other. Thus, here we consider a two-phase mixture composite with piezoelectric and piezomagnetic fractions.

We denote by $\boldsymbol{\varepsilon}$ the stain tensor; $\boldsymbol{\sigma}$ is the stress tensor; $\mathbf{E}$ is the electric field intensity vector or the electric field vector; $\mathbf{D}$ is the electric flux density vector or the electric displacement vector; $\mathbf{H}$ is the magnetic field intensity vector or magnetic field vector; $\mathbf{B}$ is the magnetic flux density vector. The fields $\boldsymbol{\varepsilon}$, $\mathbf{E}$ and $\mathbf{H}$ are determined from the functions $\mathbf{u}$, φ and ϕ by the following form:

$$\boldsymbol{\varepsilon} = (\nabla\mathbf{u} + \nabla\mathbf{u}^*)/2\,, \quad \mathbf{E} = -\nabla\varphi\,, \quad \mathbf{H} = -\nabla\phi\,, \qquad (1)$$

where superscript "*" denotes the transposition operation.

As usual, we will denote the volume-averaged quantities in the broken brackets as:

$$\langle(...)\rangle = \frac{1}{|\Omega|}\int_\Omega (...)\, d\Omega\,. \qquad (2)$$

We will formulate the auxiliary statements for electromagnetoelastic body following the proof of the effective moduli method for elastic medium. These statements also hold for the techniques used for elastic and piezoelectric bodies [13,14].

Lemma 1. These representations take place for the field characteristics (1) averaged on the volume Ω by means of the appropriate values on the boundary Γ:

(a) $\forall \boldsymbol{\varepsilon} = \boldsymbol{\varepsilon}^*$, $\langle \boldsymbol{\varepsilon} \rangle = \dfrac{1}{2|\Omega|}\int_\Gamma (\mathbf{un}^* + \mathbf{nu}^*)\, d\Gamma$;

(b) $\forall$ $\mathbf{E}, \mathbf{H}$, $\langle \mathbf{E} \rangle = -\dfrac{1}{|\Omega|}\int_\Gamma \mathbf{n}\varphi\, d\Gamma$, $\langle \mathbf{H} \rangle = -\dfrac{1}{|\Omega|}\int_\Gamma \mathbf{n}\phi\, d\Gamma$.

Lemma 2. For $\mathbf{x} \in \Gamma$ we have the following relations:

(a) if $\mathbf{u} = \mathbf{x} \cdot \boldsymbol{\varepsilon}_0$, where $\boldsymbol{\varepsilon}_0 = \overset{*}{\boldsymbol{\varepsilon}_0} = \text{const}$, i.e., $\boldsymbol{\varepsilon}_0$ is the arbitrary symmetric second rank tensor independent on $\mathbf{x}$, then $\langle \boldsymbol{\varepsilon} \rangle = \boldsymbol{\varepsilon}_0$;

(b) if $\varphi = -\mathbf{x} \cdot \mathbf{E}_0$, where $\mathbf{E}_0 = \text{const}$, i.e., $\mathbf{E}_0$ is the arbitrary vector independent on $\mathbf{x}$, then $\langle \mathbf{E} \rangle = \mathbf{E}_0$;

(c) if $\phi = -\mathbf{x} \cdot \mathbf{H}_0$, where $\mathbf{H}_0 = \text{const}$, i.e., $\mathbf{H}_0$ is the arbitrary vector independent on $\mathbf{x}$, then $\langle \mathbf{H} \rangle = \mathbf{H}_0$.

Lemma 3. If for $\mathbf{x} \in \Gamma$:

(a) $\mathbf{u} = \mathbf{x} \cdot \boldsymbol{\varepsilon}_0$, $\boldsymbol{\varepsilon}_0 = \overset{*}{\boldsymbol{\varepsilon}_0} = \text{const}$, and the equilibrium equation $\nabla \cdot \boldsymbol{\sigma} = 0$ takes place for any given symmetric second rank stress tensor $\boldsymbol{\sigma}$, then we have $\langle \boldsymbol{\sigma} : \boldsymbol{\varepsilon} \rangle = \langle \boldsymbol{\sigma} \rangle : \langle \boldsymbol{\varepsilon} \rangle$, where symbol ":" denotes the double scalar product operation between two tensors ($\boldsymbol{\sigma} : \boldsymbol{\varepsilon} = \sigma_{ij}\varepsilon_{ji}$);

(b) $\varphi = -\mathbf{x} \cdot \mathbf{E}_0$, $\mathbf{E}_0 = \text{const}$, and the equation of quasi-electrostatics $\nabla \cdot \mathbf{D} = 0$ takes place for any given electric flux density vector $\mathbf{D}$, then we have $\langle \mathbf{D} \cdot \mathbf{E} \rangle = \langle \mathbf{D} \rangle \cdot \langle \mathbf{E} \rangle$;

(c) $\phi = -\mathbf{x} \cdot \mathbf{H}_0$, $\mathbf{H}_0 = \text{const}$, and the equation of quasi-magnetostatics $\nabla \cdot \mathbf{B} = 0$ takes place for any given magnetic flux density vector $\mathbf{B}$, then we have $\langle \mathbf{B} \cdot \mathbf{H} \rangle = \langle \mathbf{B} \rangle \cdot \langle \mathbf{H} \rangle$

In accordance with fundamental form of constitutive equations we shall introduce the moduli of the magnetoelectric medium:

$$\boldsymbol{\sigma} = \mathbf{c}^{E,H} : \boldsymbol{\varepsilon} - \mathbf{e}^{H*} \cdot \mathbf{E} - \mathbf{q}^{E*} \cdot \mathbf{H}, \tag{3}$$

$$\mathbf{D} = \mathbf{e}^{H} : \boldsymbol{\varepsilon} + \boldsymbol{\kappa}^{S,H} \cdot \mathbf{E} + \boldsymbol{\alpha}^{S} \cdot \mathbf{H}, \tag{4}$$

$$\mathbf{B} = \mathbf{q}^{E} : \boldsymbol{\varepsilon} + \boldsymbol{\alpha}^{S*} \cdot \mathbf{E} + \boldsymbol{\mu}^{S,E} \cdot \mathbf{H}. \tag{5}$$

Here, in Equations (3) – (5) $\mathbf{c}^{E,H}$ is the fourth rank tensor of the elastic stiffness moduli, $\mathbf{e}^{H}$ is the third rank tensor of the piezoelectric moduli, $\mathbf{q}^{E}$ is the third rank tensor of the magnetostriction moduli, $\boldsymbol{\kappa}^{S,H}$ is the second rank tensor of the dielectric permittivity moduli, $\boldsymbol{\alpha}^{S}$ is the second rank tensor of the magnetoelectric coupling coefficients, $\boldsymbol{\mu}^{S,E}$ is the second rank tensor of the magnetic permittivity moduli, and the upper indexes indicate the constant fields under which these moduli are calculated (*S* denotes the strains, *T* denotes the stresses).

For inhomogeneous two-phase magnetoelectric body these moduli will be functions of coordinates $\mathbf{x}$: $\mathbf{c}^{E,H} = \mathbf{c}^{E,H}(\mathbf{x})$; $\mathbf{e}^{H} = \mathbf{e}^{H}(\mathbf{x})$ etc., and $\boldsymbol{\alpha}^{S} = 0$, $\forall \mathbf{x} \in \Omega$; $\mathbf{q}^{E} = 0$, $\forall \mathbf{x} \in \Omega_e$; $\mathbf{e}^{H} = 0$, $\forall \mathbf{x} \in \Omega_m$.

Note that for the case of the piezoelectric medium we do not take into account the magnetic fields $\mathbf{H}$, $\mathbf{B}$, and we use the reduced forms of the constitutive equations:

$$\boldsymbol{\sigma} = \mathbf{c}^{E} : \boldsymbol{\varepsilon} - \mathbf{e}^{*} \cdot \mathbf{E},$$

$$\mathbf{D} = \mathbf{e} : \boldsymbol{\varepsilon} + \boldsymbol{\kappa}^{S} \cdot \mathbf{E},$$

where $\mathbf{c}^{E} \equiv \mathbf{c}^{E,H}$, $\mathbf{e} \equiv \mathbf{e}^{H}$, $\boldsymbol{\kappa}^{S} \equiv \boldsymbol{\kappa}^{S,H}$, since the upper index "*H*" is not required.

Let Ω be the representative volume of the heterogeneous magnetoelectric materials with piezoelectric and piezomagnetic phases. We will determine the effective moduli $\tilde{\mathbf{c}}^{E,H}$, $\tilde{\mathbf{e}}^{H}$, $\tilde{\mathbf{q}}^{E}$, $\tilde{\boldsymbol{\kappa}}^{S,H}$, $\tilde{\boldsymbol{\alpha}}^{S}$, $\tilde{\boldsymbol{\mu}}^{S,E}$ by the following technique, similar to the well-known procedures for purely elastic and piezoelectric composites [13,14]. We consider the static magnetoelectric (piezomagnetoelectric or magnetoelectroelastic) problem for representative volume Ω:

$$\nabla \cdot \boldsymbol{\sigma} = 0, \quad \nabla \cdot \mathbf{D} = 0, \quad \nabla \cdot \mathbf{B} = 0, \quad \mathbf{x} \in \Omega, \tag{6}$$

$$\mathbf{u} = \mathbf{x} \cdot \boldsymbol{\varepsilon}_0, \quad \varphi = -\mathbf{x} \cdot \mathbf{E}_0, \quad \phi = -\mathbf{x} \cdot \mathbf{H}_0, \quad \mathbf{x} \in \Gamma. \tag{7}$$

From the solution of the problem (6), (7) and (1), (3) – (5) we find the fields $\boldsymbol{\varepsilon}$, $\mathbf{E}$, $\mathbf{H}$, $\boldsymbol{\sigma}$, $\mathbf{D}$ and $\mathbf{B}$. We note from Lemma 2, that for the problem (1), (3) – (7) the following equalities hold: $\langle \boldsymbol{\varepsilon} \rangle = \boldsymbol{\varepsilon}_0$, $\langle \mathbf{E} \rangle = \mathbf{E}_0$ and $\langle \mathbf{H} \rangle = \mathbf{H}_0$.

Let us put some "equivalent" homogeneous medium with the effective moduli $\tilde{\mathbf{c}}^{E,H}$, $\tilde{\mathbf{e}}^{H}$, $\tilde{\mathbf{q}}^{E}$, $\tilde{\boldsymbol{\kappa}}^{S,H}$, $\tilde{\boldsymbol{\alpha}}^{S}$, $\tilde{\boldsymbol{\mu}}^{S,E}$ into correspondence with initial heterogeneous medium. The constitutive equations for "equivalent" medium, similar to (3) – (5), are given in the forms:

$$\boldsymbol{\sigma}_0 = \tilde{\mathbf{c}}^{E,H} : \boldsymbol{\varepsilon}_0 - \tilde{\mathbf{e}}^{H*} \cdot \mathbf{E}_0 - \tilde{\mathbf{q}}^{E*} \cdot \mathbf{H}_0, \tag{8}$$

$$\mathbf{D}_0 = \tilde{\mathbf{e}}^{H} : \boldsymbol{\varepsilon}_0 + \tilde{\boldsymbol{\kappa}}^{S,H} \cdot \mathbf{E}_0 + \tilde{\boldsymbol{\alpha}}^{S} \cdot \mathbf{H}_0, \tag{9}$$

$$\mathbf{B}_0 = \tilde{\mathbf{q}}^E : \boldsymbol{\varepsilon}_0 + \tilde{\boldsymbol{\alpha}}^{S*} \cdot \mathbf{E}_0 + \tilde{\boldsymbol{\mu}}^{S,E} \cdot \mathbf{H}_0 . \tag{10}$$

For problem (1), (3) – (7) we accept the following equations such as relations for determination of effective moduli from (8) – (10):

$$\langle \boldsymbol{\sigma} \rangle = \boldsymbol{\sigma}_0 , \quad \langle \mathbf{D} \rangle = \mathbf{D}_0 , \quad \langle \mathbf{B} \rangle = \mathbf{B}_0 . \tag{11}$$

Note that due to Lemma 3 the average energies are equal for both heterogeneous and "equivalent" homogeneous magnetoelectric media:

$$\langle \boldsymbol{\sigma} : \boldsymbol{\varepsilon} + \mathbf{D} \cdot \mathbf{E} + \mathbf{B} \cdot \mathbf{H} \rangle / 2 = \langle \boldsymbol{\sigma}_0 : \boldsymbol{\varepsilon}_0 + \mathbf{D}_0 \cdot \mathbf{E}_0 + \mathbf{B}_0 \cdot \mathbf{H}_0 \rangle / 2 .$$

Now by using Equations (8) – (10), we can select such boundary conditions that enable to obtain obvious expressions for the effective moduli. For example, we consider the problem (1), (3) – (7) with

$$\boldsymbol{\varepsilon}_0 = \varepsilon_0 (\mathbf{e}_k \mathbf{e}_m + \mathbf{e}_m \mathbf{e}_k) , \quad \mathbf{E}_0 = 0 , \quad \mathbf{H}_0 = 0 , \tag{12}$$

where k , m are some fixed numbers $(k, m = 1, 2, 3)$; $\mathbf{e}_k$ are the unit vectors of Cartesian basis; $\boldsymbol{\varepsilon}_0 = \boldsymbol{\varepsilon}_0^* = \mathrm{const}$. Then from Equations (8) – (12) we obtain:

$$\tilde{c}_{ijkm}^{E,H} = \langle \sigma_{ij} \rangle / (2\varepsilon_0) , \quad \tilde{e}_{jkm}^{H} = \langle D_j \rangle / (2\varepsilon_0) , \quad \tilde{q}_{jkm}^{E} = \langle B_j \rangle / (2\varepsilon_0) . \tag{13}$$

If in Equation (7) we accept ($\mathbf{E}_0 = \mathrm{const}$)

$$\boldsymbol{\varepsilon}_0 = 0 , \quad \mathbf{E}_0 = E_0 \mathbf{e}_k , \quad \mathbf{H}_0 = 0 , \tag{14}$$

then from (8) – (11), (14) we find

$$\tilde{e}_{kij}^{H} = -\langle \sigma_{ij} \rangle / E_0 , \quad \tilde{\kappa}_{kj}^{S,H} = \langle D_j \rangle / E_0 , \quad \tilde{\alpha}_{kj}^{E} = \langle B_j \rangle / E_0 . \tag{15}$$

Finally, if in Equation (7) we accept ($\mathbf{H}_0 = \mathrm{const}$)

$$\boldsymbol{\varepsilon}_0 = 0 , \quad \mathbf{E}_0 = 0 , \quad \mathbf{H}_0 = H_0 \mathbf{e}_k , \tag{16}$$

then from (8) – (11), (16) we find

$$\tilde{q}_{kij}^{E} = -\langle \sigma_{ij} \rangle / H_0 , \quad \tilde{\alpha}_{jk}^{S} = \langle D_j \rangle / H_0 , \quad \tilde{\mu}_{kj}^{S,E} = \langle B_j \rangle / H_0 . \tag{17}$$

Note, that the quantities σ_{ij}, D_j and B_j in (13), (15) and (17) are different, since they are calculated from the solutions of the problems (1), (3) – (7) with various values $\boldsymbol{\varepsilon}_0$, $\mathbf{E}_0$ and $\mathbf{H}_0$ at the boundary conditions (7), (12), (14) and (16), respectively.

Equations (2), (12) – (17) allow us obtaining the full set of the effective moduli for magnetoelectric composite media with arbitrary anisotropy class. In this connection we can consider piezoelectric composite medium as a special case of magnetoelectric composite, where Ω_m is the empty set and $\Omega = \Omega_e$ is the region occupied by a heterogeneous body with only piezoelectric properties.

3. Finite Element Approximations

To solve the piezomagnetoelectric problems (1), (3) – (7) with corresponding boundary conditions for heterogeneous two-phase composite material in the representative volume Ω we can pass to their weak settings and use classical technique of finite element approximations. Let Ω_h be a region filled by the finite element mesh $\Omega_h \subseteq \Omega$, $\Omega_h = \bigcup_k \Omega^{ek}$, where Ω^{ek} is a separate finite element with the number k. For static problems we can find the approximate solution $\{\mathbf{u}_h \approx \mathbf{u},\ \varphi_h \approx \varphi,\ \phi_h \approx \phi\}$ at the finite element mesh Ω_h in the form:

$$\mathbf{u}_h(\mathbf{x}) = \mathbf{N}_u^*(\mathbf{x}) \cdot \mathbf{U}, \quad \varphi_h(\mathbf{x}) = \mathbf{N}_\varphi^*(\mathbf{x}) \cdot \boldsymbol{\Phi}, \quad \phi_h(\mathbf{x}) = \mathbf{N}_\phi^*(\mathbf{x}) \cdot \mathbf{A}, \tag{18}$$

where $\mathbf{N}_u^*$ is the matrix of the basic functions for displacements; $\mathbf{N}_\varphi^*$, $\mathbf{N}_\phi^*$ are the row vectors of the basic functions for the electric and magnetic potentials; $\mathbf{U}$, $\boldsymbol{\Phi}$, $\mathbf{A}$ are the vectors of nodal displacements, electric potentials and magnetic potentials.

Here we give the finite element statements without specific definition of the finite element types. Theoretically we can use any classical Lagrangian or serendipity isoparametric finite element approximations for mechanical, electric and magnetic fields (for example, see those described in [15]). Thus, for numerical computation we will use the 8-node hexahedral solid elements.

In accordance with conventional finite element technique we approximate the continual weak formulation by the problem in finite-dimensional spaces. Substituting (18) and similar representations for project functions into the weak formulation of the piezomagnetoelectric problem on Ω_h, we obtain the following finite element system:

$$\mathbf{K} \cdot \mathbf{a} = \mathbf{F}, \tag{19}$$

$$\mathbf{K}=\begin{bmatrix}\mathbf{K}_{uu} & \mathbf{K}_{u\varphi} & \mathbf{K}_{u\phi}\\ \mathbf{K}_{u\varphi}^{*} & -\mathbf{K}_{\varphi\varphi} & 0\\ \mathbf{K}_{u\phi}^{*} & 0 & -\mathbf{K}_{\phi\phi}\end{bmatrix}, \tag{20}$$

$$\mathbf{a}=\begin{Bmatrix}\mathbf{U}\\ \mathbf{\Phi}\\ \mathbf{A}\end{Bmatrix},\ \mathbf{F}=\begin{Bmatrix}\mathbf{F}_{u}\\ -\mathbf{F}_{\varphi}\\ -\mathbf{F}_{\phi}\end{Bmatrix}. \tag{21}$$

Here we take into account the principal boundary conditions, and therefore the vectors $\mathbf{F}_u$, $\mathbf{F}_\varphi$, $\mathbf{F}_\phi$ in (21) are not equal to zero. Block matrices $\mathbf{K}_{uu}=\sum^a \mathbf{K}_{uu}^{ek}$, $\mathbf{K}_{u\varphi}=\sum^a \mathbf{K}_{u\varphi}^{ek}$, $\mathbf{K}_{u\phi}=\sum^a \mathbf{K}_{u\phi}^{ek}$, $\mathbf{K}_{\varphi\varphi}=\sum^a \mathbf{K}_{\varphi\varphi}^{ek}$, $\mathbf{K}_{\phi\phi}=\sum^a \mathbf{K}_{\phi\phi}^{ek}$ are the global finite element matrices, obtained from the element matrices as a result of the assembling process ($\sum^a$); the block matrix $\mathbf{K}_{\varphi\phi}=0$, because $\boldsymbol{\alpha}^S=0$ in (4), (5) for the representative volume Ω of magnetoelectric heterogeneous two-phase composite material; and the element matrices are in the forms:

$$\mathbf{K}_{uu}^{ek}=\int_{\Omega^{ek}}\mathbf{B}_u^{e*}:\mathbf{c}^{E,H}:\mathbf{B}_u^{e}\,d\Omega,$$

$$\mathbf{K}_{u\varphi}^{ek}=\int_{\Omega^{ek}}\mathbf{B}_u^{e*}:\mathbf{e}^{H*}\cdot\mathbf{B}_\varphi^{e}\,d\Omega,\ \mathbf{K}_{u\phi}^{ek}=\int_{\Omega^{ek}}\mathbf{B}_u^{e*}:\mathbf{q}^{E*}\cdot\mathbf{B}_\phi^{e}\,d\Omega,$$

$$\mathbf{K}_{\varphi\varphi}^{ek}=\int_{\Omega^{ek}}\mathbf{B}_\varphi^{e*}\cdot\boldsymbol{\kappa}^{S,H}\cdot\mathbf{B}_\varphi^{e}\,d\Omega,\ \mathbf{K}_{\phi\phi}^{ek}=\int_{\Omega^{ek}}\mathbf{B}_\phi^{e*}\cdot\boldsymbol{\mu}^{S,E}\cdot\mathbf{B}_\phi^{e}\,d\Omega,$$

$$\mathbf{B}_u^e=\nabla\mathbf{N}_u^{e*},\ \mathbf{B}_\varphi^e=\nabla\mathbf{N}_\varphi^{e*},\ \mathbf{B}_\phi^e=\nabla\mathbf{N}_\phi^{e*},$$

where $\mathbf{N}_u^{e*}$, $\mathbf{N}_\varphi^{e*}$, $\mathbf{N}_\phi^{e*}$ are the matrices and the row vectors of approximate shape functions, respectively, defined on separate finite elements.

Note that here global and local stiffness matrices $\mathbf{K}_{uu}$, $\mathbf{K}_{uu}^{ek}$ are formed similarly, as in the case of pure elasticity, the matrices $\mathbf{K}_{u\varphi}$, $\mathbf{K}_{u\varphi}^{ek}$, $\mathbf{K}_{\varphi\varphi}$, $\mathbf{K}_{\varphi\varphi}^{ek}$ are the usual finite element matrices for piezoelectric body, and the matrices $\mathbf{K}_{u\phi}$, $\mathbf{K}_{u\phi}^{ek}$, $\mathbf{K}_{\phi\phi}$, $\mathbf{K}_{\phi\phi}^{ek}$ are the finite element matrices for piezomagnetic body. The global matrix $\mathbf{K}$ is the symmetric matrix, and its main blocks $\mathbf{K}_{uu}$, $\mathbf{K}_{\varphi\varphi}$, $\mathbf{K}_{\phi\phi}$ are the positive definite matrices, although the blocks $\mathbf{K}_{\varphi\varphi}$, $\mathbf{K}_{\phi\phi}$ are contained in $\mathbf{K}$ with the minus sign.

For the case of piezoelectric composite material in Equations (19) – (21) we have

$$\mathbf{K}=\begin{bmatrix}\mathbf{K}_{uu} & \mathbf{K}_{u\varphi}\\ \mathbf{K}_{u\varphi}^{*} & -\mathbf{K}_{\varphi\varphi}\end{bmatrix},\quad \mathbf{a}=\begin{Bmatrix}\mathbf{U}\\ \mathbf{\Phi}\end{Bmatrix},\quad \mathbf{F}=\begin{Bmatrix}\mathbf{F}_{u}\\ -\mathbf{F}_{\varphi}\end{Bmatrix},$$

and similarly for piezomagnetic composite material with replacement of φ by ϕ.

We can note that the matrices $\mathbf{K}$ are symmetric and have identical saddle point structure for all these composites. Further we will describe some effective approaches for solving the system (19) with symmetric saddle point matrix.

4. Cholesky Factorization for Symmetric Quasidefinite Matrices

In this section, we describe some generalized bordering methods for equations with saddle point symmetric quasidefinite matrices [16–19].

Definition 1. For the matrices $\mathbf{B}$ and $\mathbf{G}$ of ($n\times n$) and ($m\times m$) orders, respectively, the matrix $\mathbf{K}$ of the saddle point type (n, m) is called symmetric quasidefinite (sqd [20]) matrix with respect to the block decomposition (α, α'), if this matrix has the following block form:

$$\mathbf{K}=\begin{bmatrix}\mathbf{B} & \mathbf{H}\\ \mathbf{H}^{*} & -\mathbf{G}\end{bmatrix}, \tag{22}$$

and $\mathbf{B}$, $\mathbf{G}$ are positive definite matrices.

Here the block decomposition (α, α'), for the matrix $\mathbf{K}$ in the form (22) can be denoted by the following formulae: $\alpha=\{\alpha_1,\alpha_2,...,\alpha_n\}$; $n+m=N$; $J=\{1,2,...,N\}$; $\alpha_1<\alpha_2<...<\alpha_n$; $\alpha\subset J$; $\alpha'=J\setminus\alpha$; $\mathbf{K}(\alpha,\alpha)=\mathbf{K}(\alpha)=\mathbf{B}$; $\mathbf{K}(\alpha',\alpha')=\mathbf{K}(\alpha')=-\mathbf{G}$; $\mathbf{K}(\alpha,\alpha')=\mathbf{H}$; $\mathbf{K}(\alpha',\alpha)=\mathbf{H}^*$; where $\mathbf{K}(\alpha,\beta)$ is the submatrix composed of elements $K_{\alpha_i\beta_j}$; $\alpha_i\in\alpha$; $\beta_j\in\beta$.

Then the following lemmas and theorem hold.

Lemma 4. For the saddle point type (n, m) sqd matrix $\mathbf{K}$ we have $\operatorname{sign}(\det\mathbf{K})=(-1)^m$.

(Note that by Lemma 4 the saddle point sqd matrix $\mathbf{K}$ and any its principal submatrices are invertible.)

Lemma 5. Let the saddle point type (n, m) sqd matrix $\mathbf{K}$ ($n+m=N$) be the edging result of the sqd matrix $\mathbf{M}$ of the order ($N-1$) by the vector $\mathbf{v}\in\mathrm{R}^{N-1}$ and the quantity $s\in\mathrm{R}$:

$$\mathbf{K} = \begin{bmatrix} \mathbf{M} & \mathbf{v} \\ \mathbf{v}^* & s \end{bmatrix}. \tag{23}$$

Therefore, if $s > 0$, then the matrix $\mathbf{M}$ is the sqd matrix of the saddle point type ($n-1$, m) and $s > \mathbf{v}^* \cdot \mathbf{M}^{-1} \cdot \mathbf{v}$; and if $s < 0$, then the matrix $\mathbf{M}$ is the sqd matrix of the saddle point type (n, $m-1$) and $s < \mathbf{v}^* \cdot \mathbf{M}^{-1} \cdot \mathbf{v}$.

Proof. To prove Lemmas 4 and 5 we use the formula for the determinant of a block matrix with decomposition (α, α') [21]:

$$\det \mathbf{K} = \det \mathbf{K}(\alpha) \det[\mathbf{K}(\alpha') - \mathbf{K}(\alpha', \alpha) \cdot \mathbf{K}^{-1}(\alpha) \cdot \mathbf{K}(\alpha, \alpha')]. \tag{24}$$

Then, by using this formula for the matrix $\mathbf{K}$ in the form (22), we have

$$\det \mathbf{K} = \det \mathbf{B} \det(-\mathbf{G} - \mathbf{H}^* \cdot \mathbf{B}^{-1} \cdot \mathbf{H}) = (-1)^m \det \mathbf{B} \det(\mathbf{G} + \mathbf{H}^* \cdot \mathbf{B}^{-1} \cdot \mathbf{H}).$$

Since $\mathbf{B}$ and $\mathbf{G}$ are the positive definite matrices, then $\det(\mathbf{G} + \mathbf{H}^* \cdot \mathbf{B}^{-1} \cdot \mathbf{H}) > 0$, and therefore we obtain the statement of Lemma 4: $\text{sign}(\det \mathbf{K}) = (-1)^m$.

By using again formula (24) for the matrix $\mathbf{K}$ in the form (23), we have

$$\det \mathbf{K} = \det \mathbf{M} \det(s - \mathbf{v}^* \cdot \mathbf{M}^{-1} \cdot \mathbf{v}) = \det \mathbf{M}(s - \mathbf{v}^* \cdot \mathbf{M}^{-1} \cdot \mathbf{v}).$$

Therefore,

$$\text{sign}(\det \mathbf{K}) = \text{sign}(\det \mathbf{M}) \text{sign}(s - \mathbf{v}^* \cdot \mathbf{M}^{-1} \cdot \mathbf{v}). \tag{25}$$

If $s > 0$, then $\mathbf{M}$ is the sqd matrix of the saddle point type ($n-1$, m), and by Lemma 4 it follows that $\det \mathbf{M} = (-1)^m = \text{sign}(\det \mathbf{K})$. Therefore from (25) we obtain: $(s - \mathbf{v}^* \cdot \mathbf{M}^{-1} \cdot \mathbf{v}) > 0$. If $s < 0$, then $\mathbf{M}$ is the sqd matrix of the saddle point type ($n-1$, m), and so $\det \mathbf{M} = (-1)^{m-1}$, $\det \mathbf{K} = (-1)^m$. Then from (25) at $s < 0$ we have $(s - \mathbf{v}^* \cdot \mathbf{M}^{-1} \cdot \mathbf{v}) < 0$.

Theorem 1. The saddle point type (n, m) sqd matrix $\mathbf{K}$ has one and only one matrix decomposition

$$\mathbf{K} = \mathbf{L} \cdot \mathbf{D} \cdot \mathbf{L}^*, \ \mathbf{D} = \begin{bmatrix} \mathbf{I}_n & 0 \\ 0 & -\mathbf{I}_m \end{bmatrix}. \tag{26}$$

where $\mathbf{L}$ is the low triangular matrix with positive diagonal elements; $\mathbf{I}_n$, $\mathbf{I}_m$ are the identity matrices of the order n and m, respectively.

Proof. We prove this theorem by the method of induction by the order N of the matrix $\mathbf{K}$, similarly to the case of the positive definite matrices [22]. For $N-1$ the statement of Theorem 1 is trivial.

Let Theorem 1 be correct for the saddle point sqd matrix of the order $N-1$. We present the matrix $\mathbf{K}$ in the form (23). For the matrix $\mathbf{M}$ of the order $N-1$ by hypothesis of induction and by Lemma 5 the following factorization exists

$$\mathbf{M} = \mathbf{L}_M \cdot \mathbf{D}_M \cdot \mathbf{L}_M^*, \tag{27}$$

Then, the required decomposition for the matrix $\mathbf{K}$ can be written in the form

$$\mathbf{K} = \begin{bmatrix} \mathbf{L}_M & 0 \\ \mathbf{w}^* & z \end{bmatrix} \cdot \begin{bmatrix} \mathbf{D}_M & 0 \\ 0 & d \end{bmatrix} \cdot \begin{bmatrix} \mathbf{L}_M^* & \mathbf{w} \\ 0 & z \end{bmatrix}, \tag{28}$$

where

$$\mathbf{w} = \mathbf{D}_M^{-1} \cdot \mathbf{L}_M^{-1} \cdot \mathbf{v},\ z = \sqrt{|\, s - \mathbf{w}^* \cdot \mathbf{D}_M \cdot \mathbf{w}\,|}, \tag{29}$$

$$d = \operatorname{sign}(s - \mathbf{w}^* \cdot \mathbf{D}_M \cdot \mathbf{w}). \tag{30}$$

Note that $\mathbf{L}_M^{-1}$ exists since by Lemma 4 $\det \mathbf{M} \neq 0$. In addition, since by (29), (30), (27) $\mathbf{w}^* \cdot \mathbf{D}_M \cdot \mathbf{w} = \mathbf{v}^* \cdot \mathbf{M}^{-1} \cdot \mathbf{v}$, and by Lemma 5 $\operatorname{sign}(s - \mathbf{v}^* \cdot \mathbf{M}^{-1} \cdot \mathbf{v}) = \operatorname{sign} s$, then

$$d = \operatorname{sign} s \tag{31}$$

that actually proves the Theorem.

This proof is constructive, and the formulae (29), (31) are the base for the algorithm of the bordering method [22] for factorization (26) in the loop by the line number on the scheme (28). In this algorithm, the matrix $\mathbf{L}$ is formed by rows.

By Theorem 1 and Lemma 5 it is easy to construct other algorithms of the square-root method in the inner-product form and in the outer-product form that are close to the known algorithms [22] for the positive definite matrices.

For example, the algorithm in the inner-product form, which produces the matrix $\mathbf{L}$ in columns, is defined as follows:

For $j := 1$ to n do

$$d_{jj} := \operatorname{sign}(k_{jj}), \tag{32}$$

$$l_{jj} := \sqrt{| k_{jj} - \sum_{k=1}^{j-1} d_{kk} l_{jk}^2 |},$$

For $i := j+1$ to n do

$$l_{ij} := (k_{ij} - \sum_{k=1}^{j-1} d_{kk} l_{jk} l_{ik}) / (d_{jj} l_{jj}).$$

In this form, the algorithm of the inner-product is close to the ordinary algorithm of the inner-product for the square-root method for Hermitian matrices that differs only by simple formula (32). In addition, by Theorem 1 for the saddle point sqd matrix in this algorithm the permutations of rows and columns are not required, as *LDL**-factorization (26) exists, and therefore, $d_{jj} = \pm 1$; $l_{jj} \neq 0$.

The generalized bordering method (28) – (31) for determining the *LDL**-factorizations of the saddle point sqd matrices has been implemented in the finite element analysis package ACELAN, with the Newmark method for integrating the Cauchy acoustopiezoelectric transient problem [23, 24]. All procedures that we need in the finite element manipulations (the degree of freedom rotations, mechanical and electric boundary condition settings, etc.) we can provide also in symmetric form. Here we propose this strategy for finite element design of piezoelectric and magnetoelectric composite materials.

Note that for a composite material with complicated internal structure it is necessary to construct the models of representative volumes with very large number of finite elements. In these cases we obtain the finite element systems with large symmetric saddle point sqd matrices, where the special iterative method can be recommended for solving the saddle point problems [25 – 29].

5. Numerical Example

With the help of the above-mentioned formulae for the calculation of the effective moduli we can find the solution of the corresponding piezoelectric problems by the Cholesky factorization algorithms for sqd matrices in the region Ω, which should be a representative volume of porous piezoceramics. Ideally, for a representative volume it is necessary to take areas, which are big enough compared with the sizes of heterogeneity (i.e., pore), but small enough in comparison with the distances where the averaged variables vary significantly.

The model of the skeleton cohesion (3-0 or 3-3 connectivity type [30]) can be constructed in the following manner [13,31]. The cube obtained by translation of geometrically identical basic cells along three directions is considered as a representative volume. Each basic cell in its turn also represents a cube consisting of $10 \times 10 \times 10$ cubic piezoelectric or porous finite elements with 8 nodes each. A connected skeleton with piezoelectric finite elements in the cube corners always exists in the cell. The skeleton consists of a spanning parallelepiped represented by its edges (linear dimensions are pointed out by the randomizer) as well as of piezoelectric elements chains connecting the corners of the parallelepiped with the corners of the main cell. The connecting piezoelectric finite elements chains are also generated by the randomizer. The skeleton occupies 10 % of the cell volume (see Figue 1). So, the maximum possible porosity that can be achieved in this model runs up to 90 %.

The representative volumes with large number of elements can be obtained by the repetition of procedures for creating cubic structures of $10 \times 10 \times 10$ size along the three axes. In this case each structure of $10 \times 10 \times 10$ size is generated randomly in the frameworks of the procedures considered above. Additionally, in order to ensure smaller porosity (up to zero) we have applied the following algorithm. Two finite elements are chosen randomly in the representative volume constructed at the previous step. If at least one of these elements is not a pore, then these cells can be linked by some arbitrary connected path. The resulting set is then added to the previously built frame consisting of solid piezoelectric elements. After that we calculate the porosity and compare it to the given porosity. If the calculated porosity is less than the given porosity, the step of the algorithm should be repeated, i.e., again two elements should be randomly chosen in the representative volume, and so on.

One version of the representative volume Ω of 3-3 connectivity for 80 % porosity is shown in Figure 2. The size of this volume is $20 \times 20 \times 20$. Thus, here the cubic structures of $10 \times 10 \times 10$ size generally were generated in different ways $20^3/10^3 = 8$ times.

By using this model and the models of other types [13,32], we can design nanostructured properties of porous piezoceramics. Thus, after solving the static piezoelectric problems with different boundary conditions (12), (14) for these representative volumes numerically, we will find the full set of the effective moduli for porous piezoceramics. Such an approach allows for determining the properties of the intrinsic structure of pores and its distributions. It is important to note that for different types of the representative volumes we obtain similar dependencies on the porosity for the series of major properties of piezoceramics.

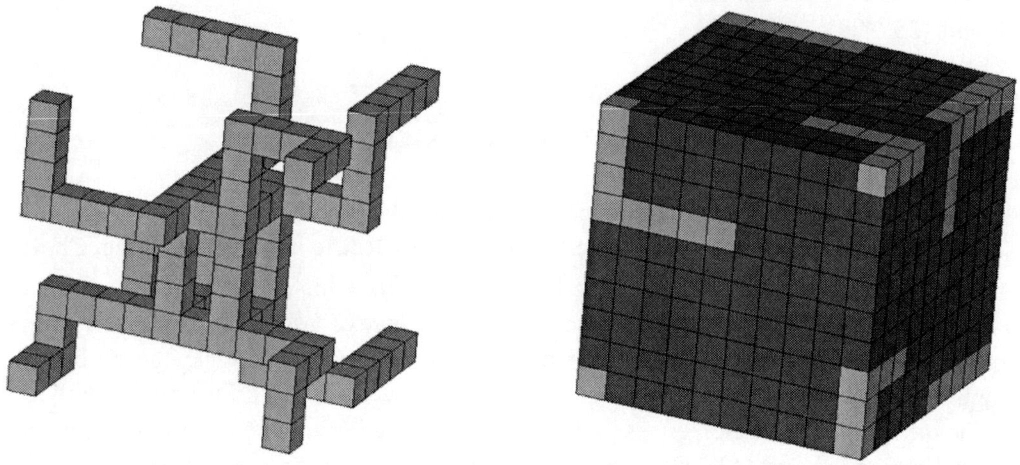

a) b)

Figure 1. Model of 3-3 connectivity type with 90 % porosity: (a) basic cell ($10 \times 10 \times 10$ elements) without pores; (b) basic cell ($10 \times 10 \times 10$ elements) with pores.

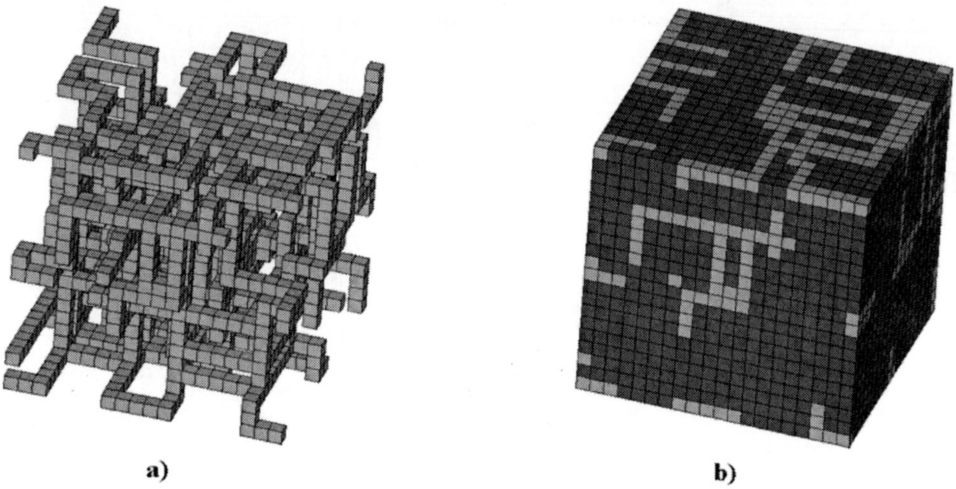

Figure 2. Model of 3-3 connectivity type with 80 % porosity: (a) basic cell ($20 \times 20 \times 20$ elements) without pores; (b) basic cell ($20 \times 20 \times 20$ elements) with pores.

Thus, the full set of non-zero material moduli in the vector-matrix forms ($\tilde{c}^E_{\alpha\beta} = \tilde{c}^E_{ijkl}$, $\tilde{e}_{i\beta} = \tilde{e}^E_{ikl}$, $\tilde{\kappa}^S_{ij} = \tilde{\varepsilon}^S_{ij}$, $\alpha, \beta = 1, 2, ..., 6$, $i, j, k, l = 1, 2, 3$ with the correspondence law $\alpha \Leftrightarrow (ij)$, $\beta \Leftrightarrow (kl)$, $1 \Leftrightarrow (11)$, $2 \Leftrightarrow (22)$, $3 \Leftrightarrow (33)$, $4 \Leftrightarrow (23) = (32)$, $5 \Leftrightarrow (13) = (31)$, $6 \Leftrightarrow (12) = (21)$) for porous piezoceramics PZT-4 is given in Table 1 ($\varepsilon_0 = 8.85 \cdot 10^{-12}$ F/m). This finite element results are obtained for the model of 3-3 connectivity type with $20 \times 20 \times 20$ cells.

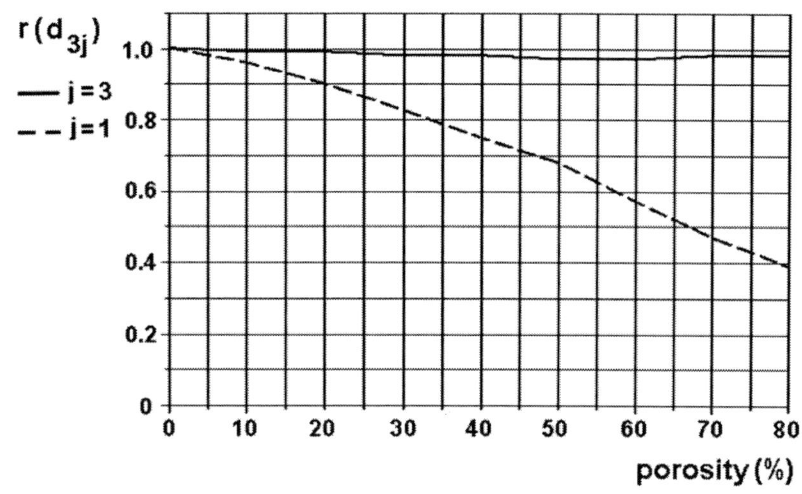

Figure 3. Dependences of effective moduli $r(d_{3j})$ on porosity.

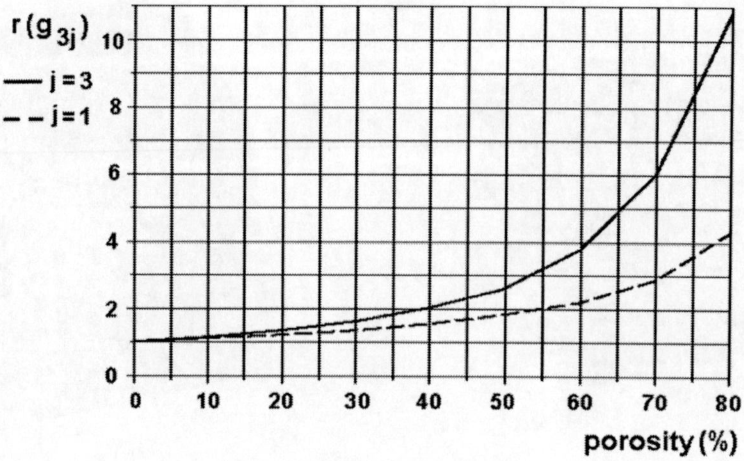

Figure 4. Dependences of effective moduli $r(g_{3j})$ on porosity.

Table 1. Effective moduli of porous piezoceramics PZT-4

p (%)	0	10	20	30	40	50	60	70	80
ρ, kg/m^3	7500	6750	6000	5250	4500	3750	3000	2250	1500
$\tilde{c}_{11}^E$, 10^{10} N/m^2	13.9	11.57	9.23	6.85	5.05	3.31	2.08	1.27	0.68
$\tilde{c}_{12}^E$, 10^{10} N/m^2	7.78	6.21	4.66	3.14	2.09	1.17	0.63	0.28	0.13
$\tilde{c}_{13}^E$, 10^{10} N/m^2	7.74	5.85	4.28	2.84	1.85	1.06	0.52	0.24	0.10
$\tilde{c}_{33}^E$, 10^{10} N/m^2	11.5	9.44	7.29	5.38	3.90	2.73	1.63	0.91	0.47
$\tilde{c}_{44}^E$, 10^{10} N/m^2	2.56	2.21	1.84	1.45	1.09	0.73	0.44	0.23	0.10
$\tilde{e}_{33}$, C/m^2	15.1	13.4	11.45	9.58	7.71	5.97	3.89	2.32	1.24
$\tilde{e}_{31}$, C/m^2	-5.2	-4.2	-3.14	-2.09	-1.32	-0.74	-0.43	-0.21	-0.10
$\tilde{e}_{15}$, C/m^2	12.7	10.86	8.89	6.86	5.04	3.30	1.96	0.99	0.44
$\tilde{\kappa}_{11}^S / \varepsilon_0$	730	657	588	511	438	350	265	191	121
$\tilde{\kappa}_{33}^S / \varepsilon_0$	635	564	488	415	343	271	198	130	75

In Figures 3, 4 we can see the dependencies on the porosity for the relative values of the main piezomoduli for porous ceramics $r(d_{3j}) = \tilde{d}_{3j} / d_{3j}$ and $r(g_{3j}) = \tilde{g}_{3j} / g_{3j}$ ($j = 1, 3$), where d_{3j} and g_{3j} are the corresponding values for dense ceramics. These dependencies are obtained from data in Table 1 and corresponding expressions for d_{3j}, g_{3j} from $c_{\alpha\beta}^E$, $e_{i\beta}$, $\kappa_{ij}^S = \varepsilon_{ij}^S$ [33]. Note that the charge coefficient $\tilde{d}_{33}$ does not decrease with the porosity increment, but at the same time the piezomodule $\tilde{d}_{31}$ decreases sufficiently fast (Figure 3). Such properties of different porous piezoceramics are also obtained from experimental data [34]. On the contrary the voltage coefficients $\tilde{g}_{33}$ and $\tilde{g}_{31}$ increase with the porosity increment (Figure 4). These dependencies of the piezomoduli determine the practical application of porous piezoceramics in many respects [34].

As a result, from the finite element solutions we can conclude that high-porous piezoceramics is a high-performance material for quasistatic and low-frequency applications in piezoelectric energy generators, because its piezo-sensitivity rapidly increases at the porosity growth. We can also note that porous ceramics is preferred for different hydro-acoustic ultrasonic applications [32,34].

Conclusion

In this chapter the theory of the effective moduli method was developed for magnetoelectric composites with piezoelectric and piezomagnetic phases. The basic statements were formulated for the average field characteristics that generalize the approaches developed for the elastic and piezoelectric media. The special boundary electromagnetoelastic problems for representative volume and necessary equations for the determination of the full set of the effective moduli were obtained for magnetoelectric media with arbitrary anisotropy. The similar techniques were applied to the modeling of effective properties for porous piezoelectric materials. The finite element design of magnetoelectric and piezoelectric composites was done with the help of the collection of special efficient algorithms for the saddle point symmetric quasidefinite matrices.

Acknowledgments

This work is supported by grants of the Southern Federal University and of the Russian Foundation for Basic Research.

References

[1] Huang, GY; Wang, BL; Mai, YW. *Mech. Res. Commun.*, 2009, vol. 36, 563-572.
[2] Lu, XY; Li, H; Wang, B. *J. Mech. Phys. Solids.*, 2011, vol. 59, 1966-1977.
[3] Nan, CW; Bichurin, MI; Dong, S; Viehland, D; Srinivasan, G. *J. Appl. Physics.*, 2008, vol. 103, 031101-1-35.
[4] Tang, T; Yu, W. *Int. J. Eng. Science.*, 2008, vol. 46, 741-757.
[5] Zhang, ZK; Soh, AK. *Europ. J. Mech. A/Solids.*, 2005, vol. 24, 1054-1067.
[6] Banno, H. *Jpn. J. Appl. Phys.*, 1993, vol. 32(1), 4214-4217.
[7] Berger, H; Gabbert, U; Koppe, H; Rodriguez-Ramos, R; Bravo-Castillero, J; Guinovart-Diaz, R; Otero, JA; Maugin, GA. *Comput. Mech.*, 2003, vol. 33, 61-67.
[8] Dunn, H; Taya, M. *J. Am. Ceram. Soc.*, 1993, vol. 76, 1697-1706.

[9] Getman, I; Lopatin, S. *Ferroelectrics*., 1996, vol. 186, 301-304.

[10] Silva, ECN; Fonseca, JSO; Kikuchi, N. *Comput. Mech*., 1997, vol. 19, 397-410.

[11] Silva, ECN; Fonseca, JS. O; Kikuchi, N. *Comput. Meth. Appl. Mech. Eng*., 1998, vol. 159, 49-77.

[12] Wersing, W; Lubitz, K; Moliaupt, J. *Ferroelectrics*., 1986, vol. 68, 77-97.

[13] Nasedkin, AV; Shevtsova, MS. In: *Ferroelectrics and Superconductors: Properties and Applications*. IA. Parinov (Ed.); Nova Science Publishers: NY., 2011, 231-254.

[14] Pobedria, BE. *Mechanics of Composite Materials*. Moscow State University Press, 1984 (in Russian).

[15] Kaltenbacher, M. *Numerical Simulation of Mechatronic Sensors and Actuators*. Springer, 2004.

[16] Akopov, ON; Belokon, AV; Nadolin, KA; Nasedkin, AV; Skaliukh, AS; Solovyev, AN. *Math. Modeling*., 2001, vol. 13(2), 51-60.

[17] George, A; Ikramov, K; Kucherov, B. *SIAM J. Matrix Anal. Appl*., 2000, vol. 21, 1318-1323.

[18] Nasedkin, AV. In: *Piezoceramic Materials and Devices*. I. A. Parinov (Ed.); Nova Science Publishers: NY., 2010, 177-218.

[19] Vanderbei, RJ. *SIAM J. Optim*., 1995, vol. 5, 100-113.

[20] Gill, P; Saunders, MA; Shinnerl, JR. *SIAM J. Matrix Anal. Appl*., 1996, vol. 17, 35-46.

[21] Horn, RA; Johnson, CR. *Matrix Analysis*., Cambridge University Press, 1985.

[22] George, A; Liu, JW. *Computer Solution of Large Sparse Positive Definite Systems*. Prentice-Hall, Englewood Cliffs, 1981.

[23] Belokon, AV; Eremeyev, VA; Nasedkin, AV; Soloviev, AN. *J. Appl. Math. Mech. (PMM)*., 2000, vol. 64, 367-377.

[24] Belokon, AV; Nasedkin, AV; Soloviev, AN. *J. Appl. Math. Mech. (PMM)*. 2002, vol. 66, 481-490.

[25] Benzi, M; Golub, GH; Liesen, J. *Acta Numerica*., 2005, vol. 14, 1-137.

[26] Benzi, M; Wathen, AJ. In: *Model Order Reduction: Theory, Research Aspects and Applications*. WHA. Schilders, HA. van der Vorst; J. Rommes (Eds.); *Mathematics in Industry*., 2008, vol. 13, 195-211.

[27] Gantner, A; Hoppe, RHW; Koster, D; Siebert, K; Wixforth, A. *Comput. Vis. Sci*., 2007, vol. 10, 145-161.

[28] Miao, SX; Cao, Y. *Appl. Math. Comput*., 2014, vol. 230, 27-34.

[29] Zhou, YY; Zhang, GF. *Appl. Math. Comput*., 2009, vol. 215, 599-607.

[30] Newnham, RE; Skinner, DP; Cross, LE. *Mater. Res. Bull*., 1978, vol. 13, 525-536.

[31] Iovane, G; Nasedkin, AV. *J. Coupled Syst. Multiscale Dyn*., 2013, vol. 1, 393-403.

[32] Nasedkin, AV; Shevtsova, MS. In: *Physics and Mechanics of New Materials and Their Applications*., IA. Parinov; SH. Chang (Eds.); Nova Science Publishers: NY., 2013, 185-202.

[33] Berlincourt, DA; Curran, DR; Jaffe, H. *Piezoelectric and Piezomagnetic Materials*. In: *Physical Acoustics*, vol. 1, *part A*, Academic Press, 1964, 233-256.

[34] Rybyanets, AN. In: *Piezoceramic Naterials and Devices*. I. A. Parinov; Ed; Nova Science Publishers: NY., 2010, 115-174.

In: Advanced Materials - Studies and Applications
Editors: I.A. Parinov, S.-H. Chang, S.Theerakulpisut
ISBN: 978-1-63463-749-7

Chapter 10

ELASTIC LOSSES AND SPATIAL DISPERSION IN POROUS PIEZOCERAMICS AND CERAMIC MATRIX COMPOSITES

A. N. Rybyanets* and A. A. Naumenko
Research Institute of Physics, Southern Federal University,
Rostov-on-Don, Russia

ABSTRACT

A comprehensive study of different composites structures with strong spatial dispersion and high losses (porous piezoceramics, composites of ceramics/ceramics and ceramics/crystals types) including finite-difference 3D simulations, impedance spectroscopy characterization and ultrasonic measurements were carried out in this chapter. The Wave 3000 Pro computer software package based on finite-differences method was used for full time-domain solution of the 3D viscoelastic wave equations. Sets of complex elastic, dielectric, and piezoelectric parameters of the porous piezoceramics and ceramic matrix piezocomposites were determined by impedance spectroscopy method using Piezoelectric Resonance Analysis (PRAP) software. Microstructure of polished, chemically etched, and chipped surfaces of porous ceramics and piezocomposite samples was observed with optical and scanning electron microscopies. The results of simulation were compared with the experimental data obtained by ultrasonic pulse-echo and through-transmit methods. Microstructural and physical mechanisms of elastic, piezoelectric and dielectric losses and spatial dispersion in ceramic matrix piezocomposites were discussed.

1. INTRODUCTION

Large-scale production of composite materials and their intensive use in ultrasonic transducers requires not only overcoming technological problems, but also a psychological

* arybyanets@gmail.com.

revolution in fundamental views of technologists and designers of piezoceramic material. The change from prevailing industry inclination to manufacture high-density, chemically perfect and structurally flawless ceramic materials to the fabrication and use of "imperfect", low density and "defective" materials (lead-metaniobate ceramics, porous ceramics and composites) has taken a long time [1–4]. Commercialization of composite materials has also led to the development of new concepts of material and ultrasonic transducers designing [5–8].

Demands on medical and NDT ultrasonic transducer performance have increased in recent years. Low-Q piezoceramics and piezocomposite materials are widely used for wide-band NDT ultrasonic transducers with high sensitivity and resolution. Some of these advanced materials are lossy, and direct use of IEEE Standards for material constant determination leads to significant errors.

The modeling and design of piezoelectric devices by finite element methods, among others, rely on the accuracy of the dielectric, piezoelectric and elastic coefficients of the active material used which is, commonly, an anisotropic ferroelectric polycrystal. The accurate description of piezoceramics must include the evaluation of the dielectric, piezoelectric and mechanical losses, accounting for the out-of-phase material response to the input signal.

Porous piezoceramics based on different piezoceramic compositions were proposed as a promising candidate for lead metaniobate replacement in wide-band ultrasonic transducers [9, 10]. Intensive research and technological work, as well as improvements in fabrication methods, have allowed large-scale manufacture of porous piezoelectric ceramics with reproducible and controllable porosity and properties [11,12].

Over the past years, considerable advances have been made to improve the mechanical properties of ceramics using ceramic composite approaches. Numerous technologies based on incorporation of functional ceramics into structural ones and *vice versa* have been developed, and novel design ideas applied in the field of functional ferroelectric ceramics [13,14].

Ceramic composites are very complex objects for theoretical modeling, NDT inspection, and ultrasonic measurements. Changes in chemical composition on the phase interfaces as well as microporosity appearance during co-firing of composite components could alter elastic and mechanical properties of composites. Spatial dispersion and scattering can distort ultrasonic pulse characteristics and make ultrasonic measurements ambiguous. Different methods were proposed for composite structures modeling and evaluation [15–17].

A comprehensive study including finite-difference 3D-simulation, impedance spectroscopy characterization, and pulse-echo ultrasonic measurements of different composite structures is described in this chapter.

The chapter is divided, somewhat arbitrarily, into three informative sections. Section 2 describes impedance spectroscopy method and software characterization of lossy piezoelectric materials. Section 3 briefly describes structure peculiarities, fabrication methods and properties of porous piezoceramics and ceramic matrix piezocomposites. Section 4 describes measurement methods and presents the results and discussion of 3D-simulation, impedance spectroscopy and pulse-echo ultrasonic measurements for porous ceramics and ceramic/crystal composites.

2. Impedance Spectroscopy Characterization of Highly Attenuating Piezocomposites

The basic techniques for finding material constants of piezoelectric materials have been present by the IEEE Standard on Piezoelectricity (1987) [18]. These methods work for many of the most widely used commercial piezoceramics based on lead-titanate-zirconate (PZT) compositions that are the high-Q_M and high-coupling coefficient piezoelectric materials. However, there is a general agreement that their use in many new piezoelectric materials such as porous ceramics, piezoelectric polymers or piezoelectric composites may lead to significant errors. Furthermore, the current IEEE Standard does not comprehensively account for the complex nature of material coefficients, as it uses only the dielectric loss factor (tan δ) and the mechanical quality factor (Q_M) to account for loss.

Numerous techniques using complex material constants have been proposed to take into account losses in low-Q_M materials and to overcome limitations in the IEEE Standard [19–21]. Iterative methods [22,23] provide means of correct determination of the complex coefficients in the linear range of poled piezoceramics based on the resonance measurements of the complex impedance. The Piezoelectric Resonance Analysis Program (PRAP) automatic iterative method [24] has been applied to the full set of standard geometries and resonance modes needed to complete complex characterization in a wide range of materials with very high and moderate loss factors [25–27].

2.1. Piezoelectric Resonance Analysis Methods

In the direct piezoelectric effect, charge generates from applying a stress to a specimen. In the reverse piezoelectric effect, the specimen will undergo strain in response to an electrical field. When four possible thermodynamic potentials are considered for adiabatic, isothermal and low amplitude conditions [28, 29], the linear equations of piezoelectricity give the relationship between electric field (E), dielectric displacement (D), strain (S) and stress (T):

$$\begin{aligned} S_p &= s^D_{pq} T_q + g_{pm} D_m \\ E_m &= \beta^T_{mn} D_n - g_{pm} T_p \end{aligned} \quad (1)$$

$$\begin{aligned} S_p &= s^E_{pq} T_q + d_{pm} E_m \\ D_m &= \varepsilon^T_{mn} E_n + d_{pm} T_p \end{aligned} \quad (2)$$

$$\begin{aligned} T_p &= c^E_{pq} S_q - e_{pm} E_m \\ D_m &= \varepsilon^S_{mn} E_n + e_{pm} S_p \end{aligned} \quad (3)$$

$$\begin{aligned} T_p &= c^D_{pq} S_q - h_{pm} D_m \\ E_m &= \beta^S_{mn} D_n - h_{pm} S_p \end{aligned} \tag{4}$$

In equations (1 – 4) d, e, g, and h are the piezoelectric constants, s and c are the elastic compliance and stiffness at constant E and D, and ε and β are the permittivity and inverse permittivity at constant S and T. The mechanical variables have three additional shear directions (4, 5, 6) while the electrical variables have only three standard directions (1, 2, 3).

Each of the sets of equations above is typically represent as a piezoelectric matrix with the transpose of the piezoelectric constants in the top right corner. Symmetry in materials means that many of the elements of the piezoelectric matrix are either equal, zero, or some function of themselves. Poled polycrystalline materials are of C_∞ or 6mm symmetry, and similar to the other sets of equations; equation set (2) reduces to a form of the piezoelectric matrix that includes only the following 10 independent constants:

$$\begin{bmatrix} S_1 \\ S_2 \\ S_3 \\ S_4 \\ S_5 \\ S_6 \\ D_1 \\ D_2 \\ D_3 \end{bmatrix} = \begin{bmatrix} s^E_{11} & s^E_{12} & s^E_{13} & 0 & 0 & 0 & 0 & 0 & d_{31} \\ s^E_{12} & s^E_{11} & s^E_{13} & 0 & 0 & 0 & 0 & 0 & d_{31} \\ s^E_{13} & s^E_{13} & s^E_{33} & 0 & 0 & 0 & 0 & 0 & d_{33} \\ 0 & 0 & 0 & s^E_{55} & 0 & 0 & 0 & d_{15} & 0 \\ 0 & 0 & 0 & 0 & s^E_{55} & 0 & d_{15} & 0 & 0 \\ 0 & 0 & 0 & 0 & 0 & 2\left(s^E_{11} - s^E_{12}\right) & 0 & 0 & 0 \\ 0 & 0 & 0 & 0 & d_{15} & 0 & \varepsilon^T_{11} & 0 & 0 \\ 0 & 0 & 0 & d_{15} & 0 & 0 & 0 & \varepsilon^T_{11} & 0 \\ d_{31} & d_{31} & d_{33} & 0 & 0 & 0 & 0 & 0 & \varepsilon^T_{33} \end{bmatrix} \begin{bmatrix} T_1 \\ T_2 \\ T_3 \\ T_4 \\ T_5 \\ T_6 \\ E_1 \\ E_2 \\ E_3 \end{bmatrix} \tag{5}$$

In general, the material constants are temperature (relative to the Curie temperature for ferroelectrics), time (logarithmic relaxation after poling) and frequency dependant. In reality, the properties may also be field dependant, which would require a re-derivation of the piezoelectric equations for higher order field terms.

Impedance spectroscopy characterization stems from the relation for impedance Z:

$$\mathbf{Z} = V/\mathbf{I} = \int \mathbf{E} \cdot d\mathbf{x} \Big/ A\frac{d\mathbf{D}}{dt}, \tag{6}$$

where V is the voltage, I is the current, and A is the surface area over which current is measured. The piezoelectric equations give the relationship between D and E by using mechanical stress (T) and strain (S). For simple one-dimensional resonance modes, many of the electric and mechanical field variables go to zero and the sets of piezoelectric equations decouple.

Solving the wave equation for one-dimensional steady-state displacement in the specimen and evaluating S from the derivative of displacement, together with the boundary conditions of an unclamped resonator, the piezoelectric equations allow us to evaluate the relationship between D and E as a function of frequency and material properties. The impedance of a thickness extensional (TE) resonator reduces to [30]:

$$\mathbf{Z} = \frac{\boldsymbol{\beta}_{33}^{S} l + \frac{\mathbf{h}_{33}^{2}}{\sqrt{\rho \mathbf{c}_{33}^{D}}\,\omega} \tan\left(\frac{\omega l \sqrt{\rho}}{2\sqrt{\mathbf{c}_{33}^{D}}} \right)}{i\omega A}, \tag{7}$$

where ρ is the specimen density, ω is the angular frequency and l is the thickness of the sample in the direction of poling. For the TE resonator, the lateral dimensions must be at least 10 times the thickness to assure a clean, one-dimensional TE mode. This can be relaxed slightly in low Q materials. In impedance spectroscopy characterization of piezoelectrics, we can adjust the material properties of equation (7) until the impedance fits a measured impedance spectrum as a function of frequency.

The common one-dimensional 6mm modes include the Length Extensional (LE), Thickness Extensional (TE), Length-Thickness Extensional (LTE), Thickness Shear (TS), Length Shear (LS), and Radial modes. Each of these modes provides electrical, elastic and piezoelectric properties of the material. Table 1 lists the properties of one form of the reduced piezoelectric matrix with the modes that will provide those properties.

Table 1. Properties provided by the common modes

Property	ε^{T}_{11}	ε^{T}_{33}	s^{E}_{11}	s^{E}_{12}	s^{E}_{13}	s^{E}_{33}	s^{E}_{55}	d_{13}	d_{33}	d_{15}
Mode	TS, LS	LE, LTE, Radial	LTE, Radial	Radial	-	LE	LS, TS	LTE, Radial	LE	TS, LS

By analyzing specimens inducing the TS, LE and Radial modes, we can obtain all but s^{E}_{13} of the reduced 6mm piezoelectric matrix. The values of s^{E}_{13} can be found from the properties determined with the TE, LTE and LE modes for other representations of the piezoelectric linear equations [8], indicating that all modes are required to construct the piezoelectric matrix. Alternatively, using an initial guess for s^{E}_{13}, the reduced matrix inverts and c^{D}_{13} compares to that found for the TE mode, eliminating the need for the LTE mode.

In practice, material properties are dispersive (i.e., they vary with frequency). This means that the range of frequency used to fit forms of equation (6) to measure spectra should be limited. It is worth noting that only a few points in the spectrum are required to determine properties. For the TE mode, only three points are required.

The IEEE Standard on Piezoelectricty (1987) [18] describes two approaches to obtaining material properties using either the difference between parallel (f_p) and series (f_s) resonance frequencies, or the frequency ratio between different order resonance peaks. These approaches can be problematic when material properties are dispersive or when a low electromechanical coupling constant results in a small difference between f_p and f_s.

Smits [23] proposed a general method for the TE, TS, LE and LTE resonators using three points around resonance. This method does not rely on critical frequencies in the spectrum and so does not lose sensitivity for low-Q materials. At resonance and anti-resonance, currents can be very large and small and this presents an instrumentation problem. Though less of an issue with low-Q materials, Smits's method is less demanding of instrumentation than the IEEE methods. Finally, because the choice of analysis points is somewhat arbitrary, noisy spectra and samples with mode coupling may still be analyzed using Smits's method. The IEEE Standard in Piezoelectricity (1987) treats all material properties as real, evaluating mechanical and electrical loss independently through Q_M and tanδ. We assumed that losses in the electromechanical coupling constant were equal to zero. It was acceptable for piezoelectric materials with low losses (such as PZT), but can lead to significant errors, when materials have large losses. In real materials, complex impedance results from complex material properties [31]. The impedance is directly associated with dielectric conduction losses [26] and attenuation of the displacement wave in the sample.

The PRAP software [24] analyses impedance spectra to determine complex material properties. This software uses a generalized form of the Smits's method to determine material properties for any common resonance mode, and a generalized ratio method for the radial mode [22] valid for all material Q's. By analyzing on each harmonic, PRAP can determine complex material properties as a function of frequency. The software always generates an impedance spectrum from the determined properties to indicate validity of the results.

3. Porous Piezoceramics and Ceramic Matrix Composites

3.1. Porous Piezoelectric Ceramics

Porous piezoelectric ceramics are important for a variety of applications such as medical ultrasonic devices, nondestructive testing, underwater acoustics, and others [32–38]. Despite the long observation regarding pore effects on material properties, the porosity-property interrelation seems not yet clear, especially for porous piezoceramics and relative composites. To describe real porous piezoceramics, it was necessary to introduce models based upon the concepts of "connectivity", "percolation", "tortuosity", and more recently, "fractal geometry" [39]. It is necessary to note that, despite rather good conformity of some modeling parameters to experimental data, none of the theoretical models truly describes a real porous piezoceramics. The situation becomes even more complicated taking into account the inconsistent experimental data for different porous ceramics available in the literature [9, 39–41]. This uncertainty is caused by objective technological and experimental problems (reproducibility of structure and properties, measurements of low-Q material parameters), as well as the use of different and frequently imperfect methods of porous ceramics fabrication and/or incorrect porosity estimations [9,41].

3.1.1. Porosity Origin and Microstructure

Generally, porous ceramics are present by two types: reticulated or open cell and foam or closed cell [42]. Open cell ceramics have interconnected pores, while in closed cell ceramics pores are usually isolated in a continuous ceramic matrix.

Porous piezoceramic is composite material composed of two phases: a piezoelectrically active (piezoceramic) phase and a passive (porous) phase. The piezoelectrically active phase forms a three-dimensional matrix, in which the passive phase exists in the form of isolated and/or interconnected inclusions. The case of isolated inclusions corresponds to 3-0 connectivity, while the interconnected inclusions case relates to 3-3 connectivity type (closed and open porosity, respectively) [2,9].

If the sizes of the passive phase inclusions are lesser than the sizes of the piezoelement, and if their shape approximates spherical, then porous piezoceramics can be considered as homogeneous [9]. The microstructure of porous ceramics is unequivocally defined by the fabrication method and by the initial chemical composition of the piezoceramic phase. The structure and properties of porous ceramics are controlled by their processing.

It is necessary also to note a qualitative structural aspect that distinguishes porous ceramics from other commixture and periodic composites, namely, the presence of a rigid 3D ceramic skeleton persisting up to relative porosity values of 75 – 95%, depending on manufacturing technology.

Figure 1 shows micrographs illustrating a micro-structural variety of porous ceramic materials. The resulting elastic, dielectric, and piezoelectric properties of porous piezoceramics are defined by the properties of its piezoceramic matrix: that is, the porosity, type of connectivity, shape, and size of pores (and, in the case of irregular shaped pores, by the degree of their primary orientation and arising anisotropy) [2,10,12,40].

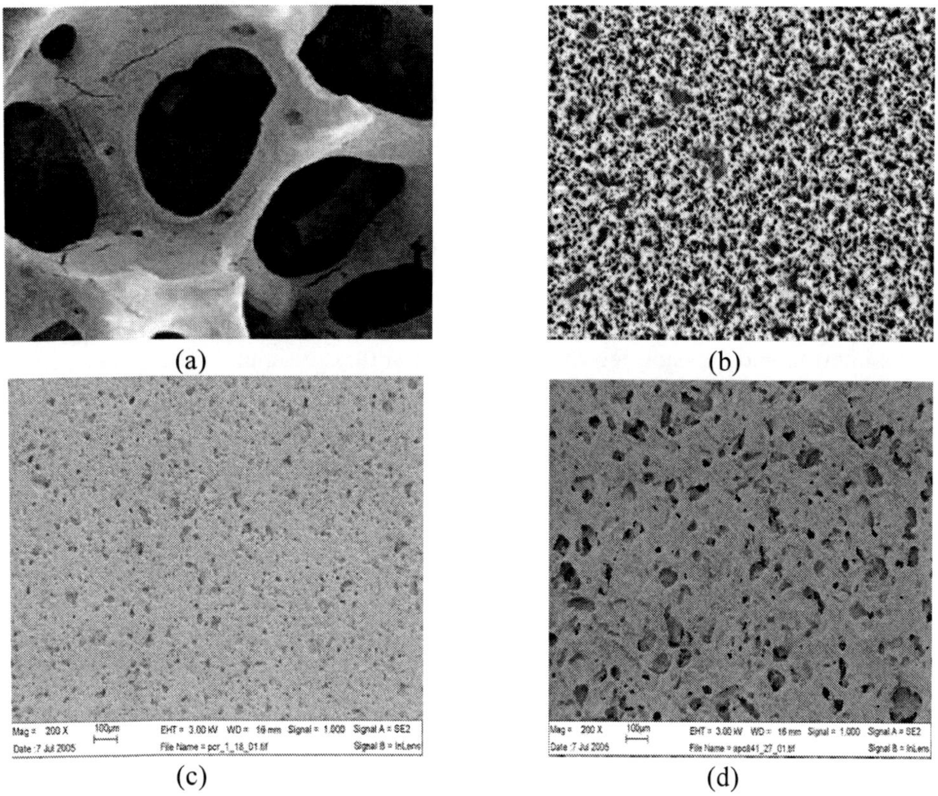

Figure 1. SEM micrographs of different porous ceramic structures: (a) open cell PZT ceramics, fabricated by polymer foam method; (b) closed cell PZT ceramics, obtained by burning out polymer granules, (c) ground surface of "soft" PZT ceramics, relative porosity of 18%, (d) ground surface of "hard" PZT ceramics, relative porosity of 21%.

3.1.2. Fabrication Methods

All existing methods of porous ceramics fabrication can be divided into two basic approaches: subtractive processes and processing limitations. In subtractive processes, certain elements of an original structure are selectively removed to create pores. In processing limitations, technological processing regimes are modified to form porous structures [39].

Porous ceramics with an open cell structure are usually formed by replicating a polymer foam/sponge. Closed cell porous ceramics are usually formed by adding porosifiers to ceramics powders.

First samples of porous piezoceramics and piezoceramic/polymer composites with 3-3 connectivity were prepared using a lost-wax process with coral as the starting material (also known as the coral replamine process) [9]. However, the replamine process is not industrially favorable because of its complicated procedures, requirement for replica storage, and low reproducibility.

A simpler technique for production of highly porous compacts was proposed in [2,9,12] with sintered mixtures of burnable plastic spheres (PMMA) and PZT powders. A series of further process for the preparation of porous ceramics and composites with 3-3 connectivity were subsequently introduced [4,7,10,11]:

- a method based on washing-out of water-soluble granules;
- the preparation of hole spheres of PZT by spray drying of alcooxides;
- the polymer foam method;
- the reaction synthesis method;
- the cryochemical method;
- the thermal decomposition of carbonates, nitrates or oxalates, as well as chemical etching and carbon activatiion;
- methods using differences in shrinkage coefficients of ceramic components, special sintering and pressing regimes, excessing of plasticiser and other processing limiters.

The majority of these methods are updates of methods [9] developed earlier for fabrication of non-piezoelectric porous materials. A universal manufacturing method suitable for mass production of a wide range of porous piezoceramics and polycrystalline composite materials was developed on the basis of a microstructure design concept for polycrystalline materials [2–4,11,12].

3.2. Ceramic Matrix Piezocomposites PZT/α-Al_2O_3

New polymer-free ceramic matrix piezocomposites with properties combining better parameters of PZT, PN type ceramics and 1-3 composites for wide-band NDT ultrasonic transducer applications were developed [6,14,16]. These new “damped by scattering” ceramic piezocomposites PZT/α-Al_2O_3 are characterized by previously unachievable low $Q_M^t = 2 - 3$, combined with high piezoelectric ($d_{33} = 250 - 350$) and electromechanical ($k_t = 0.45 - 0.5$) parameters, high Curie point (340 °C), and low acoustic impedance (15 – 20 MRayl) in a wide working frequency range (0.5 – 10 MHz). These composites were used to develop a line of wide-band NDT and medical diagnostic ultrasonic transducers with high sensitivity and resolution, which were manufactured and tested [4,16].

3.2.1. Methods of Design and Fabrication

The original microstructure design concept (MSD) for ceramic matrix composites, proposed in [6], is based on controllable substitution during composite formation of separate crystallites or crystallite groups making a polycrystal by pores, crystallites with other composition and/or structure or amorphous substances. The MSD concept offers a brand-new range of polymer-free polycrystalline piezocomposite materials with parameters that are adjustable over a wide range [2–4,10]. Starting from this point, a novel "damping by scattering" approach was proposed in [3,16] for the development of a new family of low-Q ceramic piezocomposites for wide-band ultrasonic transducer applications.

The main idea of the damping by scattering approach is introducing the scattering particles as a secondary phase into the piezoceramic matrix during the same technological process. Preliminary optimization of composite properties was fulfilled using the results of single and multiple scattering theories, composite micromechanics and effective medium models, as well as FEM models [15,16]. Material acoustic properties, along with the shape, size, and size distribution, as well as volume fraction of the scattering particles, were the main optimization parameters. Preliminary FEM calculations based on effective moduli method [6, 16] give results that take into account real composite microstructure and agree well with the preliminary experiments.

General conclusions deduced from the preliminary optimization of damped by scattering ceramic piezocomposite are follows:

- scattering particles must be irregularly shaped and anisotropic;
- size distribution of scatterers must be broad;
- piezoceramic matrix must be microporous to prevent cracking at co-firing stage, give additional attenuation, and provide mechanical flexibility and property adjustment;
- the acoustic properties of inclusions and the matrix must have a high contrast;
- particle size-to-wavelength ratio must correspond to the stochastic region to intensify the scattering.

Methods of powder preparation, ceramic matrix composite sintering and experimental samples fabrication, as well as technological regimes are described in detail in [3,4,16].

4. Experimental Scenario

The following two types of ceramic composites [3,4] with strong spatial dispersion and high losses were chosen as model samples for simulation of ultrasonic wave propagation and comparison with piezoelectric resonance analysis (PRAP) and ultrasonic measurements (Figure 2):

1. Ceramic matrix composites A850L consisting of soft PZT matrix with randomly distributed α-Al_2O_3 crystals with a mean particle diameter of ~ 200 μm and volume fraction from 9 up to 26 vol.%.

2. Porous PZT piezoceramics PCR-1 with different porosity and pore sizes (porosity of 18 %, pore size of 30 μm and porosity of 21 %, pore size of 75 μm).

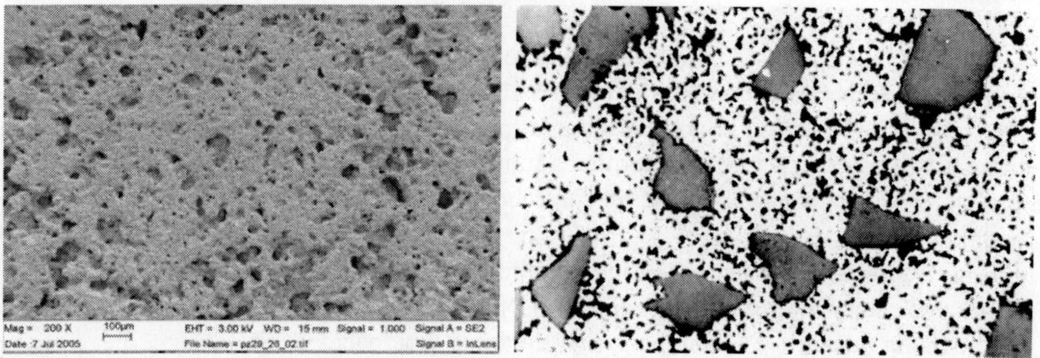

(a)

(b)

Figure 2. Optical and SEM micrographs of porous PCR-1 piezoceramic (a) and A850L composite (b) samples.

Experimental samples of porous piezoceramics with porosity range of 15 – 25 % were prepared by burning out of organic powders using specially developed technology [3,4,16]. Mixtures of organic crystal powders with different particle shapes, decomposition temperatures and grain-size distribution (from 20 up to 30 μm) were used for producing the pore volumes. The porous ceramic elements with vacuum deposited Cr/Ni electrodes were poled on air at cooling from 380 °C to 90 °C for 1 min by applying a dc-electric field of 1 – 1.5 kV/mm.

Ceramic matrix composite samples for the present study were fabricated using standard ceramic technology [3]. Mixing of the powders was carried out using a specially developed method to ensure homogeneous mixture of the components and to prevent destruction and disintegration of powders particles. Powders mixed in different proportions were granulated and cold pressed into cylinders of Ø23×20 mm^2. Special pressing and firing regimes as well as porosity agents were used for the formation of microporous piezoceramic matrices. Sintering of the green bodies was carried out at special thermal profiles to prevent cracking because of shrinkage and thermal expansion of the composite components.

For electrical measurements, Ag-paste was printed on both sides of grinded composite elements and then fired at 700 °C for 5 min. Poling treatment was carried out in silicone oil at the same temperature and electrical field as for the corresponding dense piezoceramics.

Specimens for optical microscopy were ground and polished using a specially designed procedure for composites comprising components with different hardness.

4.1. Measurement Methods

Complex elastic, dielectric and piezoelectric coefficients of porous ceramics and ceramic matrix piezocomposite elements were determined by impedance spectroscopy method using Piezoelectric Resonance Analysis (PRAP) software [24,25]. Measurements were made on Solartron Impedance/Gain-Phase Analyzer SL 1260 and Agilent 4291 Impedance Analyzer.

Sound velocity and attenuation of longitudinal and shear waves in different directions for polarized and non-polarized samples were measured by pulse-echo and through-transmit ultrasonic methods in the frequency range of 1 – 10 MHz using digital oscilloscope LeCroy Wave Surfer 422, Olympus 5800, 5077 pulser/receivers using standard Olympus transducers.

The microstructure of polished, chemically etched, and chipped surfaces of composite samples was observed with optical (NeoPphot-21) and scanning electron microscopes (SEM, Karl Zeiss) [4]. Density and shrinkage coefficients of the sintered porous ceramics and composite bodies were measured by weighing and measuring the volume, as well as by the hydrostatic weighing method.

4.2. Simulations Algorithm

The Wave 3000 Pro software package [43] was used for simulation. The program, besides simulating the complete spatial and time-dependent acoustic solution, allows simulation of ultrasound measurements in a variety of source and receiver configurations. The 3D composite objects for simulation were generated both internally, using Wave3000 Pro "Geometry" routines, and externally using optical and SEM slice data for real composites elements. Wave 3000 Pro computes an approximate solution to the 3D visco-elastic wave equation. The numerical solution is based on a finite difference algorithm [44]. The algorithm was adapted to include viscous losses. The specific acoustic equation simulated in a Wave3000 Pro simulation is as follows:

$$\rho\frac{\partial^2 w}{\partial t^2}=\left[\mu+\eta\frac{\partial}{\partial t}\right]\Delta^2 w+\left[\lambda+\mu+\phi\frac{\partial}{\partial t}+\frac{\eta}{3}\frac{\partial}{\partial t}\right]\Delta(\Delta\bullet w)$$

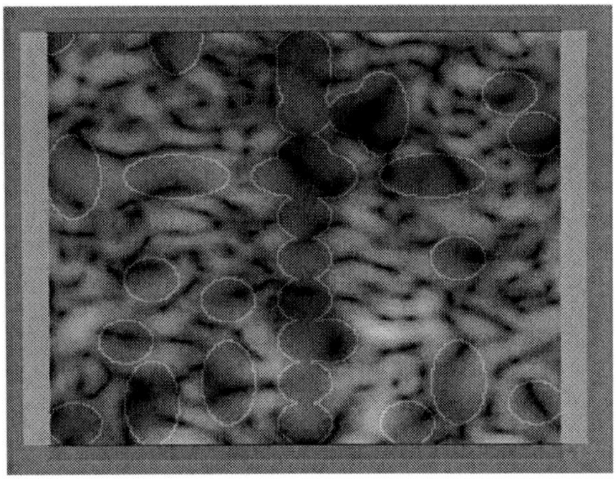

Figure 3. Wave 3000 Pro models of PZT/α-Al_2O_3 composite.

In the equation above, which applies in an isotropic elastic region, the standard designations of variables are used [44].

Figures 3 – 6 show some examples of composite and porous ceramic models and corresponding through-transmit pulse characteristics simulated by Wave 3000 Pro software.

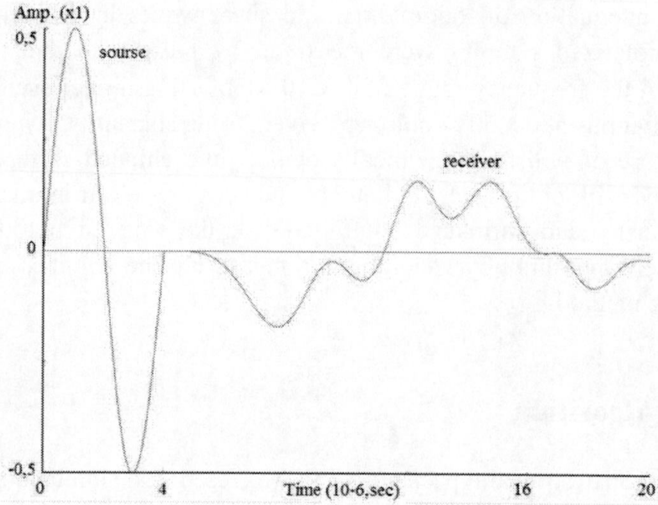

Figure 4. Through-transmit pulse characteristics for PZT/α-Al_2O_3 composite simulated by Wave 3000 Pro.

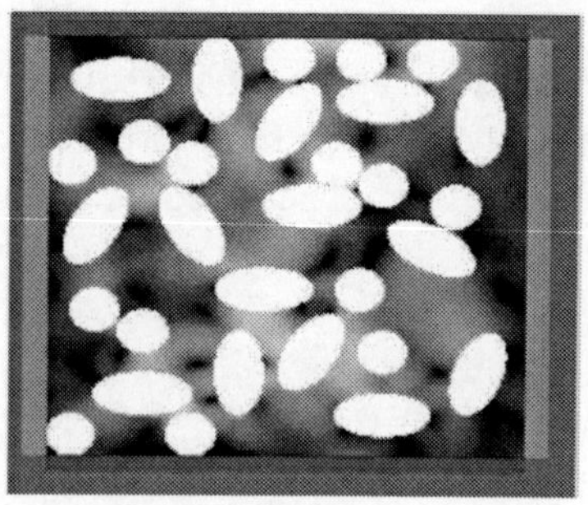

Figure 5. Wave 3000 Pro models of porous ceramics.

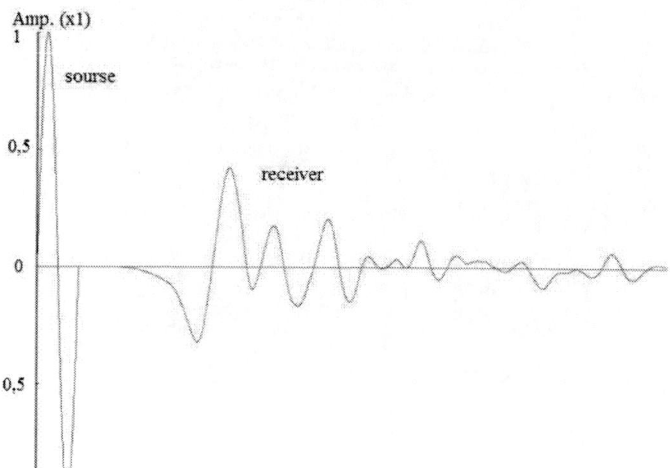

Figure 6. Through-transmit pulse characteristics for porous ceramics simulated by Wave 3000 Pro.

4.3. Results and Discussion

4.3.1. Ultrasonic Measurements

Figures 7 – 10 show echo-pulse characteristics for porous ceramics and composite elements (∅ 25 mm, thickness of 8 mm) measured at transmitted pulse frequencies of 2.25 MHz and 5 MHz.

It is readily seen from Figures 7 – 10 that because of ultrasonic wave scattering and spatial dispersion, the received echo-pulses are strongly distorted and shifted down in frequency. Thus, these pulse-echo measurements of elastic properties of ceramic composites as function of frequency are inaccurate and ambiguous.

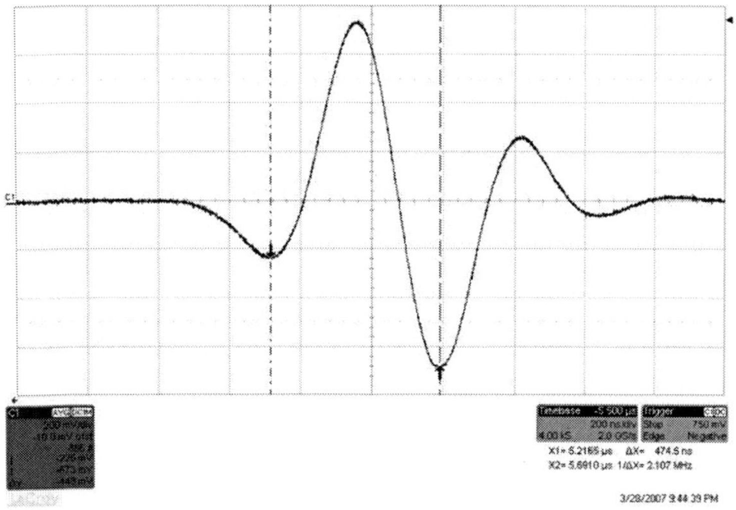

Figure 7. Echo-pulse characteristics for porous ceramics element: transmitted pulse frequency of 2.25 MHz; received pulse frequency of 2.1 MHz.

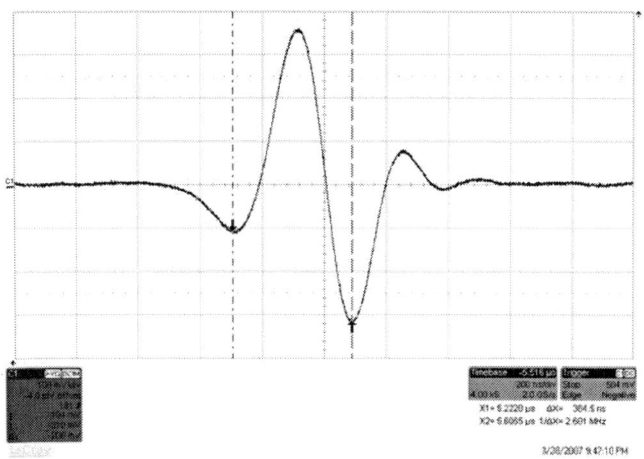

Figure 8. Echo-pulse characteristic for porous ceramics element: transmitted pulse frequency of 5 MHz; received pulse frequency of 2.6 MHz.

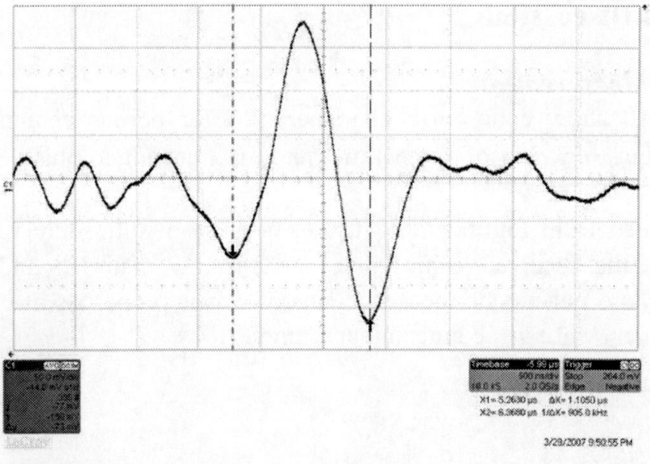

Figure 9. Echo-pulse characteristic for A850L composite element: transmitted pulse frequency of 2.25 MHz; received pulse frequency of 0.9 MHz.

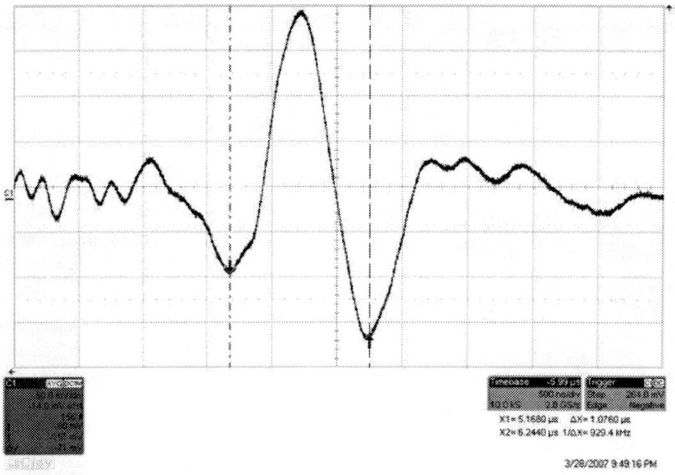

Figure 10. Echo-pulse characteristic for A850L composite element: transmitted pulse frequency of 5 MHz; received pulse frequency of 0.93 MHz.

4.3.2. Resonance Measurements and Wave Pro Simulations

Figure 11 shows an example of impedance spectra and PRAP approximations for the thickness extensional (TE) mode of PCR-1 porous ceramic disks (Ø10 × 0.5 mm^2).

Table 2 summarizes complex constants for a PCR-1 disk (with porosity of 18 %) obtained using PRAP analysis for thickness and radial resonance modes, along with corresponding IEEE Standard results. Additional physical parameters of PCR-1 porous ceramics are as follows: $Z_A = 25.5$ MRayl, $\rho = 6.6$ g/cm^3, $V_t = 3875$ m/c, $Q'_M = 25$, $T_C = 340$°C [2].

It is readily observed in Tables 2 that for moderately low-Q porous ceramics, the results obtained by PRAP and the IEEE Standard methods are practically similar. However, even here, PRAP gives sets of complex constants, thereby allowing an accounting for and an analysis of the losses.

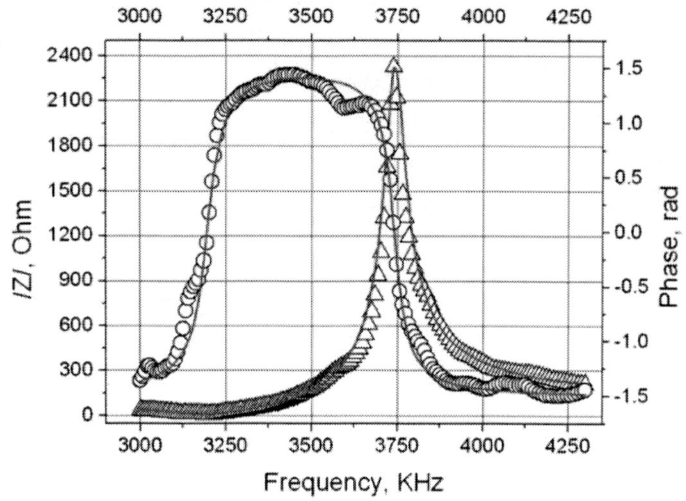

Figure 11. Impedance spectrum and PRAP approximation for TE mode of PCR-1 disk.

PRAP can be used also to determine material properties of ceramics from higher order harmonics [24].

Figure 12 shows results of PRAP analysis of material properties along with the simulation results of Wave 3000 Pro program for complex c_{33}^{D} and c_{33}^{E} obtained from the higher harmonics of the TE mode of a PCR-1 porous ceramic disk.

Table 2. PRAP analysis results for different vibration modes of PCR-1 disk

Parameter	Real	Imaginary	IEEE Standard	Error, %
Radial Extensional (RE) Mode				
f_p (Hz)	222222	754.256	221340	0.4
s_{11}^{E} (m²/N)	$1.75 \cdot 10^{-11}$	$-8.91 \cdot 10^{-14}$	$1.74 \cdot 10^{-11}$	0.6
s_{12}^{E} (m²/N)	$-4.99 \cdot 10^{-12}$	$1.48 \cdot 10^{-13}$	$-4.87 \cdot 10^{-12}$	2.4
$-d_{31}$ (C/N)	$7.92 \cdot 10^{-11}$	$-4.36 \cdot 10^{-12}$	$7.76 \cdot 10^{-11}$	2.0
ε_{33}^{T} (F/m)	$4.35 \cdot 10^{-9}$	$-1.97 \cdot 10^{-10}$	$4.27 \cdot 10^{-9}$	1.8
k_p	0.479629	− 0.01101	0.47302	1.4
σ^{P}	0.284483	0.009907	0.278866	2.0
e_{31} (C/m²)	6.32055	− 0.22792	6.16393	2.5
s_{66}^{E} (m²/N)	$4.5 \cdot 10^{-11}$	$1.18 \cdot 10^{-13}$	$4.46 \cdot 10^{-11}$	0.9
c_{66}^{E} (N/m²)	$2.22 \cdot 10^{10}$	$5.841 \cdot 10^{7}$	$2.24 \cdot 10^{10}$	0.9
Thickness Extensional (TE) Mode				
f_p *(Hz)*	$3.73 \cdot 10^{6}$	37771.9	$3.73 \cdot 10^{6}$	0.0
k_t	0.566103	− 0.00555	0.56646	0.1
c_{33}^{D} (N/m²)	$9.91 \cdot 10^{10}$	$2.01 \cdot 10^{9}$	$9.93 \cdot 10^{10}$	0.2
c_{33}^{E} (N/m²)	$6.74 \cdot 10^{10}$	$1.99 \cdot 10^{9}$	$6.75 \cdot 10^{10}$	0.2
e_{33} (C/m²)	8.3419	− 0.25107	-	-
h_{33} (V/m)	$3.81 \cdot 10^{9}$	$-1.17 \cdot 10^{8}$	-	-
ε_{33}^{S} (F/m)	$2.19 \cdot 10^{-9}$	$-1.33 \cdot 10^{-10}$	-	-

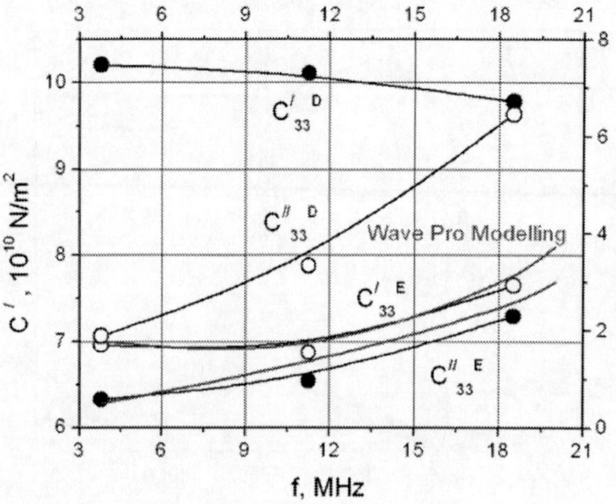

Figure 12. Frequency dependence of c'^{D}_{33}, c'^{E}_{33} (real) and c''^{D}_{33}, c''^{E}_{33} (imaginary) obtained by PRAP and Wave 3000 Pro program for higher orders peaks of TE mode of PCR-1 porous ceramic disk.

It is readily observed that c'^{E}_{33} (real), as well as c''^{E}_{33} (imaginary), increase with the frequency because of the Rayleigh scattering (dispersion) of high frequency ultrasonic waves on pores. In its turn, c'^{D}_{33} (real) decreases and c''^{D}_{33} increases significantly with the frequency as a result of the electromechanical contribution to c^{D}_{33}. Note that k_t value measured on higher harmonics drops drastically with the harmonics number.

Figures 13 and 14 show impedance spectra and PRAP approximations for thickness modes of A850L composite disks at 10 % and 20 % volume fraction of α-Al_2O_3 particles with a mean size of ~ 200 μm, respectively.

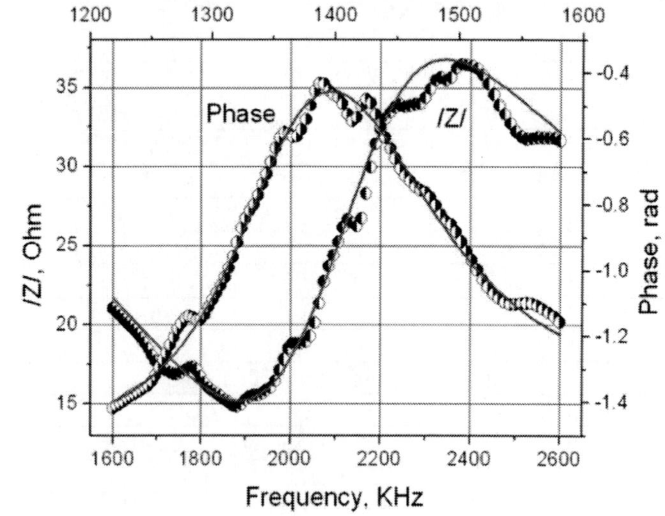

Figure 13. Impedance spectra and PRAP approximations for thickness extensional mode of A850L-10 vol. % disk (Ø19.95 × 0.8 mm^2).

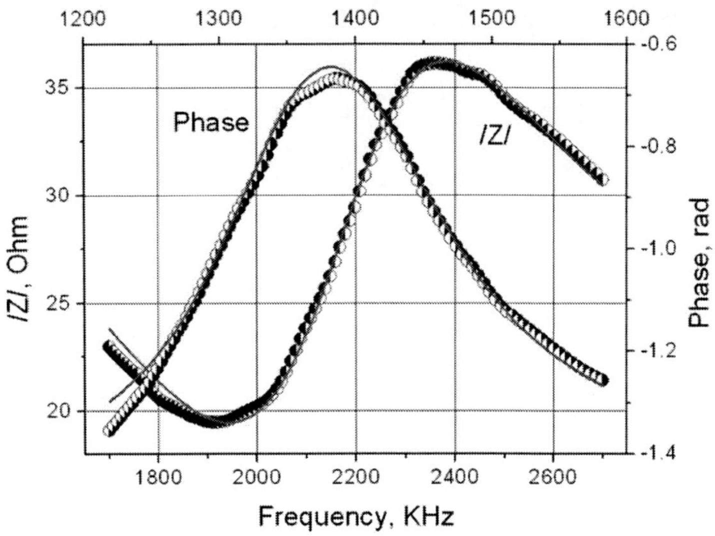

Figure 14. Impedance spectra and PRAP approximations for thickness extensional mode of A850L-20 vol. % disks (Ø19.95 × 0.8 mm^2).

Figure 13 and 14 shows clearly that at low concentrations of scattering particles, the measured impedance spectrum is modulated by individual scattering events with the period corresponding to the average distance between the scattering particles. Increase in scattering particles fraction up to 20 vol. % leads to smoothing of impedance spectra as a result of multiple scattering averaging.

Complex constants of A850L-20 vol. % disk (Ø19.95 × 0.8 mm^2) obtained by using PRAP analysis for thickness extensional and radial modes as well as corresponding IEEE Standard results, are summarized in Tables 3 and 4. Additional physical parameters of A850L-20 are defined as follows: Z_A = 22.8 MRayl, ρ = 6.4 g/cm^3, V_t = 3568 m/s, d_{33}^{quasi} = 300 pC/N, T_C = 340 °C [3].

Figure 15 shows frequency dependencies of real parts of elastic moduli c'^D_{33} and corresponding loss tan $\delta = c''^D_{33}/c'^D_{33}$ for ceramic composites PZT/α-Al_2O_3 with different volume fractions of α-Al_2O_3 measured on standard disks samples by piezoelectric resonance (PRAP) method and simulated by Wave 3000 Pro software.

Table 3. PRAP analysis results for TE mode of A850L-20 disk

PRAP Parameter	Real	Imaginary	IEEE Standard	% Error
f_p (Hz)	$2.26 \cdot 10^6$	340289	$2.29 \cdot 10^6$	1.3
f_s (Hz)	$2.01 \cdot 10^6$	304601	$2.01 \cdot 10^6$	0.0
k_t	0.492734	− 0.00138	0.513653	4.3
c_{33}^D (N/m²)	$8.15 \cdot 10^{10}$	$2.52 \cdot 10^{10}$	$8.58 \cdot 10^{10}$	5.3
c_{33}^E (N/m²)	$6.17 \cdot 10^{10}$	$1.92 \cdot 10^{10}$	$6.31 \cdot 10^{10}$	2.3
e_{33} (C/m²)	10.6185	1.31335	-	
h_{33} (V/m)	$1.91 \cdot 10^9$	$3.29 \cdot 10^8$	-	
ε_{33}^S (F/m)	$5.52 \cdot 10^{-9}$	$-2.64 \cdot 10^{-10}$	-	

Table 4. PRAP analysis results for radial mode A850L-20 disk

PRAP Parameter	Real	Imaginary	IEEE Standard	% Error
f_s^1 (Hz)	105834	1737.27	106037	0.2
f_p^1 (Hz)	110947	1623.38	111192	0.2
f_s^2 (Hz)	275536	4522.35	275194	0.1
s_{11}^E (m²/N)	$1.72 \cdot 10^{-11}$	$-5.66 \cdot 10^{-13}$	$1.75 \cdot 10^{-11}$	1.7
s_{12}^E (m²/N)	$-5.84 \cdot 10^{-12}$	$1.92 \cdot 10^{-13}$	$-6.14 \cdot 10^{-12}$	5.1
$-d_{31}$ (C/N)	$7.35 \cdot 10^{-11}$	$-3.59 \cdot 10^{-12}$	$7.38 \cdot 10^{-11}$	0.4
ε_{33}^T (F/m)	$8.27 \cdot 10^{-9}$	$-2.38 \cdot 10^{-10}$	$8.29 \cdot 10^{-9}$	0.2
k_p	0.339052	− 0.00613	0.340468	0.4
σ^P	0.339174	$8.34 \cdot 10^{-6}$	0.351532	3.6
e_{31} (C/m²)	6.46184	− 0.10351	6.51219	0.8
s_{66}^E (m²/N)	$4.61 \cdot 10^{-11}$	$-1.51 \cdot 10^{-12}$	$4.72 \cdot 10^{-11}$	2.4
c_{66}^E (N/m²)	$2.16 \cdot 10^{10}$	$7.10 \cdot 10^{8}$	$2.12 \cdot 10^{10}$	1.9

It is readily seen that the elastic modulus $c'_{33}{}^D$ increases with frequency because of frequency dispersion caused by the scattering of ultrasonic waves. It also can be seen that the frequency dependencies of the scattering loss tan δ are linear, that corresponds to stochastic scattering type tan $\delta \sim Df$ at $\lambda \sim D$. The decrease in the slope of $c'_{33}{}^D$ versus frequency dependencies at α-Al_2O_3 volume fraction growth is caused by the decrease in the spatial heterogeneity scale and the shift of the dispersion maxima to the high-frequency region. It is obvious from Figures 12 and 15 that Wave 3000 Pro simulation results are in good agreement with experimental results.

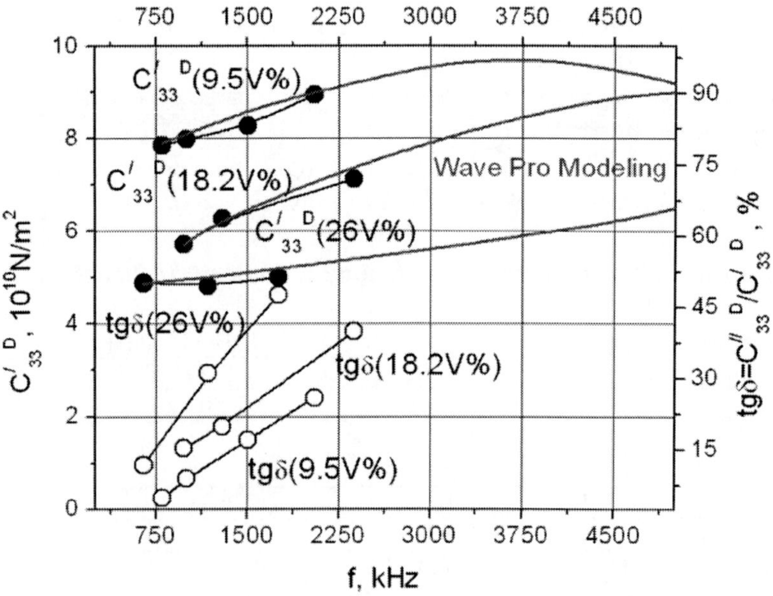

Figure 15. Measured and simulated frequency dependencies of elastic moduli $c'_{33}{}^D$ and corresponding loss tan $\delta = c''_{33}{}^D / c'_{33}{}^D$ for ceramic composites PZT/Al_2O_3 at different volume fractions of α-Al_2O_3.

CONCLUSION

A comprehensive study including finite-difference 3D-simulation, impedance spectroscopy, and pulse-echo ultrasonic measurements of porous ceramics and ceramic matrix composites with strong spatial dispersion and high losses was carried out.

Piezoelectric resonance analysis methods for automatic, iterative evaluation of complex material parameters were presented, along with the full sets of complex constants for porous ceramics and ceramic matrix piezocomposites.

It was shown that pulse-echo measurements of frequency dependencies of elastic properties for dispersive and lossy ceramic composites are inaccurate and ambiguous. In its turn, piezoelectric resonance measurements give accurate and reproducible results that well agree with the results of 3D finite-difference simulations.

Microstructural and physical mechanisms of losses and dispersion in porous ceramics and ceramic matrix piezocomposites were also considered.

In conclusion, the advantages of low-Q ceramic matrix composites were defined with reference to wide-band ultrasonic transducers applications.

REFERENCES

[1] Wersing, W. In: *Piezoelectric Materials in Devices*; Setter, N.; Swiss Federal Institute of Technology: Lausanne, Switzerland, 2002; pp. 29-66.

[2] Rybyanets, A.N. *IEEE Trans.UFFC*, 2011, *vol. 58*, 1492-1507; Rybyanets, A. N. *Ferroelectrics*, 2011, *vol. 419*, 90-96.

[3] Rybyanets, A. N.; Rybyanets, A. A. *IEEE Trans. UFFC*, 2011, *vol. 58*, 1757-1774.

[4] Rybyanets, A. N. In: *Piezoceramic Materials and Devices*, Ivan A. Parinov (Ed.). Nova Science Publishers Inc. NY, 2010, Chapter 3, pp. 113-174.

[5] Rybyanets, A. N. In: *Piezoelectrics and Related Materials: Investigations and Applications*, Ivan A. Parinov (Ed.). Nova Science Publishers Inc. NY, 2012, Chapter 5, pp.143-187.

[6] Rybjanets, A. N.; Nasedkin, A. V.; Turik A. V. *Integrated Ferroelectrics*, 2004, *vol. 63*, 179-182.

[7] Rybyanets, A. N.; Naumenko, A. A. In: *Physics and Mechanics of New Materials and Their Applications*, Ivan A. Parinov, Shun Hsiung-Chang (Eds.). Nova Science Publishers Inc., NY, 2013, Chapter 1, pp. 3-18.

[8] Rybyanets, A. N.; Zakharov, Y. N.; Raevskii, I. P.; Akopjan, V. A.; Rozhkov E. V.; Parinov I. A. In: *Physics and Mechanics of New Materials and Their Applications*, Ivan A. Parinov, Shun Hsiung-Chang (Eds.). Nova Science Publishers Inc. NY, 2013, Chapter 22, pp. 275-308.

[9] Wersing, W.; Lubitz, K.; Moliaupt, *Ferroelectrics,* 1986, *vol. 68*, 77-79.

[10] *Rybyanets, A. N.* Ferroelectrics, *2011,* vol.419*, 90-96.*

[11] Rybyanets, A. N.; Naumenko, A.A., *Journal of Modern Physics*. 2013, *vol. 4*, 1041-1049. http://dx.doi.org/10.4236/jmp.2013.48140

[12] Rybjanets, A. N.; Razumovskaja, O. N.; Reznitchenko, L. A.; Komarov, V. D.; Turik, A. V. *Integrated Ferroelectrics*, 2004, *vol. 63*, 197-200.

[13] Xiang, P.-H.; Dong, X.-L.; Chen, H.; Zhang, Z.; Guo, J.-K. *Ceramics International*, 2003, *vol. 29*, 499-503.

[14] Rybianets, A. N. *Ferroelectrics*, 2007, *vol. 360*, 84-89.

[15] Nasedkin, A.; Rybjanets, A.; Kushkuley, L.; Eshel, Y., *Proc. 2005 IEEE Ultrason. Symp.* 2005, pp. 1648-1651.

[16] Rybianets, A.; Kushkuley, L.; Eshel, Y.; Nasedkin A., *Proc. 2006 IEEE Ultrason. Symp.* 2006, pp. 1533-1536.

[17] Rybyanets, A. N.; Naumenko, A. A.; Shvetsova, N. A., In.: *Nano- and Piezoelectric Technologies, Materials and Devices*, Ivan A. Parinov (Ed.). Nova Science Publishers Inc. NY, 2013, Chapter 1, pp. 1-50.

[18] *IEEE Standard on Piezoelectricity*. ANSI/IEEE Std. 1987, 176.

[19] Alguero, M.; Alemany, C.; Pardo L.; Gonzalez A. M., *J. Am. Ceram. Soc.*, 2004, *vol. 87,* 209.

[20] Sherrit, S.; Wiederick, H. D.; Mukherjee, B. K. *SPIE Proceedings*. 1997, *vol. 3241*, 327.

[21] Rybianets, A. N.; Tasker, R., *Ferroelectrics*, 2007, *vol. 360*, 90-95.

[22] Alemany, C.; Gonzalez, A.M.; Pardo, L.; Jimenez, B.; Carmona F.; Mendiola, J. *J. Phys. D: Appl. Phvs.,* 1995, *vol. 28*, 945.

[23] Smits, J. G. *IEEE Trans. Sonics Ultrason.*, 1976, *vol. SU-23,* 393-402.

[24] *PRAP (Piezoelectric Resonance Analysis Program)*. TASI Technical Software Inc. www.tasitechnical.com.

[25] *Rybianets, A. N.; Nasedkin A. V.* Ferroelectrics*, 2007,* vol. 360*, 57-62.*

[26] Sherrit, S.; Wiederick, H. D.; Mukherjee, B.K. *Ferroelectrics.* 1992, *vol. 134*, 111.

[27] Pardo, L.; Alemany, C.; Ricote, J.; Moure, A.; Poyato, R.; Alguero, M. *Proc. Intern. Conf. Material Technology and Design of Integrated Piezoelectric Devices.* Couurmayeur, Italy, 2004, p. 145.

[28] Devonshire, A. F. *Philosophical Magazine Supplement*. 1954, *vol. 3*, 85-130.

[29] Mason, W. P. *Physical Acoustics and the Properties of Solids;* D. Van Nostrand Co. Inc.: Princeton, 1958.

[30] Berlincourt, D. A.; Curran, D. R.; Yaffe, H. In: *Physical Acoustics I, Part A;* Mason, W. P.; Academic Press: NY, 1964.

[31] Holland, R. *IEEE Trans.Sonics Ultrason.*, 1967, *vol. SU-14*, 18.

[32] Lopatin, S. S.; Lupeiko, T. G. *Izv. AN USSR. Ser. Neorg. Mat.* 1991, *vol. 27*, 1948-1951.

[33] Sarvazyan, Armen; Ostrovsky, Lev A.; Rubyanets, Andrey *Acoustical Society of America. Proceedings on Meeting on Acoustics,* 2009, *vol.6*.

[34] Rybyanets, Andrey; Nasedkin, Andrey; Domashenkina, Tamara; Rybyanets, Anastasia; Lugovaya, Maria, *Proc. 2009 IEEE Ultrason. Symp.*, 2009, pp. 1699-1702.

[35] Rybyanets, Andrey; Nasedkin, Andrey; Rybyanets, Anastasia; Lugovaya, Maria; Domashenkina, Tamara, *Proc. 2009 IEEE Ultrason. Symp.,* 2009, pp. 1765-1768.

[36] Rybianets, A.; Eshel, Y.; Kushkuley, L., *Proc. 2006 IEEE Ultrason. Symp.*, 2006, pp. 1911-1914.

[37] Rybyanets, A., *Proc. 2007 IEEE Ultrason. Symp.*, 2007, pp. 1909-1912.

[38] Goland, V.; Kushkuley, L.; Mirman, S.; Zadok, Y.; Ben-Ezra, S.; Shalgi, A.; Rybianets, A., *Proc. 2007 IEEE Ultrason. Symp.*, 2007, pp. 1305-1308.

[39] Lee, C. K.; Chiang, A. S. T.; Tsay, C. S. In: *Key Engineering Materials*; Dean-Mo Liu (Ed.); Trans. Tech. Publications: Switzerland, *vol. 115*, 1996, 21-44.

[40] Kara, H.; Ramesh, R.; Stevens, R.; Bowen, C. R., *IEEE Trans. UFFC,* 2003, *vol. 50*, 289-296.

[41] Lethiecq, M.; Levassort, F.; Trans-Huu-Hue L.P., *IEEE Ultrason. Symp. Proc.*, 2004, 1153-1156.

[42] Lee, C. K.; Chiang, A. S. T.; Tsay, C. S. In: *Key Engineering Materials*; Dean-Mo Liu (Ed.); Trans. Tech. Publications: Switzerland, *vol. 115*, 1996, 21-44.

[43] *Wave 3000 Pro (Software for 3D Ultrasound Simulation)*. CyberLogic Inc. www.cyberlogic.org.

[44] Delsanto, P. P.; Schechter, R. S.; Mignogna, R. B. *Journal Wave Motion.* 1997, *vol. 26*, 329-339.

In: Advanced Materials - Studies and Applications
Editors: I.A. Parinov, S.-H. Chang, S.Theerakulpisut
ISBN: 978-1-63463-749-7

Chapter 11

OPTIMIZATION OF FINITE ELEMENT MODELS FOR POROUS CERAMIC PIEZOELEMENTS BY PIEZOELECTRIC RESONANCE ANALYSIS METHOD

A. N. Rybyanets[1,*], A. V. Nasedkin[2,†], A. A. Naumenko[1], N. A. Shvetsova[1], M. A. Lugovaya[1] and E. I. Petrova[1]
[1]Physics Research Institute, Southern Federal University, Rostov-on-Don, Russia
[2]Vorovich Mathematics, Mechanics and Computer Sciences Institute, Southern Federal University, Rostov-on-Don, Russia

ABSTRACT

Theoretical aspects of the effective moduli method for an inhomogeneous piezoelectric media were examined. Different models of representative volume were considered. Corresponding equations for calculation of effective moduli of piezoelectric media with arbitrary anisotropy were derived. Based on these equations and using finite element method (FEM) the full set of effective moduli for PZT porous ceramics having wide porosity range was calculated. A novel approach for optimization of finite element modeling (FEM) of lossy piezoceramic elements has been proposed. The procedure of optimization has consisted in sequential and iterative application of FEM and piezoelectric resonance analysis (PRAP) to complex electric impedance spectra of piezoceramic elements. For validation of proposed optimization procedure FEM calculations of standard shape piezoelements (disks, bars and rods) made from dense and porous PZT-type piezoceramics were fulfilled using FEM ANSYS package. Comparison shows a good agreement between initial and calculated complex sets of material constants including losses ($Q = c^{/}/c^{//}$, where $c^{/}$ and $c^{//}$ real and imaginary parts of related material constants). In the case of very low-Q materials, additional corrections of material constants (effective Q for ANSYS) with recurring of optimization procedure can be fulfilled.

* Corresponding author: A. N. Rybyanets. Physics Research Institute, Southern Federal University, 194, Stachki Ave, 344090, Rostov-on-Don, Russia. E-mail: arybyanets@gmail.com.
† A. V. Nasedkin: Vorovich Mathematics, Mechanics and Computer Sciences Institute, Southern Federal University, 8a, Milchakova Street, 344090, Rostov-on-Don, Russia.

1. Introduction

In recent years low-Q piezoceramics and piezocomposite materials are widely used for wide-band medical and NDT ultrasonic transducers with high sensitivity and resolution [1–7].

Porous piezoceramics based on different piezoceramic compositions were proposed as a promising candidate for lead metaniobate ceramics replacement in wide-band ultrasonic transducers and sensors [2,3]. Porous piezoceramics have great advantage in low acoustic impedance and higher efficiency compared to conventional dense piezoceramics. Despite their long history and unique characteristics, so far porous piezoelectric ceramics have not been commercialized and their application in transducers and devices is limited.

Intensive research and technological work, as well as improvements in fabrication methods, have allowed large-scale manufacture of porous piezoelectric ceramics with reproducible and controllable porosity and properties [3,8].

The material properties of porous piezoceramics with mixed connectivities (3–0, 3–3 and 0–3) have been evaluated using different theoretical models [9–11].

After the long observation regarding pore effects on material properties, the porosity-property interrelation seems not yet clear. To describe real porous piezoceramics it was necessary to introduce models based upon the concepts of "connectivity", "percolation", "tortuosity" and, more recently, "fractal geometry" [10,12]. It is necessary to note, that despite of rather good conformity of some modeling parameters to experimental data, none of theoretical models truly describes a real porous piezoceramics [2,10]. The situation becomes even more complicated taking into account the inconsistent experimental data for different porous ceramics available in the literature [10,13–16]. It is caused by an objective technological and experimental problems (reproducibility of structure and properties, measurements of low-Q material parameters), as well as using different and frequently imperfect methods of porous ceramics fabrication and/or incorrect porosity estimations [10,17].

The majority of new advanced ceramics and composite materials are lossy and direct use of IEEE Standards [18] for material constant determination can lead to significant errors. Furthermore, the current IEEE Standard does not comprehensively account for the complex nature of material coefficients, as it uses only the dielectric loss factor (tan δ) and the mechanical quality factor (Q_M) to account for loss. Numerous techniques using complex material constants have been proposed to take into account losses in low-Q_M materials and to overcome limitations in the IEEE Standard [19–24]. Iterative methods [20,21,23] provide a means to accurately determine the complex coefficients in the linear range of poled piezoceramics from complex impedance resonance measurements. The piezoelectric resonance analysis method and program (PRAP) has been proposed for the full set of standard geometries and resonance modes needed to complete complex characterization in a wide range of materials with very high and moderate losses [25–28].

The modeling and design of piezoelectric devices by finite element methods, among others, relies on the accuracy of the dielectric, piezoelectric and elastic coefficients of the active material used – commonly an anisotropic ferroelectric polycrystal. The accurate description of piezoceramics must include the evaluation of the dielectric, piezoelectric and mechanical losses, accounting for the out-of-phase material response to the input signal [19].

Standard FEM packages that are widely used for modeling of piezoelements and devices don't take into account losses (mechanical losses can be included in calculations by implicit manner). Sets of material constants used for FEM calculations also do not contain losses data, except of Q_M for the radial mode of vibrations.

As a result, FEM calculations of real piezoelements and devices can give inadequate results for lossy materials (composites, porous ceramics etc.) [26].

This chapter presents a novel approach for optimization of finite element modeling of lossy piezoceramic elements.

The chapter is divided into three informative sections. Section 2 describes different approaches to finite element modeling of porous piezoceramics. Section 3 presents piezoelectric resonance analysis method and software for characterization of lossy piezoelectric materials. The final Section 4 introduces a novel approach for optimization of FEM modeling of lossy piezoceramic elements as well as experimental results for different porous ceramics.

2. Different Approaches to Finite Element Modeling of Effective Moduli of Porous Piezoceramics

The most promising and powerful method currently available for porous ceramics modeling is the finite element method [15,26,29], which provides not only adequate description and prediction of the properties of porous ceramics, but also the modeling of the electrophysical parameters of porous piezoelements and devices [26]. Commercial software packages (ANSYS and PiezoFlex) can be used for these calculations. Different approaches to finite element modeling of porous piezoceramics and composites were proposed, one of which is effective moduli method [15].

2.1. Mathematical Aspects of the Effective Moduli Method

Let Ω be a region occupied by the heterogeneous body with piezoelectric properties, $\Gamma = \partial\Omega$ is the boundary of the region, $\mathbf{n(x)}$ is the vector of the external unit normal to Γ ($\mathbf{x} \in \Gamma$), $\mathbf{u(x)}$ is the displacement vector-function, $\varphi(\mathbf{x})$ is the electric potential function. We denote by $\boldsymbol{\varepsilon}$ the deformation tensor, $\boldsymbol{\sigma}$ the stress tensor, $\mathbf{E}$ the electric field vector, $\mathbf{D}$ the electric displacement vector. The fields ε and $\mathbf{E}$ are determined from the functions $\mathbf{u}$ and φ by the following form:

$$\boldsymbol{\varepsilon} = (\nabla\mathbf{u} + \nabla\mathbf{u}^*)/2, \quad \mathbf{E} = -\nabla\varphi. \tag{1}$$

On the region Γ we will consider the mechanical stress vector $\mathbf{p}$ and surface density of electric charges q:

$$\mathbf{p} = \mathbf{n} \cdot \boldsymbol{\sigma}, \tag{2}$$

$$q = -\mathbf{n} \cdot \mathbf{D} . \tag{3}$$

As usual, we will denote the volume–averaged quantities in the broken brackets:

$$\langle ... \rangle = \frac{1}{\Omega} \int_{\Omega} (...) \, d\Omega . \tag{4}$$

In concordance with four equivalent basic forms of constitutive equations [30] we will introduce the moduli of piezoelectric medium $\mathbf{c}^{E}$, $\mathbf{e}$, $\boldsymbol{\varepsilon}^{S}$ and so on:

$$\boldsymbol{\sigma} = \mathbf{c}^{E} \cdot\cdot \boldsymbol{\varepsilon} - \mathbf{e}^{*} \cdot \mathbf{E} , \ \mathbf{D} = \mathbf{e} \cdot\cdot \boldsymbol{\varepsilon} + \mathbf{э}^{S} \cdot \mathbf{E} ; \tag{5}$$

$$\boldsymbol{\varepsilon} = \mathbf{s}^{E} \cdot\cdot \boldsymbol{\sigma} + \mathbf{d}^{*} \cdot \mathbf{E} , \ \mathbf{D} = \mathbf{d} \cdot\cdot \boldsymbol{\sigma} + \mathbf{э}^{T} \cdot \mathbf{E} ; \tag{6}$$

$$\boldsymbol{\sigma} = \mathbf{c}^{D} \cdot\cdot \boldsymbol{\varepsilon} - \mathbf{h}^{*} \cdot \mathbf{D} , \ \mathbf{E} = -\mathbf{h} \cdot\cdot \boldsymbol{\varepsilon} + \boldsymbol{\beta}^{S} \cdot \mathbf{E} ; \tag{7}$$

$$\boldsymbol{\varepsilon} = \mathbf{s}^{D} \cdot\cdot \boldsymbol{\sigma} + \mathbf{g}^{*} \cdot \mathbf{D} , \ \mathbf{E} = -\mathbf{g} \cdot\cdot \boldsymbol{\sigma} + \boldsymbol{\beta}^{T} \cdot \mathbf{D} ; \tag{8}$$

For inhomogeneous body these moduli will be functions of coordinates $\mathbf{x}$: $\mathbf{c}^{E} = \mathbf{c}^{E}(\mathbf{x})$, $\mathbf{e} = \mathbf{e}(\mathbf{x})$, $\boldsymbol{\varepsilon}^{S} = \boldsymbol{\varepsilon}^{S}(\mathbf{x})$ etc.

Let Ω be the representative volume of heterogeneous piezoelectric materials. We will determine the effective moduli $\mathbf{c}^{Eeff}$, $\mathbf{e}^{eff}$, $\boldsymbol{\varepsilon}^{Seff}$ by the following technique [15,31].

Consider the static piezoelectric problem for representative volume Ω:

$$\nabla \cdot \boldsymbol{\sigma} = 0 , \ \nabla \cdot \mathbf{D} = 0 , \ \mathbf{x} \in \Omega ; \tag{9}$$

$$\mathbf{u} = \mathbf{x} \cdot \boldsymbol{\varepsilon}_{0} , \ \varphi = -\mathbf{x} \cdot \mathbf{E}_{0} , \ \mathbf{x} \in \Gamma = \partial\Omega . \tag{10}$$

We shall name the problem (1), (5), (9), (10) as $u\phi$–problem and denote the solution of this problem by $\mathbf{u}^{u\varphi}$, $\varphi^{u\varphi}$. From obtained solution and (1), (5) we find $\boldsymbol{\varepsilon}^{u\varphi}$, $\mathbf{E}^{u\varphi}$, $\boldsymbol{\sigma}^{u\varphi}$ and $\mathbf{D}^{u\varphi}$, where $\boldsymbol{\varepsilon}^{u\varphi} = \boldsymbol{\varepsilon}(\mathbf{u}^{u\varphi})$ etc. Note [9], that for $u\phi$–problem $\langle \boldsymbol{\varepsilon}^{u\varphi} \rangle = \boldsymbol{\varepsilon}_{0}$ and $\langle \mathbf{E}^{u\varphi} \rangle = \mathbf{E}_{0}$.

Put in conformity to the initial heterogeneous medium some "equivalent" homogeneous medium with effective moduli $\mathbf{c}^{Eeff}$, $\mathbf{e}^{eff}$ and $\boldsymbol{\varepsilon}^{Seff}$. The constitutive equations for the "equivalent" medium analogous (5) are in the forms:

$$\boldsymbol{\sigma}_{0} = \mathbf{c}^{Eeff} \cdot\cdot \boldsymbol{\varepsilon}_{0} - \mathbf{e}^{eff*} \cdot \mathbf{E}_{0} , \ \mathbf{D}_{0} = \mathbf{e}^{eff} \cdot\cdot \boldsymbol{\varepsilon}_{0} + \mathbf{э}^{Seff} \cdot \mathbf{E}_{0} . \tag{11}$$

For $u\phi$–problem we accept the following equations as relations for definition of effective moduli from (11):

$$\langle \boldsymbol{\sigma}^{u\varphi} \rangle = \boldsymbol{\sigma}_0, \ \langle \mathbf{D}^{u\varphi} \rangle = \mathbf{D}_0. \qquad (12)$$

The moduli that were found from these conditions are marked by superscrips "$u\phi$". This variant of equivalent moduli method for piezoelectric media follows from [31]. However, for piezoelectric media it is possible to offer other ways to introduce the effective moduli considering the problems with other mechanical and electric boundary conditions. Namely, we can consider the following problems [15]:

(i) $p\varphi$–problem with boundary condition for stress vector $\mathbf{p}$ and electric potential φ:

$$\mathbf{p} = \mathbf{n} \cdot \boldsymbol{\sigma}_0, \ \varphi = -\mathbf{x} \cdot \mathbf{E}_0, \ \mathbf{x} \in \Gamma; \qquad (13)$$

(ii) uq–problem with boundary condition for displacement $\mathbf{u}$ and surface density of electric charges q:

$$\mathbf{u} = \mathbf{x} \cdot \boldsymbol{\varepsilon}_0, \ q = -\mathbf{n} \cdot \mathbf{D}_0, \ \mathbf{x} \in \Gamma; \qquad (14)$$

(iii) pq–problem with boundary condition for stress vector $\mathbf{p}$ and surface density of electric charges q

$$\mathbf{p} = \mathbf{n} \cdot \boldsymbol{\sigma}_0, \ q = -\mathbf{n} \cdot \mathbf{D}_0, \ \mathbf{x} \in \Gamma. \qquad (15)$$

From the set of solutions for these problems it is possible to define all effective moduli of porous piezoceramics. For piezoceramics homogeneously polarized along the axis x_3 for each type of problems it is enough to consider five various tasks. So, for $u\varphi$–problem we have the following tasks and formulas for calculation of moduli in tensor and two-index [30] symbols:

I-$u\phi$ $\qquad \boldsymbol{\varepsilon}_0 = \varepsilon_0 \mathbf{e}_1 \mathbf{e}_1, \ (\varepsilon_{011} = \varepsilon_0), \ \mathbf{E}_0 = 0$

$$\varepsilon_{011} \neq 0 \Rightarrow C_{11}^{E\,\text{eff},u\varphi} = \langle \sigma_{11}^{u\varphi} \rangle / \varepsilon_{011}; \ C_{12}^{E\,\text{eff},u\varphi} = \langle \sigma_{22}^{u\varphi} \rangle / \varepsilon_{011}$$

$$C_{13}^{E\,\text{eff},u\varphi} = \langle \sigma_{33}^{u\varphi} \rangle / \varepsilon_{011}; \ e_{31}^{\text{eff},u\varphi} = \langle D_3^{u\varphi} \rangle / \varepsilon_{011}$$

II-$u\phi$ $\qquad \boldsymbol{\varepsilon}_0 = \varepsilon_0 \mathbf{e}_3 \mathbf{e}_3, \ (\varepsilon_{033} = \varepsilon_0), \ \mathbf{E}_0 = 0$

$$\varepsilon_{033} \neq 0 \Rightarrow C_{33}^{E\,\text{eff},u\varphi} = \langle \sigma_{33}^{u\varphi} \rangle / \varepsilon_{033}; \ e_{33}^{\text{eff},u\varphi} = \langle D_3^{u\varphi} \rangle / \varepsilon_{033}$$

$$C_{13}^{E\,\mathrm{eff},u\varphi} = \left\langle \sigma_{11}^{u\varphi} \right\rangle / \varepsilon_{033} = \left\langle \sigma_{22}^{u\varphi} \right\rangle / \varepsilon_{033}$$

III-$u\phi$ $\quad \boldsymbol{\varepsilon}_0 = \varepsilon_0 (\mathbf{e}_2\mathbf{e}_3 + \mathbf{e}_3\mathbf{e}_2)$, $(\varepsilon_{023} = \varepsilon_{032} = \varepsilon_0)$, $\mathbf{E}_0 = 0$

$$\varepsilon_{023} \neq 0 \Rightarrow C_{44}^{E\,\mathrm{eff},u\varphi} = \left\langle \sigma_{33}^{u\varphi} \right\rangle / (2\varepsilon_{023});$$

$$e_{15}^{\mathrm{eff},u\varphi} = \left\langle D_2^{u\varphi} \right\rangle / (2\varepsilon_{023})$$

IV-$u\phi$ $\quad \boldsymbol{\varepsilon}_0 = 0$, $\mathbf{E}_0 = E_0 \mathbf{e}_1$ $(E_{01} = E_0)$

$$E_{01} \neq 0 \Rightarrow e_{15}^{\mathrm{eff},u\varphi} = -\left\langle \sigma_{13}^{u\varphi} \right\rangle / E_{01};\ Э_{11}^{S\,\mathrm{eff},u\varphi} = \left\langle D_1^{u\varphi} \right\rangle / E_{01}$$

V-$u\phi$ $\quad \boldsymbol{\varepsilon}_0 = 0$, $\mathbf{E}_0 = E_0 \mathbf{e}_3$ $(E_{03} = E_0)$

$$E_{03} \neq 0 \Rightarrow e_{31}^{\mathrm{eff},u\varphi} = -\left\langle \sigma_{11}^{u\varphi} \right\rangle / E_{03} = -\left\langle \sigma_{22}^{u\varphi} \right\rangle / E_{03}$$

$$e_{33}^{\mathrm{eff},u\varphi} = -\left\langle \sigma_{33}^{u\varphi} \right\rangle / E_{03};\ Э_{33}^{S\,\mathrm{eff},u\varphi} = \left\langle D_3^{u\varphi} \right\rangle / E_{03}$$

It is easy to receive similar formulas for $p\varphi$, uq and pq–problems (13), (14), (15).

Using of different constitutive equations from four types of problems can be useful for determination of effective moduli of the inhomogeneous structures working in one-dimensional mode of vibrations, for example, piezoceramic rods, plates and disks with piezoeletrically "soft" and "hard" modes, etc.

2.2. Models of Representative Volumes and Non-Uniform Polarization

Above mentioned formulas for calculation of effective moduli assume the solution of corresponding boundary problems of electroelasticity in region Ω that should be representative volume of porous piezoceramics. As an ideal representative volume it is necessary to take areas, big enough compared to the sizes of heterogeneity (i.e., pores), but small compared to the distances on which averaged variables varies essentially.

Nevertheless, as the first approximation it is possible to take area Ω in the form of a cube with one cubic pore inside. We shall name this model as the model 1. For the analysis of corners influence on the calculation results it is possible to consider model without angular points: spherical volume with a spherical cavity, for example, or as more general case - ellipsoid with ellipsoidal cavity. We shall name this model as the model 2.

More adequate model for the real porous piezoceramics structure at low relative porosity is a model of piezoceramic cube evenly divided into lesser cubes, part of which randomly declared as pores (model 3). Note that the similar model was considered in [31].

Obviously that even in simplest cases (models 1 and 2) solving of the boundary problems of electroelasticity (9), (10) or similar problems using an analytical methods is exceedingly difficult and for the model 3 these boundary problems can be solved only numerically. In accordance with modern standards the most effective method for numerical solution of the electroelasticity problems for representative volumes of heterogeneous media is finite element method (FEM).

It is possible to examine more complex models 3 taking into account partial polarization of the ceramics in pores vicinity [15]. In this case in first stage the process of partial polarization of porous ceramics in z-axis direction is modeled.

For this at first a quasielectrostatic problem for inhomogeneous dielectric is solved **Ω**:

$$\nabla \cdot \mathbf{D} = 0,\ \mathbf{D} = \mathbf{э} \cdot \mathbf{E},\ \mathbf{E} = -\nabla\varphi,\ \mathbf{x} \in \Omega; \tag{16}$$

$$\varphi = V_j,\ \mathbf{x} \in \Gamma_{\varphi j}; \tag{17}$$

$$\mathbf{n} \cdot \mathbf{D} = 0,\ \mathbf{x} \in \Gamma_q, \tag{18}$$

where $\Gamma = \Gamma_{\varphi 1} \cup \Gamma_{\varphi 2} \cup \Gamma_q$; $\Gamma_{\varphi 1}$, $\Gamma_{\varphi 2}$ are the electrode $z = L$ and $z = 0$, respectively, $V_1 = -E_p L$, $V_2 = 0$, E_p is the value of polarization field.

Heterogeneity of the dielectric can be defined by the following way. At fixed relative porosity **p** number of cubic finite elements n_p representing pores is calculated for the model 3 using the formula: $n_p = n_{el} p/100$.

Then the material properties with dielectric constant of vacuum are randomly assigned to n_p elements from macrocube ε_0. Using electrostatic finite elements with different material properties, problem (16) – (18) is solved. Then the calculated average values of the field E_{zm} for each finite element that is not a pore are analyzed. Marks of polarization degree are assigned to the elements if $E_{zm} > E_c$, where $E_c = k_c E_p$, $k_c \leq 1$ is the some fixed value.

It is possible also propose a "hyper-polarization" hypothesis for individual elements for which polarization field E_{zm} exceeds some fixed value of $E_{zm} > E_s$, where $E_s = k_s E_p$, $k_s \geq 1$ is the fixed value of the "hyper-polarization" field. Additional factor m_s is assigned to the "hyper-polarization" elements and used for piezoceramics moduli calculation by formulas: $c^{Es}_{\alpha\beta} = (c^E_{\alpha\beta} - c^I_{\alpha\beta})m_s + c^I_{\alpha\beta}$, $e^s_{i\alpha} = e_{i\alpha} m_s$, $\varepsilon^{Ss}_{ij} = (\varepsilon^E_{ij} - \varepsilon^I_{ij})m_s + \varepsilon^I_{ij}$, where $c^I_{\alpha\beta}$, ε^I_{ij} are the moduli of non-polarized ceramics.

In the second modeling stage the electrostatic finite elements are modified to the elements with piezoelectric analysis possibility. Four types of material properties are assigned to the new elements, namely: (i) properties of polarized piezoceramics for polarized elements, (ii) omissible piezoelectric moduli for pores, (iii) properties of isotropic elastic dielectric for non-polarized ferroelectric ceramics, (iv) properties of "hyper-polarization" piezoceramics for "hyper-polarization" elements. Then for effective moduli definition the electroelasticity problem for representative volume is solved using methods described in previous section.

To describe fully closed 3–0 type porosity it is possible to complicate the model 3 using partially controllable porosity structure. For example, it is possible to subdivide elementary compound cube on piezoelectric skeleton and lesser cube. The last one can be pore or medium with the same piezoelectric properties as the skeleton ones. To obtain given porosity, the part of lesser cubes randomly declared as pores. As a result we receive new model 4 with fully closed porosity (3–0 connectivity type).

For 3-3 connectivity type modeling, two additional models were developed. The model 5 named 3–0 - 3–3 connectivity model is a future development of the model 4. In this model despite large cubes in skeleton thin parallelepipeds also can be pores. The model 6 (3–3 connectivity type) constructed by the following manner. The cube constructed by translation of equal structured cells along three directions is considered as a representative volume. Each cell in its turn also represents cube consisting of 101010 cubic elements. Connected skeleton with the elements in the cube corners always exists in the cell. The skeleton consists of parallelepiped represented by own edges (linear dimensions are pointed out by the randomizer) as well as of elements chains connecting their corners with the corners of the main cell. Connecting elements chains are generated by the randomizer also. The skeleton occupied 10 % of cell volume. So, maximum possible porosity that can be achieved in this model runs to 90 % (see Figure 1).

3. Impedance Spectroscopy Characterization of Highly Attenuating Piezocomposites

The basic techniques for finding material constants of piezoelectric materials are outlined in the IEEE Standard on Piezoelectricity (1987) [18]. These methods work for many of the most widely used commercial piezoceramics based on lead-titanate-zirconate (PZT) compositions that are high-Q_M and high-coupling coefficient piezoelectric materials.

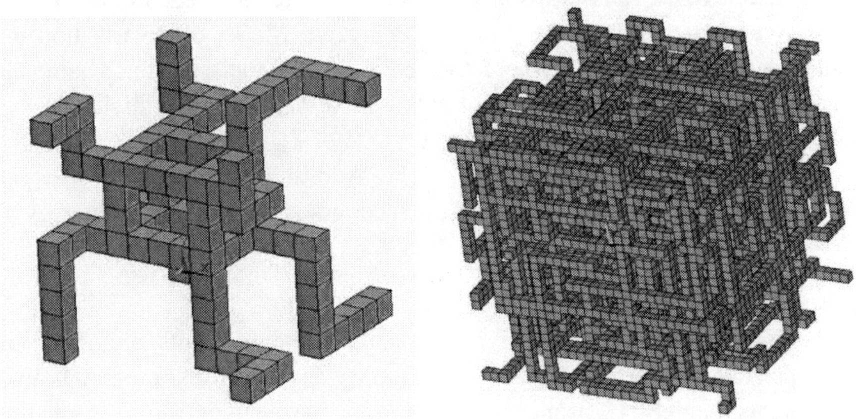

Figure 1. Skeleton and representative volume example for 3-3 connectivity model at relative porosity of 80 %.

However, there is a general agreement that their use in many new piezoelectric materials such as porous ceramics, piezoelectric polymers or piezoelectric composites may lead to significant errors. Furthermore, the current IEEE Standard does not comprehensively account for the complex nature of material coefficients, as it uses only the dielectric loss factor (tan δ) and the mechanical quality factor (Q_M) to account for loss.

Numerous techniques using complex material constants have been proposed to take into account losses in low-Q_M materials and to overcome limitations in the IEEE Standard [19–24]. Iterative methods [20–23] provide a means to accurately determine the complex coefficients in the linear range of poled piezoceramics from complex impedance resonance measurements. The Piezoelectric Resonance Analysis Program (PRAP) automatic iterative method [25] has been applied to the full set of standard geometries and resonance modes needed to complete complex characterization in a wide range of materials with very high and moderate loss factors [26–28].

3.1. Piezoelectric Resonance Analysis Methods

In the direct piezoelectric effect, charge is generated from applying a stress to a specimen. In the converse piezoelectric effect, the specimen will undergo strain in response to an electrical field. When four possible thermodynamic potentials are considered for adiabatic, isothermal and low amplitude conditions [32,33], the linear equations of piezoelectricity give the relationship between electric field (E), dielectric displacement (D), strain (S) and stress (T):

$$\begin{aligned} S_p &= s_{pq}^D T_q + g_{pm} D_m \\ E_m &= \beta_{mn}^T D_n - g_{pm} T_p \end{aligned} \tag{19}$$

$$\begin{aligned} S_p &= s_{pq}^E T_q + d_{pm} E_m \\ D_m &= \varepsilon_{mn}^T E_n + d_{pm} T_p \end{aligned} \tag{20}$$

$$\begin{aligned} T_p &= c_{pq}^E S_q - e_{pm} E_m \\ D_m &= \varepsilon_{mn}^S E_n + e_{pm} S_p \end{aligned} \tag{21}$$

$$\begin{aligned} T_p &= c_{pq}^D S_q - h_{pm} D_m \\ E_m &= \beta_{mn}^S D_n - h_{pm} S_p \end{aligned} \tag{22}$$

In Equations (19–22) d, e, g, and h are the piezoelectric constants, s and c are the elastic compliance and stiffness at constant E and D, and ε and β are the permittivity and inverse permittivity at constant S and T. The mechanical variables have three additional shear directions (4, 5, 6) while the electrical variables have only three standard directions (1, 2, 3).

Each of the sets of equations above is typically represent as a piezoelectric matrix with the transpose of the piezoelectric constants in the top right corner. Symmetry in materials means that many of the elements of the piezoelectric matrix are either equal, zero, or some function of themselves. Poled polycrystalline materials are of C_∞ or 6mm symmetry, and similar to the other sets of equations; Equation set (20) reduces to a form of the piezoelectric matrix that includes only the following 10 independent constants:

$$\begin{bmatrix} S_1 \\ S_2 \\ S_3 \\ S_4 \\ S_5 \\ S_6 \\ D_1 \\ D_2 \\ D_3 \end{bmatrix} = \begin{bmatrix} s_{11}^E & s_{12}^E & s_{13}^E & 0 & 0 & 0 & 0 & 0 & d_{31} \\ s_{12}^E & s_{11}^E & s_{13}^E & 0 & 0 & 0 & 0 & 0 & d_{31} \\ s_{13}^E & s_{13}^E & s_{33}^E & 0 & 0 & 0 & 0 & 0 & d_{33} \\ 0 & 0 & 0 & s_{55}^E & 0 & 0 & 0 & d_{15} & 0 \\ 0 & 0 & 0 & 0 & s_{55}^E & 0 & d_{15} & 0 & 0 \\ 0 & 0 & 0 & 0 & 0 & 2\left(s_{11}^E - s_{12}^E\right) & 0 & 0 & 0 \\ 0 & 0 & 0 & 0 & d_{15} & 0 & \varepsilon_{11}^T & 0 & 0 \\ 0 & 0 & 0 & d_{15} & 0 & 0 & 0 & \varepsilon_{11}^T & 0 \\ d_{31} & d_{31} & d_{33} & 0 & 0 & 0 & 0 & 0 & \varepsilon_{33}^T \end{bmatrix} \begin{bmatrix} T_1 \\ T_2 \\ T_3 \\ T_4 \\ T_5 \\ T_6 \\ E_1 \\ E_2 \\ E_3 \end{bmatrix} \tag{23}$$

In general, the material constants are temperature (relative to the Curie temperature for ferroelectrics), time (logarithmic relaxation since poling) and frequency dependant. In reality, the properties may also be field dependant, which would require a re-derivation of the piezoelectric equations for higher order field terms.

Impedance spectroscopy characterization stems from the relation for impedance Z:

$$\mathbf{Z} = V/\mathbf{I} = \int \mathbf{E} \cdot d\mathbf{x} \Big/ A\frac{d\mathbf{D}}{dt}, \tag{24}$$

where V is the voltage, I is the current, and A is the surface area over which the current is measured. The piezoelectric equations give the relationship between D and E through stress and strain. For simple one-dimensional resonance modes, many of the electric and mechanical field variables go to zero and the sets of piezoelectric equations decouple.

Solving the wave equation for one-dimensional steady-state displacement in the specimen and evaluating S from the derivative of displacement, together with the boundary conditions of an unclamped resonator, the piezoelectric equations can be used to evaluate the relationship between D and E as a function of frequency and material properties. The impedance of a thickness extensional (TE) resonator reduces to [34]:

$$\mathbf{Z} = \frac{\boldsymbol{\beta}_{33}^S l + \dfrac{\mathbf{h}_{33}^2}{\sqrt{\rho \mathbf{c}_{33}^D}\,\omega} \tan\left(\dfrac{\omega l \sqrt{\rho}}{2\sqrt{\mathbf{c}_{33}^D}}\right)}{i\omega A}, \tag{25}$$

where ρ is the specimen density, ω is the angular frequency and l is the sample thickness in the direction of poling.

For the TE resonator, the lateral dimensions must be at least 10 times the thickness to assure a clean, one-dimensional TE mode. This can be relaxed slightly in low-Q materials. In impedance spectroscopy characterization of piezoelectrics, the material properties of Equation (25) can be adjusted until the impedance fits a measured impedance spectrum as a function of frequency.

The common one-dimensional 6mm modes include the Length Extensional (LE), Thickness Extensional (TE), Length-Thickness Extensional (LTE), Thickness Shear (TS), Length Shear (LS), and Radial modes. Each of these modes provides electrical, elastic and piezoelectric properties of the material. Table 1 lists the properties of one form of the reduced piezoelectric matrix with the modes that will provide these properties.

By analyzing specimens inducing the TS, LE and Radial modes, all but s^E_{13} of the reduced 6mm piezoelectric matrix can be obtained. The s^E_{13} can be found from the properties determined with the TE, LTE and LE modes for other representations of the piezoelectric linear equations [8], indicating that all modes are required to construct the piezoelectric matrix. Alternatively, using an initial guess for s^E_{13}, the reduced matrix can be inverted and c^D_{13} compared to that found for the TE mode, eliminating the need for the LTE mode.

In practice, material properties are dispersive (i.e., they vary with frequency). This means that the range of frequency used to fit forms of Equation (24) to measured spectra should be limited. It is worth noting that only a few points in the spectrum are required to determine properties. For the TE mode, only three points are required.

The IEEE Standard on Piezoelectricty (1987) [18] describes two approaches to obtaining material properties using either the difference between parallel (f_p) and series (f_s) resonance frequencies, or the frequency ratio between different order resonance peaks. These approaches can be problematic when material properties are dispersive or when a low electromechanical coupling constant results in a small difference between f_p and f_s.

Smits [24] proposed a general method for the TE, TS, LE and LTE resonators using three points around resonance. This method does not rely on critical frequencies in the spectrum and so does not lose sensitivity for low-Q materials. At resonance and anti-resonance, currents can be very large and small and this presents an instrumentation problem. Though less of an issue with low-Q materials, Smits's method is lesser demanding of instrumentation than the IEEE methods.

Finally, because the choice of analysis points is somewhat arbitrary, noisy spectra and samples with mode coupling may still be analyzed using Smits's method. The IEEE Standard in Piezoelectricity (1987) treats all material properties as real, evaluating mechanical and electrical losses independently through Q_M and tan δ. Losses in the electromechanical coupling constant were assumed zero. This was acceptable for low-loss piezoelectric materials such as PZT, but can lead to significant errors when materials have large losses.

In real materials, complex impedance results from complex material properties [31]. The impedance is directly associated with dielectric conduction losses [27,28] and attenuation of the displacement wave in the sample.

The PRAP software [25] analyses impedance spectra to determine complex material properties. This software uses a generalized form of Smits's method to determine material properties for any common resonance mode, and a generalized ratio method for the radial mode [24] valid for all material Q's. By analyzing on each harmonic, PRAP can determine complex material properties as a function of frequency. The software always generates an impedance spectrum from the determined properties to indicate validity of the results.

Table 1. Properties provided by the common modes

Property	εT_{11}	εT_{33}	sE_{11}	sE_{12}	sE_{13}	sE_{33}	sE_{55}	d_{13}	d_{33}	d_{15}
Mode	TS, LS	LE, LTE, Radial	LTE, Radial	Radial		LE	LS, TS	LTE, Radial	LE	TS, LS

4. Optimization of Finite Element Models for Porous Ceramic Piezoelements

Procedure of optimization has consisted in sequential and iterative application of FEM and Piezoelectric Resonance Analysis (PRAP) to complex electric impedance spectra of piezoceramic elements. For validation of proposed optimization procedures, FEM calculations of standard shape piezoelements (disks, shear plates, bars, and rods) made from porous PZT-type piezoceramics were fulfilled using FEM ANSYS software package [15].

4.1. FEM Calculations and Optimization Procedure

Finite element modeling of the effective constants of porous piezoelectric ceramics based on the effective moduli method was performed using ANSYS software package [15]. Different models of representative volume were considered: piezoelectric cubes with one cubic and one spherical pore inside, cubic volume evenly divided in partial cubic volumes, with a portion randomly declared as pores, and the like ones. For modeling porous piezoceramics with 3-0/3-3 connectivity type we used the representative volume with rigid skeleton structure (Figure 2) [56]. For the models 1 and 2 finite-element calculations of effective moduli of porous ceramics APC-841 were conducted for all four types of problems: $u\phi$, $p\phi$, uq and pq. It was found that the influence of mechanical boundary conditions on the effective moduli is stronger than the electrical ones. The models $u\phi$ and uq in which fixation conditions are pointed out, are more rigid than the models $p\phi$ and pq in which mechanical stresses are fixed. Dielectric constants from the models $u\phi$ and uq are lesser than corresponding values from the models $p\phi$ and pq. The models $u\phi$ and uq gives overvalued stiffness elastic moduli, ultrasonic velocities and electromechanical coupling factors. On the contrary, piezoelectric moduli d_{33} and d_{31} are overvalued for the models $p\phi$ and pq. Dielectric constant $\varepsilon^T{}_{33}$ is less dependent from the problem type.

Comparison of calculated and measured results [15,16] for different porosities has shown that $p\phi$-problem gave more reasonable values of ultrasonic velocities, stiffness moduli and piezoelectric moduli than the problem $u\phi$. Note that in all cases piezoelectric modulus d_{31} decreases with the porosity no so fast as in the experiment. By the way, the problem $u\phi$ is most convenient for FEM calculations particularly for models with complex boundaries. From the results of the models 1 and 2 follows, that corner points enters minor distortions to integral characteristics of porous ceramics. Piezoelectric modulus d_{31} in the model 3 at homogeneous polarization practically does not decrease with the porosity. Decrease of the piezoelectric modulus d_{31} can be reached using partial polarization models however in this case piezoelectric modulus d_{33} also decreases. Serious disadvantage of the model 3 compared to more complex models 4 – 6 is a dislocation of ceramic skeleton connectivity.

The results of these models are much better than for the model 3. At the same time, the 3-3 connectivity type model gives calculated values of d_{33} that are very close to experimental ones for broad range of porosity. Calculated values of d_{31} for porous ceramics are in lesser conformity with the experiments. However, the model 6 for the 3-3 connectivity type also gives more reasonable results (see Figure 2).

Optimization procedures were following:

1. direct FEM calculations of complex electric impedance spectra using sets of material constants for standard piezoceramic elements;
2. analysis of the calculated impedance spectra by using the piezoelectric resonance analysis program (PRAP) and by deriving complex sets of material constants related to specific piezoceramic element;
3. correction of initial material constants, including losses ($Q = c'/c''$, where c' and c'' are the real and imaginary parts of relative material constant).

Validation of the optimization procedures was performed using the FEM ANSYS package. FEM-generated complex impedance spectra for standard-shaped piezoelements (disks, bars, shear plates and rods) made from "hard" porous PZT piezoceramics were processed by PRAP-software to derive a complex set of material constants.

4.2. Method of Measurements

For the experiments, we applied the PRAP automatic iterative method [25, 26] to the full set of standard geometries and resonance modes needed to complete a characterization of hard PZT porous piezoceramics with moderate loss factors.

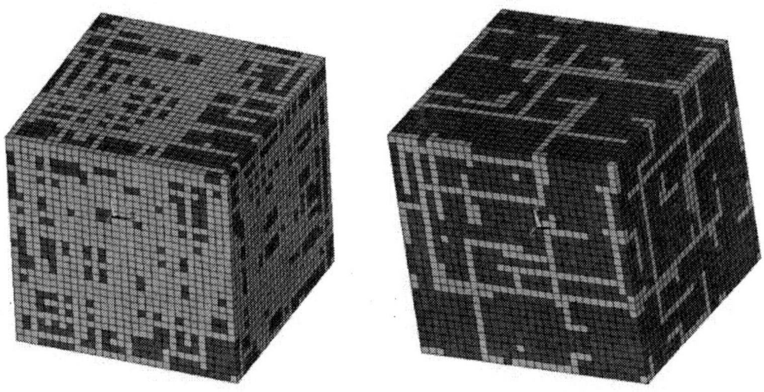

Figure 2. Examples of representative volumes used for FEM calculations of porous piezoceramics: (a) 3-3 connectivity, porosity of 20 %; (b) 3-0 connectivity, porosity of 80 %.

This software uses a generalized form of Smits's method [24] to determine material properties for any common resonance mode, and a generalized ratio method for the radial mode [34] valid for all material Q's. The software routinely generates an impedance spectrum from the determined properties to indicate the validity of the results. Measurements of electric parameters were conducted by using the Solartron Impedance/Gain-Phase Analyzer SL 1260 and Agilent 4294 Impedance Analyzer.

4.3. Results and Discussion

Figures 3 – 6 show measured complex impedance spectra and PRAP approximations for the thickness (TE), radial (RE), length extensional (LE), and thickness shear (ST) modes of "hard" porous piezoceramics with relative porosity of 21 % and average pore size of 50 μm.

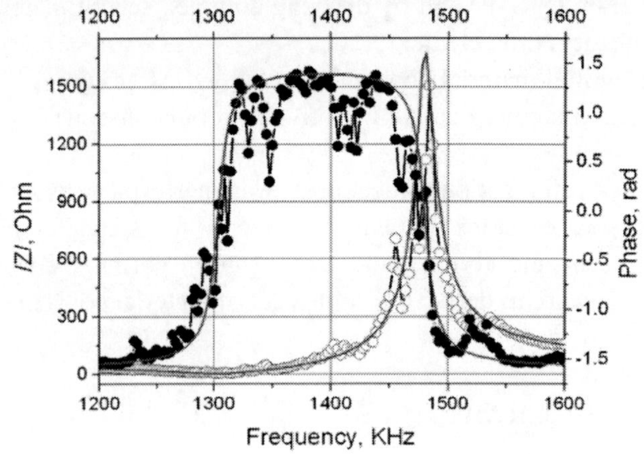

Figure 3. Measured impedance spectra and PRAP approximations for TE-mode of porous piezoceramic disk Ø20 × 1 mm^2.

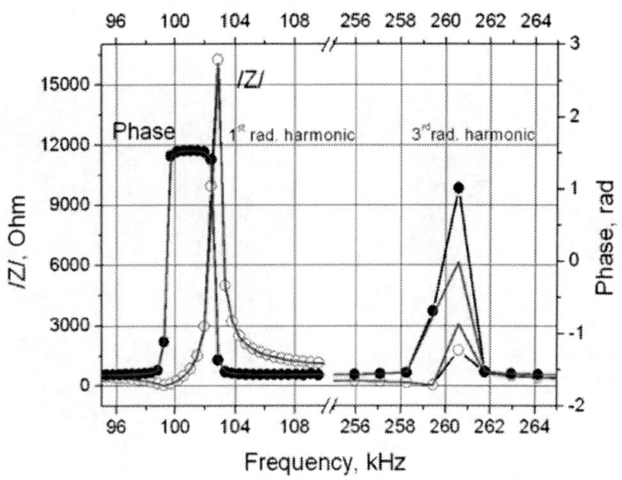

Figure 4. Measured impedance spectra and PRAP approximations for RE-mode of porous piezoceramic disk Ø20 × 1 mm^2.

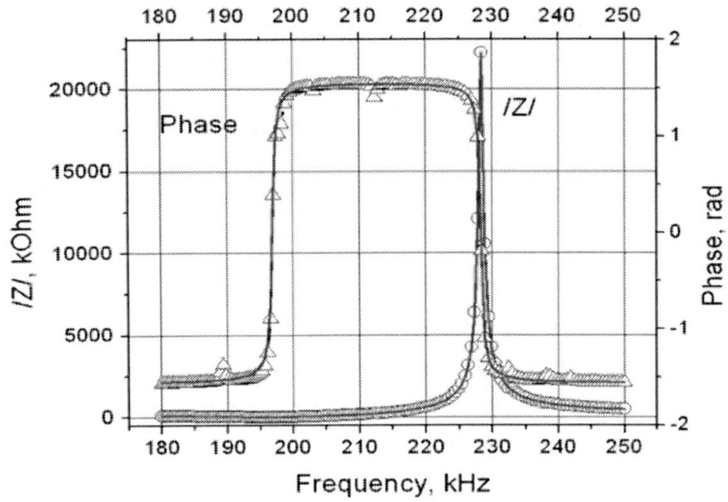

Figure 5. Measured impedance spectra and PRAP approximations for LE-mode of porous piezoceramic rod 1.5 × 1.5 × 6 mm^3.

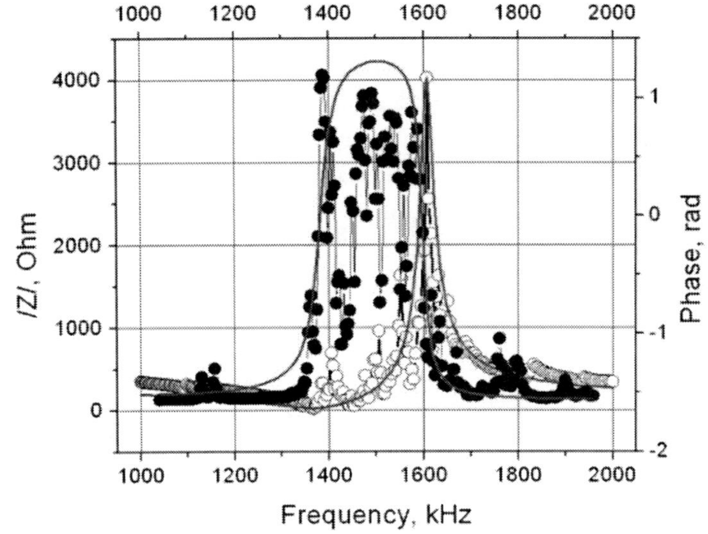

Figure 6. Measured impedance spectra and PRAP approximations for ST-mode of porous piezoceramic square 0.6 × 0.6 × 6 mm^3.

4.4. Full Set of Complex Material Constants

Table 1 summarizes the complex constants of the porous ceramics obtained using PRAP analysis for a full set of standard geometries and resonance modes (Figures 3 – 6). Additional physical parameters of porous ceramics measured by ultrasonic and hydrostatic weighting methods are by following: Z_A = 18.5 MRayl, ρ = 6.25 g/cm^3, V_t = 2960 m/s, Q_M^t = 150, (T_C = 340° C) [2, 26]. Experimental aspects of porous ceramics application in ultrasonic transducers and devices are considered in [35–40].

Table 1. Complex constants measured for various resonance modes of "hard" porous ceramics

Parameter	Real	Imaginary	Parameter	Real	Imaginary
Shear Thickness (ST) Mode			Length Extensional (LE) Mode		
f_p (Hz)	$1.60 \cdot 10^6$	–	f_p (Hz)	$2.30 \cdot 10^5$	–
k_{15}	0.56	– 0.013	k_{33}	0.55	$-8.8 \cdot 10^{-4}$
C^E_{55} (N/m²)	$1.60 \cdot 10^{10}$	$5.80 \cdot 10^8$	S^D_{33} (m²/N)	$2.12 \cdot 10^{-11}$	$-5.92 \cdot 10^{-14}$
C^D_{55} (N/m²)	$2.32 \cdot 10^{10}$	$3.50 \cdot 10^8$	S^E_{33} (m²/N)	$3.03 \cdot 10^{-11}$	$-1.26 \cdot 10^{-14}$
S^D_{55} (m²/N)	$4.31 \cdot 10^{-11}$	$-6.49 \cdot 10^{-13}$	d_{33} (C/N)	$2.54 \cdot 10^{-10}$	$-1.35 \cdot 10^{-12}$
S^E_{55} (m²/N)	$6.25 \cdot 10^{-11}$	$-2.27 \cdot 10^{-12}$	g_{33} (V·m/N)	0.04	$-7.1 \cdot 10^{-5}$
e_{15} (C/m²)	7.51	– 0.14	ε^T_{33} (F/m)	$6.85 \cdot 10^{-9}$	$-1.5 \cdot 10^{-11}$
h_{15} (V/m)	$9.65 \cdot 10^8$	$-1.32 \cdot 10^7$	Radial Extensional (RE) Mode		
d_{15} (C/N)	$4.69 \cdot 10^{-10}$	$-2.56 \cdot 10^{-11}$	f_{p1} (Hz)	$1.03 \cdot 10^5$	–
g_{15} (Vm/N)	0.04	– 0.001	S^E_{11} (m²/N)	$1.96 \cdot 10^{-11}$	$-3.10 \cdot 10^{-14}$
ε^T_{11} (F/m)	$1.13 \cdot 10^{-8}$	$-2.92 \cdot 10^{-10}$	S^E_{12} (m²/N)	$-6.37 \cdot 10^{-12}$	$2.30 \cdot 10^{-14}$
ε^S_{11} (F/m)	$7.77 \cdot 10^{-9}$	$-3.60 \cdot 10^{-11}$	d_{31} (C/N)	$5.92 \cdot 10^{-11}$	$-5.19 \cdot 10^{-14}$
Thickness Extensional (TE) Mode			ε^T_{33} (F/m)	$6.44 \cdot 10^{-9}$	$-3.09 \cdot 10^{-12}$
f_p (Hz)	$1.48 \cdot 10^6$		k_p	0.29	$-8.84 \cdot 10^{-5}$
k_t	0.52	– 0.04	σ^P	0.33	$-6.3 \cdot 10^{-4}$
C^D_{33} (N/m²)	$5.49 \cdot 10^{10}$	$3.70 \cdot 10^8$	S^E_{66} (m²/N)	$5.19 \cdot 10^{-11}$	$-1.08 \cdot 10^{-13}$
C^E_{33} (N/m²)	$4.02 \cdot 10^{10}$	$2.66 \cdot 10^8$	C^E_{66} (N/m²)	$1.93 \cdot 10^{10}$	$4.0 \cdot 10^7$
e_{33} (C/m²)	10.08	– 1.07	Physical Parameters		
h_{33} (V/m)	$1.47 \cdot 10^9$	$-7.10 \cdot 10^7$	ρ (kg/m³)	$6.25 \cdot 10^3$	–
ε^S_{33} (F/m)	$6.87 \cdot 10^{-9}$	$-4.00 \cdot 10^{-10}$	Z_A (MRayl)	18.5	–

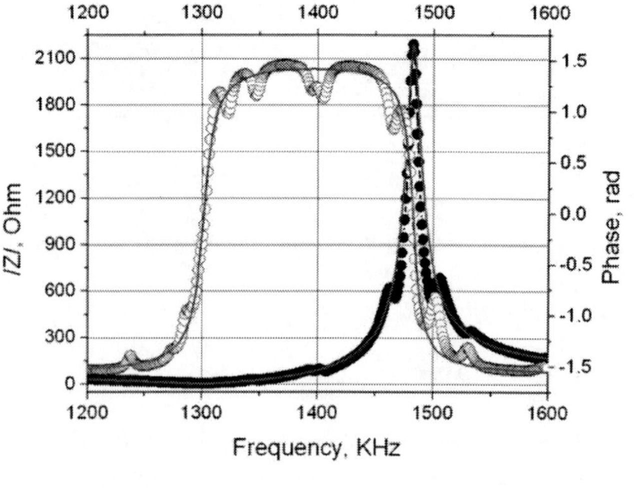

Figure 7. Impedance spectra calculated by FEM and PRAP approximations for TE-mode of porous piezoceramic disk Ø20 × 1 mm².

4.5. FEM Results

Figures 7–10 show complex impedance spectra generated by FEM using the constants shown in Table 1 and PRAP approximations for TE, RE, LE, and ST modes of porous piezoceramics.

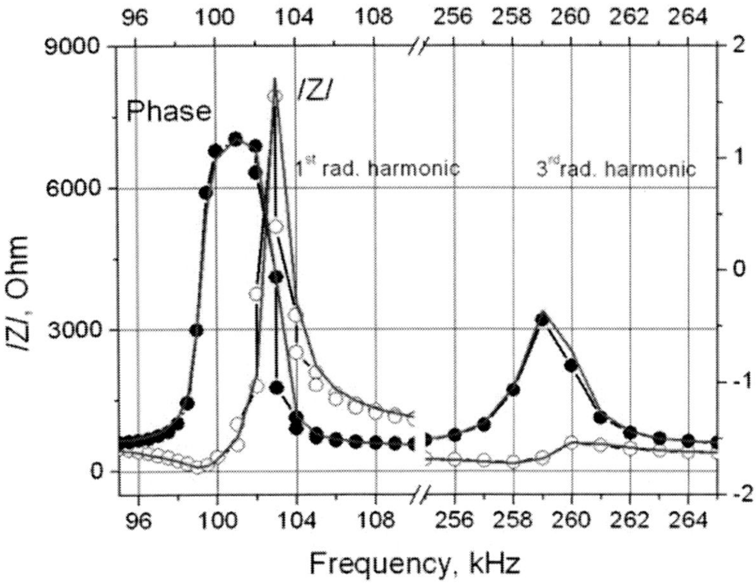

Figure 8. Impedance spectra calculated by FEM and PRAP approximations for RE-mode of porous piezoceramic disk Ø20 × 1 mm^2.

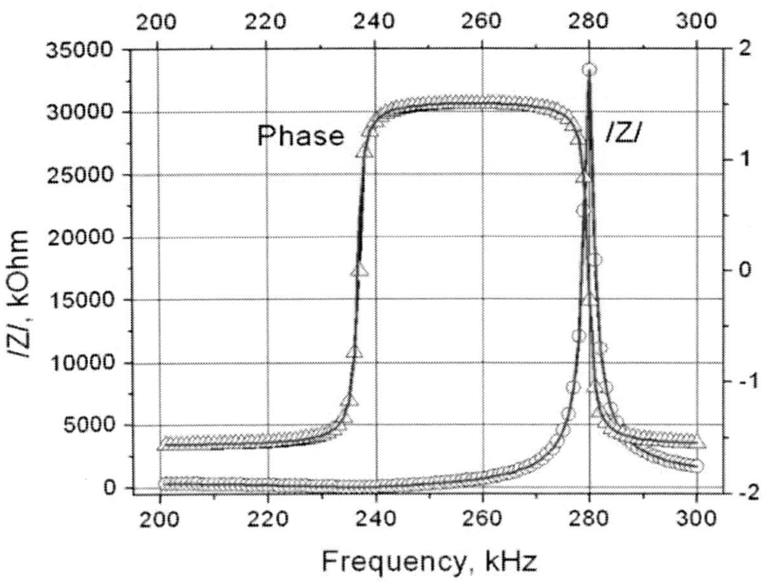

Figure 9. FEM calculated impedance spectra and PRAP approximations for LE-mode of porous piezoceramic rod 1 × 1 × 5 mm^3.

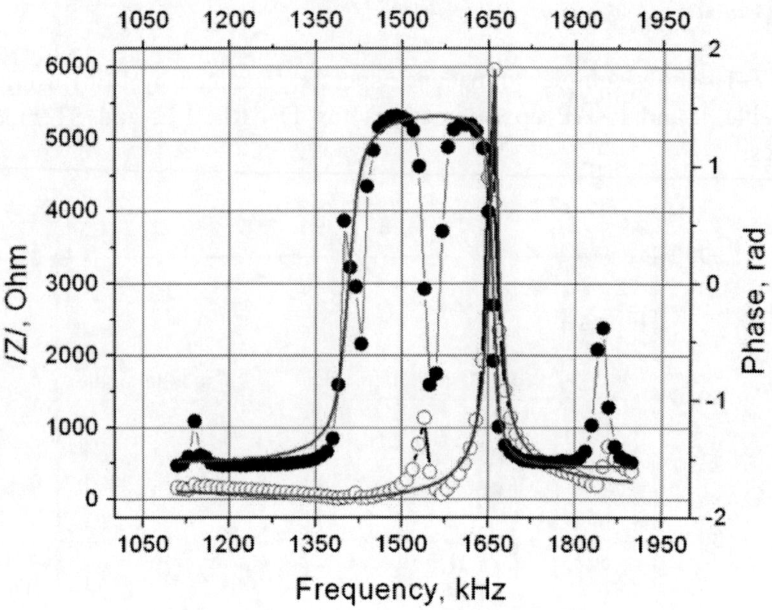

Figure 10. Impedance spectra calculated by FEM and PRAP approximations for ST-mode of porous piezoceramic square 0.6 × 0.6 × 6 mm^3.

The complex constants of "hard" porous piezoceramics, obtained using FEM generated spectra (Figures 7 – 10), are summarized in Table 2.

It is readily seen from Tables 1 and 2 that the complex material constants resulting from FEM-generated impedance spectra are very close to measured ones, including imaginary parts (losses). For other low-*Q* materials, additional corrections of material constants (*Q* effective in ANSYS) with recurring optimization procedure should be performed.

Conclusion

A novel approach for optimization of finite element modeling (FEM) of lossy piezoceramic elements was considered.

Different FEM models of porous piezoceramics based on effective moduli method were developed and considered. For the models confirmation, the porous piezoceramics with well controllable porosity were manufactured and measured. The results of FEM modeling shows that the representative volume model for 3-3 connectivity type with rigid skeleton gives better results compared to other examined models and describes experimental results for porous ceramics in the relative porosity range of 0 – 70 % very well.

Piezoelectric resonance analysis methods for automatic, iterative evaluation of complex material parameters were described, along with full sets of complex constants for various ceramic piezocomposites. Complex sets of material constants for "hard" porous piezoceramics were obtained by PRAP analysis of electric impedance spectra measured by using standard porous piezoceramic elements.

Electric impedance spectra were calculated with help of measured sets of material constants using FEM ANSYS software package.

Table 2. Complex constants obtained using PRAP analysis of FEM-generated spectra for hard porous piezoceramics

Parameter	Real	Imaginary	Parameter	Real	Imaginary
Shear Thickness (ST) Mode			Length Extensional (LE) Mode		
f_p (Hz)	$1.66 \cdot 10^{6}$	–	f_p (Hz)	$2.8 \cdot 10^{5}$	–
k_{15}	0.57	– 0.048	k_{33}	0.57	$-7.02 \cdot 10^{-4}$
C^{E}_{55} (N/m²)	$1.67 \cdot 10^{10}$	$1.48 \cdot 10^{9}$	S^{D}_{33} (m²/N)	$2.04 \cdot 10^{-11}$	$-1.038 \cdot 10^{-13}$
C^{D}_{55} (N/m²)	$2.47 \cdot 10^{10}$	$1.77 \cdot 10^{8}$	S^{E}_{33} (m²/N)	$3.04 \cdot 10^{-11}$	$-1.9 \cdot 10^{-13}$
S^{D}_{55} (m²/N)	$4.03 \cdot 10^{-11}$	$-2.89 \cdot 10^{-13}$	d_{33} (C/N)	$2.55 \cdot 10^{-10}$	$-1.27 \cdot 10^{-12}$
S^{E}_{55} (m²/N)	$5.93 \cdot 10^{-11}$	$-5.25 \cdot 10^{-12}$	g_{33} (V·m/N)	0.04	$-1.48 \cdot 10^{-4}$
e_{15} (C/m²)	9.49	– 1.26	ε^{T}_{33} (F/m)	$6.48 \cdot 10^{-9}$	$-7.9 \cdot 10^{-12}$
h_{15} (V/m)	$8.5 \cdot 10^{8}$	$-2.4 \cdot 10^{7}$	Radial Extensional (RE) Mode		
d_{15} (C/N)	$5.56 \cdot 10^{-10}$	$-1.25 \cdot 10^{-11}$	f_{p1}(Hz)	$1.03 \cdot 10^{5}$	–
g_{15} (Vm/N)	0.035	– 0.0012	S^{E}_{11} (m²/N)	$1.99 \cdot 10^{-11}$	$-8.83 \cdot 10^{-14}$
ε^{T}_{11} (F/m)	$1.63 \cdot 10^{-8}$	$-3.063 \cdot 10^{-9}$	S^{E}_{12} (m²/N)	$-6.62 \cdot 10^{-12}$	$2.53 \cdot 10^{-14}$
ε^{S}_{11} (F/m)	$1.12 \cdot 10^{-8}$	$-1.17 \cdot 10^{-9}$	d_{31} (C/N)	$6.37 \cdot 10^{-11}$	$-8.42 \cdot 10^{-13}$
Thickness Extensional (TE) Mode			ε^{T}_{33} (F/m)	$6.64 \cdot 10^{-9}$	$-2.19 \cdot 10^{-11}$
f_p (Hz)	$1.48 \cdot 10^{6}$	–	k_p	0.3	– 0.002
k_t	0.52	– 0.006	σ^{P}	0.33	0.0028
C^{D}_{33} (N/m²)	$5.49 \cdot 10^{10}$	$3.145 \cdot 10^{8}$	S^{E}_{66} (m²/N)	$5.29 \cdot 10^{-11}$	$-1.26 \cdot 10^{-13}$
C^{E}_{33} (N/m²)	$4.03 \cdot 10^{10}$	$5.92 \cdot 10^{8}$	C^{E}_{66} (N/m²)	$1.89 \cdot 10^{10}$	$4.49 \cdot 10^{7}$
e_{33} (C/m²)	9.26	– 0.35	Physical Parameters		
h_{33} (V/m)	$1.57 \cdot 10^{9}$	$-2.9 \cdot 10^{7}$	ρ (kg/m³)	$6.25 \cdot 10^{3}$	–
ε^{S}_{33} (F/m)	$6.87 \cdot 10^{-9}$	$-3.29 \cdot 10^{-10}$	Z_A (MRayl)	18.5	–

Complex sets of material constants were derived from FEM generated complex impedance spectra using PRAP analysis. Comparison shows a good agreement between initial and calculated complex sets of material constants including losses ($Q = c^{//}c^{//}$, where $c^{/}$ and $c^{//}$ are the real and imaginary parts of related material constants).

For the case of more low-Q materials, additional corrections of material constants (Q effective for ANSYS) with recurring of optimization procedure were proposed.

REFERENCES

[1] Wersing, W. In: *Piezoelectric Materials in Devices*; N. Setter (Ed.); Swiss Federal Institute of Technology: Lausanne, Switzerland, 2002; pp. 29-66.

[2] Rybyanets, A. N. *IEEE Trans.UFFC*, 2011, vol. 58, 1492-1507; Rybyanets, A. N. *Ferroelectrics*, 2011, vol. 419, 90-96.

[3] Rybyanets, A. N. *Ferroelectrics,* 2011, vol. 419, 90-96.

[4] Rybyanets, A. N., Rybyanets, A. A. *IEEE Trans.* UFFC, 2011, vol. 58, 1757-1774.

[5] Rybyanets, A. N. In: *Piezoceramic Materials and Devices*; Ivan A. Parinov (Ed.). Nova Science Publishers Inc. NY, 2010, Chapter 3, pp. 113-174.

[6] Rybyanets, A. N. In: *Piezoelectrics and Related Materials: Investigations and Applications*, Ivan A. Parinov (Ed.). Nova Science Publishers Inc. NY, 2012, Chapter 5, pp. 143-187.

[7] Rybyanets, A. N., Zakharov, Y. N., Raevskii, I. P., Akopjan, V. A., Rozhkov E. V., Parinov, I. A. In: *Physics and Mechanics of New Materials and Their Applications*, Ivan A. Parinov, Shun Hsiung-Chang (Eds.). Nova Science Publishers Inc. NY, 2013, Chapter 22, pp. 275-308.

[8] Rybyanets, A. N., Naumenko, A. A. In: *Physics and Mechanics of New Materials and Their Applications*, Ivan A. Parinov, Shun Hsiung-Chang (Eds.). Nova Science Publishers Inc., NY, 2013, Chapter 1, pp. 3-18.

[9] Rybjanets, A. N., Nasedkin, A. V., Turik, A. V. *Integrated Ferroelectrics*, 2004, vol. 63, pp. 179-182.

[10] Wersing, W., Lubitz, K., Moliaupt. *Ferroelectrics*, 1986, vol. 68, 77-79.

[11] Banno, H. *Jpn. J. Appl. Phys.*, 1993, vol. 32(1), 4214–4217.

[12] Xiang, P.-H., Dong, X.-L., Chen, H., Zhang, Z., Guo, J.-K. *Ceramics International*. 2003, 29, 499-503.

[13] Rybjanets, A. N., Razumovskaja, O. N., Reznitchenko, L. A., Komarov, V. D., Turik, A. V. *Integrated Ferroelectrics*, 2004, vol. 63, 197-200.

[14] Rybianets, A. N. *Ferroelectric*, 2007, vol. 360, 84-89.

[15] Nasedkin, A., Rybjanets, A., Kushkuley, L., Eshel, Y. *Proc. 2005 IEEE Ultrason. Symp.*, 2005, pp. 1648-1651.

[16] Rybianets, A., Kushkuley, L., Eshel, Y., Nasedkin, A. *Proc.2006 IEEE Ultrason. Symp.*, 2006, pp. 1533-1536.

[17] Lethiecq, M., Levassort, F., Trans-Huu-Hue, L. P., *IEEE Ultrason. Symp. Proc.*, 2004, 1153-1156.

[18] *IEEE Standard on Piezoelectricity*. ANSI/IEEE Std. 1987, 176.

[19] Rybyanets, A. N., Naumenko, A. A., Shvetsova, N. A. In: *Nano- and Piezoelectric Technologies, Materials and Devices*, Ivan A. Parinov (Ed.). Nova Science Publishers Inc. NY, 2013, Chapter 1, pp. 275-308.

[20] Alguero, M., Alemany, C., Pardo, L., Gonzalez, A. M. *J. Am. Ceram. Soc.*, 2004, vol. 87, 209.

[21] Sherrit, S., Wiederick, H. D., Mukherjee, B. K. *SPIE Proceedings*, 1997, vol. 3241, 327.

[22] Rybianets, A. N., Tasker, R. *Ferroelectrics*, 2007, vol. 360, 90-95.

[23] Alemany, C., Gonzalez, A. M., Pardo, L., Jimenez, B., Carmona, F., Mendiola, J. *J. Phvs. D: Appl. Phvs.*, 1995, vol. 28, 945.

[24] Smits, J. G. *IEEE Trans. Sonics Ultrason.*, 1976, vol. SU-23, 393-402.

[25] *PRAP (Piezoelectric Resonance Analysis Program)*. TASI Technical Software Inc. www.tasitechnical.com.

[26] Rybianets, A. N., Nasedkin, A. V. *Ferroelectrics,* 2007, vol. 360, 57-62.

[27] Sherrit, S., Wiederick, H. D., Mukherjee, B. K. *Ferroelectrics,* 1992, vol. 134, 111.

[28] Pardo, L., Alemany, C., Ricote, J., Moure, A., Poyato, R., Alguero, M. *Proc. Intern. Conf. Material Technology and Design of Integrated Piezoelectric Devices.* Couurmayeur, Italy. 2004, p. 145.

[29] Kara, H., Ramesh, R., Stevens, R., Bowen, C. R. *IEEE Trans. UFFC,* 2003, vol. 50, 289-296.
[30] Berlincourt, D. A., Curran, D. R., Jaffe, H. In: *Physical Acoustics I, Part A;* Mason, W. P.; Academic Press: NY, 1964.
[31] Getman, I., Lopatin, S. *Ferroelectrics*, 1996, vol. 186, 301-304.
[32] Devonshire, A. F. *Philosophical Magazine Supplement*, 1954, vol. 3, 85-130.
[33] Mason, W. P. *Physical Acoustics and the Properties of Solids;* D. Van Nostrand Co. Inc.: Princeton, 1958.
[34] Holland, R. *IEEE Trans. Sonics Ultrason*. 1967, vol. SU-14, 18.
[35] Sarvazyan, Armen; Ostrovsky, Lev; Rubyanets, Andrey. *Acoustical Society of America. Proceedings on Meeting on Acoustics.* 2009, vol. 6.
[36] Rybyanets, Andrey, Nasedkin, Andrey; Domashenkina, Tamara, Rybyanets, Anastasia; Lugovaya, Maria. *Proc. 2009 IEEE Ultrason. Symp.*, 2009, pp. 1699-1702.
[37] Rybyanets, Andrey. Nasedkin, Andrey; Rybyanets, Anastasia; Lugovaya, Maria; Domashenkina, Tamara. *Proc. 2009 IEEE Ultrason. Symp.,* 2009, pp. 1765-1768.
[38] Rybianets, A., Eshel, Y., Kushkuley, L. *Proc. 2006 IEEE Ultrason. Symp.*, 2006, pp. 1911-1914.
[39] Rybyanets, A. *Proc. 2007 IEEE Ultrason. Symp.*, 2007, pp. 1909-1912.
[40] Goland, V., Kushkuley, L., Mirman, S., Zadok, Y., Ben-Ezra, S., Shalgi, A., Rybianets, A. *Proc. 2007 IEEE Ultrason. Symp.*, 2007, pp. 1305-1308.

In: Advanced Materials - Studies and Applications
Editors: I.A. Parinov, S.-H. Chang, S.Theerakulpisut
ISBN: 978-1-63463-749-7

Chapter 12

MODELING OF PIEZOELECTRIC ELEMENTS WITH INHOMOGENEOUS POLARIZATION BY USING ACELAN

***A. N. Soloviev*[*,1,2,3], *A. S. Skaliukh*[2] and *P. A. Oganesyan*[1,3,4]**

[1]Department of Theoretical and Applied Mechanics, Don State Technical University, Rostov-on-Don, Russia

[2]Department of Mathematics, Mechanics and Computer Sciences, Southern Federal University, Rostov-on-Don, Russia

[3]Department of Mechanics of Active Materials, Southern Scientific Centre of Russian Academy of Sciences, Rostov-on-Don, Russia

[4]Aircraft Technology Department, Southern Scientific Centre of Russian Academy of Sciences, Rostov-on-Don, Russia

ABSTRACT

We consider the problem on influence of a heterogeneous remnant polarization on physical characteristics of piezoceramic transducers. With this aim, we study the following problems: (i) mathematical modeling of the polarization process and its finite element implementation; (ii) development of two-dimensional finite element models of piezoceramic transducers in the form of rods with transverse polarization; (iii) calculation of physical characteristics of the piezotransducers with heterogeneous remnant polarization. In particular, it is shown that electromechanical coupling factor could be increased by choosing a specific distribution of preliminary polarization in the transducer. On the other hand, piezotransducer can be developed, which is excited only on a certain frequency mode.

[*] solovievarc@gmail.com.

1. Introduction

Piezoceramic transducers during their fabrication are subject to a number of technological operations and studies, including sintering of ceramics, its polarization, machining, measurement of characteristics, etc. Naturally, designing the piezoceramic transducers is based on modeling of above processes, but only some of them can be described by mathematical models with a required accuracy. To fabricate piezoceramic transducer, it is necessary preliminary to model a mechanical destruction of ferroelectric crystal, processed by using a series of the following technological operations: dry mixing of precursors, pressing, sintering and cooling of green sample, applying the electrodes and the specimen polarization in intense electric field. Sufficient mathematical models have not been developed for description of these technological procedures. In particular, numerous mathematical models have been developed for study of polarization processes. However they allow us to describe these processes correctly only in some details and require further investigations.

Better results have been obtained in mathematical modeling (on the base of finite element methods) of transducers, operating under certain conditions, taking into account the mechanical and electric fields of low intensity. However, there are numerous difficulties of a computational nature even in solution of linear problems, describing these devices.

Definition of thermal properties into framework of a mathematical model complicates the problem dramatically, even if full set of characteristics into framework of the thermo-electro-elasticity model is not required.

Above-mentioned circumstances define the following stages of our study. First we consider a polarization model of polycrystalline ferroelectric material under fields of high intensity. Then we develop numerical models, based on finite-element method, to calculate physical characteristics of piezoelectric samples are subject to fields of low intensity, when the field of preliminary polarization does not change. The main emphasis in this analysis is made on characteristics of the heterogeneous remnant polarization field. Here we study the following interesting problem with viewpoint of practical application, when above characteristics change significantly, compared to transducer with uniform polarization.

Electrical and mechanical fields of different intensities can induce both reversible and irreversible processes in a polycrystalline specimen, which do not alter the structure of the material or change it, respectively.

Mathematical modeling of reversible processes in piezoelectric materials is based on the governing equations of continuum mechanics and electrodynamics, which are defined by linear constitutive relations for the strain tensor and polarization vector, on the one hand, and the mechanical stress tensor and electric field vector, on the other hand. This situation changes dramatically for ferroelectric materials, when the electrical and mechanical fields reach certain threshold values. The processes of polarization and deformation become *irreversible*. In such a case, the constitutive relations become complex nonlinear relations of the hysteresis type. Note that hysteresis operators take place in many problems, describing the irreversible processes, for example, in problems of plasticity. Structure and investigation methods of the hysteresis operators in the plasticity problems can be found in classical works [1,2]. Then, it has been developed a method that was used for solving the problems of polarization of the polycrystalline ferroelectrics [3–5]. Introduction of mathematical element "hysteron" allowed one to describe the phenomena of magnetization in ferromagnetic materials [6] and plasticity in crystals [7]. Further studies have shown that it was impossible to create a simple qualitative mathematical model applicable for practical purposes without taking into account internal structure of material. Some works have appeared, in which were considered both as domain structure of the material as its polycrystalline structure. A hysteretic operator has been constructed in the form of an ordinary differential equation to describe the magnetization under the magnetic field of polycrystalline ferromagnetic [8]. An auxiliary function of limit magnetization is no positive definite. Integral representation of the distribution function of domains in magnetic field has been constructed, based on the statistical Boltzmann law. Then this theory was used in one-dimensional case in the theory of polarization of polycrystalline ferroelectrics [9]. Its generalization to three-dimensional case is given in [10]. In the present paper, the hysteresis type operators were constructed for three-dimensional cases. It allowed us to determine the polarization vector in dependence on value of electric field with the help of calculating successive increments of the polarization vector for given increments of the electric field vector. In our study, one-dimensional Jiles – Atherton model is generalized to three-dimensional case at certain assumptions on behavior of the remnant polarization vector, when electric field changes its direction.

2. Subject of Study

Let us consider a representative volume of polycrystalline ferroelectric continuum, which contains a great number of crystallites. Each crystallite consists of a great number of domains, and each of the domains has a set of cells with the same direction of spontaneous polarization, as shown in Figure. 1.

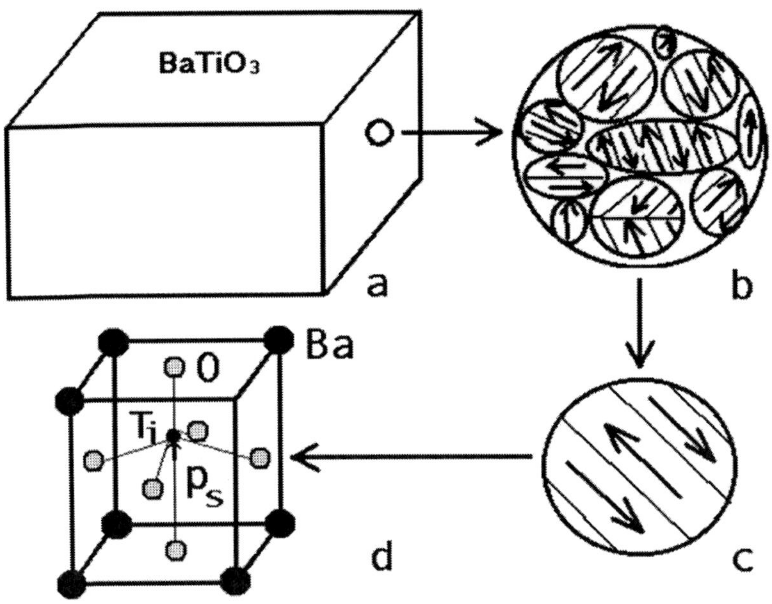

Figure 1. Structure of perovskite-type polycrystalline ferroelectric material on example of barium titanate ceramic: a - ceramic sample; b - representative volume; c - dipoles in the crystallite; d - nuclear cell.

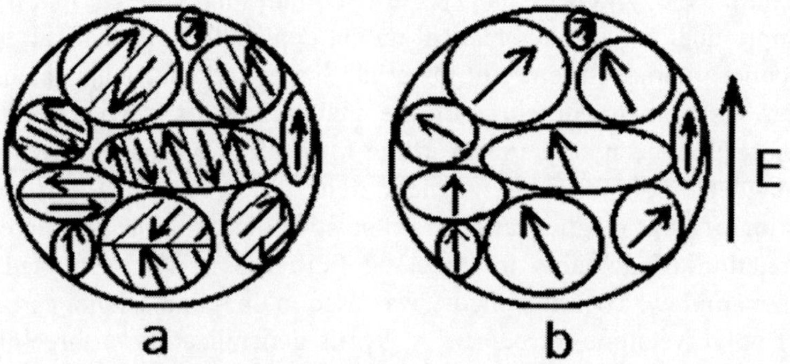

Figure 2. Directions of the spontaneous polarization vectors in domains: a - before polarization; b - after polarization.

Each of the domains in ferroelectric phase has a spontaneous polarization $\mathbf{p}_S$ and spontaneous deformation ε_S, one of the principal axes of which coincides with a direction of the spontaneous polarization. Hereafter we shall denote vector and tensor fields by bold type. A reorientation process of domains begins inside of all crystallites, when an electric field $\mathbf{E}$ is applied to sample. All domains are arranged in a direction close to direction of the electric field in accordance with internal crystalline structure. Figure 2a shows a representative volume in depolarized state, and Figure 2b demonstrates the domain distribution after polarization. It was established in experiments that the domain reorientation began only when the electric field reached a certain "threshold" value.

3. Physical Problem

Let $\Omega \subset \mathbf{R}^3$ presents a macroscopic body of "pure" (thermo-depolarized) ceramic, which is parameterized in $\mathbf{x}$ (Figure 3).

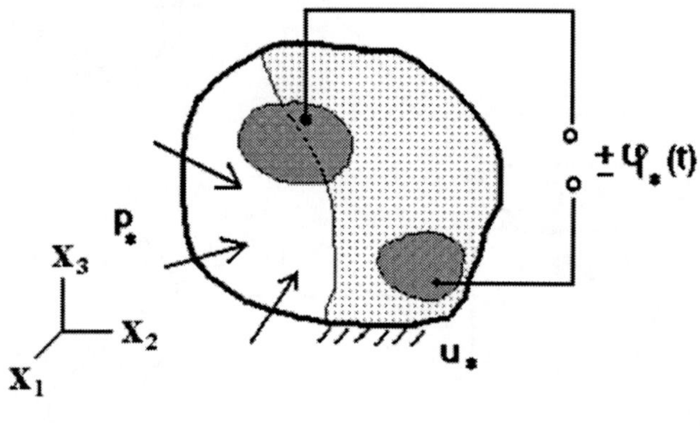

Figure 3. Macroscopic ceramic body.

Furthermore, let $\mathbf{u}$ is the macroscopic displacement field, φ is the macroscopic scalar potential, $\boldsymbol{\sigma}$ is the macroscopic mechanical stress and $\mathbf{D}$ is the electric displacement. In order to treat an electromechanical boundary-value problem, a surface S of the considered body is divided into mechanical and electrical parts: $S = S_\sigma \cup S_u$ and $S = S_D \cup S_\varphi$, respectively, with $S_\sigma \cap S_u = 0$ and $S_D \cap S_\varphi = 0$. The given boundary conditions on S_u and S_φ as well as the natural boundary conditions on S_σ and S_D are defined by $\mathbf{u} = \mathbf{u}_*$ on S_u; $\varphi = \varphi_*$ on S_φ; $\mathbf{n} \cdot \boldsymbol{\sigma} = \mathbf{p}_*$ on S_σ; $Q = \mathbf{D} \cdot \mathbf{n}$ on S_D.

It is required to identify the fields of displacements $\mathbf{u}$ and stresses $\boldsymbol{\sigma}$, electric field $\mathbf{E}$, electric displacement $\mathbf{D} = \varepsilon_0 \mathbf{E} + \mathbf{P}_e + \mathbf{P}_0$ as well as the fields of remnant polarization $\mathbf{P}_0$ and residual strain $\boldsymbol{\varepsilon}_0$ for all values of the boundary functions, i.e., displacement vector $\mathbf{u}_*$, stress vector $\mathbf{p}_*$, electric potential φ_* and electric charge Q.

4. MODEL OF POLARIZATION

Let unit vector is applied to each of the domains of representative volume. So these domains have one common original point inside the representative volume. After applying an electric field, the domains begin to reorient, changing directions of their unit vectors, but the vector ends locate on common unit surface (see Figure 4). Then, each of the domains can be characterized by local coordinates (ρ, φ, ψ). It is easy to see that domains with different values of coordinate ρ, but the same values of coordinates φ, ψ contribute equally to the total polarization of the representative volume. This allows one to perform an averaging only on the latter two coordinates. It is sufficient to introduce a unit sphere, centered at (x^1, x^2, x^3), to calculate a polarization. We assume that applied field direction coincides with x^3-axis.

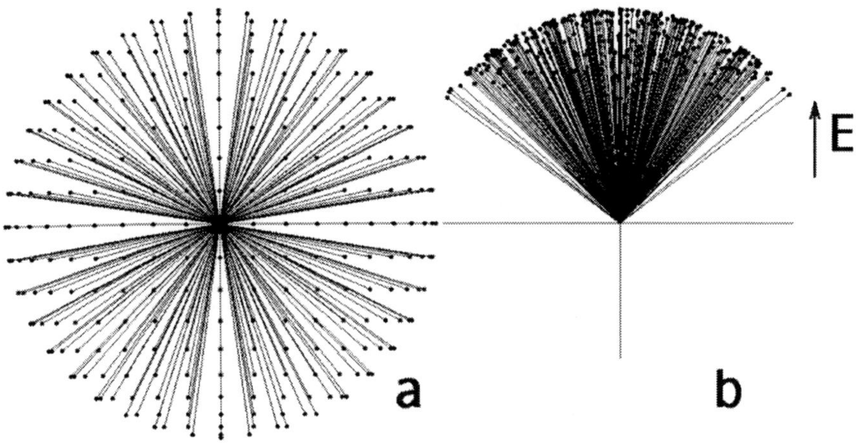

Figure 4. Directions of unit vectors: a – non-polarized state, b – state, polarized by electric field.

As a basis, we take a model scheme, presented in [9, 10]. The common research scheme can be summarized as the following sequence of procedures:

(i) statement of representative volume with greatest possible polarization, using the statistical Boltzmann law for domain reorientations;
(ii) definition of the domain rotation energy, taking into account that part of the energy spends on destruction of pinning the domain walls;
(iii) calculation of work of the electric field in the polarization process;
(iv) determination of energy losses for ideal and real processes of polarization;
(v) consideration of the energy balance equation, taking into account the energy, required to destruct the pinning of the domain walls and energy losses in actual process of polarization;
(vi) derivation of the equation set in finite-difference form, defining the polarization vector increment in dependence on the increment of electric field vector as

$$-\mathbf{P}_0 = -\mathbf{P}_\infty + k\frac{d\mathbf{P}_0}{|d\mathbf{E}^{\mathbf{ef}}|}.$$

For one-dimensional case this equation can be rewritten as

$$dP_0 = \frac{|dE^{ef}|}{k}(P_\infty - P_0). \tag{1}$$

The equation involves unknown parameters p_s, a, α, k, E_c. For their determination, we must use experimental data and achieve the best coincidence of calculated and test data.

5. Solution of the Equation and Determination of Model Parameters

Equation (1) has been reduced to an ordinary differential equation in [9, 10]. It could be solved numerically by using, for example, the Runge-Kutta method of fourth order. We solve this equation by the method of successive approximations. Indeed, it is easy to see that the left and right sides of Equation (1) depend on the increment dP_0. Moreover, $P_\infty = P_\infty(P_0)$. Let us choose a certain set of the parameters p_s, a, α, k, E_c, and let the value of remnant polarization $P_0^{(i-1)}$ for a given state $C^{(i-1)}$ and electric field $E^{(i-1)}$ are known. Then let transition to next state is given by an increment of the electric field $dE^{(i-1)}$. We define the iteration scheme (index " m ") as follows:

$$\left(dP_0^{(i-1)}\right)_m = \frac{|dE^{(i-1)} + \alpha\left(dP_0^{(i-1)}\right)_{m-1}|}{k}\left(P_\infty[(P_0^{(i-1)})_{m-1}] - (P_0^{(i-1)})_{m-1}\right), \qquad m = 1,2,\ldots$$

$$(P_0^{(i-1)})_m = (P_0^{(i-1)})_{m-1} + \left(dP_0^{(i-1)}\right)_m$$

For the first iteration step, we take

$$\left(dP_0^{(i-1)}\right)_0 = 0, \quad (P_0^{(i-1)})_0 = P_0^{(i-1)}.$$

The iteration process begins after fulfillment of the condition:

$$\left|\left(dP_0^{(i-1)}\right)_m - \left(dP_0^{(i-1)}\right)_{m-1}\right| < \varepsilon .$$

In practice, a number of iterations does not exceed of 10, at $\varepsilon = 10^{-6}$.

After definition of the remnant polarization increment in the transition from state $C^{(i-1)}$ to state $C^{(i)}$, we can determine the remnant polarization $P_0^{(i)}$ in the state $C^{(i)}$.

Then we define an induced polarization as a difference between "limit" and remnant parts of the polarization [9], and introduce a factor of proportionality to the electric field as

$$P_e = c[P_\infty(P_0) - P_0] + \beta E . \tag{2}$$

The physical meaning of second addendum in right side of Equation (2) is yet unclear at this stage, but it may be, for example, electronic polarization. It is easy to see that if the remnant part of the polarization is known, then calculation of the induced polarization is reduced to computing of the "limit" part $P_\infty(P_0)$ by using Equation (2). However, there are two additional scalar parameters: c, β. Once the induced part of the polarization $P_e^{(i)}$ is found in the state $C^{(i)}$, a whole polarization $P^{(i)} = P_e^{(i)} + P_0^{(i)}$ can be easily determined in this state. So, all required quantities for the state $C^{(i)}$ are found and we can consider a new state.

Equation (1) presents in actuality the equation in finite differences. The solution of Equation (1) by the method of successive approximations can be formally written in the form of the following integral relation:

$$P_0^{(i)} = P_0^{(i-1)} + \int_{(E^{(i-1)},P_0^{(i-1)})}^{dE^{(i-1)}} f(E^{(i-1)}, P_0^{(i-1)}, |\Delta E^{(i-1)}|) dE . \tag{3}$$

Before application of the proposed approximations into numerical scheme of the main problem, its solution should be "configured" to a certain material. This is achieved by appropriate choice of the model parameters $p_s, a, \alpha, k, E_c, c, \beta$ so that experimental and calculation data have the best coincidence. Experimental data are usually absent as for remnant polarization, as for a whole polarization. But as it has been shown above, the proposed model allowed us to find a whole polarization. The problem of defining the parameters $p_s, a, \alpha, k, E_c, c, \beta$ reduces to solution of inverse problems. To solve this auxiliary problem, the methods of minimization of a discrepancy functional are applied.

The discrepancy functional is constructed as follows. Let us suppose we have experimental data, obtained with the help of uniaxial tests. For example, it may be the dielectric loop of small or large hysteresis. These test data for specific material are present usually either in graphic or tabular form. In any case, we can choose always a set of "n" points $\{E_k, P_k^{\exp}\}_{k=1}^{n}$ on the base of them. For numerical computations, we can take into account a change of electric field, which is described in experiment. By using the proposed model for a certain set of parameters $p_s, a, \alpha, k, E_c, c, \beta$, we obtain corresponding calculated (theoretical) values of polarization, based on the given values of electric field $\{E_k, P_k^{theor}\}_{k=1}^{n}$. Comparison of experimental and theoretical data leads to the quadratic functional of discrepancy:

$$F(p_s, a, \alpha, k, E_c, c, \beta) = \sum_{k=1}^{n} \left(P_k^{\mathbf{exp}} - P_k^{theor}\right)^2. \tag{4}$$

The main condition for choice of the model parameters is a minimization of the functional: $F(p_s, a, \alpha, k, E_c, c, \beta) \rightarrow \min$.

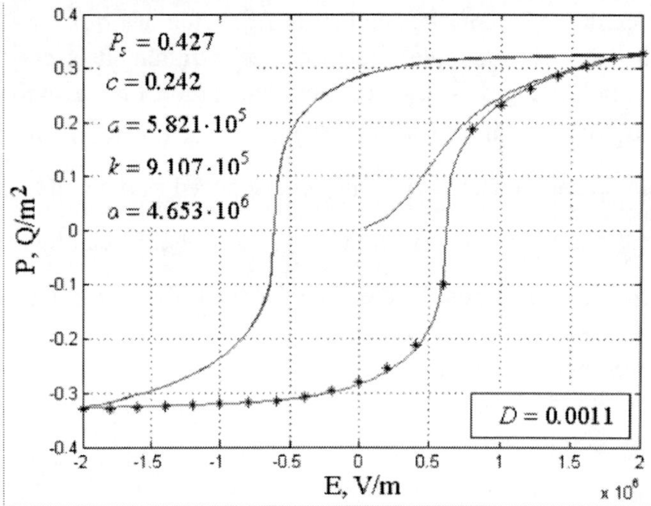

Figure 5. Theoretical and experimental data.

There are many works, devoted to the minimization of functional (4). The main difficulty consists in that functionals of this type have a complex structure. It complicates in a significant measure search of the functional minimum. Among recent works, devoted to this problem, we can note the studies of genetic algorithms [11–14]. Leaving aside a detailed description of the computational process, we note only that the numerical analysis was performed by using the genetic algorithms in MATLAB software. We use experimental data, taken of works [15,16], and obtain comparative results of the theoretical and experimental data, shown in Figure 5, where continuous lines present experimental data, and points show calculated values.

Note that the Jiles – Atherton model describes quite well full sloops of the hysteresis. However, there are significant shortcomings, when we try to describe a small hysteresis loop. We have done some work on modernization of the method, as noted above. Recall that main element of our study consisted in introduction of a new parameter – the coercive field (that required changing the calculation scheme of the model), and an additional factor in definition of the induced polarization. As a result of numerical experiments, the small and large loops of the hysteresis have been investigated. We compared the experimental small hysteresis loops, obtained in [3], with our results of numerical experiments, found by using the proposed computational scheme. The comparison results are present in Figures 6 and 7. Figure 6 shows data, obtained in [17], and Figure 7 shows similar curves, defined numerically by using the next set of parameters:

$$p_s = 0.55\ \text{K/m}^2,\quad a = 8.79\cdot 10^5\ \text{C/m}^2,\quad \alpha = 6.1\cdot 10^6\ \text{V}\cdot\text{m/C},$$
$$k = 6.9\cdot 10^5\ \text{C/m}^2,\quad E_c = 6.0\cdot 10^5\ \text{V/m},\quad c = 0.15,\quad \beta = 0.8\cdot 10^6\ \text{V}\cdot\text{m/C}$$

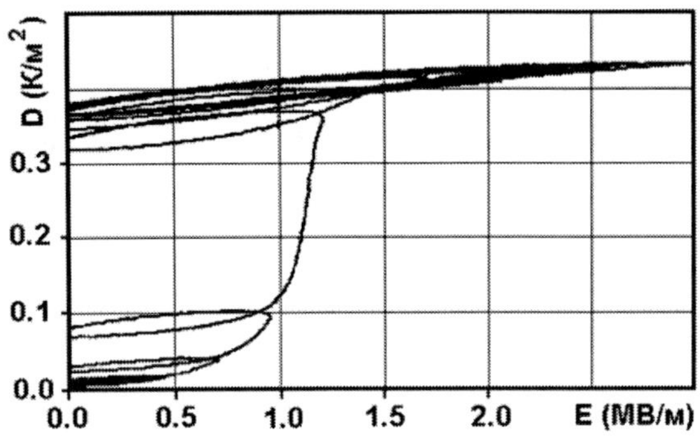

Figure 6. Experiment (Salten M. et al. [17]).

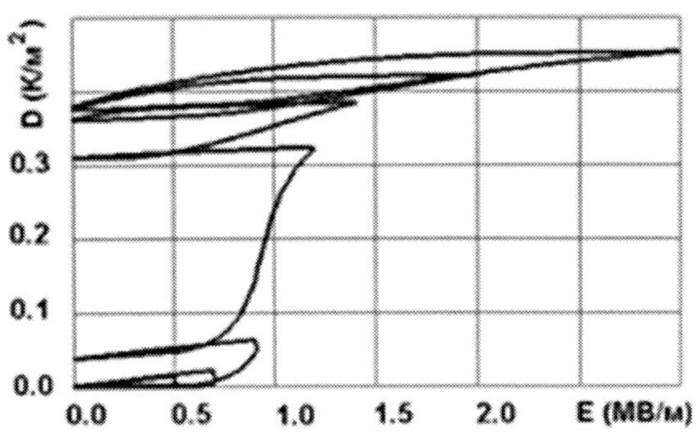

Figure 7. Numerical results of our model.

It is easy to see that there is not only qualitative but also quantitative coincidence.

6. Generalization of Model to Three-dimensional Case

The next step in constructing an algorithm of general solution consists in calculation of the remnant polarization vector, i.e., all of its components in three-dimensional case. With this aim, we developed an algorithm, using the modified one-dimensional model.

Let us suppose that vector of the remnant polarization $\mathbf{P}_0^{(i-1)}$ is known for a state $C^{(i-1)}$ at some point M of the bulk material. In transition to the next state, electric field gets an increment $\Delta\mathbf{E}^{(i-1)}$ and becomes $\mathbf{E}^{(i)} = \mathbf{E}^{(i-1)} + \Delta\mathbf{E}^{(i-1)}$. Now we need to find vector of the remnant polarization $\mathbf{P}_0^{(i)}$ in a state $C^{(i)}$.

We modify a local coordinate system $M\xi\eta\zeta$ so that $M\xi$-axis and $M\eta$-axis lie into plane of the vectors $\mathbf{E}^{(i)}$ and $\mathbf{P}_0^{(i-1)}$. Then $M\zeta$-axis is perpendicular to the plane, forming with these axes a right-handed system with basis vectors $\{\mathbf{e}_1, \mathbf{e}_2, \mathbf{e}_3\}$. In the case of a situation when $\mathbf{P}_0^{(i-1)} \uparrow\uparrow \mathbf{E}^{(i)}$, the choice of the $M\eta$-axis and $M\zeta$-axis can be arbitrary, but so they form a right-handed system. For simplicity, we can choose the $M\eta$-axis parallel to one of the planes of the global system of coordinates, for example, passing through the x_1-axis and x_2-axis.

The choice of the local system can uniquely determine a transition matrix $\mathbf{H}$ from the global coordinate system to the local one. The first vector is expressed through vector $\mathbf{E}^{(i)}$ as $e_1 = \{h_{11}, h_{12}, h_{13}\} = \{E_\xi^{(i)}/E, E_\eta^{(i)}/E, E_\zeta^{(i)}/E,\}; E = \sqrt{(E_\xi^{(i)})^2 + (E_\eta^{(i)})^2 + (E_\zeta^{(i)})^2}$. If $\mathbf{E}^{(i)} = 0$, then $\mathbf{e}_1 = \mathbf{i}_1$, that can be identified with the first unit vector of the global system. We choose the second vector $\mathbf{e}_2$ from the following conditions: (i) co-planarity of the vectors $\mathbf{E}^{(i)}$ and $\mathbf{P}_0^{(i-1)}$; (ii) orthogonality of the vectors $\mathbf{E}^{(i)}$ and $\mathbf{e}_2$; (iii) the normalization condition, $|\mathbf{e}_2| = 1$ (if $\mathbf{P}_0^{(i-1)} = 0$, then $\mathbf{e}_2 = \mathbf{i}_2$). Thus, to determine the vector $\mathbf{e}_2$, we should solve the set of equations:

$$\begin{cases} h_{21}(E_\eta^{(i)} P_{0\zeta}^{(i-1)} - E_\zeta^{(i)} P_{0\eta}^{(i-1)}) + h_{22}(E_\xi^{(i)} P_{0\zeta}^{(i-1)} - E_\zeta^{(i)} P_{0\xi}^{(i-1)}) + h_{23}(E_\xi^{(i)} P_{0\eta}^{(i-1)} - E_\eta^{(i)} P_{0\xi}^{(i-1)}) = 0, \\ h_{21}E_\xi^{(i)} + h_{22}E_\eta^{(i)} + h_{23}E_\zeta^{(i)} = 0, \\ h_{21}^2 + h_{22}^2 + h_{23}^2 = 1. \end{cases}$$

Since first equation in this set is quadratic, the equation set has two families of solutions. Let us choose the solution, which corresponds to the situation, when the vectors $\mathbf{e}_2$ and $\mathbf{P}_0^{(i-1)}$ constitute an acute angle. The third vector is found from the orthogonality condition as $\mathbf{e}_3 = \mathbf{e}_1 \times \mathbf{e}_2$. By using the obtained transition matrix, the components of vectors $\mathbf{E}^{(i-1)}$ and $\mathbf{P}_0^{(i-1)}$ in the local coordinate system are defined as

$$\begin{pmatrix} E_\xi^{(i-1)} \\ E_\eta^{(i-1)} \\ E_\zeta^{(i-1)} \end{pmatrix} = \begin{pmatrix} h_{11} & h_{12} & h_{13} \\ h_{21} & h_{22} & h_{23} \\ h_{31} & h_{32} & h_{33} \end{pmatrix} \begin{pmatrix} E_1^{(i-1)} \\ E_2^{(i-1)} \\ E_3^{(i-1)} \end{pmatrix}, \quad \begin{pmatrix} P_{0\xi}^{(i-1)} \\ P_{0\eta}^{(i-1)} \\ P_{0\zeta}^{(i-1)} \end{pmatrix} = \begin{pmatrix} h_{11} & h_{12} & h_{13} \\ h_{21} & h_{22} & h_{23} \\ h_{31} & h_{32} & h_{33} \end{pmatrix} \begin{pmatrix} P_{01}^{(i-1)} \\ P_{02}^{(i-1)} \\ P_{03}^{(i-1)} \end{pmatrix}.$$

Note that $P_{0\zeta}^{(i-1)} = 0$ in the chosen coordinate system, but in the case, when $\mathbf{P}_0^{(i-1)} \uparrow\uparrow \mathbf{E}^{(i)}$, we have the conditions $P_{0\eta}^{(i-1)} = 0, \quad P_{0\zeta}^{(i-1)} = 0$. By using new coordinate system, components of the electric field are defined in the state $C^{(i)}$ as $\mathbf{E}^{(i)} = \{E^{(i)}, \ 0, \ 0\}$, and components of the electric field increment have the forms:

$$\Delta E_{\xi}^{(i-1)} = E_{\xi}^{(i)} - E_{\xi}^{(i-1)}, \Delta E_{\eta}^{(i-1)} = -E_{\eta}^{(i-1)}, \Delta E_{\zeta}^{(i-1)} = -E_{\zeta}^{(i-1)}.$$

We take the following assumptions:

(i) increments of components of the remnant polarization vector are caused only by the corresponding components of the electric field increment;
(ii) absolute values of the remnant polarization vector cannot exceed value of the polarization saturation, i.e., it should be always satisfied the inequality $| P_0^{(i)} | \leq p_{sat}$.

The next treatment analyzes changes of absolute value of electric field and its direction. Let the absolute value of the electric field increases, but its direction changes a little. Then growth of polarization is possible only along $M\xi$-axis, while along other axes the polarization can only decrease. Now suppose that the electric field changes its direction, almost without changing its meaning. Again we have a situation, when growth of polarization takes place only along the $M\xi$-axis, and along other axes the polarization can reduce, too. The case, when the electric field decreases, while remaining almost unchanged in its direction, corresponds also to the case considered.

This analysis with the introduced assumptions allows us to apply above-mentioned one-dimensional model. Only we obtain three one-dimensional models in this case, each of which has a primary value and increment of the electric field. Let us first discuss an arising situation for each of the components. There are obvious increments of the electric field on the $M\xi$-axis and $M\eta$-axis, which cause corresponding increment of the polarization. Further experiments show that change in direction of the electric field can lead to a rotation of the remnant polarization vector, corresponding to the field direction. It means that the polarization increment along $M\zeta$-axis is absent. So we obtain obvious equality $P_{0\zeta}^{(i-1)} = 0$. Based on this treatment and taking into account integral relation (3), we have

$$P_{0\xi}^{(i)} = P_{0\xi}^{(i-1)} + \int_{(E_{\xi}^{(i-1)}, P_{0\xi}^{(i-1)})}^{\Delta E_{\xi}^{(i-1)}} f(E_{\xi}^{(i-1)}, P_{0\xi}^{(i-1)}, | \Delta E_{\xi}^{(i-1)} |) dE,$$

$$P_{0\eta}^{(i)} = P_{0\eta}^{(i-1)} + \int_{(E_{\eta}^{(i-1)}, P_{0\eta}^{(i-1)})}^{\Delta E_{\eta}^{(i-1)}} f(E_{\eta}^{(i-1)}, P_{0\eta}^{(i-1)}, | \Delta E_{\eta}^{(i-1)} |) dE, \quad (5)$$

$$P_{0\zeta}^{(i)} = 0.$$

Note that the one-dimensional model has been built in such a way that remnant polarization, derived from it, could not exceed saturation polarization. After that we find two components of polarization. The first component does not exceed a value of the saturation polarization in all cases, but the second component can only decrease at increasing the first component. However, this decreasing may be insufficient to satisfy the assumption $|P_0^{(i)}| \leq p_{sat}$. It is caused by the fact that one-dimensional models describe only reorientation of 180°-domains, but no reorientation of 90°-domains. To overcome this shortcoming, it is required to verify additionally the condition:

$$|P_0^{(i)}| = \sqrt{\left((P_{0\xi}^{(i)})^2 + (P_{0\eta}^{(i)})^2 + (P_{0\zeta}^{(i)})^2\right)} = \sqrt{\left((P_{0\xi}^{(i)})^2 + (P_{0\eta}^{(i)})^2\right)} \leq p_{sat}.$$

Case 1. If the inequality is satisfied, then components of the remnant polarization are defined in local coordinates of the state $C^{(i)}$.

Case 2. If the inequality is violated, then we perform additional calculations. First, note that a violation of the condition can occur only due to the component $P_{0\eta}^{(i)}$, since the first component cannot exceed the saturation polarization. It is obvious that the component $P_{0\eta}^{(i)}$ cannot exceed value of $|P_\tau| = \sqrt{(p_{sat})^2 - (P_{0\xi}^{(i)})^2}$. Therefore, instead of values, calculated by using Equation (5), we choose values of $P_{0\eta}^{(i)} = \text{sign}(P_{0\eta}^{(i)})\,|P_\tau|$. As a result, all components of the remnant polarization are calculated in local coordinates of the state $C^{(i)}$.

By using transformation matrix for transition to global coordinates, vector components of the remnant polarization in the state $C^{(i)}$ are found as

$$\begin{pmatrix} P_{01}^{(i)} \\ P_{02}^{(i)} \\ P_{03}^{(i)} \end{pmatrix} = \begin{pmatrix} h_{11} & h_{21} & h_{31} \\ h_{12} & h_{22} & h_{32} \\ h_{13} & h_{23} & h_{33} \end{pmatrix} \begin{pmatrix} P_{0\xi}^{(i)} \\ P_{0\eta}^{(i)} \\ P_{0\zeta}^{(i)} \end{pmatrix}. \tag{6}$$

The next important step is to determine a residual strain in the state $C^{(i)}$. Consider the case, when mechanical stresses are sufficiently small. In this case, we can assume that deformation in the local coordinates is defined by incompressibility condition of the considered material. It means that tension deformation acts along direction of the remnant polarization, and compression strain arises along transverse direction, i.e., we have the following relation:

$$\boldsymbol{\varepsilon}_0^{(i)} = \frac{\varepsilon_{sat}\mathbf{P}_0^{(i)}}{p_{sat}}\left(\mathbf{e}_1 \otimes \mathbf{e}_1 - \frac{1}{2}\mathbf{e}_2 \otimes \mathbf{e}_2 - \frac{1}{2}\mathbf{e}_3 \otimes \mathbf{e}_3\right).$$

To determine the strain components in the global coordinates, we use known formulae of transition. First, by using components of the transition matrix **H**, we construct a transformation matrix **Q** in the form:

$$\mathbf{Q} = \begin{pmatrix} h_{11}^2 & h_{11}^2 & h_{11}^2 & h_{12}h_{13} & h_{11}h_{13} & h_{11}h_{12} \\ h_{21}^2 & h_{22}^2 & h_{23}^2 & h_{22}h_{23} & h_{21}h_{23} & h_{21}h_{12} \\ h_{31}^2 & h_{32}^2 & h_{33}^2 & h_{32}h_{33} & h_{31}h_{33} & h_{31}h_{32} \\ 2h_{21}h_{31} & 2h_{21}h_{31} & 2h_{21}h_{31} & h_{22}h_{33}+h_{23}h_{32} & h_{22}h_{33}+h_{23}h_{32} & h_{22}h_{33}+h_{23}h_{32} \\ 2h_{11}h_{31} & 2h_{12}h_{32} & 2h_{31}h_{33} & h_{12}h_{33}+h_{13}h_{32} & h_{11}h_{33}+h_{13}h_{31} & h_{11}h_{32}+h_{12}h_{31} \\ 2h_{11}h_{21} & 2h_{12}h_{22} & 2h_{13}h_{23} & h_{12}h_{23}+h_{13}h_{22} & h_{11}h_{23}+h_{13}h_{21} & h_{11}h_{22}+h_{12}h_{21} \end{pmatrix}$$

Then we introduce vectors, formed by components of the tensor $\boldsymbol{\varepsilon}_0^{(i)}$ in the global and local coordinate systems:

$$\boldsymbol{\varepsilon}_{0g}^{(i)} = \{\varepsilon_{11}, \varepsilon_{22}, \varepsilon_{33}, 2\varepsilon_{23}, 2\varepsilon_{13}, 2\varepsilon_{12}\}^T,$$

$$\boldsymbol{\varepsilon}_{0l}^{(i)} = \{\varepsilon_{11}^l, \varepsilon_{22}^l, \varepsilon_{33}^l, 2\varepsilon_{23}^l, 2\varepsilon_{13}^l, 2\varepsilon_{12}^l\}^T.$$

Finally, we write the transformation formula:

$$\boldsymbol{\varepsilon}_{0g}^{(i)} = \mathbf{Q}^T \cdot \boldsymbol{\varepsilon}_{0l}^{(i)}. \tag{7}$$

Expressions (5), (6) determine completely vector of remnant polarization $\mathbf{P}_0^{(i)}$ and tensor of residual strain $\boldsymbol{\varepsilon}_0^{(i)}$ in the state $C^{(i)}$. Below, we shall also use increments of these parameters, which are easily found as $\Delta\mathbf{P}_0^{(i-1)} = \mathbf{P}_0^{(i)} - \mathbf{P}_0^{(i-1)}$, $\Delta\boldsymbol{\varepsilon}_0^{(i-1)} = \boldsymbol{\varepsilon}_0^{(i)} - \boldsymbol{\varepsilon}_0^{(i-1)}$.

7. Determination of Elastic and Electric Characteristics in the Case of Heterogeneous Polarization in Some Direction

Since elastic and electric characteristics depend on the remnant polarization vector, which changes its direction and length in transition from one point of the body to other, it is necessary to obtain formulae for calculation of these characteristics.

It is convenient for these calculations to introduce matrix representations of the parameters in the Voigt forms:

$$\Delta\boldsymbol{\varepsilon}_e^{(i)} = \begin{pmatrix} \Delta\varepsilon_{e11}^{(i)} \\ \Delta\varepsilon_{e22}^{(i)} \\ \Delta\varepsilon_{e33}^{(i)} \\ 2\Delta\varepsilon_{e23}^{(i)} \\ 2\Delta\varepsilon_{e13}^{(i)} \\ 2\Delta\varepsilon_{e23}^{(i)} \end{pmatrix}, \quad \Delta\boldsymbol{\sigma}^{(i)} = \begin{pmatrix} \Delta\sigma_{11}^{(i)} \\ \Delta\sigma_{22}^{(i)} \\ \Delta\sigma_{33}^{(i)} \\ \Delta\sigma_{23}^{(i)} \\ \Delta\sigma_{13}^{(i)} \\ \Delta\sigma_{12}^{(i)} \end{pmatrix}, \quad \Delta\mathbf{D}_e^{(i)} = \begin{pmatrix} \Delta D_{e1}^{(i)} \\ \Delta D_{e2}^{(i)} \\ \Delta D_{e3}^{(i)} \end{pmatrix}, \quad \Delta\mathbf{E}^{(i)} = \begin{pmatrix} \Delta E_1^{(i)} \\ \Delta E_2^{(i)} \\ \Delta E_3^{(i)} \end{pmatrix}, \quad \mathbf{P}_0^{(i)} = \begin{pmatrix} P_{01}^{(i)} \\ P_{02}^{(i)} \\ P_{03}^{(i)} \end{pmatrix}, \dots$$

$$\mathbf{S}(\mathbf{P}_0^{(i)}) = \begin{pmatrix} S_{11} & S_{12} & S_{13} & S_{14} & S_{15} & S_{16} \\ S_{12} & S_{22} & S_{23} & S_{24} & S_{25} & S_{26} \\ S_{13} & S_{23} & S_{33} & S_{34} & S_{35} & S_{36} \\ S_{14} & S_{24} & S_{34} & S_{44} & S_{45} & S_{46} \\ S_{15} & S_{25} & S_{35} & S_{45} & S_{55} & S_{56} \\ S_{16} & S_{26} & S_{36} & S_{46} & S_{56} & S_{66} \end{pmatrix}, \quad \mathbf{d}(\mathbf{P}_0^{(i)}) = \begin{pmatrix} d_{11} & d_{12} & d_{13} & d_{14} & d_{15} & d_{16} \\ d_{21} & d_{22} & d_{23} & d_{24} & d_{25} & d_{26} \\ d_{31} & d_{32} & d_{33} & d_{34} & d_{35} & d_{36} \end{pmatrix},$$

$$\text{э}(\mathbf{P}_0^{(i)}) = \begin{pmatrix} \text{э}_{11} & \text{э}_{12} & \text{э}_{13} \\ \text{э}_{12} & \text{э}_{22} & \text{э}_{23} \\ \text{э}_{13} & \text{э}_{23} & \text{э}_{33} \end{pmatrix}.$$

We assume that partially polarized ceramic belongs to a class of locally anisotropic materials. By this, the remnant polarization vector determines a direction of anisotropy axis of material in each point of the body. We introduce in each point a new local coordinate system $Mz_1z_2z_3$, assuming that Mz_3-axis coincides with direction of the remnant polarization vector $\mathbf{P}_0^{(i)}$ in the state $C^{(i)}$. Then third basis vector can be defined in the form: $\mathbf{v}_3 = \dfrac{\mathbf{P}_0^{(i)}}{|\mathbf{P}_0^{(i)}|} = l_{31}\mathbf{i} + l_{32}\mathbf{j} + l_{33}\mathbf{k}$ (if $\mathbf{P}_0^{(i)} = 0$, then $l_{31} = 0, l_{32} = 0, l_{33} = 1$). The condition of transverse isotropy of material implies that others two vectors of the local basis can be defined arbitrarily in a plane perpendicular to $\mathbf{v}_3$. The simplest possibility is choice of direction of the vector $\mathbf{v}_2$ in parallel to the coordinate plane Ox_1x_2 of the global coordinate system. By defining the basis vectors of the local basis as $\mathbf{v}_1 = l_{11}\mathbf{i} + l_{12}\mathbf{j} + l_{13}\mathbf{k}, \mathbf{v}_2 = l_{21}\mathbf{i} + l_{22}\mathbf{j} + 0\mathbf{k}$, we obtain the following relations:

$$|\mathbf{v}_2| = 1, \qquad \mathbf{v}_3 \cdot \mathbf{v}_2 = 0, \qquad \mathbf{v}_1 = \mathbf{v}_2 \times \mathbf{v}_3.$$

By denoting components of the remnant polarization vector $\mathbf{P}_0^{(i)}$ through $\{P_1, P_2, P_3\}$, we can write directing cosines of the local system as

$$l_{11} = \frac{P_1}{\sqrt{P_1^2 + P_2^2 + P_3^2}}, \quad l_{12} = \frac{P_2}{\sqrt{P_1^2 + P_2^2 + P_3^2}}, \quad l_{13} = \frac{P_3}{\sqrt{P_1^2 + P_2^2 + P_3^2}},$$

$$l_{21} = \frac{P_2}{\sqrt{P_1^2 + P_2^2}}, \quad l_{22} = -\frac{P_1}{\sqrt{P_1^2 + P_2^2}}, \quad l_{23} = 0,$$

$$l_{11} = \frac{P_1 P_3}{\sqrt{P_1^2 + P_2^2}\sqrt{P_1^2 + P_2^2 + P_3^2}}, \quad l_{12} = -\frac{P_2 P_3}{\sqrt{P_1^2 + P_2^2}\sqrt{P_1^2 + P_2^2 + P_3^2}}, \quad l_{13} = \frac{\sqrt{P_1^2 + P_2^2}}{\sqrt{P_1^2 + P_2^2 + P_3^2}}.$$

Thus, the transformation matrix takes the form:

$$\mathbf{L} = \begin{pmatrix} l_{11} & l_{12} & l_{13} \\ l_{21} & l_{22} & l_{23} \\ l_{31} & l_{32} & l_{33} \end{pmatrix}.$$

By using the components of this matrix, we introduce auxiliary matrix $\mathbf{R}$, required to convert the physical units:

$$\mathbf{R} = \begin{pmatrix} l_{11}^2 & l_{12}^2 & l_{13}^2 & 2l_{12}l_{13} & 2l_{11}l_{13} & 2l_{11}l_{12} \\ l_{21}^2 & l_{22}^2 & l_{23}^2 & 2l_{22}l_{23} & 2l_{21}l_{23} & 2l_{21}l_{22} \\ l_{31}^2 & l_{32}^2 & l_{33}^2 & 2l_{32}l_{33} & 2l_{31}l_{33} & 2l_{31}l_{32} \\ l_{21}l_{31} & l_{22}l_{32} & l_{23}l_{33} & l_{22}l_{33} + l_{23}l_{32} & l_{21}l_{33} + l_{23}l_{31} & l_{21}l_{32} + l_{22}l_{31} \\ l_{11}l_{31} & l_{12}l_{32} & l_{13}l_{33} & l_{12}l_{33} + l_{13}l_{32} & l_{11}l_{33} + l_{13}l_{31} & l_{11}l_{32} + l_{12}l_{31} \\ l_{11}l_{21} & l_{12}l_{22} & l_{13}l_{23} & l_{12}l_{23} + l_{13}l_{22} & l_{11}l_{23} + l_{13}l_{21} & l_{11}l_{22} + l_{12}l_{21} \end{pmatrix}.$$

Now we can define physical characteristics of material in the local coordinate system. Let isotropic elastic properties of non-polarized ceramics are elastic modulus E and Poisson's ratio ν. Let $\mathbf{S}_0$ is the matrix of elastic compliances of an isotropic material, and $\mathbf{S}_{sat}$ is the compliance matrix of the ceramic polarized up to saturation. Then we have in the local coordinate system the following matrix representations:

$$\mathbf{S}_0 = \frac{1}{E}\begin{pmatrix} 1 & -\nu & -\nu & 0 & 0 & 0 \\ -\nu & 1 & -\nu & 0 & 0 & 0 \\ -\nu & -\nu & 1 & 0 & 0 & 0 \\ 0 & 0 & 0 & 2(1+\nu) & 0 & 0 \\ 0 & 0 & 0 & 0 & 2(1+\nu) & 0 \\ 0 & 0 & 0 & 0 & 0 & 2(1+\nu) \end{pmatrix},$$

$$\mathbf{S}_{sat} = \begin{pmatrix} S_{11} & S_{12} & S_{13} & 0 & 0 & 0 \\ S_{12} & S_{11} & S_{13} & 0 & 0 & 0 \\ S_{13} & S_{13} & S_{33} & 0 & 0 & 0 \\ 0 & 0 & 0 & S_{44} & 0 & 0 \\ 0 & 0 & 0 & 0 & S_{44} & 0 \\ 0 & 0 & 0 & 0 & 0 & \frac{1}{2}(S_{11}-S_{12}) \end{pmatrix}$$

Then we determine elastic properties of partially polarized ceramic in the form:

$$\overset{0}{\mathbf{S}} = \mathbf{S}_0 + \frac{|\mathbf{P}_0^{(i)}|}{p_{sat}}(\mathbf{S}_{sat} - \mathbf{S}_0) = \mathbf{S}_0 + \frac{|\mathbf{P}_0^{(i)}|}{p_{sat}}\mathbf{S}_1 .$$

Let isotropic dielectric properties of non-polarized ceramic are described by coefficient of permittivity $э$ and permittivity of polarized ceramic is characterized by values of $э_{11}$, $э_{33}$. Then the matrices of dielectric properties of the non-polarized and polarized ceramics can be written as

$$\mathbf{э}_0 = \begin{pmatrix} э & 0 & 0 \\ 0 & э & 0 \\ 0 & 0 & э \end{pmatrix}, \quad \mathbf{э}_{sat} = \begin{pmatrix} э_{11} & 0 & 0 \\ 0 & э_{11} & 0 \\ 0 & 0 & э_{33} \end{pmatrix}.$$

Dielectric properties of partially polarized ceramic in the local system can be described as

$$\overset{0}{\mathbf{э}} = \mathbf{э}_0 + \frac{|\mathbf{P}_0^{(i)}|}{p_{sat}}(\mathbf{э}_{sat} - \mathbf{э}_0) = \mathbf{э}_0 + \frac{|\mathbf{P}_0^{(i)}|}{p_{sat}}\mathbf{э}_1 .$$

Piezoelectric properties of polarized piezoelectric ceramic are characterized by three moduli d_{15}, d_{31}, d_{33}, and matrix of piezoelectric properties has the form:

$$\mathbf{d}_{sat} = \mathbf{d}_1 = \begin{pmatrix} 0 & 0 & 0 & 0 & d_{15} & 0 \\ 0 & 0 & 0 & d_{15} & 0 & 0 \\ d_{31} & d_{31} & d_{33} & 0 & 0 & 0 \end{pmatrix}.$$

Partially polarized ceramics will be characterized in the local system by the matrix:

$$\overset{0}{\mathbf{d}} = \frac{|\mathbf{P}_0^{(i)}|}{p_{sat}}\mathbf{d}_1 .$$

By using the transition matrix and the matrices of material characteristics in the local coordinates, we can determine characteristics of material in the global coordinate system in the state $C^{(i)}$ as

$$\mathbf{S} \equiv \mathbf{S}(\mathbf{P}_0^{(i)}) = \mathbf{R}^T \cdot \overset{0}{\mathbf{S}} \cdot \mathbf{R} = \mathbf{S}_0 + \frac{|\mathbf{P}_0^{(i)}|}{p_{sat}} \mathbf{R}^T \cdot \mathbf{S}_1 \cdot \mathbf{R},$$

$$\mathbf{d} \equiv \mathbf{d}(\mathbf{P}_0^{(i)}) = \mathbf{L}^T \cdot \overset{0}{\mathbf{d}} \cdot \mathbf{R} = \frac{|\mathbf{P}_0^{(i)}|}{p_{sat}} \mathbf{L}^T \cdot \mathbf{d}_1 \cdot \mathbf{R},$$

$$\text{э} \equiv \text{э}(\mathbf{P}_0^{(i)}) = \mathbf{L}^T \cdot \overset{0}{\text{э}} \cdot \mathbf{L} = \text{э}_0 + \frac{|\mathbf{P}_0^{(i)}|}{p_{sat}} \mathbf{L}^T \cdot \text{э}_1 \cdot \mathbf{L}.$$

Thus, the developed mathematical model of the polarization allows us to find the remnant polarization field and determine physical characteristics of heterogeneously polarized ceramic in dependence on coordinates of points of the body. Further study of properties of specific transducers is conducted by using finite element method, in which the physical characteristics of the considered material are already certain functions of the coordinates, and they do not change, regardless of electrical and mechanical fields.

8. Finite Element Model of Heterogeneous Electro-Elastic Material in ACELAN

Let us examine a piezoelectric transducer Ω, presented by a set of areas $\Omega_j = \Omega_{pk}$; $k = 1, 2,..., N_p$; $j = k$ with properties of piezoelectric materials, and a set of areas $\Omega_j = \Omega_{em}$; $m = 1,2,..., N_e$; $j = N_p + m$ with properties of elastic materials. It is appropriate to describe physical-mechanical processes, taking place in the media Ω_{pk} and Ω_{em} into framework of the piezoelectricity (electric elasticity) and elasticity theory.

We assume that the following constitutive equations are satisfied (the piezoelectric medium is present by areas $\Omega_j = \Omega_{pk}$):

$$\rho_{pk}\ddot{\mathbf{u}} + \alpha_{dj}\rho_j\dot{\mathbf{u}} - \nabla \cdot \boldsymbol{\sigma} = \mathbf{f}_j;\ \nabla \cdot \mathbf{D} = 0, \tag{8}$$

$$\boldsymbol{\sigma} = \mathbf{c}_j^E \cdot\cdot(\boldsymbol{\varepsilon} + \beta_{dj}\dot{\boldsymbol{\varepsilon}}) - \mathbf{e}_j^T \cdot \mathbf{E};\mathbf{D} + \varsigma_d\dot{\mathbf{D}} = \mathbf{e}_j \cdot\cdot (\boldsymbol{\varepsilon} + \varsigma_d\dot{\boldsymbol{\varepsilon}}) + \text{э}_j^S \cdot \mathbf{E}, \tag{9}$$

$$\boldsymbol{\varepsilon} = (\nabla\mathbf{u} + \nabla\mathbf{u}^T)/2;\mathbf{E} = -\nabla\varphi \tag{10}$$

where ρ(x, t) is the continuous function of coordinates (material density), u(x) is the displacement vector-function, σ is the stress tensor, f are the mass forces, D is the electric induction vector; c_j^E are the components of the elastic constant tensor; e_j are the piezoelectric stress components, ε is the strain tensor, E is the electric field vector; φ(x) is the electric potential function; $э_j^S$ are the components of the dielectric permittivity tensor; α_{dj}, β_{dj}, ζ_d are non-negative damping coefficients, and the other symbols have standard designations for theory of electroelasticity with exception of index j, corresponding to area Ω_j (the piezomoduli e_j are equal to zero for the elastic media $\Omega_j = \Omega_{ek}$).

Only stress fields would be considered for the media $\Omega_j = \Omega_{em}$ with pure elastic properties. The constitutive relationships, similar to Equations (8) – (10), are used with negligible electric fields and piezoelectric connectivity effects. Equations (8) – (10) are added to mechanical and electrical boundary conditions, as well as initial conditions in the case of non-stationary problem. Numerical modeling of devices that can be described by Equations (8) – (10) is performed by using finite element method.

In addition to previous equations, all material properties are handled as functions of coordinates:

$$\rho_k = \rho_{pk}(\mathbf{x}),\ \mathbf{c}_j^E = \mathbf{c}_j^E(\mathbf{x}),\ \boldsymbol{\varepsilon}_{\mathbf{a}\,j}^{\ S} = \boldsymbol{\varepsilon}_{\mathbf{a}\,j}^{\ S}(\mathbf{x}),\ \mathbf{e}_j^T = \mathbf{e}_j^T(\mathbf{x}) \tag{11}$$

$$g = g^i + |P|(g^a - g^i) \text{ for tensors } \mathbf{c}_j^E \text{ and } э_j^S;\ g = |P| g^a \text{ for tensor } \mathbf{e}_j^T, \tag{12}$$

where g is the corresponding tensor component, i marks isotropic state, a marks anisotropic state. Tensor of piezoconstants $\mathbf{e}_j^T$ will be zero for isotropic bodies.

9. Definition of Heterogeneous Polarization in ACELAN

There are two ways to define heterogeneous polarization in ACELAN software. The polarization field can be obtained by using the finite-element model, described in Section 8 or given preliminary. In the case of known distribution of the polarization field, either analytical or tabular function can be used. To keep the tabular data relevant to different meshes, ACELAN provides interpolation utilities, based on spline algorithms. Figure 8 shows: (a) estimated polarization field, (b) its transformation to a refined mesh with a spline for rotation angle, and (c) two independent splines for rotation angle and modulus.

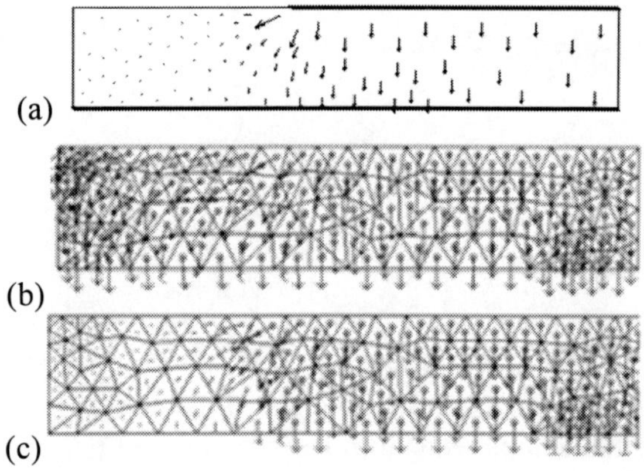

Figure 8. Using obtained polarization field on refined mesh.

We use thin-plate spline and smoothing spline, created by using Meinguet algorithm [18]. Idea of the thin-plate spline method consists in minimizing the full free energy of infinite elastic thin plate, deformed by bending in some points:

$$J_2(\varphi) = \int_{-\infty}^{+\infty}\int_{-\infty}^{+\infty}\left[\left(\frac{\partial^2\varphi}{\partial x^2}\right)^2 + 2\left(\frac{\partial^2\varphi}{\partial x\partial y}\right)^2 + \left(\frac{\partial^2\varphi}{\partial y^2}\right)^2\right]dx\,dy \to \min. \tag{13}$$

The solution of this problem can be described as $\varphi(x_i, y_i) = f_i,\quad i = \overline{1,N}$.

$$\varphi(x, y) = \sum_{i=1}^{N} c_i r_i^2 \ln r_i^2 + c_{N+1} + c_{N+2}\, x + c_{N+3}\, y,\; r_i^2 = (x - x_i)^2 + (y - y_i)^2. \tag{14}$$

The set of linear equations, presented below, must be solved to find coefficients c_i:

$$\left.\begin{aligned} &\varphi(x_j, y_j) = \sum_{i=1}^{N} c_i r_{ij}^2 \ln r_{ij}^2 + c_{N+1} + c_{N+2}x_j + c_{N+3}y_j = f_j;\\ &r_{ij}^2 = (x_j - x_i)^2 + (y_j - y_i)^2,\quad j = \overline{1,N};\\ &\sum_{i=1}^{N} c_i = 0;\ \sum_{i=1}^{N} x_i c_i = 0;\ \sum_{i=1}^{N} y_i c_i = 0. \end{aligned}\right\} \tag{15}$$

This equation set is no positively definite and has zeros on main diagonal. More effective way to build a spline was described by J. Meinguet, considering the smoothing splines. The minimizing functional differs from one of the previous method:

$$J_m(\varphi) \to \min_{\varphi\in\Theta},\quad \Theta = \left\{\varphi \in W_2^m\left(\mathrm{R}^n\right) \mid \varphi(t^i) = f^i, i = \overline{1,N}\right\}, \tag{16}$$

$$J_m(\varphi) = \left\|\nabla^m\varphi\right\|_{L_2(\Omega)} = \left(\int_{\Omega}\sum_{|\alpha|=m}\frac{m!}{\alpha!}(D^\alpha\varphi)^2\, d\Omega\right)^{1/2}$$

where $t^i\ \ (i = \overline{1,N})$ are the interpolation nodes, $f_i\ \ (i = \overline{1,N})$ are the known values. Here m presents required smoothness degree; the method uses a number of input points is greater than $2m$. Obviously, it can be easily achieved in finite-element case. Numerical solution has the following form:

$$\varphi(t) = \sum_{i=1}^{N} c_i E_m\left(t, t^i\right) + p(t),\quad t \in \mathrm{R}^n; \tag{17}$$

$$E_m(u,v) = \begin{cases} (-1)^{n/2-1}\,|u - v|^{2m-n}\ln|u - v|, & \text{at even } n;\\ (-1)^{(n-1)/2}\,|u - v|^{2m-n}, & \text{at odd } n; \end{cases}$$

$$p(t) = \sum_{\nu=1}^{M} d_\nu \phi_\nu(t) \in P_{m-1},$$

where $\phi_\nu(t)$, $\nu = \overline{1,M}$, $M = C^n_{n+m-1}$ are the different monomials with powers are lesser or equal to m. The coefficients can be obtained from the set of equations:

$$\left.\begin{aligned} &\sum_{j=1}^{N} c_j E_m(t^i, t^j) + \sum_{\nu=1}^{M} d_\nu \phi_\nu(t^i) = f^i, \quad i = \overline{1,N}; \\ &\sum_{j=1}^{N} c_j \phi_\nu(t^j) = 0, \qquad \nu = \overline{1,M}. \end{aligned}\right\} \tag{18}$$

Idea of the method is to divide input points to define system of fundamental polynomials with power m. It leads to two sets of equations with different orders. The equation set with smaller order can be solved by any method, and the equation set with larger $(M - m)$ order is symmetric and positively definite. So we can use effective method of square root.

10. Plain Piezoelectric Transducer Analysis

Numerical experiments were conducted with a plain piezoelectric transducer (see Figure 9) with free boundaries and electrodes on upper and lower surfaces. Material was PZT-4, the model included a study of damping with coefficients $\alpha = 2.7 \cdot 10^2, \beta = \zeta = 3.84 \cdot 10^{-8}$. Length of the rod was 1 cm, its thickness changed in numerical experiments from 0.025 to 0.200 cm. Oscillations were excited by a voltage of 200 V.

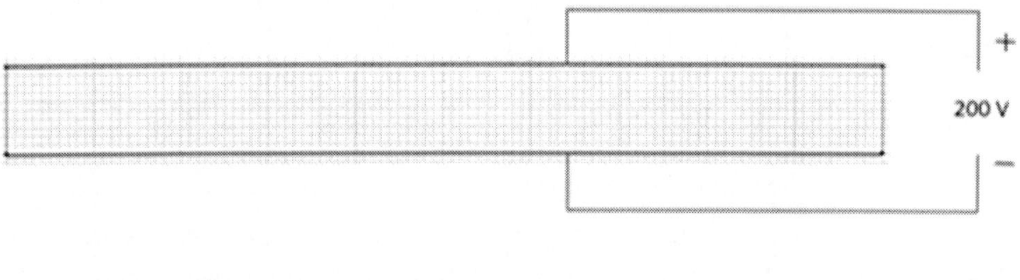

Figure 9. Scheme of piezoelectric rod.

First part of the study consisted in analyzing eigenforms of the rod. In this research, longitudinal oscillations with coupled electro-mechanical fields were observed. Three of them are present in Figure 10.

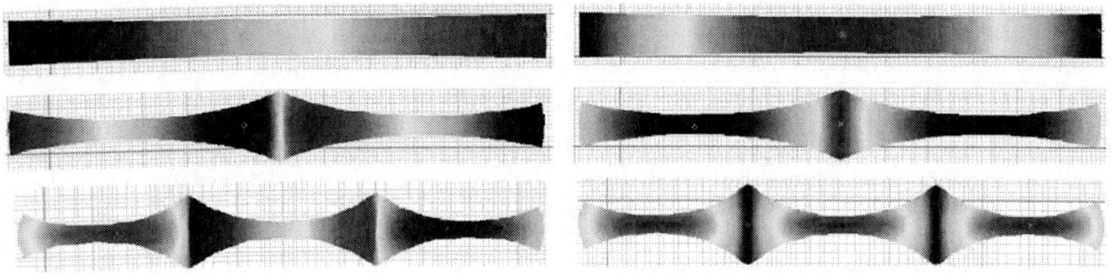

Figure 10. First three eigenforms of the rod for x-component distribution (left) and its strain (right).

We suggested that polarization field responded for strain field. The strain distribution of x-component on the middle line of the rod is present in Figure 11.

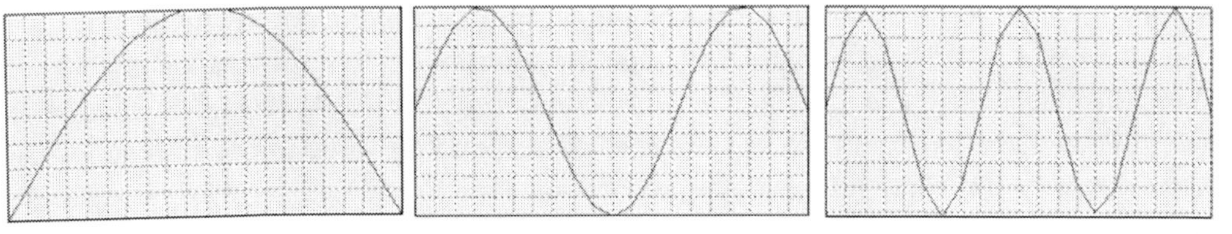

Figure 11. Distribution of dimensionless strain of x-component on the middle line for three eigenforms.

This type of functions can be described as $F(x) = \sin(\pi kx / l)$, where l is the length of the rod. Efficiency of piezo-device can be estimated by electro-mechanical coupling coefficient. In this study, we used the following representation:

$$K^2 = \frac{\varphi_a^2 - \varphi_r^2}{\varphi_a^2}, \tag{19}$$

where φ_r is the resonance frequency and φ_a is the anti-resonance frequency. Some of observed polarization fields are present in Figure 12.

(a)

(b)

(c)

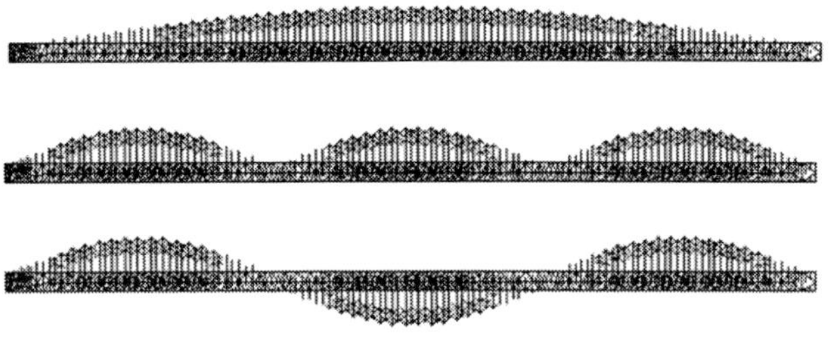

Figure 12. Some inhomogeneous polarization fileds.

Three eigenmodes were measured for each of the resonance and anti-resonance frequencies of the polarization field. We estimated efficiency by comparing the electro-mechanical coupling coefficient of the heterogeneously polarized specimen with uniformly polarized sample. Frequency differences between different cases were lesser than 5%, but growth of the electro-mechanical coupling coefficient was large in some cases. A better scheme than uniform polarization was not found for the first eigenmode. Figure 12(b) shows the most efficient polarization field, obtained for the second eigenmode, and Figure 12(c) shows the best field for the third eigenmode. Specimens of different thickness were analyzed. Figure 13 shows how the electro-mechanical coupling coefficient differs for the second and third eigenmodes; K_n^2 is the electro-mechanical coupling coefficient for the best heterogeneous field, K_0^2 is the same factor for uniform field, h is the thickness of the rod, l is the length of the rod.

Other important parameter of piezodevice is a frequency response. Bandwidth expansion of a device leads to more effective usage. In the case of longitudinal oscilations, dispalcemnet of x-compoment can be used for measurements. The frequency response analysis (FRA) was also performed by using ACELAN. We constructed plots (see Figure 14) and evaluated bandwidths for all uniform fields and corresponding optimum heterogenious fields.

Usual scheme of the bandwidth evaluation determines this parameter at 50% of maximum displacement. The possible way of comparing the FRA results is to take the bandwidth value from uniform case and use it as corresponding value in heterogeneous case. Figure 15 shows how bandwidth for the second eigenmode differs for various values of thickness and polarization.

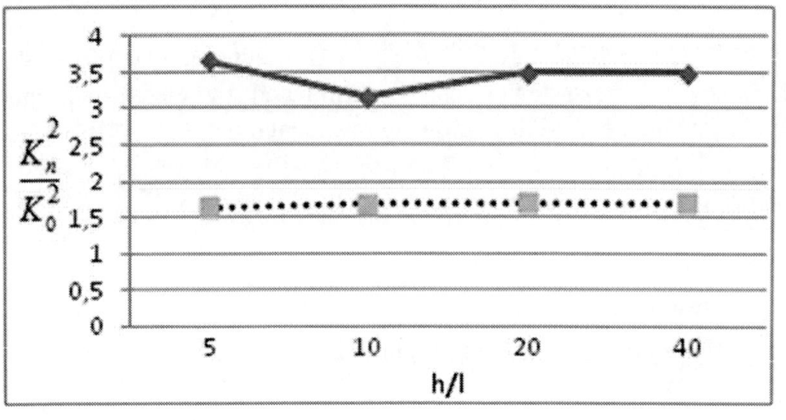

Figure 13. Changes of electro-mechanical coupling coefficient for second (continuous line) and third (dotted line) eigenmodes.

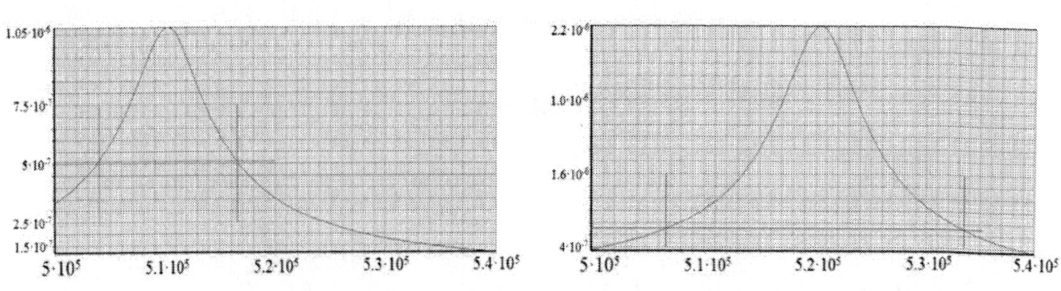

Figure 14. Analysis of frequency response for second eigenmode; axes present longitudinal displacement modulus (m) and frequency (Hz) for uniform (left) and heterogeneous (right) fields.

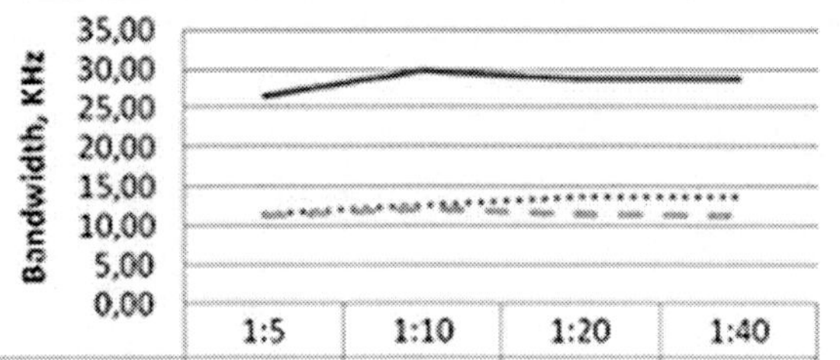

	1:5	1:10	1:20	1:40
······· Uniform	11,50	12,60	13,70	13,70
—— Polarization 3b	26,50	29,80	28,60	28,60
– – Polarization 3c	11,20	12,20	11,50	11,30

Figure 15. Bandwith for second eigenmode; dimensionless parameter *h/l* presents on *x*-axis.

The results of FRA show that the bandwith increases almost 3 times and this increasing is close to growth of the electro-mechanical coupling coefficient. The same trend is observed for the third eigenmode.

ACKNOWLEDGMENTS

The investigation was supported by the Russian Foundation for Basic Research (grants Nos. 12-01-00829-a, 13-08-01094-a) and the project No. 1334 of the base part of task No. 2014/174 to fulfill state works in the field of scientific activities of the Department of Education and Science of Russia.

REFERENCES

[1] Ilyushin, A. A., *Plasticity*, Publishing House of Academy of Sciences of the USSR, Moscow, 1963, pp. 1-272 (In Russian).

[2] Hill, R. *Mathematical Theory of Plasticity*, Gostekhizdat, Moscow, 1956, pp. 1-407 (In Russian).

[3] Huber, J. E.; Fleck, N. A. *Journal of the Mechanics and Physics of Solids*, 2001, *vol. 49*, 785-811.

[4] McMeeking, R. M.; Landis, C. M. *International Journal of Engineering Science*, 2002, *vol. 40*, 1553-1577.

[5] Landis, C. M. *Journal of the Mechanics and Physics of Solids*, 2002, *vol. 50*, 127-152.

[6] Preisach, F. *Zs. f. Phys*., *vol. 94*, 1935, 277-302.

[7] Akulov, N. S., *Dislocations and Plasticity*, Publishing House of the Academy of Sciences of the BSSR, Minsk, 1961, pp. 1-112 (In Russian).

[8] Jiles, D. C.; Atherton, D. L. *Journal of Magnetism and Magnetic Materials*, 1986, *vol.* 61, 48-60.

[9] Smith, R. C.; Ounaies, Z. A. *Journal of Intelligent Material Systems and Structures*, 2000, *vol. 11(1)*, 62-79.

[10] Belokon, A. V.; Skaliukh, A. S. *Mathematical Modeling of Irreversible Processes of Polarization*. FIZMATLIT, Moscow, 2010, pp. 1-328 (In Russian).

[11] Jiaju, Zheng; Shuying, Cao; Hongli, Wang; Wenmei, Huang, *Neurocomputing*, 2007, *vol. 70(4-6)*, 749-761.

[12] Che-Hang, Chan; Guangjun, Liu, *Mechatronics*, 2007, *vol. 17(7)*, 391-402.

[13] Krzysztof, Chwastek; Jan, Szczyglowski, *Mathematics and Computers in Simulation*, 2006, *vol. 71(3)*, 206-211.

[14] Kerry, Gallagher; Malcolm, Sambridge, *Computers & Geosciences*, 1994, *vol. 20(7-8)*, 1229-1236.

[15] Yan, B.; Waechter, D.; Blacow, R.; Prasad, S. E. In: *Proc. Int. Conf. on Smart Materials and Structures, Montreal, Quebec, Canada*, 2002, pp. 33-39.

[16] Dunmin, Lin; Kwok, K. W.; Chan, H. L. W. *Appl. Phys. Let.*, 2007, *vol. 90*, 3.

[17] Salten, M.; Schneider, G. A.; Knoblauch, V.; McMeeking, R. M. *Int. Journal of Solids and Structures*, 2005, *vol. 42*, 3953-3966.

[18] Meinguet, J. *Polynomial and Spline Approximation. Proc. NATO Adv. Study Inst. Calgary*, 1978; Kluwer Academic Publishers, Dordrecht, Boston, London, 1979. pp. 163-190.

In: Advanced Materials - Studies and Applications
Editors: I.A. Parinov, S.-H. Chang, S.Theerakulpisut
ISBN: 978-1-63463-749-7

Chapter 13

ACOUSTIC EMISSION DIAGNOSTICS OF THE KINETICS OF PHYSICOCHEMICAL PROCESSES IN LIQUID AND SOLID MEDIA

***S. I. Builo*[1,*]*, D. M. Kuznetsov*[2,†] *and V. L. Gaponov*[2,‡]**

[1]Vorovich Mathematics, Mechanics and Computer Sciences Institute, Southern Federal University, Rostov-on-Don, Russia
[2]Don State Technical University, Rostov-on-Don, Russia

ABSTRACT

It has been established that such processes as dissolution, melting, electrolysis, and homogeneous chemical reactions are accompanied by the emission of acoustic waves in the sound and ultrasonic frequency ranges. The objective of this study was to investigate the features of acoustic emission (AE) during growth, dissolution, crystallization and melting of crystals and to try to understand the mechanisms of accompanying AE. We discovered, that in the most general case, histograms of the distribution of time intervals between AE events during ice crystallization group well around an exponent corresponding to the density of time intervals between events of a Poisson stream. Thus, the methods of quantitative reconstruction of the parameters of structural transformations on the basis of AE signals that have been developed in our previous investigations by using the Poisson model can be used in studies of crystallization processes.

The data from our experiments show that the AE method can be successfully applied to study quantitatively the dynamics of the physicochemical processes of dissolution and crystallization of ionic compounds in a liquid medium.

* S. I. Builo, e-mail: bsi@math.sfedu.ru.
† D. M. Kuznetsov, e-mail: kuznetsovdm@mail.ru.
‡ V. L. Gaponov, e-mail: v.gaponov6591@yandex.ru.

1. Introduction

The method of acoustic emission (AE) analysis has ever-increasing application in the physics of condensed media and strength, fracture mechanics, and problems of nondestructive testing and diagnostics [1–9]. Recently, it has been revealed that the AE method also allows one to obtain important (often unique) information on the kinetics of chemical and physicochemical processes in liquid media. It has been established that such processes as dissolution, melting, electrolysis, and homogeneous chemical reactions are also accompanied by the emission of acoustic waves in the sound and ultrasonic frequency ranges [10–15].

There is no generally accepted viewpoint on the nature of the AE phenomenon during the dissolution, crystallization, and melting of substances. For example, AE can be caused by an exothermal or endothermic character of the crystallization and dissolution processes [17]. In fact, the release (absorption) of latent crystallization energy during phase transitions of the first kind in a liquid may be quite substantial. A change in the temperature of a subsurface liquid layer causes a density perturbation and, as a result, the generation of acoustic waves. However, if the crystal dissolution process is considered to result from gradually overcoming an energy barrier by individual ions and molecules, which is distributed over a volume, the acoustic signals generated in this process will be fully compensated by thermal fluctuations of particles of a medium. In this case, the dissolution process cannot be detected in any way.

Some authors treat the hypothesis that during dissolution of a solid in a liquid that has a narrow zone in which molecules of a dissolved substance are detached and introduced the solution serves as a perturbation zone of the liquid density. The change in the liquid density in this zone propagates over the entire volume in the form of sound signals. According to this theory, the AE amplitude increases to a detectable level during dissolution of substances owing to the effect of resonance coincidence of the forced vibration of a liquid with the frequency of natural crystal vibrations.

However, based on this assumption, one could expect that, first, the frequency range of AE waves during dissolution of crystals with dimensions of, e.g., 10 and 1 mm must have differed tenfold; second, as crystals are being dissolved and their dimensions correspondingly decrease, the resonance frequency must increase.

In the course of these experiments, we studied the AE spectrum of various crystals and the results unambiguously showed that the AE frequency range virtually does not depend on the dimensions of crystals during their dissolution or melting. In addition, as a crystal is being dissolved, no increase in the frequency of a detected signal is observed. In our opinion, this indicates the incorrect interpretation of the AE phenomenon during crystal dissolution, melting, and growth as an effect of the coincidence of forced vibrations of a liquid with the natural vibration frequency of the crystal.

2. Mechanisms of Acoustic Emission, Accompanying Growth, Dissolution and Melting of Crystals

The objective of this study was to investigate the features of AE during growth, dissolution, and melting of crystals and to try to understand the mechanisms of accompanying AE.

A simplified scheme of the experiment is shown in Figure 1 and the experimental results are plotted in Figures 2, 3. AE signals were received with a highly sensitive piezoelectric transducer mounted on the bottom part of a vessel with a liquid under study. The vessel was conical and its narrow part was attached to the working surface of the AE transducer through a contact grease layer. AE signals were received and statistically processed using АП-71Э (our design) [3, 5] and A-Line32D (Interyunis Co.) digital diagnostic AE systems. The signal shapes were recorded and spectrally analyzed with an A-Line32D AE system.

Figures 2 and 3 show typical parameters of the recorded AE signals arising during crystal dissolution. It is seen, as a substance is being dissolved, acoustic emission with amplitude sufficient for detection is generated and its parameters obviously correlate with dissolution process dynamics. The data in Figure 3b correspond well to exponential decay (owing to a decrease in the substance mass) of a dissolution process, the kinetics of which can be easily quantitatively determined by measuring stored AE pulses.

In the course of these our experiments, we studied the AE spectrum of various crystals and the results unambiguously showed that the AE frequency range virtually does not depend on the dimensions of crystals during their dissolution or melting. In addition, as a crystal is being dissolved, no increase in the frequency of a detected signal is observed. In our opinion, this indicates the incorrect interpretation of the AE phenomenon during crystal dissolution, melting, and growth as an effect of the coincidence of forced vibrations of a liquid with the natural vibration frequency of a crystal.

We believe that the observed discrepancy between the expected and experimental results can be explained on the basis of the phenomenon of "beats", i.e., quasiperiodic changes in the amplitude of vibrations formed upon addition of several harmonic vibrations with similar frequencies ω_1, ω_2, ... ω_n [11].

In the classical treatment of beats, a low-frequency wave with high amplitude is formed as a result of the summation of harmonic vibrations with an equal frequency difference. In the case of a stochastic set of harmonic vibrations, at first glance, the phenomenon of beats cannot be expected. To simulate this process, we developed a program that allows variation of the frequencies of the summed waves, the number of waves, and their phases and, depending on the conditions, specifies a fixed frequency difference or an arbitrary set of frequencies.

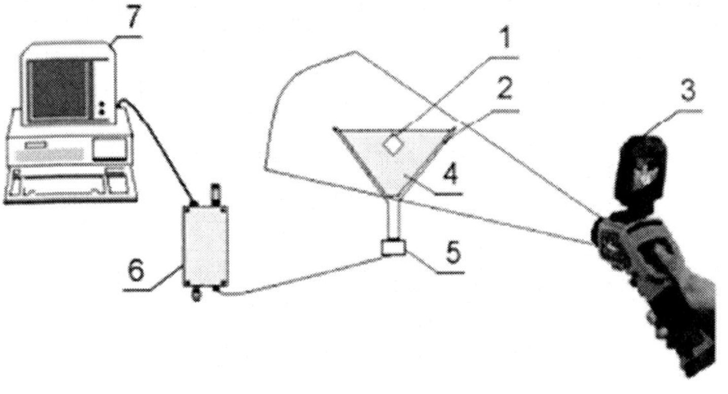

Figure 1. Simplified scheme of the experiment: 1 - specimen of the studied component, 2 - conical vessel, 3 - infrared imager, 4 - solvent, 5 - AE transducer, 6 - preamplifier, 7 - digital AE diagnostic block.

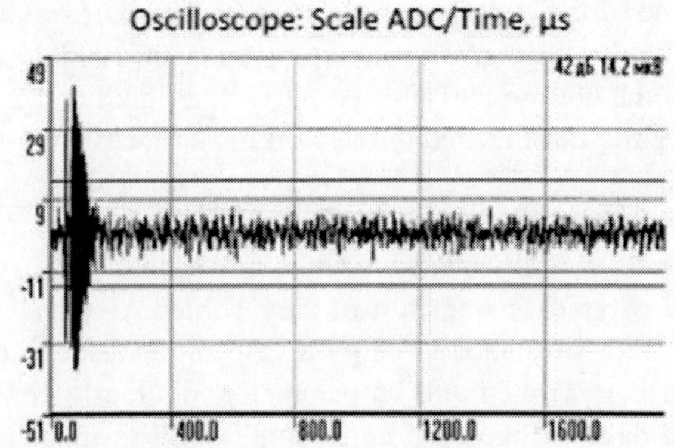

a

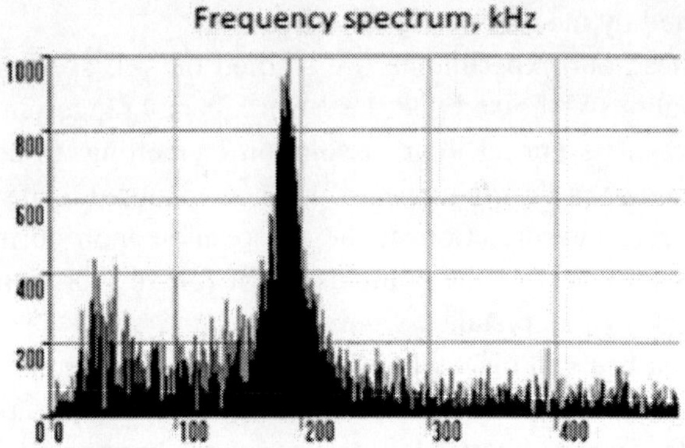

b

Figure 2. (a) shape and (b) frequency spectrum of an AE pulse during dissolution of $NiSO_4$ crystal. The time from the start of the process is 3 min 5 s (screenshots from the monitor screen A-Line32D).

Because we assume that beats between close vibration frequencies of dissolved substance units (clusters, molecules, or ions) are the actual AE source during crystal dissolution and growth in liquid media, we chose a Gaussian distribution of the summed frequencies. The frequency is varied in the program using the parameters of the distribution function of a random value of the wave frequency spread.

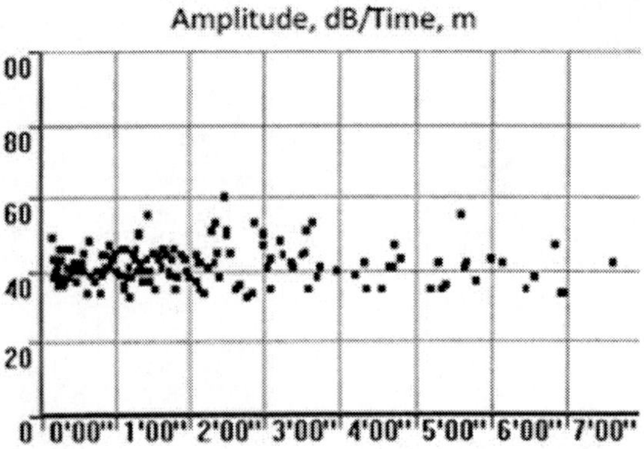

a

Figure 3. (Continued).

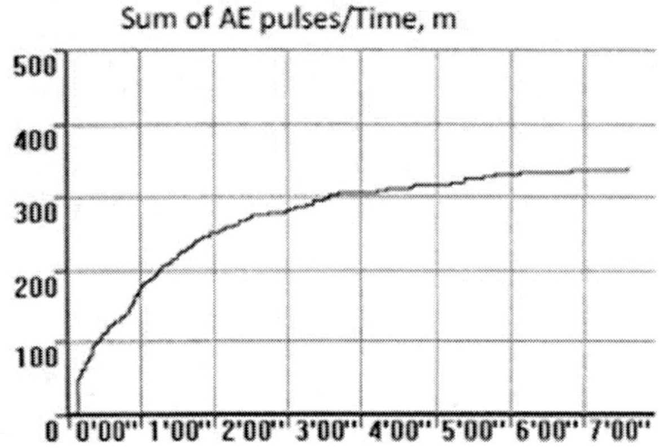

b

Figure 3. (a) amplitude and (b) storage kinetics of AE pulses during dissolution of $NiSO_4$ crystal (screenshots from the monitor screen A-Line32D).

This program helped us to reveal and predict the formation of low-frequency waves with pulsating amplitude upon summation of a large number of waves with a random set of similar frequencies with a preset mode and a preset standard deviation. Since we used the hypothesis that beats between close frequencies of atomic vibrations ($10^{12} - 10^{13}$ Hz) can really be an AE source in liquids, it follows from physical considerations that a single mode frequency distribution must be observed.

3. Applicability of the Model of Random Poisson Process to Problems of Fracture and Crystallization of Solids

Figure 4a shows the obtained histograms of the distribution of time intervals between AE events during ice crystallization. As is seen, this distribution groups well around an exponent corresponding to the density of time intervals between events of a Poisson stream of intensity $\dot{N}_a$. For this stream, the probability $P_r(\Delta t)$ of the appearance of *r* events within a time interval Δt obeys the distribution:

$$P_r(\Delta t) = [(\dot{N}_a \Delta t)^r / r!]\exp(-\dot{N}_a \Delta t) \tag{1}$$

Thus, the methods of quantitative reconstruction of the parameters of structural transformations on the basis of AE signals that have been developed in our previous investigations by using the Poisson model [2–9,16], can be used in studies of crystallization processes.

Interesting results were also obtained in the course of an amplitude analysis of AE signals (Figure 4b). It is known that the AE amplitude u_{ae} is related to the emission energy E_{ae}. Hereinafter, we consider *u* as meaning the maximum (peak u_p) amplitude of the AE pulse envelope.

The emission energy E_{ae} is proportional to the amplitude squared u_{ae}^2 and proportional to the area of the radiation source, which is usually proportional to the square of the average linear size of a microsource.

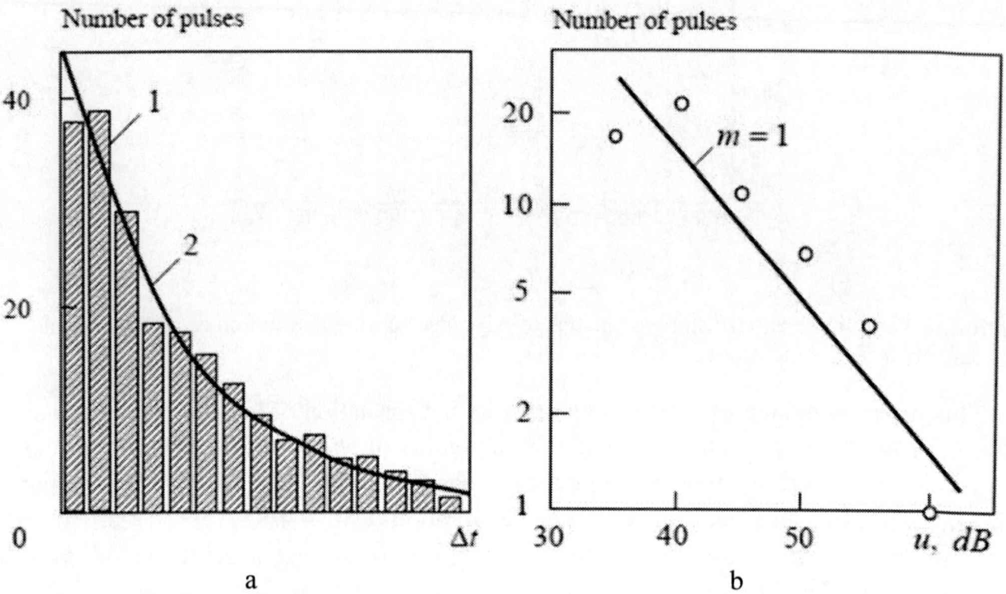

Figure 4. (a) Experimental histograms of the distribution of time ranges between AE events during ice crystallization and (b) amplitude distribution of AE signals during dissolution of $NiSO_4$ crystal in water (from 2nd to 8th min). The sample size is 202 events; the number of analyzer channels is 15; the channel width is 0.45 ms: 1 -experimental values and 2 - approximating Poisson stream.

Analysis of the form of this distribution shows that the amplitude distribution density $w_n(u/u_t)$ normalized to the unit probability of the sum of all events is properly described by the power dependence:

$$w_n(u/u_t) = (m-1)(u/u_t)^{-m}, \tag{2}$$

where u is the amplitude of an AE pulse and u_t is the discrimination threshold (noise cutoff level) [11].

The power form of the AE amplitude distribution (2) allows us to suggest that the distribution in the linear dimensions of the microscopic dissolution regions of studied crystals has similar form (at least, beginning with the second minute of dissolution).

In the region of large amplitudes (60 dB or more), a slight deviation $w(u)$ from the power dependence is observed. This is probably associated with a finite number of real AE microsources [3–5].

Thus, the AE effect is also observed in liquid media and there is a fundamental possibility of diagnosing crystallization, dissolution, and melting processes on the basis of AE data. The results of our recent studies show that the mechanism of beats may cause the recorded AE events, i.e., the formation of low-frequency (up to 10^4–10^7 Hz) acoustic waves during crystal dissolution (melting) and growth in a liquid, electrolysis, and in homogeneous chemical reactions of liquid reagents.

The physicochemical nature of vibrations composing the process of beats is beyond the framework of this study. At the same time, it is obvious that this mechanism is conditioned by the discreteness of the dissolution process, in which, e.g., in the case of dissolution of an ionic compound (e.g., NaCl crystals) in a polar solvent (water), the permittivity of the solvent plays the main role and the dissolution process occurs at the level of individual ions. In some cases, a liquid may display a property of structuredness.

Thus, the presence of suprapremolecular water complexes detected with optical methods may lead to the generation of acoustic waves at a level of vibrations of clusters that include up to 10^9 molecules, rather than at a level of individual molecules.

On the example of $CaSO_4 \cdot 2H_2O$ crystallization the connection of curing astringent building mixtures with the reaction volume temperature and AE parameters were investigated. Curing of gypsum is calcium sulfate hemihydrate transition when mixing with its water dihydrate:

$$CaSO_4 \cdot 0.5H_2O + 1.5H_2O \rightarrow CaSO_4 \cdot 2H_2O \qquad (3)$$

In the crystallization process is carried gypsum phase transition of a substance in a crystalline compound with a lower energy. The excess energy is released in the form of AE and latent heat. AE parameters recorded by us in the course of the process (3) are shown in Figure 5. The release of this latent energy of crystallization leads to heating of the solution on the boundary of the growing crystal.

It is established that the crystallization latent energy process passes most intensively in a fairly narrow time range (20–30 minutes), which coincides with the most intense AE radiation (see Figure 5b).

The final stage of curing taking place during 1 – 2 hours consists in the formation of crystalline concretion of sufficiently large crystals of gypsum dihydrate (see Figure 6).

These results allow us to propose a new AE method for the diagnosis of curing of the materials. The results of this study can be used in the development of new methods for diagnosing physicochemical processes in liquid media and by controlling and diagnosing the dynamics of dissolution and crystallization into furnaces for growing single crystals.

4. Acoustic Emission Monitoring of the Kinetics of Dissolution of Substances

The purposes of our works were to analyze the dynamics of changes in the number of AE events during the dissolution of crystals of ionic substances and to determine the possibility of establishing the correlation relationships between the AE parameters and the amount of a dissolved compound.

NaCl and $NiSO_4$ crystals were taken as tested components. The selection of these materials was made based on the fact that the dissolution of these crystals has been well studied by conventional methods. A simplified scheme of the experiment is shown in Figure 1 and the experimental results are plotted in Figure 7.

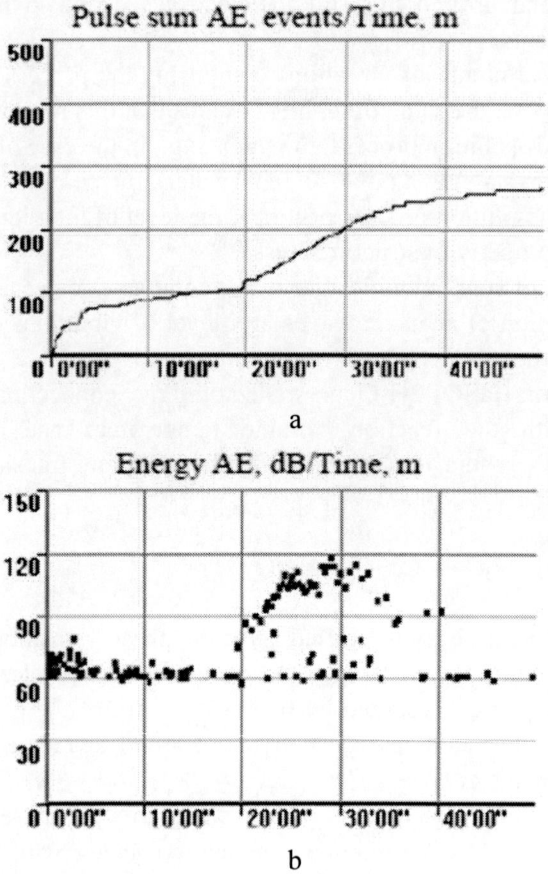

Figure 5. Changes of the parameters of AE during the early stages of $CaSO_4 \cdot 2H_2O$ crystallization.

Let us consider the dynamics of changes in the intensity of an AE event stream (number of AE events per unit time) $\dot{N}_a = dN_a/dt$ and in the total number of AE events N_a with time. We assume that the nature of the induction of AE signals during the dissolution of ionic crystals is associated with considerable energy expenditures on the disruption of ionic bonds. This must lead to the formation of local solvent density gradients at the crystal–solvent interface and, as a consequence, to the emission of recordable elastic waves into a liquid [10–13]. Moreover, we considered the fact that the AE signal distortion and overlap ratio in the described experiments was much lower than unity and the intensity of an AE event stream $\dot{N}_a$ was quantitatively estimated directly by measuring the activity $\dot{N}_\Sigma$ (intensity of a detected AE pulse stream) without using the reconstruction procedure (i.e., $\dot{N}_a \approx \dot{N}_\Sigma$) [2–6].

Let *m* is the initial mass of dissolving crystal and *M* the mass of the crystal part that has been already dissolved. Then, at any time moment, the mass of the crystal is equal to the difference $m - M$ and the crystal mass dissolution rate is dM/dt.

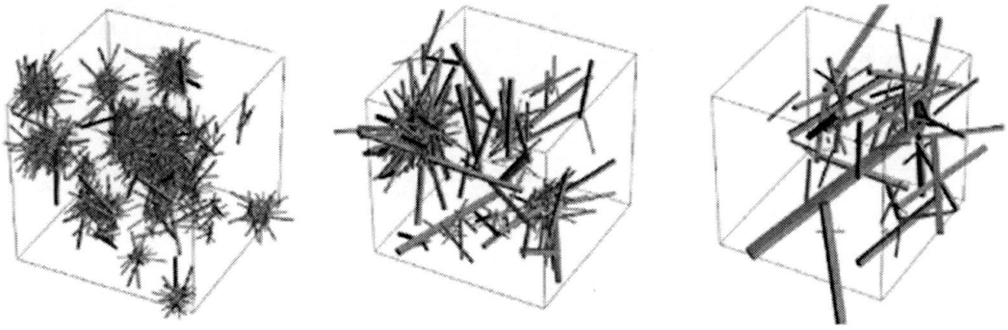

a

Pulse sum AE, events/Time, h

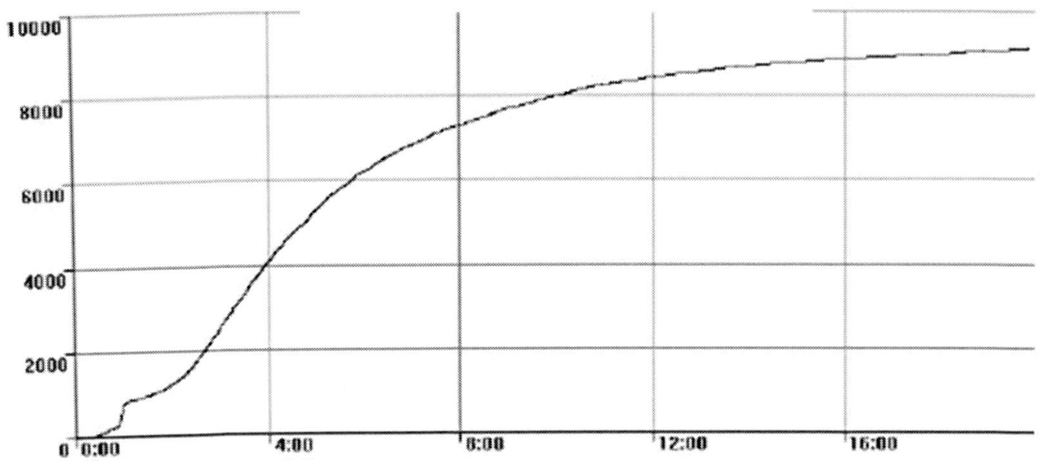

b

Figure 6. (a) Stages of growth of $CaSO_4 \cdot 2H_2O$ according to [14] and (b) sum of AE pulses stream in the process of curing the gypsum mixture.

Assuming that the intensity of AE event stream (number of AE events per unit time) dN_a/dt is proportional to the dissolved component's mass change rate dM/dt, we can write the equation:

$$\frac{dN_a}{dt} = K\frac{dM}{dt}, \tag{4}$$

where K is a certain proportionality coefficient.

The crystal mass at any instant is $m - M$ and the crystal's mass dissolution rate is dM/dt. Assuming that the dissolution rate is proportional to the current crystal mass, we obtain the following differential equation of the process:

$$\frac{dM}{dt} = k(m - M), \tag{5}$$

where k is a certain proportionality coefficient.

Integrating this equation, we obtain the solution with respect to the current crystal mass $m - M$ at the initial condition: $t = 0$, $M = 0$:

$$m - M = me^{-t/\tau}, \tag{6}$$

where τ is the relaxation time constant determined by the dissolution rate.

Substituting Equation (6) into Equation (4) and separating the variables, we obtain the following equation for the dynamics of changes in the total number of AE events with the dissolution of a salt crystal:

$$N_a = Km(1 - e^{-t/\tau}), \tag{7}$$

where N_a is the total number of AE events at the current time moment t, K is a certain coefficient that depends on the sensitivity of AE instruments and the acoustic signal recording method, m is the mass of the crystal at the initial time moment, and τ is the relaxation time constant of dissolution process, which is determined from the emission data shown in Figure 7.

Using Equation (7), we can calculate the total number of AE events at any instant for the given mass of a crystal.

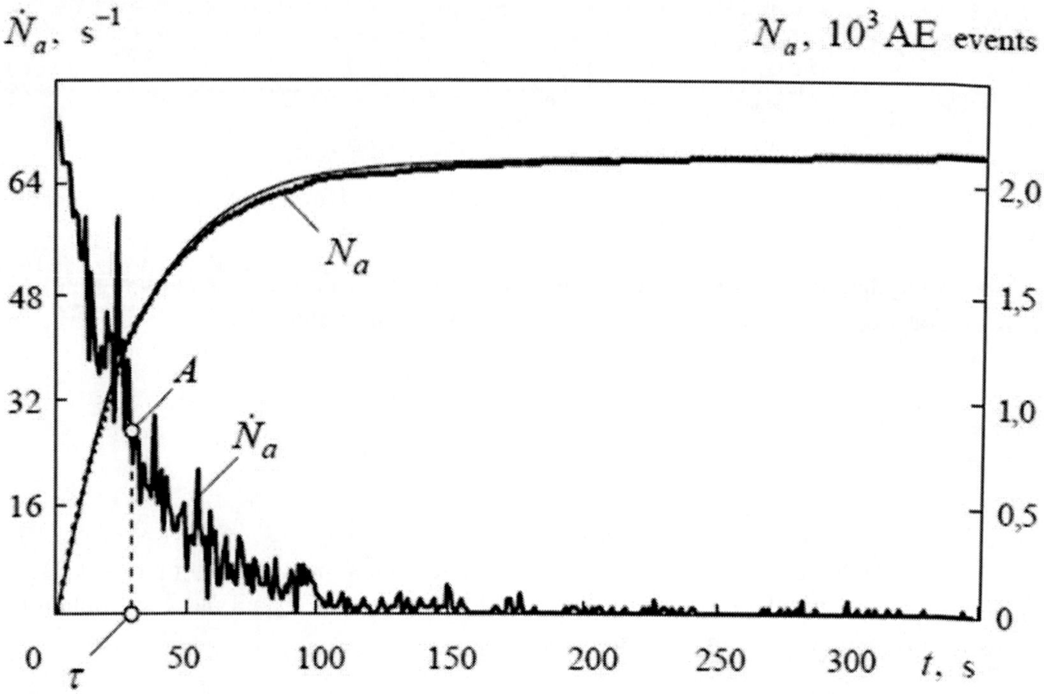

Figure 7. Dynamics of changes in intensity of AE event stream and total number of AE events during dissolution of 0.17 g NaCl crystal.

In Figure 7, the solid line is the theoretical dependence expressed by Equation (7), and the microtriangles denote the experimental data. We can observe very good agreement (with an accuracy of better than 3 – 5%) between the curve of Equation (7) and the experimental data for NaCl, as well as the previously obtained AE data on the dissolution of $NiSO_4$ crystals. This allows us to propose Equation (7) for the experimental estimation of crystal dissolution dynamics from the data of AE measurements.

For example, after certain values of K and τ have been preliminarily determined from the data of AE experiments, we can then experimentally estimate the dissolution rate and mass of the remaining substance from the recorded values of N_a. Certain values of the dissolution time constant τ can be determined from the decrease of dN_a/dt curve's values in e times (point A in Figure 7). After determining τ and substituting experimental values of N_a at times $t \geq (3–5)\tau$ and the initial value of the component mass m into Equation (7), we can easily obtain a certain value of the coefficient K from Equation (7).

The above described results were obtained at room temperature. The AE studies of the dissolution of NaCl within a range of temperatures showed that, according to our results, the relaxation time constant changes slightly with the temperature of the solution. The higher the temperature of the water is, the lower the relaxation time constant and the higher dissolution rate of the crystal is, as fully confirmed by the physical chemistry of the dissolution of ionic compounds. An increase in the temperature of a solution increases the rate of diffusion of solvated particles into water and, correspondingly, reduces the period of dissolution [17].

Hence, the data from these our experiments show that the AE method can be successfully applied for the quantitative study of the dynamics of the physicochemical processes of dissolution in a liquid medium.

5. Acoustic Emission Monitoring of the Impregnation of Porous Medium

The objectives of our works were to analyze the time history of the quantity of AE events in the process of capillary liquid flow in porous media for the interdisciplinary expansion of the application area of the effects of emissions and the creation of a physical nondestructive testing method for the impregnation depth and completeness of materials from parameters of the accompanying acoustic emissions. Until the present time, experimental studies of the dynamics of liquid flows in capillaries were basically performed via optical methods using transparent materials. The known Haines jumps effect can act as a theoretical substantiation of the possibility of using the AE method for the determination of the dynamics of capillary liquid flows. The physical nature of Haines jumps is as follows. When the capillary flow in a corrugated capillary is volumetric, the meniscus must periodically stretch and shrink in the capillary expansions and contractions, respectively. The liquid penetration in the porous object generates a growing percolation cluster, whose expansions occur in irregular jumps that are related to filling the pockets of large radius pores (as compared with the radius of the narrow neck, which connects the pocket to the cluster). In this case, the meniscus assumes a form that is far from equilibrium. The meniscus passes these places very rapidly. This phenomenon is called Haines jumps.

It is specified by the pressure fluctuations in a liquid during impregnation and drainage, accompanied by the corresponding energy dissipation and thus should be accompanied by acoustic radiation of the recorded level.

A simplified scheme of the experiment is shown in Figure 8 and the experimental results are plotted in Figure 9.

The ceramic specimen (in our case, a porous aluminooxide carrier, Al_2O_3) is hung in a liquid so that the specimen edge adjoins the liquid level, leading to capillary impregnation. During the impregnation, the induced AE waves reached reservoir walls and were observed by highly sensitive piezoelectric transducer. The AE signals that were received by the transducer were amplified by a preamplifier and next applied along the cable to the input of the A-Line32D digital measuring complex A-Line32D (Interyunis Co.). Before and after the impregnation, the specimen was weighed on a precise scale. The temperature of the solution and the specimen were monitored using an infrared imager.

In the course of the experiments the energy E_a of the AE signals was recorded relative to some threshold level, activity (number per time unit) $\dot{N}_\Sigma$ and the sum of the AE pulses N_Σ [2–5].

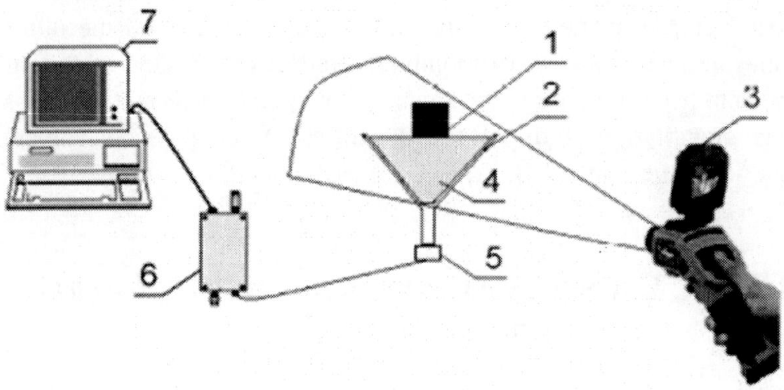

Figure 8. Simplified scheme of the experiment: 1 - specimen of the studied component, 2 - conical vessel, 3 - infrared imager, 4 - liquid, 5 - AE transducer, 6 - preamplifier, 7 - digital AE diagnostic block.

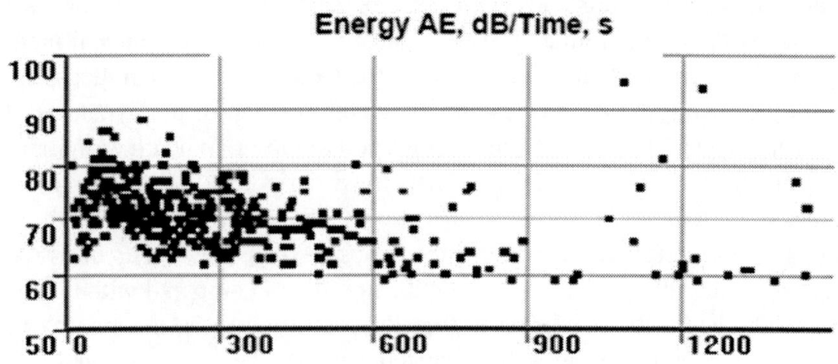

a

Figure 9. (Continued).

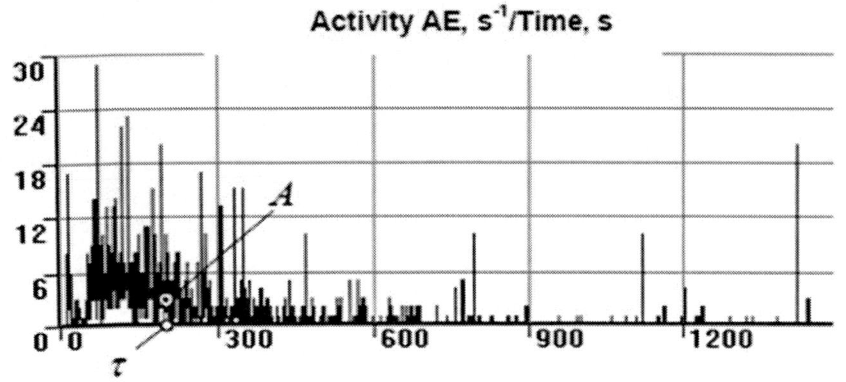

b

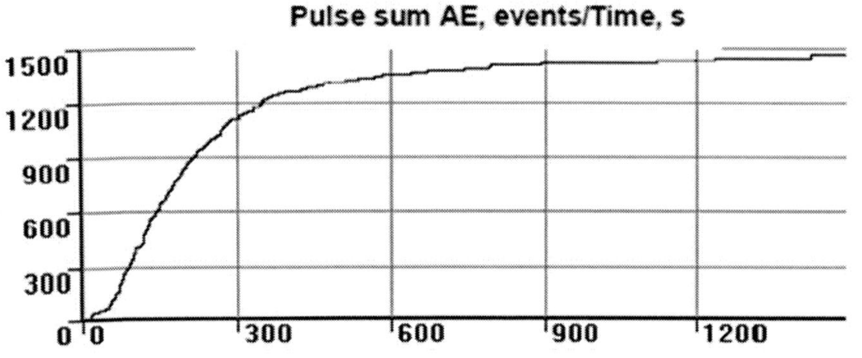

c

Figure 9. AE parameters in the impregnation process of porous Al_2O_3 specimen.

Taking into account the fact that in these experiments the distortion and overlapping factor of the AE signals was much smaller than unity, the quantitative evaluation of the stream intensity of AE events was determined directly from measurements of activity $\dot{N}_\Sigma$ without recourse to the recovery procedure (it was considered that $\dot{N}_a \approx \dot{N}_\Sigma$ and $N_a \approx N_\Sigma$) [2–5].

The statistical treatment of results showed that the recorded intensity of the AE event stream actually drops exponentially and the total number of AE events is accurately described within the framework of the exponentially decaying process model, which is similar to that considered earlier (4) – (7):

$$N_a = Km(1 - e^{-t/\tau}), \tag{8}$$

where N_a is the total number of AE events that occurred by this time t; K is a coefficient that depends on the sensitivity of the AE equipment and acoustic signal recording procedure; m is the total mass of the absorbed liquid by the capillary impregnation process completion; and τ is the relaxation time constant of the impregnation process, which is determined from the emission data in Figure 9b.

Using Equation (8), one can consider the total number of AE events or the mass of the absorbed liquid at any time moment. Unfortunately, due to the opaqueness of the studied specimens, in these experiments it did not prove a possibility to directly evaluate the time history of height of the rise of the liquid during their impregnation. Let us try to evaluate this dynamic indirectly, using the experimental results, which contains measurement data on the velocity and capillary rising height of 95% alcohol in glass capillaries of different diameters [15]. The statistical treatment of the results showed that the time history of the height of the liquid rise, h, is described with an accuracy of better than 5% by the ratio $h = h_{\max}(1 - e^{-t/\tau})$, which is a particular case of formula (8).

Here h_{max} is the established (maximal) level of the liquid rise. The application of formula (8) to the evaluation of the height of the liquid rise (the impregnation depth) also follows from the fact that when the liquid density and the average sectional area of the capillary or the specimen unchange, the time history of the height of the liquid rise should exactly repeat the time history of the liquid mass.

This allows one to use formula (8) to experimentally evaluate the dynamics of the impregnation process and capillary liquid rise from AE measurement data. For example, having preliminarily determined the particular values of K and τ from the AE experiment data, it is possible to experimentally estimate the mass m (completeness of the impregnation), of the absorbed liquid and the rise height h (depth of the impregnation) from the recorded values of N_a at any time t.

It is possible to determine particular values of impregnation time constant τ from the drop of the values of the dependence plot dN_a/dt by a factor of e (point A in Figure 9b). The value of τ can also be determined directly from the data (Figure 9c) in the course of approximation (e.g., by the least squares method) of experimental values of N_a by ratio (8). Having determined τ and substituted the experimental values of N_a at the times $t \geq (3 - 5)\tau$ and the finite mass of the absorbed liquid m into (8), from formula (8) we obtain the particular value of the coefficient K.

Similarly, having substituted the experimental values of N_a at the times $t \geq (3 - 5)\tau$ and h_{max} instead of m into (8), from formula (8) we obtain another particular value of the coefficient K for evaluating the dynamics of the capillary liquid rise (the depth of the impregnation) from AE measurement data.

The above described results were obtained at room temperature. The AE studies of the capillary impregnation process in a temperature range showed that when the solution temperature is increased to 40 – 50 °C, the total AE event number N_a and the constant time value of relaxation of the impregnation process τ are approximately halved. The reason for this is likely in the physics of the process; since the wetability of the surface by the liquid becomes higher at higher temperatures, the impregnation rate increases and it is no uniforly (total number of Haines jumps) decreases.

Hence, the data from these our experiments show that the acoustic emission method can be successfully applied for the quantitative study of dynamics of the capillary liquid flow in impregnation processes. This allows us to propose a new nondestructive inspection method of the impregnation depth and completeness of materials based on the parameters of the accompanying acoustic emissions.

Thus, the results of our experiments demonstrate that the AE method can be used not only for studying the degradation processes of solids, but also for the quantitative study of the dynamics of physico-chemical processes in liquid media.

ACKNOWLEDGMENTS

This work was supported by the Russian Ministry of Education and Science, project No. 231.01-11/2014-29, and the Russian Foundation for Basic Research, projects Nos. 09-08-00283-a, 12-08-01190-a, 14-08-90015 Bel-a, 14-08-00829-a.

REFERENCES

[1] Acoustic Emission; *ASTM Special Publication 505:* Baltimore, 1972; pp. 1-337.

[2] Builo, S. I. In: *Physics and Mechanics of New Materials and Their Applications*. Ivan A. Parinov, Shun-Hsyung Chang (Eds.). Nova Science Publishers, New York, 2013, Chapter 15, pp. 171-184.

[3] Builo, S. I. *Physicomechanical and Statistical Aspects of Increasing the Reliability of Results of Acoustic-Emission Testing and Diagnostics*, Southern Federal University Press, Rostov-on-Don, 2008, pp. 1-192 [in Russian].

[4] Builo, S. I. *Investigation of Acoustic Emission during Mechanical Deformation. Ph.D. Dissertation*; Rostov State University Press, Rostov-on-Don, 1983; pp. 1-191, [in Russian].

[5] Builo, S. I. *Physicomechanical and Statistical Aspects of Acoustic Emission Pre-destructive State Diagnostics. Doctor of Sciences in Physics and Mathematics Dissertation*; Southern Federal University Press, Rostov-on-Don, 2009, pp. 1-279 [in Russian].

[6] Builo, S. I. *Russian Journal of Nondestructive Testing*, 1995, vol. 31(7), 492-503.

[7] Builo, S. I. *Russian Journal of Nondestructive Testing*, 2004, vol. 40(8), 552-560.

[8] Builo, S. I. *Russian Journal of Nondestructive Testing*, 2007, vol. 43(3), 188-195.

[9] Builo, S. I. *Russian Journal of Nondestructive Testing*, 2008, vol. 44(8), 517-526.

[10] Kuznetsov, D. M., Gaponov, V. L., Smirnov, A. N. *Engineering Physics*, 2008, vol. 44 (1), 16-21.

[11] Builo, S. I., Kuznetsov, D. M. *Russian Journal of Nondestructive Testing*, 2010, vol. 46(9), 684-689.

[12] Kuznetsov, D. M., Builo, S. I., Ibragimova, J. A. *Chemical Technology*, 2011, vol. 6(2), 112-114.

[13] Builo, S. I., Kuznetsov, D. M., Gaponov, V. L., Trepachev, V. V. *Russian Journal of Nondestructive Testing*, 2012, vol. 48(10), 594-597.

[14] Lemarchand, A., Boudoire, F., Boucard, E., Chotard, Th., Smith, A. *Journal of Physical Chemistry C*, 2012, vol. 116, 4671-4678.

[15] Builo, S. I., Kuznetsov, D. M., Gaponov, V. L. *Russian Journal of Nondestructive Testing*, 2014, vol. 50(7), 392-395.

[16] Builo, S. I., Orlov, S. V. *Russian Journal of Nondestructive Testing*, 2014, vol. 50(8), 464-468.

[17] Builo, S.I., Kuznetsov, D.M. and Gaponov, V.L. *Abstracts of International Symposium on Physics and Mechanics of New Materials and Underwater Applications (PHENMA 2014)*, Khon Kaen, Thailand, March 27-29, 2014, pp. 23-24.

In: Advanced Materials - Studies and Applications
Editors: I.A. Parinov, S.-H. Chang, S.Theerakulpisut
ISBN: 978-1-63463-749-7

Chapter 14

CHANGE OF FERMI LEVEL IN AREA OF NEAR-SURFACE CHARGE IN MISM STRUCTURES

G. Ya. Karapetyan*, V. G. Dneprovski and I. A. Parinov
Vorovich Mathematics, Mechanics and Computer Sciences Institute,
Southern Federal University, Rostov-on-Don, Russia

ABSTRACT

We consider a behavior of Fermi level in the metal-insulator-semiconductor-metal (MISM) structures in area of near-surface charge. The Fermi level changes in a direction from semiconductor bulk to its surface, located under insulator, if we apply a voltage to gate. Bottom level of conduction band does not change, but the Fermi level bends, changing a concentration of charge carriers in the area of near-surface charge. We explain this phenomenon by that semiconductor surface is an equipotential surface and has the same electric potential, regardless of a charge density on the surface. Then the bottom of the conduction band in the area of near-surface charge should be disposed at the same level as the bottom of the conduction band on the semiconductor surface outside of the insulator (near-surface charge area). This circumstance leads to bending the Fermi level, since the charge density under the gate area differs significantly, compared to the charge density in the semiconductor bulk and on its surface outside of the gate area. The density of charge carriers becomes greater at lesser difference between the Fermi level and bottom of the conduction band.

INTRODUCTION

One of the causes of degradation of the light-emitting diodes (LEDs) characteristics is their overheating. A cooling the LED light source performs usually with the help of refrigerator of Peltier [1]. Similar to other cases of a forced cooling, it is necessary to expend an additional energy to overcome a parasitic EMF in a cooled element that leads to increased energy consumption by such lamps. Moreover, the usually used radiators, which remove a

* jorichkaka@yandex.ru.

heat from a hot edge of the cooler, complicate a design of the LED lamps and increase their cost. Therefore, we have proposed a novel method of cooling the LEDs, based on the metal-insulator-semiconductor-metal (MISM) structure, which differs from known structure of the Peltier cooler, owing to absence of the hot edge, i.e., there is only a cool edge (see Figure 1) [2]. In this case, the cooler not only removes heat from the LEDs, but also provides the lamps with electric energy, transformed from an absorbed heat.

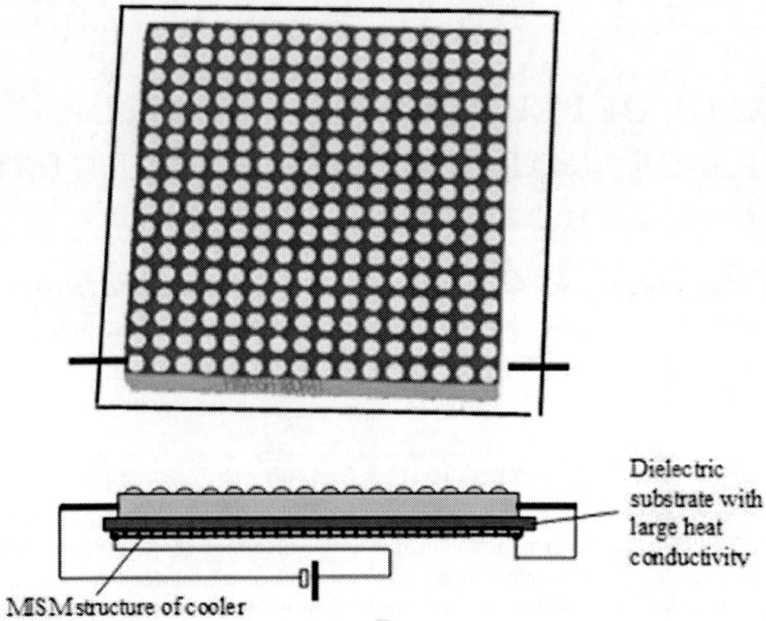

Figure 1. LEDs matrix with semiconductor thermoelectric cooler of novel type, reducing the energy consumption.

We assume that levels of the conduction and valence bands bend on the semiconductor surface under action of electric field in metal-insulator-semiconductor (MIS) structures. At the same time, the Fermi level remains constant in all points of the semiconductor. By bending the band levels, a distance between them and the Fermi level changes, leading to significant change in a concentration of charge carriers in the near-surface layer of the semiconductor, located under insulator [3]. However, the concentration of electrons depends on difference between the Fermi level and level of the conduction band or valence band. Let us assume presence of bending the Fermi level and absence of a distortion of level of the conduction band or valence band. Then the concentration distribution of charge carriers will be the same as in the case of bending the level of the conduction band or valence band at the constant Fermi level. It will mean that a count off should be performed, beginning from no Fermi level, but from bottom of the conduction band or top of the valence one.

Such an ambiguous interpretation of distortion of the Fermi level and levels of the conduction and valence bands arises from the fact that the treated MIS structure is not the metal-insulator-semiconductor-metal (MISM) structure. At the same time, to connect the MIS structure, an electrode is required on the hand of the semiconductor. Therefore, all of the MIS structures are the MISM structures, in reality.

The results of the chapter show that namely the Fermi level distorts in the MISM structures in the near-surface charge area, but no level of the conduction band or valence band. This allows one assume that a difference of electric potentials between the electrode, connected to the semiconductor surface, and the electrode, connected to near-surface charge, creates an electromotive force. Here heat energy of environment is the external source of electric energy. The structure will cool and protect high-powered LEDs from overheating.

2. METHOD

We have shown [2, 4, 5] that a count off in charged capacitor should be performed, beginning from no Fermi level, but from bottom of the conduction band. The study of a process of charging the capacitor, representing MISM structure, defined this conclusion. We showed that energy of such a charged capacitor is determined as

$$W_C = \frac{U_0^2 + 2U_{kA}U_0}{2}C, \tag{1}$$

where C is the capacitance of the MISM structure, U_{kA} is the difference of electric potentials between metal and semiconductor in presence of a transverse magnetic field or at the temperature T_1. The Peltier coefficient is equal to a difference of electric potentials at internal junction ($\Pi = U_{kA}$), if we understand as this difference of electric potentials the parameter $(E_c - F)/e$ [3], where E_c is the energy on bottom of the conduction band of semiconductor, F is the Fermi level, e is the electron charge. The Peltier coefficient decreases after removal of the magnetic field or if the temperature drops down to T_2. However the capacitor energy does not change, because we assume that dielectric constant, size of the capacitor and charge on it are independent on the changes of the magnetic field and temperature. The difference of electric potentials at internal junction decreases and riches the value of U_k, but the energy of the charged capacitor becomes equal to

$$W_C = \frac{U^2 + 2U_kU_0}{2}C. \tag{2}$$

Since capacitance of the capacitor and its charge do not change, then the capacitor energy also does not change. Then we can write

$$C\frac{U^2 + 2U_kU}{2} = C\frac{U_0^2 + 2U_0U_{kA}}{2}, \tag{3}$$

where U is the voltage on metallic electrodes of the thermoelectric capacitor, determined as

$$U = -U_k + \sqrt{U_k + U_0^2 + 2U_0U_{kA}}. \tag{4}$$

It is not difficult to see that $U = 0$ at $U_0 = 0$, and $U \approx U_0 + \Delta U$, where $\Delta U = U_{kA} - U_k$ at $U_0 >> U_{kA}$, U_k.

Figure 2 depicts a chart of electric potentials for the charged capacitor.

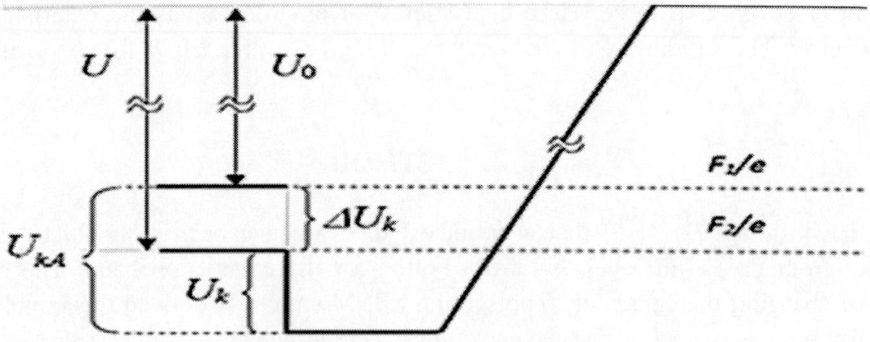

Figure 2. Chart of electric potentials for charged capacitor with MISM structure.

This chart shows clearly that difference of electric potentials at internal junction counts off, beginning from no Fermi level, but from bottom of the conduction band. It follows from that electric potential of bottom of the conduction band does not change at changing the external conditions, i.e., changes of magnetic field and/or temperature (see Figure 1).

3. Results and Discussion

Further, let us consider plate of the *n*-type semiconductor with electrode on plate bottom surface, and other electrode disposes above of the plate at some distance from it. This structure forms a capacitor, whose one plate is the electrode, which does not connect to the plate, and other electrode is the plate itself (see Figure 3). The square of surface of these electrodes is much lesser than square of the plate surface. We choose the plate thickness much greater than screening length of surface charge, i.e., charge on the upper surface does not influence on junction of the lower electrode with the plate.

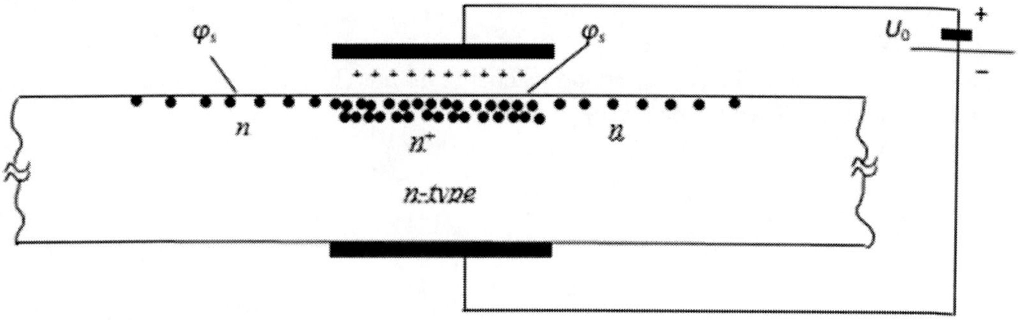

Figure 3. Semiconductor plate with electrodes.

The capacitor charges from a battery with voltage U_0. The battery connects in such a way that its positive pole connects to the electrode, which does not join with the semiconductor plate. The concentration of electrons on the semiconductor surface, located under upper electrode, can exceed significantly the electron concentration in the semiconductor bulk and on its surface outside of the electrode. It takes place due to the electrons are attracted to the semiconductor surface by positively charged electrode. However, according to the laws of electrostatics for conducting bodies, a whole surface of the plate is an equipotential surface. Therefore, the electric potential of the total surface (including area, located under upper electrode), as we can assume, is equal to φ_s. Moreover, note that the electric potentials of the conducting bulk bodies are the same, when the electric field equals zero in their bulk. So, we can choose the electric potential φ_s for a whole plate. However, the equality of the electric potentials at the body surface and in its bulk means that bottom of the conduction band has the same level in the total body. However, then the Fermi level should change in area under upper electrode, because the electron density, which depends on the difference between the Fermi level and bottom of the conduction band, differs from the electron concentration on surface and into bulk of the plate.

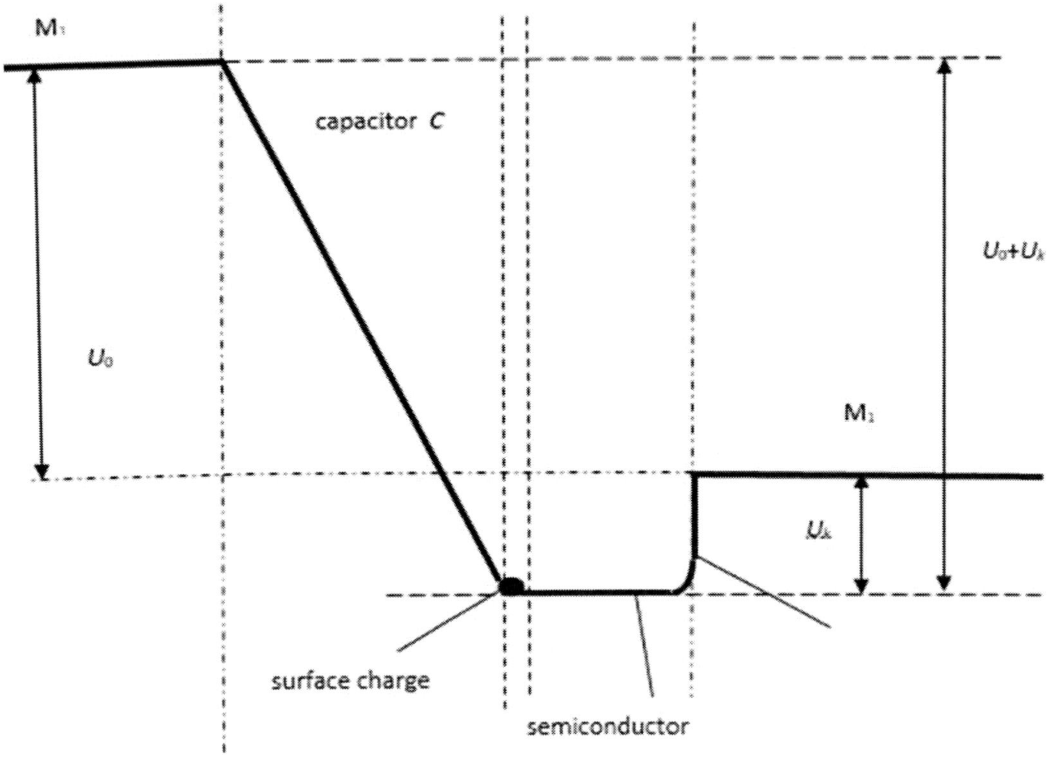

Figure 4. Chart of electric potentials of the capacitor, presented in Figure 3.

Figure 4 shows chart of electric potentials for the charged capacitor, depicted in Figure 3. As we can see in this chart, the electric potential of near-surface charge is equal to the electric potential of semiconductor plate, and creates a greater voltage on the capacitor, compared to the battery, on the value of U_k. This occurs, as shown in [2–4], because heat energy absorbs during charging the capacitor at junction of lower electrode (see Figure 3) with the semiconductor plate (i.e., the junction cools in this case). The heat (kinetic) energy transforms into potential energy of the charged capacitor and this energy determines from expression (1). Thus, electric potential of the near-surface charge equals φ_s, and concentration of electrons in the charge is much higher than in the conductor bulk. The count off is carried out, beginning from bottom of the conduction band. Then the difference of potentials at the junction between the near-surface charge and metallic electrode counts off, beginning from bottom of the conduction band. It leads to the conclusion that the Fermi level of the electrode does not coincide with the Fermi levels of lower electrode and semiconductor (see Figure 5).

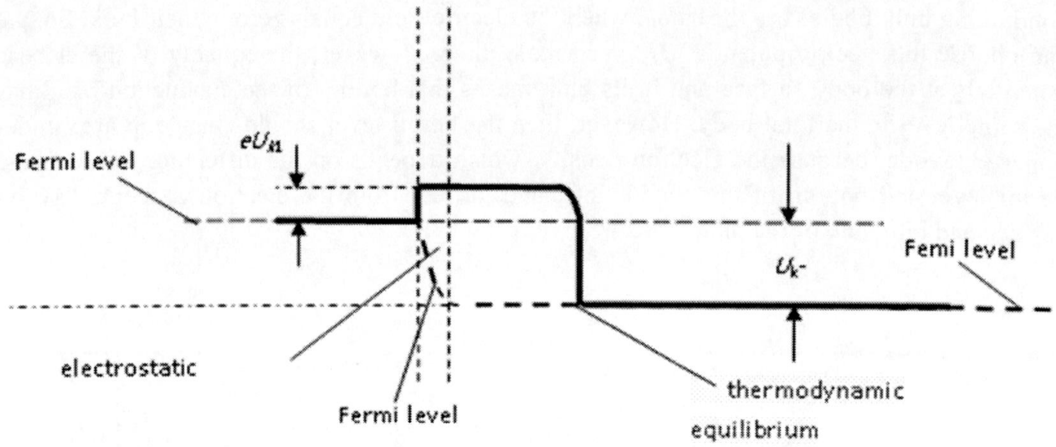

Figure 5. Chart of energy for lower plate of the capacitor, presented in Figure 3.

Since concentration of electrons in the near-surface charge is much higher than in the semiconductor bulk and taking into account that the count off performs, beginning from bottom of the conduction band then it should lead to a distortion of the Fermi level (see Figure 5). Thus electric potential difference forms between electrode, connected to the near-surface charge, and electrode, connected to the semiconductor plate; it is equal to $\Delta U_k = U_k - U_{k1}$ (see Figure 5). By assuming that the Fermi level does not bend, we should suppose that the near-surface charge has an electric potential, which is equal to the electric potential of lower electrode. Then the energy, absorbed in the process of charging, does not transform into potential energy of the charged capacitor and disappears in unknown direction that violates the law of energy conservation. The fact that surface of the semiconductor plates is an equipotential surface shows that an electrostatic equilibrium arises, in which a charge distribution exists on the surface in absence of currents. The junction between lower electrode and semiconductor plate creates thermodynamic equilibrium, in which the Fermi levels coincide in metal and semiconductor. This occurred due to the lower electrode connected before charging the capacitor, when this equilibrium was possible.

Since the Fermi level of lower electrode and upper electrode, connected to the near-surface charge, are different, then we can speak on that the difference of electric potentials at the junction between lower electrode and semiconductor plate is an electromotive force, in which non-electric force, driving charges against the forces of the electric field, is the environmental heat.

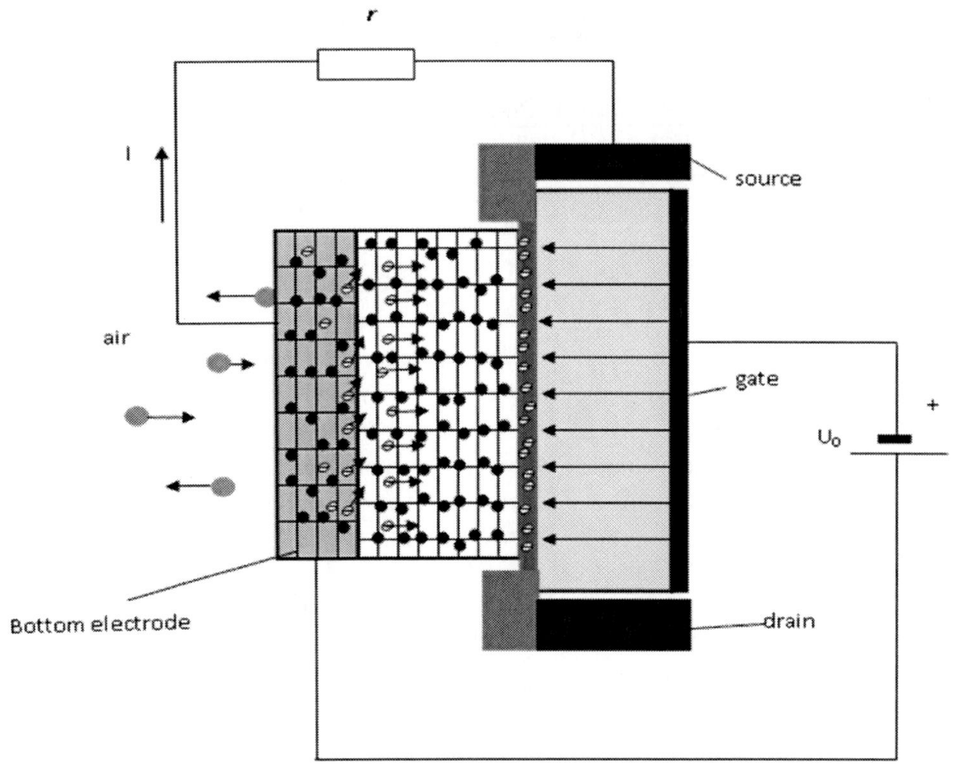

Figure 6. Explanation of heat absorption from environment by MISM structure.

Figure 6 explains how the environmental heat transforms into electric energy in the MISM structure. As is shown above, the electric potential difference $\Delta U_k = U_k - U_{k1}$ forms between lower electrode, and drain and source, connected in parallel one to other and connected to the near-surface charge. Therefore, if we connect resistor r to these electrodes then the current flows through the resistor along the direction, in which electrons should overcome a potential barrier. These electrons, reaching the near-surface charge, give their energy to electric field, maintaining the attained potential difference in each moment of time. The electrons that move from the near-surface charge take away this amount of energy. Thus, the energy of the charged capacitor remains constant in each moment of time, and the energy, absorbed at overcoming the potential barrier, evolves on the resistor r.

Air molecules affect on the MISM structure (see Figure 6), interacting with surface of metallic electrode. The air molecules influence on atoms in crystalline lattice. Then the atoms interact with electrons, causing their energy re-distribution. This process allows the electrons to overcome the potential barrier. The electrons with lesser energy, which cannot overcome the potential barrier, take away the energy of atoms of the crystalline lattice, as a result, of interaction with them. At the same time, the atoms of the crystalline lattice interact with molecules of air and take away their energy. However, the energy of the air molecules recovers in each moment of time, owing to solar radiation, which provides energy of the molecules (sun warms air, water, earth). This process leads to transformation of the solar energy into electrical one. So, the converter begins to cool, if we insulate it of heat; the energy distribution starts to decrease and aspires to zero in end of the process.

CONCLUSION

Thus, a distortion of the Fermi level observes in near-surface area of the MISM structure. It is a consequence of the electrostatic equilibrium, in which total semiconductor surface is the equipotential surface. It does not violate the law of energy conservation. If we assume, that the Fermi level does not distort then the law of energy conservation violates, since a heat, absorbed at charging the capacitor, disappears by unknown way.

The MISM structure fabricates on silicon plate [2,3]. Let us take low-resistance silicon (with specific resistance of 0.01 Ω·cm or less) for this plate and we process a layer on the plate with a thickness of 10 μm and a specific resistance being no lesser than 1 Ω·cm. If we fabricate MISM structures, having the sizes of 0.1×0.1 mm^2, on this layer, then the electric resistance between near-surface charge and low-ohms part of the plate is ~ 10 Ω. The electric resistance in the near-surface charge (where a concentration of electrons is close to the concentration of electrons in metal) is equal to 100 Ω at thickness of 5 nm, length of 100 mm and width of 100 mm. Thus, electric resistance between lower electrode and the near-surface charge equals 110 Ω for the MISM structure with sizes of 0.1×0.1 mm^2.

We can place 10,000 MISM structures in a square of 1 cm^2. If these structures connect in parallel, electric resistance between lower electrode and near-surface charge is 10,000 times less and equals 0.011 Ω. If we connect in series three such 1 cm^2-plates and place one plate under other, then 0.5 W of absorbed power can be obtained at electric load resistance of 0.1 Ω, and the Peltier coefficient of $\Pi = 0.1$ V. The thickness of this cooler is ~ 1 mm, taking into account that thickness of the silicon plate is equal to 0.35 mm. Thus, 1 cm^2-square of the proposed cooler allows one to take away up to 0.5 W of heat power, which is enough to cool sufficiently high-powered LED. If we connect 1,000 coolers in series that correspond to the silicon plate with sizes of 33×33 cm^2, then output voltage is equal to 300 V, and internal electric resistance is 33 Ω. The heat power up to 500 W evolves on the electric load resistance of 100 Ω, and we can provide an electric energy of 1,000 LEDs with power of 0.5 W for each. Thus, we can create lamp with sizes of 31.6×31.6 cm^2 from 1,000 LEDs, which absorb a heat power is equal to 500 W, and the absorbed heat transforms into electric energy, to ensure electric power for the LEDs matrix.

REFERENCES

[1] Bell, L. Accelerating the Commercialization of Promising New Thermoelectric Materials. www.asminternational.org

[2] Karapetyan, G. Ya.; Dneprovski, V. G., In: Physics and Mechanics of New Materials and their Applications, Ivan A. Parinov, Shun-Hsyung Chang (Eds.). *Nova Science Publishers,* New York, Chapter 29, 2013, pp. 405-416.

[3] Stilbans, L. S., Physics of Semiconductors. *Nauka,* Moscow. 1967, pp 1 – 451 (In Russian).

[4] Dneprovski, V. G.; Karapetyan, G. Ya.; Pustovalov, G. A.; Lysenko, M. V., *J. Microsystem Engineering,* 2003, No 9, 38 – 42 (In Russian).

[5] Lunin, L. S.; Karapet'yan, G. Ya.; Dneprovskii, V. G.; Kataev V. F., *Technical Physics,* 2013, vol. 58(11), 1619 – 1624.

In: Advanced Materials - Studies and Applications
Editors: I.A. Parinov, S.-H. Chang, S.Theerakulpisut
ISBN: 978-1-63463-749-7

Chapter 15

EXCITATION OF UN-DAMPED OSCILLATIONS IN *LC*-CIRCUIT WITH THERMOELECTRIC CAPACITOR WITH SEMICONDUCTOR PLATE

G. Ya. Karapetyan* and V. G. Dneprovski
Vorovich Mathematics, Mechanics and Computer Sciences Institute,
Southern Federal University, Rostov-on-Don, Russia

ABSTRACT

We consider process of excitation of the un-damped oscillations in a circuit, consisting of capacitor, one plate of which contains metal-semiconductor junction. This junction is under influence of transverse magnetic field, generated by ferromagnetic core of circuit coil. Moreover, the core includes a permanent magnet; therefore magnetic field contains both alternating (*ac*) and direct (*dc*) components. We write oscillation equation of the circuit and show that the un-damped (*dc*) oscillations could arise due to absorption of environment heat and further cooling the metal-semiconductor junction at certain parameters of semiconductor and value of the *dc*-field.

1. INTRODUCTION

The chapter considers a structure, consisting of permanent magnet and *LC*-circuit, in which the inductance uses coil, wounded on core of a magnetic material. Thermoelectric capacitor disposes in a gap of the magnetic circuit (see Figure 1). One of plates of the capacitor is metallic, and other consists of metallic and semiconductor materials. The semiconductor has strong dependence of thermo-EMF on magnetic field and locates under insulator [1].

The magnetic field changes a ratio of fast and slow electrons in current, owing to the fast electrons scatter on heat oscillations of crystalline lattice in lesser degree than slow electrons

---

* jorichkaka@yandex.ru.

[2]. This leads to growth of relative quantity of the fast electrons in the current with increasing the magnetic field and further increasing the Peltier factor Π = αT =Uk. Therefore thermo-EMF $\alpha = \alpha_0[1 + c_\alpha(\eta_H B)^2]$ increases also, where α0 is the thermo-EMF in absence of the magnetic field, c_α is the coefficient, depending on the character of electron scattering (for example, $c_\alpha \approx 0.154$ at the scattering on heat phonons), ηH is the Hall mobility of electrons in semiconductor, *B* is the magnetic induction of the field.

The un-damped (*dc*) oscillations excite in the circuit in this case because of the absorption of environment heat. It occurs due to the summary magnetic field (*dc*-field of permanent magnet and *ac*-field of inductance) in the gap of the magnetic circuit and the Peltier factor increases in the semiconductor-metal junction at a charging the capacitor and decreases at its discharging. As a result, the capacitor cools in the circuit per one period of oscillations, and the absorbed heat transforms into electric energy.

2. CONCEPT

Let us consider the oscillatory circuit, shown in Figure 1.

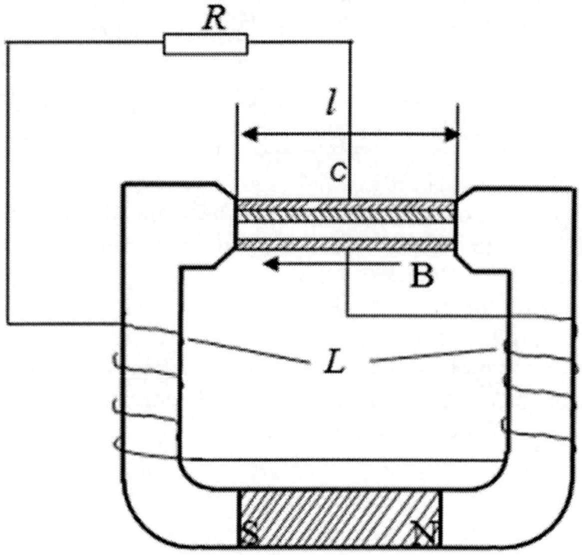

Figure 1. Circuit for study of un-damped (*dc*) oscillations.

In this circuit, the thermoelectric capacitor locates in the gap of the magnetic circuit, which consists of permanent magnet and two branches of a soft magnetic material circuit with wounded inductance *L*. The current *i*(*t*) flows through the inductance, when oscillations excite into circuit, and magnetic field appears in the gap of the magnetic circuit with the induction:

$$B(t) = B_g + \frac{\mu_0 N i(t)}{l_m/\mu + l_{mag}/\mu_{mag} + l}, \qquad (1)$$

where B_g is the magnetic induction in the gap, caused by permanent magnet; l is the width of the gap; l_m is the total length of the magnetic circuit branches; l_{mag} is the length of the permanent magnet with $\mu_0 = 12.56 \cdot 10^{-7}$ H/m; μ is the dimensionless magnetic permeability of the branches of the magnetic circuit; μ_{mag} is the dimensionless magnetic permeability of the permanent magnet; N is the number of turns of the inductance L. Since there is the next dependence [1]:

$$U_{km} = U_k \left\{1 + c_\alpha (\eta_H B(t))^2\right\}, \; U_k = \alpha_0 T, \tag{2}$$

then a differential equation, describing oscillations in such a circuit at $q = Cu_c$, has the form:

$$LC\frac{d^2u_c}{dt^2} + C\frac{du_c}{dt}\left\{R - 2c_\alpha\eta_H^2U_k\frac{B_g\mu_0N}{l_m/\mu + l_{mag}/\mu_{mag} + l} - c_\alpha\eta_H^2U_k\frac{\mu_0^2N^2i(t)}{(l_m/\mu + l_{mag}/\mu_{mag} + l)^2}\right\} + \tag{3}$$

$$+u_c = U_k + c_\alpha\eta_H^2U_kB_g^2 \equiv U_{k\text{м}},$$

here $i(t) = \dfrac{dq}{dt} = C\dfrac{du_c}{dt}$, uc is the voltage on the capacitor:

$$u_c = u + U_{k\text{м}}, \tag{4}$$

where u is the voltage on the capacitor in respect to external metallic electrodes. Then inhomogeneous differential equation (3) reduces to the homogeneous one by substituting relation (4):

$$LC\frac{d^2u}{dt^2} + C\frac{du}{dt}\left\{R - 2c_\alpha\eta_H^2U_k\frac{B_g\mu_0N}{l_m/\mu + l_{mag}/\mu_{mag} + l} - c_\alpha\eta_H^2CU_k\frac{\mu_0^2N^2}{(l_\text{м}/\mu + l_{mag}/\mu_{mag} + l)^2}\frac{du}{dt}\right\} + u = 0. \tag{5}$$

Then by using the expressions

$$2c_\alpha\eta_H^2U_k\frac{B_g\mu_0N}{l_m.\mu + l_{mag}/\mu_{mag} + l} = R_{neg}; \; r = c_\alpha\eta_H^2CU_k\frac{\mu_0^2N^2}{(l_m/\mu + l_{mag}/\mu_{mag} + l)^2},$$

we obtain the equation:

$$LC\frac{d^2u}{dt^2} + C\frac{du}{dt}\{R - R_{neg}\} + rC\left(\frac{du}{dt}\right)^2 + u = 0, \tag{6}$$

and Equation (6) reduces at $R = R_{neg}$ to the following one:

$$LC\frac{d^2u}{dt^2}+rC\left(\frac{du}{dt}\right)^2+u=0. \tag{7}$$

Equation (7) describes periodic un-damped oscillations, if value of $rC\omega^2U << 1$, where $\omega = 2\pi f$, f is the oscillation frequency, U is the oscillation amplitude of voltage on capacitor. The study of Equation (7) by using the phase plane method defines an ellipse, corresponding to the un-damped oscillations.

3. DISCUSSION

Equation (6) describes oscillations in circuit with self-excitation at $R < R_{neg}$, i.e., amplitude of oscillations increases during time.

Increasing of the oscillation amplitude in the circuit, shown in Figure 1, occurs with each stage of a charging the capacitor. The voltage on the capacitor increases by ΔU, causing an increase of current, which, in its turn, leads to increasing the magnetic field induction, and therefore growth of U_{kM}. So, an increase of voltage causes increasing the current and this growth of current leads to increasing the junction voltage. Hence, the energy, absorbed from environment, grows also for account of cooling the thermoelectric capacitor.

It is necessary to choose the current direction, so that a magnetic field induction, generated by the current, adds to the magnetic field induction of permanent magnet in the gap at charging the capacitor (the case, when current, flowing through metal-semiconductor junction, causes the junction cooling). It occurs, when thermoelectric capacitor electrode, being at junction with semiconductor, connects to inductance L on the hand of the north pole of permanent magnet and the inductance wounded counterclockwise. Otherwise, the sign before of R_{neg} changes to contrary and oscillation generation does not arise.

The intrinsic semiconductors demonstrate maximal increasing thermo-EMF, which can reach up to 100% [1]. Therefore, maximal amplitude of oscillations in the circuit, described by Equation (6), estimates from the condition:

$$c_\alpha(\mu_H B)^2=1. \tag{8}$$

By taking into account that $B(t) = B_g + \frac{\mu_0 Ni(t)}{l_m/\mu + l_{mag}/\mu_{mag} + l}$, we can write the following expression for steady-state current in the circuit as

$$I_{s-s} = \frac{l_m/\mu + l_{mag}/\mu_{mag} + l}{\eta_H N\mu_0}\left(\sqrt{\frac{1}{c_\alpha}} - \eta_H B_g\right). \tag{9}$$

The heat amount that absorbs or evolves at the metal-semiconductor junction is found as $dQ = \Pi idt = P(dq/dt)dt = \Pi dq$. Then, let us assume that heat capacities of metal, semiconductor and dielectric are the same and equal to c. Then we define a temperature increment as $dT = dQ/(cm) = \Pi dq/(cm)$, where m is the capacitor mass, and q is the charge on the capacitor. Then we obtain

$$\frac{dT}{T} = \frac{\alpha dq}{cm} \rightarrow T = T_0 \exp\left(\pm \frac{aq}{cm}\right). \tag{10}$$

The sign of plus indicates a heating and the sign of minus corresponds to a cooling of the junction. We define the total amount of heat that evolves or absorbs on the capacitor at its charging or discharging in the form:

$$Q = \int_0^{q_0} \alpha T_0 \exp\left(\pm \frac{aq}{cm}\right) dq = T_0 cm \left[1 \mp \exp\left(\pm \frac{aq_0}{cm}\right)\right], \tag{11}$$

where q_0 is the charge on capacitor, $q_0 = CU_0$. We obtain at the heating

$$Q_+ = T_0 cm\left[\exp\left(\frac{aq_0}{cm}\right) - 1\right]. \tag{12}$$

and at the cooling

$$Q_- = T_0 cm\left[\exp\left(-\frac{aq}{cm}\right) - 1\right]. \tag{13}$$

If $\frac{\alpha q_0}{cm} << 1$, then $Q_\pm = T_0 cm\left[\frac{\alpha q_0}{cm} \pm \frac{1}{2}\left(\frac{\alpha q_0}{cm}\right)^2\right]$, (14)

i.e., the capacitor heats in greater degree than cools. This difference defines as

$$\Delta T = \frac{Q_+ - Q_-}{cm} = T_0\left(\frac{\alpha q_o}{cm}\right)^2. \tag{15}$$

This effect leads to the following increase of the Peltier coefficient during heating:

$$\frac{\Delta \Pi}{\Pi} = \frac{\Delta Q}{T_0 \alpha q_0} = \frac{\alpha q_0}{cm}. \tag{16}$$

For the values of the parameters $C = 10^{-6}$ F, $U = 10$ V $q_0 = 10^{-5}$C, $\alpha = 0.2 \cdot 10^{-3}$ V/K, $c \approx 1$ J/(g·K), $m \approx 10^{-2}$ g, we obtain:

$$\frac{\Delta \Pi}{\Pi} = 0.2 \cdot 10^{-6}. \tag{17}$$

At the same time, the change of the Peltier coefficient due to the magnetic field is much greater ($\Delta\Pi/\Pi \approx 2.0$ [2 – 5]) at cooling. Therefore, additional increasing Π during heating compensates completely and change of Π for account of the heating is a neglected value, compared to the changing Π due to magnetic field. It is interesting to note, that additional losses arise in absence of the magnetic field, owing to oscillations in the circuit cause heating the capacitor in accordance with above-described effect. We shall compare a power, evolving on the capacitor in the form of heat, with a power, absorbed at the metal-semiconductor junction.

With this aim, we present the capacitor as capacitance C and resistor $R_{\tan\delta} = \tan\delta/(\omega C)$, connected in series, where $\tan\delta$ is the tangent of dielectric loss angle. The power, evolving on this resistor, is equal to $\frac{R_{\tan\delta} i^2}{2}$. The power, absorbed at the junction, determines as

$$W_k = \frac{i^2 R_{op}}{2}, \tag{18}$$

where

$$R_{neg} = 2c_\alpha \eta_H^2 U_k \frac{B_c \mu_0 N}{l_m/\mu + l_{mag}/\mu_{mag} + l}. \tag{19}$$

Then the ratio of the power, lost due to dielectric losses, to the power, absorbed at the junction, defines as

$$\frac{R_{\tan\delta}}{R_{op}} = \frac{(l_m/\mu + l_{mag}/\mu_{mag} + l)\tan\delta}{2\omega C c_\alpha \eta_H^2 U_k B_g \mu_0 N} = \frac{\left(1 + \frac{l_m}{\mu l} + \frac{l_{mag}}{\mu_{mag} l}\right) d \tan\delta}{2\omega\varepsilon\varepsilon_0 c_\alpha \eta_H^2 U_k B_g \mu_0 N}, \tag{20}$$

where $C = \frac{\varepsilon\varepsilon_0 l l_{width}}{d}$, l_{width} is the width of thermoelectric capacitor, d is the thickness of insulator.

Since the magnetic field is alternating, it produces a vortex electric field, and therefore eddy current and vortex EMF $E_{vort} = d\Phi/dt = \omega B_g S_{cs}$, where S_{cs} is the capacitor square perpendicular to direction of the magnetic field lines. Then the power, consumed by magnetic field on a heating the capacitor, is equal to $W_{vort} = (E_{vort})^2/R$, where R is the electric resistance of metallic film (as rule, having lowest resistance) of the capacitor electrode, $R = \rho_m l_{width}/d_m$, d_m is the thickness of the metallic film and $S_{cs} = d_m l_{width}$.

Owing to $B = \frac{\mu_0 N i(t)}{l_м/\mu + l_{mag}/\mu_{mag} + l}$, we have

$$W_{vort} = \frac{1}{2}\frac{\left(\omega B_g dl_{width}\right)^2}{\rho_M \dfrac{l_{width}}{dl}} = \frac{1}{2}\frac{\omega^2 \mu_0^2 N^2 d^3 l^2 i^2}{\left(l_M / \mu + l_{mag} / \mu_{mag} + l\right)^2 \rho_m l}, \tag{21}$$

where ρ_m is the specific resistance of metal, d_m is the total thickness of metallic plates of both electrodes. Then

$$\frac{W_{vort}}{W_k} = \frac{\omega^2 \mu_0 N d^3 l_{width}^2}{2\rho_m c_\alpha \eta_H^2 U_k B_g \left(l_M / \mu + l_{mag} / \mu_{mag} + l\right)}. \tag{22}$$

The power, scattered at the electric resistance of semiconductor, is equal to $W_s = i^2 R_s$. Then

$$\frac{W_s}{W_k} = \frac{R_s}{R_{neg}} = \frac{\rho_s d_s \left(l_M / \mu + l_{mag} / \mu_{mag} + l\right)}{2c_\alpha \eta_H^2 U_k B_c \mu_0 N l l_{width}}, \tag{23}$$

where ρ_s is the specific resistance of semiconductor, d_s is the thickness of semiconductor film or plate.

The square of thermoelectric capacitor in its cross-section does not exceed of 0.01 cm^2, if its width is l_{width} = 1 cm and total thickness of metallic, insulator and semiconducting films is no greater than 10 – 100 μm. Then we choose the gap square of S_g = 0.5 cm^2 in its cross-section, taking into account that the gap can include several thermoelectric capacitors. Then the cross-section square of the magnetic circuit and width of the gap in the magnetic circuit are equal to [6]:

$$S_m = \frac{B_g \sigma S_g}{B_d}; \; l_{mag} = \frac{H_g f l}{H_d} = \frac{B_g f l}{\mu_0 H_d}, \tag{24}$$

where B_g is the given value of magnetic field induction in the gap; σ is the factor of the magnetic flux scattering; B_d and H_d are the magnetic induction and magnetic field, corresponding to the point $(BH/2)_{max}$ of a demagnetization curve for a material of magnet; f is the factor, taking into account a fall of magnetic EMF in magnetic armature and junctions.

4. Some Estimates

In this section, we shall obtain some estimates, based on our model of the thermoelectric capacitor. Usually, the values of f = 1.1 – 1.4 and σ = 1.25 – 3 are used for study of the magnetic circuit, depicted in Figure 1.

We have mentioned that a change of junction voltage took place synchronously with changing the magnetic field, if the change occurred no less than per 10^{-6} s. Therefore we shall take f_0 = 60 KHz. Then, we choose a barium magnet [6] as the permanent magnet, for which specific resistance lies into range of $10^4 - 10^6$ Ω·cm, then the losses on eddy currents are near to zero. We take manganese-zinc ferrite as the branches of magnetic circuit, which has low losses at re-magnetization are equal to 2 – 4 μW/(cm^3·Hz). We have $B_d \approx B_r/2 = 0.2$ T, $H_d \approx H_c/2 = 1.2$ kA/cm for the barium magnet [11], where B_r and H_c are the residual induction and coercive force, respectively. As semiconductor material, we choose the indium antimonide, which has record-high electron mobility of $\eta_H = 10$ m^2/(V·s) at room temperature. Then we assume that *dc*-component of magnetic induction is close to *ac*-component. Let $B_c = 0.154$ T, then the *ac*-component, calculated from expression (1), is equal to 0.1 T. Now we determine the cross-section square $S_m = 1.15$ cm^2 and length of permanent magnet l_{mag} = 0.49 cm from formulae (24) by assuming that $l = 0.4$ cm.

We choose the number of turns $N = 100$ in the inductance L with a wire diameter of 1.12 mm and, as a result, calculate $l_m = 112$ mm. Let us assume $\alpha = 1$ mV/K that corresponds to the concentration of electrons in semiconductor is equal to $10^{16} - 10^{17}$ Ω·cm [1]. Then the specific resistance of semiconductor is $10^{-3} - 10^{-2}$ Ω·m and $U_k = \alpha T = 0.3$ V (at $T = 300$ K). Since permanent magnet operates in the saturation mode, its dimensionless magnetic permeability exceeds slightly of 1. Therefore, we can choose $\mu_{mag} = 1.5$ [6]. We can also select a value of $\mu = 1000$ for the circuit branches of soft magnetic material. Then we calculate $R_{neg} = 0.024$ Ω by using formula (19) and above-pointed numerical values of the required parameters.

We choose copper wire with a length of 5 m and electric resistance of 0.084 Ω for the inductance that is greater than negative electric resistance. Therefore, a generation of oscillations is absent. We define inductance as $L = \mu\mu_0 N^2 S_m / l_m$. Because the magnetic circuit includes permanent magnet with low magnetic permeability and air gap, we can estimate the effective magnetic permeability in the form: $\mu_{ef} \approx \dfrac{(l_m + l_{mag})}{l_m/\mu + l_{mag}/\mu_{mag} + l} = 15.8$. We calculate L = 203.7 μH by assuming $S_m = 1.15$ cm^2 and $l_m = 0.112$ m.

Then we define the capacitance of the thermoelectric capacitor as $C = 1/\omega^2 L = 34.6$ nF and the electric resistance, caused by dielectric losses of tan $\delta = 0.0005$, as $R_{tg\delta} = \omega L \cdot \tan\delta = 0.038$ Ω. It defines again a greater value, compared to negative electric resistance. Further we determine the dielectric permittivity of the thermoelectric capacitor as $\varepsilon = \dfrac{Cd}{\varepsilon_0 l l_{width}} = 97.7$ and estimate the eddy current loss as $\dfrac{W_{vort}}{W_k} = 1.69 \cdot 10^{-5}$ from expression (22) at $\rho_m = 10^{-8}$ Ω·m, i.e., the eddy current loss is near to zero.

We estimate the loss in the semiconductor as $\dfrac{W_s}{W_k} = 0.0041$ by using formula (23) at $\rho_s = 10^{-2}$ Ω·m and $d_s = 3$ μm; again the loss in the semiconductor is near to zero. The amplitude of steady-state oscillations of voltage on the capacitor, we determine as $I_{s-s} = 5.9$ A by using expression (9) and taking into account that $B_g = 0.154$ T.

Then we calculate the following parameter: $r = c_\alpha \eta_H^2 C U_k \dfrac{\mu_0^2 N^2}{(l_m/\mu + l_{mag}/\mu_{mag} + l)^2}$ = $8.74 \cdot 10^{-11}$ s^2/V. Because $rC\omega^2 U = r\omega I_{s-s} = 19.43 \cdot 10^{-8} \ll 1$, Equation (7) describes the undamped oscillations.

The magnetic field, created by the inductance, $H = NI_{s\text{-}s}/(l_m/\mu + l/\mu_{mag} + l) = 79.65$ kA/m, is several times lesser than value of coercive force for the barium magnet (about 200 kA/m). Therefore, the loss on hysteresis is very small. The losses in soft magnetic material, we can estimate, if know specific losses on re-magnetization in the material. As shown in [5], the loss in the soft magnetic material W-04 equals 3.3 W at the frequency of 60 kHz and $B = 0.1$ T. The electric resistance of loss $R_{loss} = 0.095\ \Omega$ inserts in electric circuit at the current of 5.9 A. Then the summary positive electric resistance in the circuit, caused by the dielectric loss, ferromagnetic loss and resistive loss in wire, is equal to $R_w = 0.22\ \Omega$. So, to create the generation of oscillations, we should instead of one thermoelectric capacitor to use several capacitors, connected in series. The junction voltages add in this case and negative electric resistance increases in n times (where n is the number of the same thermoelectric capacitors, connected in series).

The negative electric resistance increases in 20 times and equals 0.48 Ω at $n = 20$, which is greater than $R_w = 0.22\ \Omega$, leading to the circuit generation. The total capacitance does not change at choice of insulator with value of $\varepsilon = 1{,}954$ (corresponding to $n = 20$), but with double dielectric loss ($\tan\delta = 0.001$). As it follows from formulae (19) and (20), the loss on eddy currents and loss in semiconductor do not change, i.e., they are near to zero. Therefore the electric load resistance can be chosen as $R = R_{neg} - R_s = 0.225\ \Omega$ and the useful power, evolving on load resistor, is equal to $W_R = I^2R/2 = 3.9$ W. We can change the magnitude of the electric load resistance, if connect it to the circuit through additional winding, wounded on the magnetic core. In this case, each turn of additional winding induces the voltage, $\omega_0 B_g S_m =$ 4.33 V. Obviously, by choosing the number of turns, we can connect to the transducer any electric load resistance. Only, it is necessary that the power, evolving on the electric load, does not exceed of 3.9 W. Otherwise, the positive electric resistance arises in the contour, which is greater than 0.225 Ω, and the oscillation generation breaks.

REFERENCES

[1] Ure Ronald, W., *IEEE*, 1963, *vol. 51(5)*, 702-717.

[2] Wolf, R.; Smith, G., *Bull Am. Phys. Soc.*, 1961, *vol. 6(2)*.

[3] Jim, W. M.; Amith, A., *Solid State Electronics*, 1972, *vol. 15*.

[4] Weiner, A. L., *Cascade Thermoelectric Sources of Cold*. Moscow, Soviet Radio, 1976 (In Russian).

[5] Waldock, A., *TIER*, 1963, *vol. 51(5)*, 697 – 702.

[6] Reinboth, T. *Magnetic Materials and their Applications*, A. A. Preobrazhensky (Ed.), Leningrad, Energy, 1974 (In Russian).

PART III. MECHANICS OF ADVANCED MATERIALS

In: Advanced Materials - Studies and Applications
Editors: I.A. Parinov, S.-H. Chang, S.Theerakulpisut
ISBN: 978-1-63463-749-7

Chapter 16

FLAT PUNCH SLIDING ON BOUNDARY OF ELASTIC HALF-PLANE

V. B. Zelentsov*, B. I. Mitrin and S. S. Volkov
Research and Education Center "Materials",
Don State Technical University, Rostov-on-Don, Russia

ABSTRACT

The dynamic contact problem on the motion of a flat punch on the boundary of an elastic half-plane is considered.

During the movement, sliding indentation occurs such a way that the surface of the punch remains parallel to the half-plane boundary at each moment of time. In moving coordinates relative to the moving punch position, the contact problem is reduced to a 2-D integral equation.

Approximate solution of the integral equation of the problem is constructed in the form of Neumann series, whose zero term is represented by a superposition of solutions of the corresponding 2-D integral equations on a coordinate half-axis minus a solution of the integral equation on a whole axis.

This approach allows one to construct a solution of the 2-D integral equation of the problem in four velocity ranges of the punch movement, covering a whole spectrum of its velocities, and also to conduct a detailed analysis of features of the contact stresses and vertical displacements of a free surface on the boundary of the contact zone. To obtain effective solutions of the problem, containing no singular quadratures, an approximate method of solution of the integral equation is proposed, based on the special approximation of an integrand of the kernel of the integral equation in a complex plane. The considered problem initially has been studied by Galin [1] by using methods of complex analysis in a subsonic range of the punch velocity. In [2–5] and other works, similar problems have been investigated in different velocity ranges. In [6,7], the problem has been solved in the range of steady movements without taking into account inertia of the foundation. In [8], the problem has been solved by the method used here, but in a limited velocity range.

* vbzelen@gmail.com

1. The Problem Statement and Its Integral Equation

Let us consider the dynamic contact problem about a flat punch with width of $2a$ $(-a-Vt \le x \le a-Vt)$, moving parallel to y-axis with a constant velocity V on the boundary $(y=0)$ of an elastic half-plane $(|x|<\infty, y\ge 0)$ to the negative direction of x-axis. The indentation of the moving punch into the elastic half-plane is performed according to the law $y=\varepsilon(t)$ $(t\ge 0)$ in such a way that its base remains parallel to x-axis. The friction and adhesion forces between the punch base and the elastic half-plane are absent. The scheme of the problem is present in Figure 1.

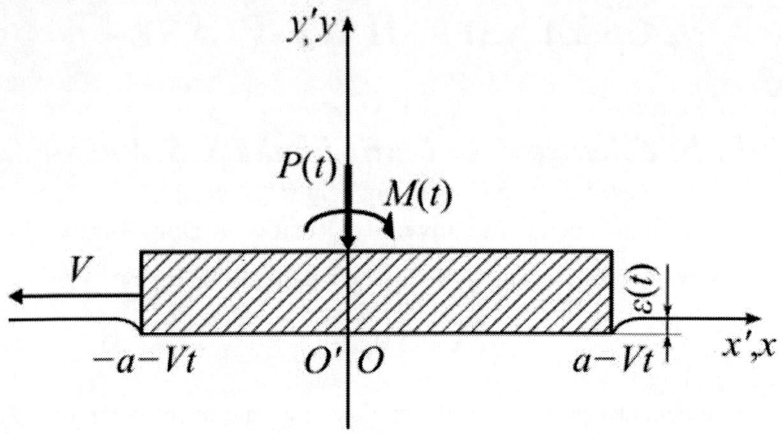

Figure 1. Flat punch, moving along x-axis in negative direction with constant velocity V, deforms elastic half-plane by the value of $\varepsilon(t)$ ($\varepsilon(t_0)=0$), while remaining parallel to x-axis at each moment. Moving coordinate system x', y' is connected with moving punch.

The half-plane surface outside of the punch contact zone with the elastic medium is free of stresses: $\sigma_{yy}=\sigma_{xy}=0$, where σ_{yy} is the normal stress and σ_{xy} is the tangential stress. The displacements u, υ and stresses σ_{yy}, σ_{xy}, σ_{xx} in the elastic half-plane aspire to zero at infinity. Due to the elastic half-plane was in steady state before time t, initial conditions of the problem are zero: $u=\upsilon=\partial u/\partial t=\partial \upsilon/\partial t=0$, at $t=0$.

The boundary conditions of the formulated contact problem in a fixed coordinate system x, y are written [1,2] as

$$y=0 \quad \sigma_{yy}(x+Vt,0,t)=\varphi(x+Vt,t) \qquad -a-Vt\le x\le a-Vt \tag{1}$$

$$\upsilon(x+Vt,0,t)=\varepsilon(t) \qquad -a-Vt\le x\le a-Vt \tag{2}$$

$$\sigma_{yy}(x+Vt,0,t)=0 \qquad -\infty<x<-a-Vt,\ a-Vt<x<\infty \tag{3}$$

$$\sigma_{xy}(x+Vt,0,t)=0 \qquad -\infty<x<\infty \tag{4}$$

At fulfillment of the conditions at infinity and initial conditions of the problem, where $\varphi(x+Vt,t)$ are the unknown normal stresses on the half-plane boundary under the punch base.

The Lamé movement equations of the plane elasticity theory in the displacement formulation in the fixed coordinate system x, y are written [9] as

$$(\lambda+\mu)\ \mathrm{grad\ div}\ \mathbf{u} + \mu\Delta\mathbf{u} = \rho\mathbf{u}''_{tt},$$

where $\mathbf{u} = \mathbf{u}(u,v)$ is the displacement vector, whose coordinates, being scalar functions $u(x,y,t)$, $v(x,y,t)$, are the horizontal and vertical displacements in the elastic medium, λ, μ are Lamé parameters, ρ is the density of the elastic medium, primes designate second derivative with respect to time t. To obtain integral equation with fixed limits of integration are independent on time, it is necessary their change to a moving coordinate system x', y', connected with the punch, moving to the left direction with the velocity V. In this case, the moving coordinates x', y' are connected with the fixed ones x, y via formulae $x' = x+Vt$, $y' = y$. The transition to the moving coordinates in (1)-(4) leads to the mixed boundary conditions:

$$y'=0 \qquad \sigma_{yy}(x',0,t) = \varphi(x',t) \qquad -a \le x' \le a \tag{5}$$

$$v(x',0,t) = \varepsilon(t) \qquad -a \le x' \le a \tag{6}$$

$$\sigma_{yy}(x',0,t) = 0 \qquad -\infty < x' < -a,\ a < x' < \infty \tag{7}$$

$$\sigma_{xy}(x',0,t) = 0 \qquad -\infty < x' < \infty \tag{8}$$

and the differential equation in partial derivatives for the elastic medium has the form:

$$(\lambda+\mu)\ \mathrm{grad\ div}\ \mathbf{u} + \mu\Delta\mathbf{u} = \rho(\mathbf{u}''_{tt} + 2V\,\mathbf{u}''_{x't} + V^2\,\mathbf{u}''_{x'x'}). \tag{9}$$

To reduce the mixed boundary-value problem (5)-(9) (primes are omitted hereinafter for simplicity) with zero initial conditions and conditions at infinity to integral equation, we use Laplace integral transform on time t (index L) and Fourier integral transform on longitudinal coordinate x (index F) [10] in the next designations:

$$\begin{Bmatrix} u^{LF} \\ v^{LF} \end{Bmatrix}(\alpha, y, p) = \int_0^\infty e^{-pt}dp \int_{-\infty}^{\infty} \begin{Bmatrix} u \\ v \end{Bmatrix}(x,y,t) e^{i\alpha x} dx, \tag{10}$$

$$\begin{Bmatrix} u \\ v \end{Bmatrix}(x,y,t) = \frac{1}{2\pi i}\int_{-i\infty}^{i\infty} e^{pt}dp \frac{1}{2\pi}\int_{-\infty}^{\infty} \begin{Bmatrix} u^{LF} \\ v^{LF} \end{Bmatrix}(\alpha, y, p) e^{-i\alpha x} d\alpha, \tag{11}$$

By using of which, the mixed boundary-value problem (6)–(9) is reduced to the next 2-D integral equation (IE):

$$\int_0^t d\tau \int_{-a}^{a} \varphi(\xi,\tau)k(\xi-x,t-\tau)d\xi = -2\pi\mu c_V^{-1}\varepsilon(t) \qquad |x|\le a,\ t>0 \tag{12}$$

$$k(x,t)=\frac{1}{2\pi i}\int_{-i\infty}^{i\infty} e^{pt}dp\int_{\Gamma} K(u)e^{iuxp/c_2}du \tag{13}$$

$$K(u)=(1-iu\beta_2)^2\sigma_2 R^{-1}(u),\quad R(u)=c_V\left[\left(2u^2+(1-iu\beta_2)^2\right)^2-4u^2\sigma_1\sigma_2\right] \tag{14}$$

$$\sigma_1=\sqrt{u^2+(1-iu\beta_2)^2},\quad \sigma_2=\sqrt{u^2+\beta^2(1-iu\beta_2)^2},$$

$$\beta_k=V/c_k,\quad k=1,2$$

$$c_1=\sqrt{\frac{\lambda+2\mu}{\rho}},\quad c_2=\sqrt{\frac{\mu}{\rho}},\quad \beta=\frac{c_2}{c_1}$$

on normal stresses $\varphi(x,t)=\sigma_{yy}(x,0,t)$, arising under the punch. Here Γ is the integration contour in a complex plane $u=\sigma+i\tau$ along the straight line under the angle of $-\arg p$ to a real plane, c_1, c_2 are the velocities of longitudinal and transverse elastic waves.

The factor c_V in each of four velocity ranges (I. $V>c_1$; II. $c_2<V<c_1$; III. $c_R<V<c_2$; IV. $V<c_R$, c_R is the velocity of Rayleigh wave) is calculated as

I. $\beta_1>1$, $\beta_2>1$

$$c_V=i\left(1+k_2^2\right)k_1/\Delta,\quad \Delta=\left(1-k_2^2\right)^2+4k_1k_2,\quad k_1=\sqrt{\beta_1^2-1},\quad k_2=\sqrt{\beta_2^2-1} \tag{15}$$

II. $\beta_1<1$, $\beta_2>1$

$$c_V=-\left(1+k_2^2\right)\bar{k}_1/\Delta_1,\ \Delta_1=\left(1-k_2^2\right)+i4\bar{k}_1k_2,\ \bar{k}_1=\sqrt{1-\beta_1^2} \tag{16}$$

III, IV. $\beta_1<1,\beta_2<1$

$$c_V=-\left(1+k_2^2\right)^2\bar{k}_1/\Delta_2,\ \Delta_2=\left(1-k_2^2\right)^2-4\bar{k}_1\bar{k}_2,\ \bar{k}_2=\sqrt{1-\beta_2^2} \tag{17}$$

The function of complex variable $K(u)$ (14) is neither even nor odd. On the real axis of the complex plane $u = \sigma + i\tau$, $K(u)$ is the complex-valued function. At the small values of $|u|$, we have

$$K(u) = K(0) + O(u) \text{ at } |u| \to 0, \; K(0) = \beta / c_V \tag{18}$$

At the great values of $|u|$, on the real axis, we have

$$K(u) = \frac{1}{|u|} + O\left(\frac{1}{u|u|}\right) \text{ at} |u| \to \infty, \arg u = 0 \tag{19}$$

and in the complex plane on a fixed Riemann surface, we have

$$K(u) = \frac{1}{u} + O\left(\frac{1}{u^2}\right) \text{ at } |u| \to \infty, 0 < \arg u < 2\pi \tag{20}$$

In the complex plane $u = \sigma + i\tau$, the function $K(u)$ has the following isolated singular points: four algebraic branching points $u = -i\gamma_k$ ($k = 1,2,3,4$), located on the imaginary axis, in which γ_k are calculated as

$$\gamma_1 = \frac{\beta}{\beta_1 - 1}, \; \gamma_2 = \frac{1}{\beta_2 - 1}, \; \gamma_3 = \frac{1}{\beta_2 + 1}, \; \gamma_4 = \frac{\beta}{\beta_1 + 1} \tag{21}$$

and two Rayleigh poles $u = -i\gamma_k$ ($k = 5,6$), that also situate on the imaginary axis:

$$\gamma_5 = \frac{\eta_R}{\beta_R - 1}, \; \gamma_6 = \frac{\eta_R}{\beta_R + 1}, \; \beta_R = \frac{V}{c_R} \tag{22}$$

where $\pm i\eta_R$ are the roots of the Rayleigh classic equation:

$$\left(2u^2 + 1\right)^2 - 4u^2\sqrt{u^2 + 1}\sqrt{u^2 + \beta^2} = 0$$

Four various cases are present for disposition of the isolated singular points $u = -i\gamma_k$ ($k = 1,2,\ldots,6$) on the imaginary axis of a complex plane, depending on the value of V:

$$\text{I. } \beta_1 > 1, \; \beta_2 > 1, \gamma_1 > \gamma_2 > \gamma_5 > \gamma_6 > \gamma_3 > \gamma_4 > 0 \tag{23}$$

(all points $u = -i\gamma_k$ ($k = 1,2,\ldots,6$) situate on the imaginary axis of lower half-plane of the complex plane $u = \sigma + i\tau$).

II. $\beta_1 < 1$, $\beta_2 > 1$, $\gamma_1 < 0$, $\gamma_2 > \gamma_5 > \gamma_6 > \gamma_3 > \gamma_4 > 0$ (24)

(the point $u = -i\gamma_1$ locates in upper half-plane and other points dispose into the lower half-plane).

III. $\beta_1 < 1$, $\beta_2 < 1$, $\gamma_2 < \gamma_1 < 0$, $\gamma_5 > \gamma_6 > \gamma_3 > \gamma_4 > 0$ (25)

(two points $u = -i\gamma_k$ ($k = 1,2$) situate in upper half-plane and other points ($k = 3,4,5,6$) dispose into lower half-plane).

IV. $\beta_1 < 1$, $\beta_2 < 1$, $\gamma_5 < \gamma_2 < \gamma_1 < 0$, $\gamma_6 > \gamma_3 > \gamma_4 > 0$ (26)

(three points $u = -i\gamma_k$ ($k = 1,2,5$) dispose into upper half-plane and other points ($k = 3,4,6$) into lower half-plane).

We make cuts in each of the four cases of disposition of the singular points $u = -i\gamma_k$ ($k = 1,2,3,4$) for a single-valued representation of $K(u)$ in the complex plane $u = \sigma + i\tau$ by the next way:

I. From branching points $u = -i\gamma_k$ ($k = 12,3,4$), disposed into lower half-plane on the imaginary axis (23), along negative part of the imaginary axis to $-i\infty$;

II. From point $u = -i\gamma_1$ $(\gamma_1 < 0)$ along positive part of the imaginary axis to $+i\infty$ into upper half-plane; from points $u = -i\gamma_k$ ($k = 2,3,4$, $\gamma_k > 0$) along negative part of the imaginary axis to $-i\infty$ into lower half-plane;

III. From points $u = -i\gamma_k$ ($k = 1,2$, $\gamma_k < 0$) along positive part of the imaginary axis to $+i\infty$ into upper half-plane; from points $u = -i\gamma_k$ ($k = 3,4$, $\gamma_k > 0$) along negative part of the imaginary axis to $-i\infty$ into lower half-plane;

IV. The same way as in the case III.

The cuts we made allow one to calculate the quadratures, included into kernel (13) of integral equation (12), and to represent the next analytical form:

$$k(x,t) = \frac{2c_2}{x}\left[K_1\left(\frac{c_2 t}{x}\right)H(x) + K_2\left(-\frac{c_2 t}{x}\right)H(-x)\right] \quad x \neq 0 \tag{27}$$

in each of the ranges of V. The formulae for the functions of real variable $K_1(\xi)$ and $K_2(\xi)$ for each of the velocity ranges have the forms:

I. $K_1(\xi) = 0$ (28)

$$K_2(\xi) = -l_{11}^{-}(\xi)H(-\gamma_2 + \xi)H(\gamma - \xi)H(-\gamma_4 + \xi)H(\gamma_3 - \xi) - l_{12}^{-}(\xi)H(-\gamma_3 + \xi)H(\gamma_2 - \xi)$$

$$l_{11}^{-}(\xi)=\left(1-\xi\beta_2\right)^2\sqrt{\xi-\gamma_4}\sqrt{\gamma_1-\xi}\sqrt{\beta_1^2-1}\eta_1(\xi)r_{-}^{-1}(\xi)$$
$$r_{-}(\xi)=c_V\left[\eta_1^2(\xi)-16\xi^4(\xi-\gamma_1)(\xi-\gamma_2)(\xi-\gamma_3)(\xi-\gamma_4)\left(\beta_1^2-1\right)\left(\beta_2^2-1\right)\right]$$
$$l_{12}^{-}(\xi)=\left(1-\xi\beta_2\right)^2\sqrt{\xi-\gamma_4}\sqrt{\gamma_1-\xi}\sqrt{\beta_1^2-1}R_{-}^{-1}(\xi)$$
$$R_{-}(\xi)=c_V\left[\eta_1(\xi)-4\xi^2\sqrt{\gamma_1-\xi}\sqrt{\gamma_2-\xi}\sqrt{\xi-\gamma_3}\sqrt{\xi-\gamma_4}\sqrt{\beta_1^2-1}\sqrt{\beta_2^2-1}\right]$$
$$\eta_1(\xi)=\left(-2\xi^2+\left(1-\xi\beta_2\right)^2\right)^2.$$

$$\text{II. } K_1(\xi)=-l_{21}^{+}(\xi)H(\gamma_1+\xi) \tag{29}$$
$$K_2(\xi)=-l_{21}^{-}(\xi)H(-\gamma_2-\xi)H(-\gamma_4-\xi)H(\gamma_3+\xi)-l_{22}^{-}(\xi)H(-\gamma_3-\xi)H(\gamma_2+\xi)$$
$$l_{21}^{\pm}(\xi)=\left(1\pm\xi\beta_2\right)^2\sqrt{\xi\pm\gamma_1}\sqrt{\xi\pm\gamma_4}\sqrt{1-\beta_1^2}\eta_2^{\pm}(\xi)r_{\pm}^{-1}(\xi)$$
$$r_{\pm}(\xi)=c_V\left[\eta_2^{\pm 2}(\xi)+16\xi^2(\xi\pm\gamma_1)(\xi\pm\gamma_2)(\xi\pm\gamma_3)(\xi\pm\gamma_4)\left(1-\beta_1^2\right)\left(\beta_2^2-1\right)\right]$$
$$l_{22}^{-}(\xi)=\left(1-\xi\beta_2\right)^2\sqrt{\xi-\gamma_1}\sqrt{\xi-\gamma_4}\sqrt{1-\beta_1^2}R_{-}(\xi)$$
$$R_{-}(\xi)=c_V\left[\eta_2^{-}(\xi)-4\xi^2\sqrt{\xi-\gamma_1}\sqrt{\gamma_2-\xi}\sqrt{\xi-\gamma_3}\sqrt{\xi-\gamma_4}\sqrt{1-\beta_1^2}\sqrt{\beta_2^2-1}\right]$$
$$\eta_2^{\pm}(\xi)=\left(-2\xi^2+\left(1\pm\xi\beta_2\right)^2\right)^2.$$

$$\text{III. } K_1(\xi)=-l_{31}^{+}(\xi)H(\gamma_2+\xi)-l_{32}^{+}(\xi)H(\gamma_1+\xi)H(-\gamma_2-\xi) \tag{30}$$
$$K_2(\xi)=-l_{31}^{-}(\xi)H(-\gamma_3+\xi)-l_{32}^{-}(\xi)H(-\gamma_3+\xi)H(\gamma_4-\xi)$$
$$l_{31}^{\pm}(\xi)=\left(1\pm\xi\beta_2\right)^2\sqrt{\xi\pm\gamma_1}\sqrt{\xi\pm\gamma_4}\sqrt{1-\beta_1^2}R_{\pm}^{-1}(\xi)$$
$$R_{\pm}(\xi)=c_V\left[\eta_3^{\pm}(\xi)-4\xi^2\sqrt{\xi\pm\gamma_1}\sqrt{\xi\pm\gamma_2}\sqrt{\xi\pm\gamma_3}\sqrt{\xi\pm\gamma_4}\sqrt{1-\beta_1^2}\sqrt{1-\beta_2^2}\right]$$
$$l_{32}^{\pm}(\xi)=\left(1\pm\xi\beta_2\right)^2\sqrt{\xi\pm\gamma_1}\sqrt{\xi\pm\gamma_4}\sqrt{1-\beta_1^2}\eta_3^{\pm}(\xi)r_{\pm}^{-1}(\xi)$$
$$r_{\pm}(\xi)=c_V\left[\eta_3^{\pm 2}(\xi)\mp 16\xi^4(\xi\pm\gamma_1)(\xi\pm\gamma_2)(\xi\pm\gamma_3)(\xi\pm\gamma_4)\left(1-\beta_1^2\right)\left(1-\beta_2^2\right)\right]$$
$$\eta_3^{\pm}(\xi)=\left(-2\xi^2+\left(1\pm\xi\beta_2\right)^2\right)^2.$$

$$\text{IV. } K_1(\xi)=-l_{41}^{+}(\xi)H(\gamma_2+\xi)-l_{42}^{+}(\xi)H(\gamma_1+\xi)H(-\gamma_2-\xi) \tag{31}$$
$$K_2(\xi)=-l_{41}^{-}(\xi)H(-\gamma_3+\xi)-l_{42}^{-}(\xi)H(-\gamma_4+\xi)H(\gamma_3-\xi)$$
$$l_{41}^{\pm}(\xi)=l_{31}^{\pm}(\xi),\ l_{42}^{\pm}(\xi)=l_{32}^{\pm}(\xi)$$

The functions $l_{12}^{-}(\xi)$, $l_{22}^{-}(\xi)$, $l_{31}^{-}(\xi)$, $l_{41}^{\pm}(\xi)$ in (28), (29), (30), (31) demonstrate, respectively, singular discontinuities at the points $\xi=\gamma_5,\gamma_6$.

It should be noted, that 2-D integral equation coincides at $V=0$ with classic one of dynamic problem for the non-stationary indentation of a flat punch into an elastic plane [11]. Moreover, if $K(u)$ is replaced in integral equation (1) by the first term of the asymptotic relation (19) and it is assumed that $\varphi(x,t)$ is independent of t, then the integral equation (12) reduces to the integral equation of the problem in the range of steady movements [6,7].

2. Approximate Solution of Integral Equation

Approximate solution of the problem or 2-D integral equation (12) is constructed in the form of Neumann series, whose zero term is represented by the superposition [12–14]

$$\varphi(x,t) = \varphi^{+}(a+x,t) + \varphi^{-}(a-x,t) - \varphi^{\infty}(x,t) \qquad 0 \le t < 2a/c_1,\ |x| \le a \qquad (32)$$

of solutions $\varphi^{\pm}(x,t)$ of 2-D integral equations on coordinate half-axis

$$\int_0^t d\tau \int_0^\infty \varphi^{\pm}(\xi,\tau) k(\pm(\xi - x), t-\tau) d\xi = -2\pi\mu c_V^{-1} \varepsilon(t) \qquad 0 \le x < \infty \qquad (33)$$

and solution $\varphi^{\infty}(x,t)$ of 2-D integral equation on a whole axis

$$\int_0^t d\tau \int_{-\infty}^\infty \varphi^{\infty}(\xi,\tau) k(\xi - x, t-\tau) d\xi = -2\pi\mu c_V^{-1} \varepsilon(t) \qquad -\infty < x < \infty \qquad (34)$$

Integral equations (33) are continued to a whole real axis on x-coordinate as

$$\int_0^t d\tau \int_0^\infty \varphi^{\pm}(\xi,\tau) k(\pm(\xi - x), t-\tau) d\xi = \begin{cases} -2\pi\mu c_V^{-1} \varepsilon(t) & 0 \le x < \infty \\ -2\pi\mu c_V^{-1} \upsilon_{\mp}(x,t) & -\infty < x < 0 \end{cases} \qquad (35)$$

where $\upsilon_{\mp}(x,t)$ are the unknown functions, represented in the operator form as

$$\upsilon_{\mp}(x,t) = \int_0^t d\tau \int_{-\infty}^0 \varphi^{\pm}(\xi,\tau) k(\pm(\xi - x), t-\tau) d\xi \qquad -\infty < x < 0$$

being vertical displacements of free surface of the elastic half-plane outside of the contact zone.

After application of Laplace integral transform to left and right parts of 2-D integral equation (35), we obtain Wiener–Hopf 1-D integral equations relatively of Laplace transforms $\varphi^{\pm L}(x,p)$ of normal stresses $\varphi^{\pm}(x,t)$, continued to a whole real axis [15]:

$$\int_0^{\infty} \varphi^{\pm L}(x,p)k(\pm(\xi - x)p/c_2)d\xi = \begin{cases} -2\pi\mu c_V^{-1}\varepsilon^L(p) & 0 \le x < \infty \\ -2\pi\mu c_V^{-1} v_{\mp}^L(x,p) & -\infty < x < 0 \end{cases} \quad (36)$$

$$k(\xi) = \int_{\Gamma} K(u)e^{iu\xi}du$$

where $K(u)$ is defined by (14), and the contour Γ has been described above.

In integral equation (36), a replacement of variables ξ and x is carried out by using formulae $\xi = \frac{c_2}{p}\xi'$, $x = \frac{c_2}{p}x'$ (primes are omitted in the following text), and integral equations, obtained on a half-axis with unknown functions $\varphi^{\pm L}\left(\frac{c_2}{p}\xi, p\right)$ and $v_{\mp}^L\left(\frac{c_2}{p}x, p\right)$, are reversed by the Wiener–Hopf method. At the first step, we apply Fourier integral transform on variable x to the obtained integral equations. This leads to corresponding functional equations in some strip $\eta_- < \mathrm{Im}(u) < \eta_+$ of the complex plane $u = \sigma + i\tau$:

$$\varphi^{\pm LF}(u,p)K(\pm u) = \frac{\mu}{c_V}\frac{p}{c_2}\frac{\varepsilon^L(p)}{iu} - \frac{\mu}{c_V}\frac{p}{c_2}\upsilon_{\mp}^{LF}(u,p) \quad (37)$$

$$\varphi^{\pm LF}(u,p) = \int_0^{\infty} \varphi^{\pm L}\left(\frac{c_2}{p}\xi, p\right)e^{i\xi u}d\xi, \qquad \upsilon_{\mp}^{LF}(u,p) = \int_{-\infty}^{0} \upsilon_{\mp}^{L}\left(\frac{c_2}{p}\xi, p\right)e^{i\xi u}d\xi$$

Factorization of functions $K(\pm u)$ in (37) is the next step of the Wiener–Hopf method, consisting in representation of $K(\pm u)$ in the form of the product [15]:

$$K(\pm u) = K_+^{\pm}(u)K_-^{\pm}(u) \quad (38)$$

where $K_+^{\pm}(u)$ are the regular functions into upper half-plane $(\mathrm{Im}\, u > \eta_-)$, and $K_-^{\pm}(u)$ are the regular functions into lower half-plane $(\mathrm{Im}\, u < \eta_+)$ of the complex plane $u = \sigma + i\tau$.

After fulfillment of factorization (38), the solution of functional equations (37) is carried out by using standard scheme [15]. After calculation of transforms $\varphi^{\pm LF}(u,p)$ and $\upsilon_{\mp}^{LF}(u,p)$, their inverse Laplace and Fourier transforms are conducted. As a result, we obtain solutions of integral equation (35) in the form of double quadratures:

$$\varphi^{\pm}(x,t) = \frac{\mu}{c_V}\frac{1}{2\pi i}\int_{-i\infty}^{i\infty} \frac{p}{c_2}e^{pt}dp\frac{1}{2\pi}\int_{-\infty+ic}^{\infty+ic} \frac{g_+^{\pm}(u,p)}{K_+^{\pm}(u)}e^{-iuxp/c_2}du \qquad 0 \le x < \infty \quad (39)$$

$$v_{\mp}(x,t)=\frac{1}{2\pi i}\int_{-i\infty}^{i\infty}e^{pt}dp\frac{1}{2\pi}\int_{-\infty+id}^{\infty+id}g_{-}^{\pm}(u,p)K_{-}^{\pm}(u)e^{-iuxp/c_2}du \qquad -\infty<x<0 \tag{40}$$

in which $\eta_{-}<c<d<\eta_{+}$ and

$$g_{+}^{\pm}(u,p)=\frac{\varepsilon^L(p)}{K_{-}^{\pm}(0)}\frac{1}{iu},\ g_{-}^{\pm}(u,p)=\frac{\varepsilon^L(p)}{iu}\left(\frac{1}{K_{-}^{\pm}(u)}-\frac{1}{K_{-}^{\pm}(0)}\right) \tag{41}$$

and functions $K_{\pm}^{\pm}(u)$ are calculated below.

The solution $\varphi^{\infty}(x,t)$ of integral equation (34) on a whole coordinate axis is defined by using Laplace and Fourier integral transforms (10), (11) and calculated as

$$\varphi^{\infty}(x,t)=-\frac{\mu}{\beta}\frac{\dot{\varepsilon}(t)}{c_2} \quad -\infty<x<\infty \tag{42}$$

3. FORMULAE OF $K_{\pm}^{\pm}(u)$ FOR FACTORIZATION OF $K(\pm u)$

For factorization of $K(\pm u)$ and determination of formulae for $K_{\pm}^{\pm}(u)$, the functions $K(\pm u)$ are represented by the following form, convenient for the factorization:

$$K(\pm u)=q^{\pm}(u)r^{\pm}(u) \quad q^{\pm}(u)=\frac{\sqrt{\gamma_4\mp iu}\sqrt{-\gamma_1\pm iu}}{(\gamma_6\mp iu)(-\gamma_5\pm iu)} \tag{43}$$

$$r^{\pm}(u)=\sqrt{1-\beta_1^2}\,(1\mp iu\beta_2)^2(\gamma_6\mp iu)(-\gamma_5\pm iu)R^{-1}(\pm u) \tag{44}$$

where $R(u)$ is defined by (14). So the constructed functions $r^{\pm}(u)$ (44) satisfy the conditions:

$$\lim_{|u|\to\infty}r^{\pm}(u)=1 \quad \eta_{-}<\mathrm{Im}(u)<\eta_{+} \tag{45}$$

The functions $r^{\pm}(u)$ (44) contain as isolated singular points in the complex plane $u=\sigma+i\tau$ four algebraic branch points $u=-i\gamma_k$ ($k=1,2,3,4$) (9) and two removable singular points $u=-i\gamma_k$ ($k=5,6$).

The factorization of functions $K(\pm u)$ (38), presented in the form (43), is divided into factorizations of functions $q^{\pm}(u)=q_{+}^{\pm}(u)q_{-}^{\pm}(u)$ and $r^{\pm}(u)=r_{+}^{\pm}(u)r_{-}^{\pm}(u)$.

The factorization of $q^{\pm}(u)$ is reached in each of velocity ranges V by using elementary operations, and the factorization of $r^{\pm}(u)$ is carried out in agreement [15] with formulae

$$r_{+}^{\pm}(u)=\exp\left(H_{+}^{\pm}(u)\right),\ H_{+}^{\pm}(u)=\frac{1}{2\pi i}\int\limits_{-\infty+ic}^{\infty+ic}\frac{\ln r^{\pm}(\zeta)}{\zeta-u}d\zeta \tag{46}$$

$$r_{-}^{\pm}(u)=\exp\left(H_{-}^{\pm}(u)\right),\ H_{-}^{\pm}(u)=-\frac{1}{2\pi i}\int\limits_{-\infty+id}^{\infty+id}\frac{\ln r^{\pm}(\zeta)}{\zeta-u}d\zeta \tag{47}$$

a calculation of quadratures in which depends on disposition of the singular points $u=-i\gamma_k$ ($k=1,2,3,4$) of the functions $r^{\pm}(u)$ (44) in the complex plane $u=\sigma+i\tau$.

By computing quadratures in (46), (47), taking into account (43) – (45), we obtain the formulae

$$K_{\pm}^{\pm}(u)=q_{\pm}^{\pm}(u)r_{\pm}^{\pm}(u),\ r_{\pm}^{\pm}(u)=\exp\left(H_{\pm}^{\pm}(u)\right) \tag{48}$$

for each velocity range V.

I. $K_{+}^{+}(u)=K(u)$, $H_{+}^{+}(u)=\ln r^{+}(u)$, $q_{+}^{+}(u)=q^{+}(u)$ (49)

$$K_{-}^{+}(u)=1,\ H_{-}^{+}(u)=0,\ q_{-}^{+}(u)=1 \tag{50}$$

$$K_{+}^{-}(u)=1,\ H_{+}^{-}(u)=0,\ q_{+}^{-}(u)=1 \tag{51}$$

$$K_{-}^{-}(u)=K(-u),\ H_{-}^{-}(u)=\ln r^{-}(u),\ q_{-}^{-}(u)=q^{-}(u) \tag{52}$$

Formulae (49) – (52) are found, taking into account that all singular points $r_{+}^{+}(u)$ for this velocity range dispose into lower half-plane (23), and all singular points $r_{-}^{-}(u)$ are located in upper half-plane of the complex plane $u=\sigma+i\tau$.

II. $q_{+}^{+}(u)=\dfrac{\sqrt{\gamma_4-iu}}{\left(-\gamma_5+iu\right)\left(\gamma_6-iu\right)}$, (53)

$$H_{+}^{+}(u)=\frac{1}{\pi}\int\limits_{\gamma_4}^{\gamma_3}\frac{\operatorname{arctg}\omega_{+}^{+}(\xi)}{\xi-iu}d\xi-\frac{1}{\pi}\int\limits_{\gamma_2}^{\infty}\frac{\operatorname{arctg}\omega_{+}^{-}(\xi)}{\xi-iu}d\xi$$

$$\omega_{+}^{\pm}(\xi)=4\xi^2\sqrt{\xi-\gamma_1}\sqrt{\pm\gamma_2\mp\xi}\sqrt{\pm\gamma_3\mp\xi}\sqrt{\xi-\gamma_4}\sqrt{1-\beta_1^2}\sqrt{\beta_2^2-1}\eta^{-1}(\xi)$$

$$q_-^+(u) = \sqrt{-\gamma_1 + iu}\,,$$

$$H_-^+(u) = -\frac{1}{\pi}\int_{-\infty}^{\gamma_1} \frac{\operatorname{arctg}\omega_-^+(\xi)}{\xi + iu} d\xi \tag{54}$$

$$\omega_-^+(\xi) = 4\xi^2\sqrt{\gamma_1 - \xi}\sqrt{\gamma_2 - \xi}\sqrt{\gamma_3 - \xi}\sqrt{\gamma_4 - \xi}\sqrt{1-\beta_1^2}\sqrt{\beta_2^2 - 1}\eta^{-1}(\xi)$$

$$q_+^-(u) = \sqrt{-\gamma_1 - iu}\,, \qquad H_+^-(u) = -\frac{1}{\pi}\int_{-\infty}^{\gamma_1} \frac{\operatorname{arctg}\omega_+^-(\xi)}{\xi + iu} d\xi \tag{55}$$

$\omega_+^-(\xi) = \omega_-^+(\xi)$, where $\omega_-^+(\xi)$ from (54).

$$q_-^-(u) = \frac{\sqrt{\gamma_4 + iu}}{(-\gamma_5 - iu)(\gamma_6 + iu)},$$

(56)

$$H_-^-(u) = \frac{1}{\pi}\int_{\gamma_4}^{\gamma_3} \frac{\operatorname{arctg}\omega_-^-(\xi)}{\xi + iu} d\xi - \frac{1}{\pi}\int_{\gamma_2}^{\infty} \frac{\operatorname{arctg}\omega_-^+(\xi)}{\xi + iu} d\xi$$

$$\omega_-^\pm(\xi) = 4\xi^2\sqrt{-\gamma_1 + \xi}\sqrt{\pm\xi \mp \gamma_2}\sqrt{\pm\xi \mp \gamma_3}\sqrt{\xi - \gamma_4}\sqrt{1-\beta_1^2}\sqrt{\beta_2^2 - 1}\eta^{-1}(\xi) = \omega_+^\mp(\xi)$$

where $\omega_+^\mp(\xi)$ is defined by (53)

$$\text{III. } q_+^+(u) = \frac{\sqrt{\gamma_4 - iu}}{(-\gamma_5 + iu)(\gamma_6 - iu)}, \qquad H_+^+(u) = \frac{1}{\pi}\int_{\gamma_4}^{\gamma_3} \frac{\operatorname{arctg}\omega_+^-(\xi)}{\xi - iu} d\xi \tag{57}$$

$$q_-^+(u) = \sqrt{-\gamma_1 + iu}\,, \qquad H_-^+(u) = \frac{1}{\pi}\int_{\gamma_2}^{\gamma_1} \frac{\operatorname{arctg}\omega_-^+(\xi)}{\xi - iu} d\xi \tag{58}$$

$$\omega_\pm^+(\xi) = 4\xi^2\sqrt{\mp\gamma_1 \pm \xi}\sqrt{-\gamma_2 + \xi}\sqrt{\gamma_3 - \xi}\sqrt{\pm\xi \mp \gamma_4}\sqrt{1-\beta_1^2}\sqrt{1-\beta_2^2}\eta^{-1}(\xi)$$

$$q_+^-(u) = \sqrt{-\gamma_1 - iu}\,, \qquad H_+^-(u) = -\frac{1}{\pi}\int_{\gamma_2}^{\gamma_1} \frac{\operatorname{arctg}\omega_+^-(\xi)}{\xi + iu} d\xi \tag{59}$$

$$q_-^-(u) = \frac{\sqrt{\gamma_4 + iu}}{(-\gamma_5 - iu)(\gamma_6 + iu)}, \qquad H_-^-(u) = \frac{1}{\pi}\int_{\gamma_4}^{\gamma_3} \frac{\operatorname{arctg}\omega_-^-(\xi)}{\xi + iu} d\xi \tag{60}$$

$$\omega_\pm^-(\xi) = 4\xi^2\sqrt{\pm\gamma_1 \mp \xi}\sqrt{-\gamma_2 + \xi}\sqrt{\gamma_3 - \xi}\sqrt{\pm\gamma_4 \mp \xi}\sqrt{1-\beta_1^2}\sqrt{1-\beta_2^2}\eta^{-1}(\xi)$$

In (53) – (60), $\eta(\xi) = \left(-2\xi^2 + (1-\xi\beta_2)^2\right)^2$

$$\text{IV. } q_+^+(u) = \frac{\sqrt{\gamma_4 - iu}}{\gamma_6 - iu}, \qquad H_+^+(u) = \frac{1}{\pi}\int_{\gamma_4}^{\gamma_3} \frac{\operatorname{arctg}\omega_+^+(\xi)}{\xi - iu} d\xi \tag{61}$$

$$q_-^+(u) = \frac{\sqrt{-\gamma_1 + iu}}{-\gamma_5 + iu}, \qquad H_-^+(u) = -\frac{1}{\pi}\int_{\gamma_2}^{\gamma_1} \frac{\operatorname{arctg}\omega_-^+(\xi)}{\xi - iu} d\xi \tag{62}$$

where $\omega_\pm^+(\xi)$ is defined by (58),

$$q_+^-(u) = \frac{\sqrt{-\gamma_1 - iu}}{-\gamma_5 - iu}, \qquad H_+^-(u) = -\frac{1}{\pi}\int_{\gamma_2}^{\gamma_1} \frac{\operatorname{arctg}\omega_+^-(\xi)}{\xi + iu} d\xi \tag{63}$$

$$q_-^-(u) = \frac{\sqrt{\gamma_4 + iu}}{\gamma_6 + iu}, \qquad H_-^-(u) = \frac{1}{\pi}\int_{\gamma_4}^{\gamma_3} \frac{\operatorname{arctg}\omega_-^-(\xi)}{\xi + iu} d\xi \tag{64}$$

where $\omega_\pm^-(\xi)$ is defined by (60).

4. The Problem Solution and its Main Properties

The solution of the stated problem is found by superposition (32), in which $\varphi^\pm(a \pm x, t)$ are defined by formulae (39), and $\varphi^\infty(x,t)$ by formula (42). The vertical displacements of free surface outside of the punch contact zone with elastic half-plane $\upsilon_\mp(a \pm x, t)$ are calculated by using formula (40). The computation of quadratures in (39), (40) for all four velocity ranges V leads to the following formulae:

$$\text{I. } \varphi^+(x,t) = \varphi^\infty(x,t)H(t) + x^{-1}\sum_{k=1}^{2} \Phi_k^+(x,t)H(t - x\tau_k)\ t>0\,, x>0 \tag{65}$$

$$\Phi_1^+(x,t) = \frac{\mu}{\pi}\frac{\partial}{\partial t}\left(\int_{t-x\tau_1}^{t-x\tau_2} + \int_{t-x\tau_3}^{t-x\tau_4}\right)\varepsilon(\tau)G_1^+(x, c_2(t-\tau))d\tau$$

$$\Phi_2^+(x,t) = \frac{\mu}{\pi}\frac{\partial}{\partial t}\int_{t-x\tau_2}^{t-x\tau_3} \varepsilon(\tau)G_2^+(x, c_2(t-\tau))d\tau\ , \tau_k = \frac{\gamma_k}{c_2},\ k=1,2,3,4$$

$$G_1^+(x,t) = \left(-2t^2 + (x - \beta_2 t)^2\right)^2 R^{-1}(x,t)$$

$$G_2^+(x,t) = G_1^+(x,t) - \widetilde{G}_1^+(x,t)R^{-1}(x,t)$$
$$\widetilde{G}_1^+(x,t) = 4t^2\sqrt{t-x\gamma_4}\sqrt{x\gamma_1 - t}\sqrt{t-x\gamma_3}\sqrt{x\gamma_2 - t}\sqrt{\beta_1^2-1}\sqrt{\beta_2^2-1}$$
$$R(x,t) = t(x-\beta_2 t)^2\sqrt{t-x\gamma_4}\sqrt{x\gamma_1 - t}\sqrt{\beta_1^2-1}$$

$$\upsilon_-(x,t) = 0 \qquad t>0,\, x<0 \tag{66}$$

$$\varphi^-(x,t) = -\frac{\mu\,\dot{\varepsilon}(t)}{\beta\ c_2} \qquad t>0,\, x>0 \tag{67}$$

$$\upsilon_+(x,t) = -\frac{1}{\pi\beta}\sum_{k=1}^{2} V_k^+(x,t)H(t+x\tau_k) \qquad t>0,\, x<0 \tag{68}$$

$$V_1^+(x,t) = \left(\int\limits_{t+x\tau_1}^{t+x\tau_2} + \int\limits_{t+x\tau_3}^{t+x\tau_4}\right)\varepsilon(\tau)B_1^+(x, c_2(t-\tau))d\tau$$

$$V_2^+(x,t) = \int\limits_{t+x\tau_2}^{t+x\tau_3}\varepsilon(\tau)B_2^+(x, c_2(t-\tau))d\tau$$

$$B_1^+(x,t) = x(x-\beta_2 t)^2\sqrt{t-x\gamma_4}\sqrt{x\gamma_1 - t}\,b_1(x,t)\sqrt{\beta_1^2-1}\,R_1^{-1}(x,t)$$
$$B_2^+(x,t) = x(x-\beta_2 t)^2\sqrt{t-x\gamma_4}\sqrt{x\gamma_1 - t}\sqrt{\beta_1^2-1}\,R_2^{-1}(x,t)$$
$$R_1(x,t) = t\left(b_1^2(x,t) - b_2^2(x,t)\right)$$
$$R_2(x,t) = t\left(b_1(x,t) - b_2(x,t)\right)$$
$$b_1(x,t) = \left(-2t^2 + (x-\beta_2 t)^2\right)^2$$
$$b_2(x,t) = 4t^2\sqrt{t-x\gamma_4}\sqrt{x\gamma_1 - t}\sqrt{t-x\gamma_3}\sqrt{x\gamma_2 - t}\sqrt{\beta_1^2-1}\sqrt{\beta_2^2-1}$$

The integral $V_2^+(x,t)$ is understood in the sense of the Cauchy principal value [15], because the under integral function $B_2^+(x,t)$ has singular discontinuities at the Rayleigh poles $\tau = t + \tau_{5,6}x$.

II. $\varphi^+(x,t) = \varphi^\infty(x,t)H(t) - x^{-3/2}\Phi^+(x,t)H(t - x\tau_4) \quad t>0,\ x>0$ (69)

$$\Phi^+(x,t) = \frac{K_+^+(0)}{\pi}\frac{\partial}{\partial t}\int\limits_0^{t-x\tau_4}\varepsilon(\tau)G_+^+(x, c_2(t-\tau))d\tau,\ G_+^+(x,t) = \frac{(-x\gamma_5 + t)(x\gamma_6 - t)}{t\sqrt{t-x\gamma_4}\,r_+^+\left(-itx^{-1}\right)}$$

where $r_+^+(u)$, $K_+^+(0)$ are defined from (48), (53).

$$\upsilon_{-}(x,t) = -(-x)^{-1/2} V_0^{-}(x,t) H(t - x\tau_1) \qquad t > 0\,,\, x < 0 \tag{70}$$

$$V_0^{-}(x,t) = \frac{1}{\pi K_{-}^{+}(0)} \int_0^{t - x\tau_1} \varepsilon(\tau) G_{-}^{+}(x, c_2(t-\tau)) d\tau\,,\; G_{-}^{+}(x,t) = c_2 t^{-1} \sqrt{-x\gamma_1 + t} r_{-}^{+}\left(-\frac{it}{x}\right)$$

where $r_{-}^{+}(u)$ is defined by (48), (54).

$$\varphi^{-}(x,t) = \varphi^{\infty}(x,t) H(t) - x^{1/2} \Phi^{-}(x,t) H(t + x\tau_1) \qquad t > 0\,,\, x > 0 \tag{71}$$

$$\Phi^{-}(x,t) = \frac{K_{+}^{-}(0)}{\pi} \frac{\partial}{\partial t} \int_0^{t + x\tau_1} \varepsilon(\tau) G_{+}^{-}(x, c_2(t-\tau)) d\tau\,,\; G_{+}^{-}(x,t) = \frac{1}{t\sqrt{x\gamma_1 + t} r_{+}^{-}\left(-itx^{-1}\right)}$$

where $r_{+}^{-}(u)$, $K_{+}^{-}(0)$ are defined by (48), (55).

$$\upsilon_{+}(x,t) = -(-x)^{3/2} V_0^{+}(x,t) H(t + x\tau_4) \qquad t > 0\,,\; x < 0 \tag{72}$$

$$V_0^{+}(x,t) = \frac{1}{\pi K_{-}^{-}(0)} \int_0^{t + x\tau_4} \varepsilon(\tau) G_{-}^{-}(x, c_2(t-\tau)) d\tau\,, G_{-}^{-}(x,t) = -\frac{c_2\sqrt{t + x\gamma_4}}{t(x\gamma_5 + t)(x\gamma_6 + t)} r_{-}^{-}\left(-\frac{it}{x}\right)$$

where $r_{-}^{-}(u)$, $K_{-}^{-}(0)$ are defined by (48), (56), and integral in (72) is understood in the sense of the Cauchy principal value, because the under integral function $G_{-}^{-}(x,t)$ has singular discontinuities at the Rayleigh poles $\tau = t + \tau_{5,6} x$.

III. $$\varphi^{+}(x,t) = \varphi^{\infty}(x,t) H(t) + x^{-3/2} \Phi^{+}(x,t) H(t - x\tau_4) \quad t > 0\,,\; x > 0 \tag{73}$$

$$\Phi^{+}(x,t) = \frac{K_{+}^{+}(0)}{\pi} \frac{\partial}{\partial t} \int_0^{t - x\tau_4} \varepsilon(\tau) G_{+}^{+}(x, c_2(t-\tau)) d\tau\,,\; G_{+}^{+}(x,t) = \frac{(-x\gamma_5 + t)(x\gamma_6 - t)}{t\sqrt{t - x\gamma_4} r_{+}^{+}\left(-itx^{-1}\right)}$$

where $r_{+}^{+}(u)$, $K_{+}^{+}(0)$ are calculated by using formulae (48), (57).

$$\upsilon_{-}(x,t) = -(-x)^{-1/2} V_0^{-}(x,t) H(t - x\tau_1) \qquad t > 0\,,\; x < 0 \tag{74}$$

$$V_0^{-}(x,t) = \frac{1}{\pi K_{-}^{+}(0)} \int_0^{t - x\tau_1} \varepsilon(\tau) G_{-}^{+}(x, c_2(t-\tau)) d\tau\,,\; G_{-}^{+}(x,t) = \frac{c_2\sqrt{-x\gamma_1 + t}}{t r_{-}^{+}\left(-itx^{-1}\right)}$$

where $r_{-}^{+}(u)$, $K_{-}^{+}(0)$ are calculated by Equations (48), (58).

$$\varphi^-(x,t)=\varphi^\infty(x,t)H(t)-x^{1/2}\Phi^-(x,t)H(t+x\tau_1) \qquad t>0\,,\ x>0 \tag{75}$$

$$\Phi^-(x,t)=\frac{K_+^-(0)}{\pi}\frac{\partial}{\partial t}\int\limits_0^{t+x\tau_1}\varepsilon(\tau)G_+^-(x,c_2(t-\tau))d\tau\,,\ G_+^-(x,t)=\frac{1}{t\sqrt{x\gamma_1+t}\,r_+^-(-itx^{-1})}$$

in which $r_+^-(u)$, $K_+^-(0)$ are calculated by Equations (48), (59).

$$\upsilon_+(x,t)=-(-x)^{3/2}V_0^+(x,t)H(t+x\tau_4) \qquad t>0\,,\ x<0 \tag{76}$$

$$V_0^+(x,t)=\frac{1}{\pi K_-^-(0)}\int\limits_0^{t+x\tau_4}\varepsilon(\tau)G_-^-(x,c_2(t-\tau))d\tau\,,G_-^-(x,t)=-\frac{c_2\sqrt{t+x\gamma_4}}{t(x\gamma_5+t)(x\gamma_6+t)}r_-^-\left(-\frac{it}{x}\right)$$

where $r_-^-(u)$, $K_-^-(0)$ are calculated by Equations (48), (60), and the integral is understood in the sense of the Cauchy principal value, because the under integral function has singular discontinuities at the Rayleigh poles $\tau=t+\tau_{5,6}x$.

$$\text{IV. } \varphi^+(x,t)=\varphi^\infty(x,t)H(t)+x^{-1/2}\Phi^+(x,t)H(t-x\tau_4) \quad t>0\,,\ x>0 \tag{77}$$

$$\Phi^+(x,t)=\frac{K_+^+(0)}{\pi}\frac{\partial}{\partial t}\int\limits_0^{t-x\tau_4}\varepsilon(\tau)G_+^+(x,c_2(t-\tau))d\tau\,,\ G_+^+(x,t)=\frac{x\gamma_6-t}{t\sqrt{t-x\gamma_4}\,r_+^+(-itx^{-1})}$$

where $r_+^+(u)$, $K_+^+(0)$ are calculated by Equations (48), (61).

$$\upsilon_-(x,t)=(-x)^{1/2}V_0^-(x,t)H(t-x\tau_1) \qquad t>0\,,\ x<0 \tag{78}$$

$$V_0^-(x,t)=\frac{1}{\pi K_-^+(0)}\int\limits_0^{t-x\tau_1}\varepsilon(\tau)G_-^+(x,c_2(t-\tau))d\tau\,,\quad G_-^+(x,t)=\frac{c_2\sqrt{-x\gamma_1+t}}{t(-x\gamma_5+t)}r_-^+\left(-\frac{it}{x}\right)$$

where $r_-^+(u)$, $K_-^+(0)$ are calculated by Equations (48), (62), and the integral is understood in the sense of the Cauchy principal value, because the under integral function has singular discontinuities at the Rayleigh pole $\tau=t-\tau_5 x$.

$$\varphi^-(x,t)=\varphi^\infty(x,t)H(t)-x^{-1/2}\Phi^-(x,t)H(t+x\tau_1) \qquad t>0\,,\ x>0 \tag{79}$$

$$\Phi^{-}(x,t)=\frac{K_{+}^{-}(0)}{\pi}\frac{\partial}{\partial t}\int_{0}^{t+x\tau_1}\varepsilon(\tau)G_{+}^{-}(x,c_2(t-\tau))d\tau\,,\; G_{+}^{-}(x,t)=\frac{x\gamma_5+t}{t\sqrt{x\gamma_1+tr_{+}^{-}(-itx^{-1})}}$$

where $r_{+}^{-}(u)$, $K_{+}^{-}(0)$ are calculated by Equations (48), (63).

$$\upsilon_{+}(x,t)=-(-x)^{1/2}V_0^{+}(x,t)H(t+x\tau_4) \qquad t>0\,,\; x<0 \tag{80}$$

$$V_0^{+}(x,t)=\frac{1}{\pi K_{-}^{-}(0)}\int_{0}^{t+x\tau_4}\varepsilon(\tau)G_{-}^{-}(x,c_2(t-\tau))d\tau\,,\; G_{-}^{-}(x,t)=-\frac{c_2\sqrt{t+x\gamma_4}}{t(x\gamma_6+t)}r_{-}^{-}\left(-\frac{it}{x}\right)$$

where $r_{-}^{-}(u)$, $K_{-}^{-}(0)$ are calculated by Equations (48), (54), and the integral is understood in the sense of the Cauchy principal value, because the under integral function has singular discontinuities at the Rayleigh pole $\tau=t+\tau_6 x$. Formulae (77), (78) for $\varphi^{+}(x,t)$ and $\upsilon_{-}(x,t)$ at $V=0$ coincide with corresponding formulae [11].

5. Asymptotic Analysis of Obtained Solutions

In this section, features of the obtained solutions (65) – (80) are determined at the boundary $(x=0)$ of punch contact with elastic half-plane, along with factors at the singularity and character of their dependence on velocity V of the punch movement.

I. By carrying out the replacement of variable

$$c_2(t-\tau)=\tau' x\,,\; c_2 d\tau=-x d\tau'$$

in quadratures in (65) and then going to limit at $x\to 0+0$, we obtain the estimates:

$$\varphi^{+}(x,t)=\varphi^{\infty}(x,t)\varphi_0^{+}(V)+O(x) \qquad x\to 0+0 \tag{81}$$

$$\varphi_0^{+}(V)=1-\frac{\beta}{\pi}\left[\left(\int_{\gamma_2}^{\gamma_1}+\int_{\gamma_4}^{\gamma_3}\right)h_1(\tau)d\tau+\int_{\gamma_3}^{\gamma_2}h_2(\tau)d\tau\right]$$

$$h_1(t)=\left(-2t^2+(1+\beta_2 t)^2\right)^2 r^{-1}(t)$$

$$h_2(t)=h_1(t)-4t^2\sqrt{t-\gamma_4}\sqrt{\gamma_1-t}\sqrt{t-\gamma_3}\sqrt{\gamma_2-t}\sqrt{\beta_1^2-1}\sqrt{\beta_2^2-1}r^{-1}(t)$$

$$r(t)=t(1+\beta_2 t)^2\sqrt{t-\gamma_4}\sqrt{\gamma_1-t}\sqrt{\beta_1^2-1}$$

Here γ_k $(k=1,2,3,4)$ are calculated in (21). Then, it follows from (66), (67) that

$$\upsilon_-(x,t)=0 \qquad x\to 0-0 \tag{82}$$

$$\varphi_-(x,t)=-\frac{\mu\,\dot{\varepsilon}(t)}{\beta\,c_2} \qquad x\to 0+0 \tag{83}$$

After the replacement of variable, pointed before formula (81), we obtain from (68) that

$$\upsilon_+(x,t)=-\frac{\varepsilon(t)}{\beta}\upsilon_0^+(V)+O(x) \qquad x\to 0-0 \tag{84}$$

$$\upsilon_0^+(V)=\frac{1}{\pi}\left[\left(\int_{\gamma_2}^{\gamma_1}+\int_{\gamma_4}^{\gamma_3}\right)q_1(t)dt+\int_{\gamma_3}^{\gamma_2}q_2(t)dt\right]$$

$$q_1(t)=l_1(t)l_2(t)r_1^{-1}(t)$$

$$q_2(t)=l_1(t)r_2^{-1}(t)$$

$$l_1(t)=(1-\beta_1 t)^2\sqrt{t-\gamma_4}\sqrt{\gamma_1-t}\sqrt{\beta_1^2-1}$$

$$l_2(t)=\left(-2t^2+(1-\beta_2 t)^2\right)^2$$

$$r_1(t)=l_2^2(t)-16t^4(t-\gamma_1)(t-\gamma_2)(t-\gamma_3)(t-\gamma_4)(\beta_1^2-1)(\beta_2^2-1)$$

$$r_2(t)=l_2(t)-4t^2\sqrt{\gamma_1-t}\sqrt{\gamma_2-t}\sqrt{t-\gamma_3}\sqrt{t-\gamma_4}\sqrt{\beta_1^2-1}\sqrt{\beta_2^2-1}$$

Integral of $q_2(t)$ is understood in the sense of the Cauchy principal value [15], because $q_2(t)$ has singular discontinuities at the Rayleigh poles $t=\gamma_{5,6}$.

Then, it follows from (81) that, when the punch moves with velocity $V>c_1$ on the boundary of the elastic half-plane, finite stress jump arises at the frontal edge of the punch. This jump differs of $\varphi^\infty(x,t)$ (42) only by factor $\varphi_0^+(V)$, depended on V, when displacement $\upsilon(x,0,t)=\varepsilon(t)$ and stress $\sigma_{xy}(x,0,t)=0$ at $|x|<\infty$ are determined on a whole boundary. The fronts of longitudinal and transverse waves, created by frontal edge of the punch due to its supersonic velocity *V*, are situated under the punch.

The values of stresses coincide at the back edge of the punch with $\varphi^\infty(x,t)$ (42), and wave fronts, created by the back edge, propagate on free surface and initiate two Rayleigh waves (68). The neighboring wave propagates toward the punch with velocity $V-c_R$, and distant wave propagates relatively of the punch with velocity $V+c_R$. The displacement jump at the back edge is proportional to the punch penetration $\varepsilon(t)$ with a factor, depended on *V*.

II. Asymptotic relations for $H_\pm^\pm(u)$ (53) – (56) into a complex plane at great values of $|u|$ have been used for obtainment of asymptotic formulae in the case:

$$H_+^{\pm}(u) = \pm\alpha \ln(-iu) + O\left(\frac{1}{-iu}\right) \qquad \text{at } |u| \to \infty,\ \operatorname{Im} u > \eta_- \tag{85}$$

$$H_-^{\pm}(u) = \mp\alpha \ln(iu) + O\left(\frac{1}{iu}\right) \qquad \text{at } |u| \to \infty,\ \operatorname{Im} u < \eta_+$$

$$\alpha = \frac{1}{\pi}\operatorname{arctg}\frac{4\sqrt{1-\beta_1^2}\sqrt{\beta_2^2-1}}{\left(-2+\beta_2^2\right)^2}, \qquad 0 < \alpha \le \frac{1}{2}, \tag{86}$$

where $\eta_\pm$ are defined by (39), (40). The obtained estimates (85) are based on study of asymptotic representations of the divergent improper singular integrals. All divergent improper integrals are understood in the sense of finite part [16], also as all other divergent quadratures in the formulae obtained:

$$\varphi^+(x,t) = -\frac{\mu}{\beta} x^{-3/2+\alpha} \varphi_0^+(t,V) + O\left(x^{-1/2+\alpha}\right) \qquad x \to 0+0 \tag{87}$$

$$\varphi_0^+(t,V) = \frac{\cos\pi\alpha}{\pi} K_+^+(0) \frac{\partial}{\partial t} \int_0^t \varepsilon(\tau)(c_2(t-\tau))^{1/2-\alpha} d\tau,\ K_+^+(0) < 0$$

$$\upsilon_-(x,t) = -(-x)^{-1/2+\alpha} \upsilon_-^0(t,V) + O\left((-x)^{1/2+\alpha}\right) \quad x \to 0-0 \tag{88}$$

$$\upsilon_-^0(t,V) = \frac{c_2 \cos\pi\alpha}{\pi K_-^+(0)} \int_0^t \varepsilon(\tau)(c_2(t-\tau))^{-1/2-\alpha} d\tau,\ K_-^+(0) > 0$$

$$\varphi^-(x,t) = \frac{\mu}{\beta} x^{1/2-\alpha} \varphi_0^-(t,V) + O\left(x^{3/2-\alpha}\right) \qquad x \to 0+0 \tag{89}$$

$$\varphi_0^-(t,V) = \frac{\cos\pi\alpha}{\pi} K_+^-(0) \frac{\partial}{\partial t} \int_0^t \varepsilon(\tau)(c_2(t-\tau))^{-3/2+\alpha} d\tau,\ K_+^-(0) > 0$$

$$\upsilon_+(x,t) = (-x)^{3/2-\alpha} \upsilon_+^0(t,V) + O\left((-x)^{5/2-\alpha}\right) \quad x \to 0+0 \tag{90}$$

$$\upsilon_+^0(t,V) = \frac{c_2 \cos\pi\alpha}{\pi K_-^-(0)} \int_0^t \varepsilon(\tau)(c_2(t-\tau))^{-5/2+\alpha} d\tau,\ K_-^-(0) < 0$$

At the frontal and back edges of punch, the normal stresses (87), (89) have the same sign with the sign of the punch penetration into elastic half-plane, but the contact stresses have opposite sign. The normal stresses (87) have jump with exponent of singularity $-3/2+\alpha$ at the boundary of frontal edge [17], witnessing on fracture of half-plane material by the frontal edge. The vertical displacements of the half-plane surface on the frontal edge have jump with exponent of singularity $-1/2+\alpha$ and aspire to $-\infty$ along negative part of y-axis. This is similar to formation of chip, cutting of which occurs for account of movement of the penetrated punch along x-axis. The front of longitudinal wave in this velocity range locates before the frontal edge.

The displacements are smooth with exponent $3/2-\alpha$ at the back edge, witnessing on bulging surface under punch. At free surface behind back edge, similar as in previous range, two Rayleigh waves arise, which are created by elastic waves, initiated by the back edge of punch. The front of longitudinal wave c_1, propagating toward the punch, disposes under punch.

III. The asymptotic estimates for $\varphi^{\pm}(x,t)$, $\upsilon_{\mp}(x,t)$ coincide in this velocity range with corresponding formulae of the previous velocity range (87) – (90) in the case of $\alpha=0$. Due to this the great stresses and displacements at the boundary of contact zone become even greater and intensity of material cutting of the half-plane material increases. The wave fronts c_1, c_2, created by frontal edge of punch and propagating along direction of V, are located before the frontal edge of the punch and the wave fronts c_1, c_2, created by the frontal edge of the punch and propagating along direction of V, dispose under punch. Two Rayleigh waves are observed behind back edge of the punch similarly to the previous ranges.

$$\text{IV. } \varphi^{+}(x,t)=-\frac{\mu}{\beta}x^{-1/2}\varphi_0^{+}(t,V)+O\left(x^{1/2}\right) \qquad x\to 0+0 \tag{91}$$

$$\varphi_0^{+}(t,V)=\frac{K_+^{+}(0)}{\pi}\frac{\partial}{\partial t}\int_0^t \varepsilon(\tau)(c_2(t-\tau))^{-1/2}d\tau,\ K_+^{+}(0)>0$$

$$\upsilon_{-}(x,t)=(-x)^{1/2}\upsilon_{-}^{0}(t,V)+O\left((-x)^{3/2}\right) \qquad x\to 0-0 \tag{92}$$

$$\upsilon_{-}^{0}(t,V)=\frac{c_2}{\pi K_{-}^{+}(0)}\int_0^t \varepsilon(\tau)(c_2(t-\tau))^{-3/2}d\tau,\ K_{-}^{+}(0)>0$$

$$\varphi^{-}(x,t)=-\frac{\mu}{\beta}x^{-1/2}\varphi_0^{-}(t,V)+O\left(x^{1/2}\right) \qquad x\to 0+0 \tag{93}$$

$$\varphi_0^{-}(t,V)=\frac{K_+^{-}(0)}{\pi}\frac{\partial}{\partial t}\int_0^t \varepsilon(\tau)(c_2(t-\tau))^{-1/2}d\tau,\ K_+^{-}(0)>0$$

$$\upsilon_{+}(x,t)=(-x)^{1/2}\upsilon_{+}^{0}(t,V)+O\left(x^{3/2}\right) \qquad x\to 0-0 \tag{94}$$

$$\upsilon_{+}^{0}(t,V)=\frac{c_2}{\pi K_{-}^{-}(0)}\int_0^t \varepsilon(\tau)(c_2(t-\tau))^{-3/2}d\tau,\ K_{-}^{-}(0)>0$$

Formulae (91) – (94) reflect classic results on the singularity exponents for stresses and smoothness for displacements [11]. Only the factors at the singularities depend on movement velocity V of punch in this velocity range. Here, the fronts of elastic waves c_1, c_2 and front of Rayleigh wave c_R, coinciding on direction with V, dispose before frontal edge of the punch, and opposite on direction with V, locate behind back edge of the punch.

In order to determine singularities of solutions $\varphi^{\pm}(x,t)$, $\upsilon_{\mp}(x,t)$ and character of their joining at the range boundary (for inter-range values of velocity V), it is used either limit transition in the obtained solutions at $V = c_1 \pm 0, c_2 \pm 0$, or solutions for these cases are constructed anew.

1. To study $\varphi^{\pm}(x,t)$, $\upsilon_{\mp}(x,t)$ at $V = c_1$, we construct solutions in the first velocity range at $V = c_1 + 0$. The function $K_+^+(u) = K(u)$ in this case takes the form:

$$K_+^+(u) = (1 - iu\beta_2)^2 \sqrt{\gamma_4 - iu}\sqrt{2\beta} R^{-1}(u), \tag{95}$$

$$R(u) = c_V\left[\left(2u^2 + (1 - iu\beta_2)^2\right)^2 - 4u^2\sqrt{\gamma_2 - iu}\sqrt{\gamma_3 - iu}\sqrt{\gamma_4 - iu}\sqrt{2\beta}\sqrt{\beta_2^2 - 1}\right]$$

where

$$\gamma_2 = \frac{\beta}{1-\beta},\ \gamma_3 = \frac{\beta}{1+\beta},\ \gamma_4 = \frac{\beta}{2},\ \gamma_5 = \frac{\eta_R}{\overline{\beta}_R - 1},\ \gamma_6 = \frac{\eta_R}{\overline{\beta}_R + 1},\ \overline{\beta}_R = \frac{c_1}{c_R} \tag{96}$$

$$c_V = -\frac{\sqrt{2\beta}}{\beta^2\left(2 - \beta^{-2}\right)^2},\ \beta = \frac{c_2}{c_1}.$$

Construction of solutions in this case, based on formula (39), taking into account (95), leads to the following formulae for $V = c_1 + 0$:

$$\varphi^+(x,t) = \varphi^\infty(x,t)H(t) + x^{-3/2}\sum_{k=1}^{2}\Phi_k^+(x,t)H(t - x\tau_k)\quad t > 0,\ x > 0 \tag{97}$$

$$\Phi_1^+(x,t) = \frac{\mu}{\pi}\frac{\partial}{\partial t}\left(\int_0^{t-x\tau_2} + \int_{t-x\tau_3}^{t-x\tau_4}\right)\varepsilon(\tau)G_1^+(x, c_2(t-\tau))d\tau$$

$$\Phi_2^+(x,t) = \frac{\mu}{\pi}\frac{\partial}{\partial t}\int_{t-x\tau_2}^{t-x\tau_3}\varepsilon(\tau)G_2^+(x, c_2(t-\tau))d\tau$$

$$G_1^+(x,t) = \left(-2t^2 + \left(x - t\beta^{-1}\right)^2\right)^2 R^{-1}(x,t)$$

$$G_2^+(x,t) = G_1^+(x,t) - 4t^2\sqrt{x\gamma_2 - t}\sqrt{t - x\gamma_3}\sqrt{t - x\gamma_4}\sqrt{2\beta}\sqrt{\beta_2^2 - 1}R^{-1}(x,t)$$

$$R(x,t) = t\left(x - t\beta^{-1}\right)^2\sqrt{t - x\gamma_4}\sqrt{2\beta}$$

The formula for $\varphi^+(x,t)$ from range II at $V = c_1$ is obtained by using limit transition in (69) at $V = c_1 - 0$:

$$\varphi^+(x,t) = \varphi^\infty(x,t)H(t) - x^{-3/2}\Phi^+(x,t)H(t - x\tau_4) \;\; t > 0 \, , \; x > 0 \tag{98}$$

where $\Phi^+(x,t)$, $G_+^+(x,t)$ are defined by (69), in which $r_+^+(u)$ is given by formula (48), and $H_+^+(u)$, included in $r_+^+(u)$, is found from (53), where

$$\omega_+^\pm(\xi) = 4\xi^2\sqrt{\pm\gamma_2 \mp \xi}\sqrt{\pm\gamma_3 \mp \xi}\sqrt{\xi - \gamma_4}\sqrt{2\beta}\sqrt{\beta^{-2} - 1}\eta^{-1}(\xi) \tag{99}$$

$$\eta(\xi) = \left(-2\xi^2 + \left(1 - \xi\beta^{-1}\right)^2\right)^2 , \; \beta = c_2 / c_1$$

A comparison of (97), obtained at $V = c_1 + 0$, with (98), determined at $V = c_1 - 0$, is conducted for boundary of the contact zone at $x \to 0+0$ and leads to the next formula for factors at the singularity $x^{-3/2}$:

$$\lim_{\substack{x\to 0+0 \\ V = c_1 \pm 0}} \varphi^+(x,t)x^{3/2} = -\frac{\mu}{\beta}\frac{K_+^+(0)}{\pi}\frac{\partial}{\partial t}\int_0^t \varepsilon(\tau)\sqrt{c_2(t-\tau)}d\tau \tag{100}$$

Here $K_+^+(0)$ is calculated for $V = c_1 + 0$ by using formula (49), and $K_+^+(0)$ found for $V = c_1 - 0$ on the base of formula (53), where $\omega_+^\pm(\xi)$ are computed by formula (99). So, the factors at the singularity $x^{-3/2}$ in (97) and (98) coincide in form with (100), but differ one of other on value.

The solutions $\varphi^-(x,t)$ (67) at $V = c_1 + 0$ and $\varphi^-(x,t)$ (71) at $V = c_1 - 0$ coincide on the back edge at $x \to 0+0$

$$\lim_{\substack{x\to 0+0 \\ V = c_1 \pm 0}} \varphi^-(x,t) = -\frac{\mu}{\beta}\frac{\dot{\varepsilon}(t)}{c_2}. \tag{101}$$

A comparison of the obtained solutions $\upsilon_+(x,t)$ (68) at $V = c_1 + 0$ and $\upsilon_+(x,t)$ (72) at $V = c_1 - 0$ on the boundary of contact zone (on the back edge) leads to the next relationship between factors at $(-x)^{3/2}$:

$$\lim_{\substack{x\to 0-0 \\ V = c_1 \pm 0}} \upsilon_+(x,t)(-x)^{-3/2} = \mp\frac{c_2}{\pi K_-^-(0)}\int_0^t \varepsilon(\tau)\left(c_2(t-\tau)\right)^{-5/2}d\tau \, . \tag{102}$$

Here $K_-^-(0)$ is calculated for the case of $V = c_1 + 0$ by using formula (52), and $K_-^-(0)$ computed for $V = c_1 - 0$ by formula (56), where $\omega_-^\pm(\xi) = \omega_+^\mp(\xi)$ are found from (99).

It should be noted, that factors at $(-x)^{3/2}$ for $\upsilon_+(x,t)$ at $V = c_1 \pm 0$ have different signs, namely: the punch (its back edge) moves down from wave (from bulging) at $V = c_1 + 0$. At the same time, the punch disposes on the wave (on bulging) at $V = c_1 - 0$.

The displacement $\upsilon_-(x,t)$ on the frontal edge $(x \to 0-0)$ is equal to zero (66) at $V = c_1 + 0$, but the displacements $v_-(x,t)$ (70) are not restricted at $V = c_1 - 0$, by this

$$\lim_{\substack{x\to 0-0 \\ V=c_1-0}} \upsilon_-(x,t)(-x)^{1/2} = -\frac{c_2}{\pi K_-^+(0)} \int_0^t \varepsilon(\tau)\left(c_2(t-\tau)\right)^{-1/2} d\tau \tag{103}$$

2. The study of solutions $\varphi^{\pm}(x,t)$, $\upsilon_{\pm}(x,t)$ is conducted at $V = c_2$ by using limit transitions in (69) – (72) at $V = c_2 + 0$ and in (73) – (76) at $V = c_2 - 0$. It should be noted that we have at $V = c_2$

$$K(u) = (1-iu)^2 \sqrt{-\gamma_1 + iu}\sqrt{\gamma_4 - iu}\sqrt{1-\beta_1^2}\, R^{-1}(u), \tag{104}$$

$$R(u) = c_V\left[\left(2u^2 + (1-iu)^2\right)^2 - 4u^2\sqrt{-\gamma_1 + iu}\sqrt{\gamma_3 - iu}\sqrt{\gamma_4 - iu}\sqrt{2}\sqrt{1-\beta^2}\right]$$

$$\gamma_1 = \frac{\beta}{\beta - 1},\ \gamma_3 = \frac{1}{2},\ \gamma_4 = \frac{\beta}{1+\beta},\ \gamma_5 = -\frac{\eta_R}{\overline{\beta}_R - 1},\ \gamma_6 = \frac{\eta_R}{\overline{\beta}_R - 1},\ \overline{\beta}_R = \frac{c_2}{c_R}.$$

By carrying out limit transitions in (69) at $V = c_2 + 0$, and in (73) at $V = c_2 - 0$, we obtain on the boundary of contact zone at $x \to 0+0$

$$\lim_{\substack{x\to 0+0 \\ V=c_2\pm 0}} \varphi^+(x,t)x^{3/2} = \frac{\mu}{\beta}\frac{K_+^+(0)}{\pi}\frac{\partial}{\partial t}\int_0^t \varepsilon(\tau)\left(c_2(t-\tau)\right)^{1/2} d\tau \tag{105}$$

where $K_+^+(0)$ is calculated for case $V = c_2 + 0$ from (48), (53), taking into account that $\gamma_2 \to +\infty$, when $V \to c_2$, and improper integral in (53) disappears and

$$\omega_+^+(\xi) = 4\xi^2\sqrt{\gamma_1 - \xi}\sqrt{\gamma_3 - \xi}\sqrt{\xi - \gamma_4}\sqrt{2}\sqrt{1-\beta^2}\,\eta^{-1}(\xi) \tag{106}$$

$$\eta(\xi) = \left(-2\xi^2 + (1-\xi)^2\right)^2$$

The value of $K_+^+(0)$ in (105) is calculated for case $V = c_2 - 0$ by using formulae (48), (57), in which

$$\omega_+^+(\xi) = 4\xi^2\sqrt{\xi-\gamma_1}\sqrt{\gamma_3-\xi}\sqrt{\xi-\gamma_4}\sqrt{2}\sqrt{1-\beta^2\eta^{-1}(\xi)} \tag{107}$$

where $\eta(\xi)$ calculated from (106). So, the factors at the singularity $x^{-3/2}$ in (69) and (73) coincide at $x \to 0+0$ in form with (105), but are no equal on value.

The stresses coincide at the back edge $\varphi^-(x,t)$ (71) for $V = c_2 + 0$, and $\varphi^-(x,t)$ (75) for $V = c_2 - 0$:

$$\lim_{\substack{x\to 0+0 \\ V=c_2\pm 0}} \varphi^-(x,t) = -\frac{\mu\,\dot{\varepsilon}(t)}{\beta\; c_2} \tag{108}$$

The displacements at the back edge $\upsilon_+(x,t)$ (72) for $V = c_2 + 0$ and $\upsilon_+(x,t)$ (76) for $V = c_2 - 0$ lead at $x \to 0-0$ to the relationship:

$$\lim_{\substack{x\to 0-0 \\ V=c_2\pm 0}} \upsilon_+(x,t)(-x)^{-3/2} = \frac{c_2}{\pi K_-^-(0)}\int_0^t \varepsilon(\tau)\left(c_2(t-\tau)\right)^{-5/2} d\tau \tag{109}$$

in which $K_-^-(0)$ is calculated for $V = c_2 + 0$ by using formulae (48), (56), where divergent integral disappears in $H_-^-(u)$, and $\omega_-^-(\xi)$ coincides with $\omega_+^+(\xi)$ from (107). The value of $K_-^-(0)$ is calculated for $V = c_2 - 0$ by using formulae (48), (60), in which $\omega_-^-(\xi)$ coincides with $\omega_+^+(\xi)$ from (107). So, the factors at $x^{3/2}$ in (72) and (76) at $x \to 0-0$ coincide in form with (109), but differ on value.

The displacements at the frontal edge of punch $\upsilon_-(x,t)$ from (70) for $V = c_2 + 0$ and $\upsilon_-(x,t)$ from (74) for $V = c_2 - 0$ lead to the relationship:

$$\lim_{\substack{x\to 0-0 \\ V=c_2-0}} \upsilon_-(x,t)(-x)^{1/2} = -\frac{c_2}{\pi K_-^+(0)}\int_0^t \varepsilon(\tau)\left(c_2(t-\tau)\right)^{-1/2} d\tau \tag{110}$$

where $K_-^+(0)$ is calculated for $V = c_2 + 0$ on the base of formulae (48), (54), in which $\omega_-^+(\xi)$ coincides with $\omega_+^+(\xi)$ from (106), the value of $K_-^+(0)$ is calculated for $V = c_2 - 0$ on the base of formulae (48), (54), in which $\omega_-^+(\xi)$ coincides with $\omega_+^+(\xi)$ from (106), and the value of $K_-^+(0)$ is calculated for $V = c_2 - 0$ on the base of formulae (48), (58), in which $\omega_-^+(\xi)$ coincides with $\omega_+^+(\xi)$ from (107). The factors at $(-x)^{-1/2}$ in (70) and (74) at $x \to 0-0$ coincide in form with (110), but differ on value.

3. The problem solution does not exist at $V = c_R$. In particular, it is connected with that what root of the Rayleigh equation $R(u) = 0$ (17), $\gamma_5 = \eta_R(\overline{\beta}_R - 1)^{-1}$ does not exist.

Above analysis shows that the problem solution with slowing V from $c_1 + 0$ down to $c_2 - 0$ at the transition of V through the range boundaries c_1, c_2 changes qualitatively and quantitatively.

This is connected with that the fronts of the elastic waves c_1 and c_2 go out from under punch before its frontal edge on free surface. At the same time, the fronts of the elastic waves c_1 and c_2 go in under the punch through its back edge.

6. Integral Characteristics of the Problem

The force $P(t)$, acting on punch, is found as

$$P(t) = -\int_{-a}^{a} \varphi(x,t)dx \qquad t > 0 \tag{111}$$

where $\varphi(x,t)$ is given by superposition (32). Formula (111) after substitution of (32) has the form:

$$P(t) = P^{\infty}(t) - P^{-}(t) - P^{+}(t) \tag{112}$$

$$P^{\pm}(t) = \int_{-a}^{a} \varphi^{\pm}(a \pm x,t)dx\,, \quad P^{\infty}(t) = \int_{-a}^{a} \varphi^{\infty}(x,t)dx$$

where $\varphi^{\pm}(x,t)$ are defined in Section 4 or by formulae (39), and $\varphi^{\infty}(x,t)$ is calculated via (42).

The moment, applied to punch and keeping the punch base parallel to x-axis, is calculated on base of the formula:

$$M(t) = -\int_{-a}^{a} x\varphi(x,t)dx \qquad t > 0 \tag{113}$$

where $\varphi(x,t)$ is calculated from (32) and after its substitution in (113), we have

$$M(t) = M^{\infty}(t) - M^{-}(t) - M^{+}(t) \tag{114}$$

$$M^{\pm}(t)=\int_{-a}^{a}x\varphi^{\pm}(a\pm x,t)dx\,,\quad M^{\infty}(t)=\int_{-a}^{a}x\varphi^{\infty}(x,t)dx$$

Below, we use formulae (39) for $\varphi^{\pm}(x,t)$, which are required for calculation of $P_{\pm}(t)$, $M_{\pm}(t)$, and formula (42) for $\varphi^{\infty}(x,t)$. By this, we obtain for all ranges in our case

$$P_{\infty}(t)=-\frac{\mu}{\beta}t_2\dot{\varepsilon}(t)\,,\quad M_{\infty}(t)=0 \qquad t>0 \tag{115}$$

$$\text{I. } P^{+}(t)=P_{+}^{\infty}(t)H(t)+\sum_{k=1}^{2}P_k^{+}(t)H(t-\gamma_k t_2)\quad t>0 \tag{116}$$

$$P_{+}^{\infty}(t)=-\frac{\mu}{\beta}\left(t_2\dot{\varepsilon}(t)-p_0^{+}\varepsilon(t)\right)$$

$$P_1^{+}(t)=-\frac{\mu}{\beta}\frac{1}{\pi t_2}\left(\int_{t-\gamma_1 t_2}^{t-\gamma_2 t_2}+\int_{t-\gamma_3 t_2}^{t-\gamma_4 t_2}\right)\varepsilon(\tau)f_1\left(\frac{t-\tau}{t_2}\right)d\tau$$

$$P_2^{+}(t)=-\frac{\mu}{\beta}\frac{1}{\pi t_2}\int_{t-\gamma_2 t_2}^{t-\gamma_3 t_2}\varepsilon(\tau)f_2\left(\frac{t-\tau}{t_2}\right)d\tau$$

$$p_0^{+}=\frac{\beta}{\pi}\left(\int_{\gamma_2}^{\gamma_1}+\int_{\gamma_4}^{\gamma_3}\right)f_1(\xi)d\xi+\frac{\beta}{\pi}\int_{\gamma_3}^{\gamma_2}f_2(\xi)d\xi\,,$$

$$f_1(\xi)=\left(-2\xi^2+(1-\xi\beta_2)^2\right)^2R^{-1}(\xi)$$

$$f_2(\xi)=f_1(\xi)-4\xi^2\sqrt{\gamma_1-\xi}\sqrt{\gamma_2-\xi}\sqrt{\xi-\gamma_3}\sqrt{\xi-\gamma_4}\sqrt{\beta_1^2-1}\sqrt{\beta_2^2-1}R^{-1}(\xi)$$

$$R(\xi)=\xi^2(1-\xi\beta_2)^2\sqrt{\xi-\gamma_4}\sqrt{\gamma_1-\xi}\sqrt{\beta_1^2-1}$$

$$P_{-}(t)=P_{\infty}(t)=-\frac{\mu}{\beta}t_2\dot{\varepsilon}(t) \tag{117}$$

$$M^{+}(t)=M_{+}^{\infty}(t)H(t)+\sum_{k=1}^{2}M_k^{+}(t)H(t-\gamma_k t_2)\quad t>0 \tag{118}$$

$$M_{+}^{\infty}(t)=-\mu\left(c_1p_0^{+}\varepsilon(t)+c_2m_0^{+}\tilde{\varepsilon}(t)\right)$$

$$M_1^+(t) = -\mu\frac{1}{\pi t_2}\left(\int_{t-\gamma_1 t_2}^{t-\gamma_2 t_2} + \int_{t-\gamma_3 t_2}^{t-\gamma_4 t_2}\right)\widetilde{\varepsilon}(\tau)\widetilde{f}_1\left(\frac{t-\tau}{t_2}\right)d\tau$$

$$M_2^+(t) = -\mu\frac{1}{\pi t_2}\int_{t-\gamma_2 t_2}^{t-\gamma_3 t_2}\widetilde{\varepsilon}(\tau)\widetilde{f}_2\left(\frac{t-\tau}{t_2}\right)d\tau$$

$$m_0^+(t) = \frac{\mu}{\pi}\left(\int_{\gamma_2}^{\gamma_1} + \int_{\gamma_4}^{\gamma_3}\right)\widetilde{f}_1(\xi)d\xi + \frac{1}{\pi}\int_{\gamma_2}^{\gamma_3}\widetilde{f}_2(\xi)d\xi$$

$$\widetilde{f}_k(\xi) = \xi^{-1}f_k(\xi),\ k=1,2,\ \widetilde{\varepsilon}(t) = \int_0^t \varepsilon(\tau)d\tau,\ t_2 = 2a/c_2.$$

$$M_-(t) = 0 \qquad t>0 \qquad (119)$$

$$\text{II. } P^+(t) = P_+^\infty(t)H(t) + P_0^+(t)H(t-\gamma_4 t_2) \qquad t>0 \qquad (120)$$

$$P_0^+(t) = -\frac{\mu}{\beta}\frac{K_+^+(0)}{\pi t_2}\int_0^{t-\gamma_4 t_2}\varepsilon(\tau)G_+^+\left(\frac{t-\tau}{t_2}\right)d\tau$$

$$G_+^+(\xi) = \frac{(-\gamma_5+\xi)(\gamma_6-\xi)}{\xi^2\sqrt{\gamma_4+\xi}\,r_+^+(-i\xi)},\ p_0^+ = \frac{K_+^+(0)}{\pi}\int_{\gamma_4}^{\infty}G_+^+(\xi)d\xi$$

where improper integral is understood in the sense of finite part, and $P_+^\infty(t)$ given in (116).

$$P^-(t) = P_-^\infty(t)H(t) + P_0^-(t)H(t+\gamma_1 t_2) \qquad t>0 \qquad (121)$$

$$P_-^\infty(t) = -\frac{\mu}{\beta}\left(t_2\varepsilon(t) - p_0^-\varepsilon(t)\right)$$

$$P_0^-(t) = -\frac{\mu}{\beta}\frac{K_+^-(0)}{\pi t_2}\int_0^{t+\gamma_1 t_2}\varepsilon(\tau)G_+^-\left(\frac{t-\tau}{t_2}\right)d\tau$$

$$G_+^-(\xi) = \frac{1}{\xi^2\sqrt{\gamma_1^-+\xi}\,r_+^-(-i\xi)},\ p_0^- = \frac{K_+^-(0)}{\pi}\int_{-\gamma_1}^{\infty}G_+^-(\xi)d\xi$$

$$M^+(t) = M_+^\infty(t)H(t) + \sum_{k=1}^{2}M_k^+(t)H(t-\gamma_4 t_2) \qquad t>0$$

$$M_+^{\infty}(t) = -\frac{\mu}{\beta}\left(ap_0^+\varepsilon(t) + c_2 m_0^+\widetilde{\varepsilon}(t)\right)$$

$$M_1^+(t) = -\frac{\mu}{\beta}\frac{K_+^+(0)}{\pi t_2}a\int_0^{t-\gamma_4 t_2}\varepsilon(\tau)G_+^+\left(\frac{t-\tau}{t_2}\right)d\tau$$

$$M_2^+(t) = \frac{\mu}{\beta}\frac{K_+^+(0)}{\pi t_2}c_2\int_0^{t-\gamma_4 t_2}\widetilde{\varepsilon}(\tau)\widetilde{G}_+^+\left(\frac{t-\tau}{t_2}\right)d\tau$$

$$m_0^+ = \frac{K_+^+(0)}{\pi}\int_{\gamma_4}^{\infty}\widetilde{G}_+^+(\xi)d\xi\,,\ \widetilde{G}_+^+(\xi) = \xi^{-1}G_+^+(\xi)$$

where $G_+^+(\xi)$ is given in (120), and p_0^+ defined by (116).

$$M^-(t) = M_-^{\infty}(t)H(t) + \sum_{k=1}^{2} M_k^-(t)H(t+\gamma_1 t_2) \qquad t>0 \tag{122}$$

$$M_-^{\infty}(t) = -\frac{\mu}{\beta}\left(ap_0^-\varepsilon(t) + c_2 m_0^-\widetilde{\varepsilon}(t)\right)$$

$$M_1^-(t) = -\frac{\mu}{\beta}\frac{K_+^-(0)}{\pi t_2}a\int_0^{t+\gamma_1 t_2}\varepsilon(\tau)G_+^-\left(\frac{t-\tau}{t_2}\right)d\tau$$

$$M_2^-(t) = -\frac{\mu}{\beta}\frac{K_+^-(0)}{\pi t_2}c_2\int_0^{t+\gamma_1 t_2}\widetilde{\varepsilon}(\tau)\widetilde{G}_+^-\left(\frac{t-\tau}{t_2}\right)d\tau$$

$$m_0^- = \frac{K_+^-(0)}{\pi}\int_{-\gamma_1}^{\infty}\widetilde{G}_+^-(\xi)d\xi\,,\ \widetilde{G}_+^-(\xi) = \xi^{-1}G_+^-(\xi)\,,$$

where $G_+^-(\xi)$, p_0^- are found by (121), and $\widetilde{\varepsilon}(t)$ is calculated by (117).

In the obtained formulae (120) – (122) for force characteristics of the problem, their finite parts are pointed, obtained as a result of integrating discontinuous solutions (69) – (72).

III. The formulae for force characteristics in this velocity range coincide with corresponding formulae, obtained in the previous range, differing only in that what the functions $r_\pm^\pm(u)$, $K_\pm^\pm(u)$ are calculated by using formulae (48), (57) – (60).

IV. $P^+(t) = P_+^{\infty}(t)H(t) + P_0^+(t)H(t-\gamma_4 t_2) \qquad t>0$ (123)

$$P_0^+(t) = -\frac{\mu}{\beta}\frac{K_+^+(0)}{\pi t_2}\int_0^{t-\gamma_4 t_2}\varepsilon(\tau)G_+^+\left(\frac{t-\tau}{t_2}\right)d\tau\,,\ G_+^+(\xi) = \frac{\gamma_6-\xi}{\xi^2\sqrt{\xi-\gamma_4}\,r_-^+(-i\xi)}\,.$$

The value of p_0^+ is calculated in (120) by using the pointed functions $G_+^+(\xi)$, and $P_+^\infty(t)$, given in (116).

$$P_-(t) = P_-^\infty(t)H(t) + P_0^-(t)H(t+\gamma_1 t_2) \qquad t>0 \qquad (124)$$

$$P_0^-(t) = \frac{\mu}{\beta}\frac{K_+^-(0)}{\pi t_2}\int_0^{t+\gamma_1 t_2}\varepsilon(\tau)G_+^-\left(\frac{t-\tau}{t_2}\right)d\tau\,,\ G_+^-(\xi) = \frac{\gamma_5+\xi}{\xi^2\sqrt{\gamma_1+\xi}\,r_+^-(-i\xi)}$$

The value of p_0^- is calculated in (121) by using the pointed function $G_+^-(\xi)$.

$$M^+(t) = M_+^\infty(t)H(t) + \sum_{k=1}^{2} M_k^+(t)H(t-\gamma_4 t_2) \qquad t>0 \qquad (125)$$

$$M_1^+(t) = -\frac{\mu}{\beta}\frac{K_+^+(0)}{\pi t_2}a\int_0^{t-\gamma_4 t_2}\varepsilon(\tau)G_+^+\left(\frac{t-\tau}{t_2}\right)d\tau$$

$$M_2^+(t) = \frac{\mu}{\beta}\frac{K_+^+(0)}{\pi t_2}c_2\int_0^{t-\gamma_4 t_2}\tilde{\varepsilon}(\tau)\tilde{G}_+^+\left(\frac{t-\tau}{t_2}\right)d\tau\,,\ \tilde{G}_+^+(\xi) = \xi^{-1}G_+^+(\xi)$$

the function $G_+^+(\xi)$ is calculated in (124), and p_0^+, m_0^+ are determined on its base by using formulae (120), (122), $M_+^\infty(t)$ is given by (122).

$$M^-(t) = M_-^\infty(t)H(t) + \sum_{k=1}^{2} M_k^-(t)H(t+\gamma_1 t_2) \qquad t>0 \qquad (126)$$

$$M_1^-(t) = -\frac{\mu}{\beta}\frac{K_+^-(0)}{\pi t_2}a\int_0^{t+\gamma_1 t_2}\varepsilon(\tau)G_+^-\left(\frac{t-\tau}{t_2}\right)d\tau$$

$$M_2^-(t) = -\frac{\mu}{\beta}\frac{K_+^-(0)}{\pi t_2}c_2\int_0^{t+\gamma_1 t_2}\tilde{\varepsilon}(\tau)\tilde{G}_+^-\left(\frac{t-\tau}{t_2}\right)d\tau\,,\ \tilde{G}_+^-(\xi) = \xi^{-1}G_+^-(\xi)\,,$$

the function $G_+^-(\xi)$ is calculated in (125), and p_0^-, m_0^- are determined on its base by using formulae (121), (123), $M_-^\infty(t)$ is given by (123).

7. Proper Approximation of Function

Except for the velocity range I, where the problem solutions (65) – (68), (116) – (119) do not contain singular quadratures in addition to Rayleigh waves, the solutions, obtained for ranges II – IV (formulae (69) – (80), (120) – (127)), include the singular quadratures, presented in $r_{\pm}^{\pm}(u)=\exp\left(H_{\pm}^{\pm}(u)\right)$, where $H_{\pm}^{\pm}(u)$ are found by singular integrals (53) – (64). This circumstance impedes in some way a numerical analysis of the obtained results and their next application. To obtain solutions without singular quadratures, it is assumed to approximate [18, 19] Fourier imagine of kernel of the integral equation (12), the function $K(u)$ (14) in the complex plane $u=\sigma+i\tau$ by an expression of special form $K_V(u)$, which satisfies the next requirements:

(i) the function $K_V(u)$ includes all isolated singular points of the function $K(u)$ (21), (22) in the complex plane, and does not include other isolated singular points;
(ii) the function $K_V(u)$ has the same asymptotical properties that the function $K(u)$ (18) – (20);
(iii) the function $K_V(u)$ allows a factorization by using elementary operations.

One of the kinds of approximation $K_V(u)$ in range II, satisfying all above requirements (i) – (iii) is the expression:

$$K_V(u)=q(u)r(iu)\,,\ q(u)=\frac{\sqrt{-\gamma_1+iu}\sqrt{\gamma_4-iu}}{(-\gamma_5+iu)(\gamma_6-iu)}\,, \tag{128}$$

$$r(u)=\exp\left(M_{14}^{\alpha}(u)+M_{14}(u)\right) \tag{129}$$

$$M_{ml}^{\alpha}(u)=-\frac{1}{4}\alpha\sum_{k=m}^{l}(k-2)(k-3)\ln\omega_*(u,\gamma_k) \tag{130}$$

$$M_{ml}(u)=\sum_{n=0}^{N}d_n\sum_{k=m}^{l}\omega_n(u,\gamma_k) \tag{131}$$

$$\omega_*(u,\gamma_k)=\begin{cases}\left(\sqrt{\gamma_k}+\sqrt{\gamma_k-u}\right)^{-2} & k=2,3,4\\ \left(\sqrt{-\gamma_k}+\sqrt{-\gamma_k+u}\right)^{-2} & k=1\end{cases} \tag{132}$$

$$\omega_n(u,\gamma_k)=\begin{cases}\left(\dfrac{\sqrt{\gamma_k}+\sqrt{\gamma_k-u}}{\sqrt{\gamma_k}}\right)^{-2(n+1)} & k=2,3,4\\ \left(\dfrac{\sqrt{-\gamma_k}+\sqrt{-\gamma_k+u}}{\sqrt{-\gamma_k}}\right)^{-2(n+1)} & k=1\end{cases} \tag{133}$$

where γ_k ($k=1,2,\ldots,6$) are found in (21), (22), and their signs in the considered range are pointed in (24), d_n are the unknown parameters of the approximation which are required to calculate, and α is present by (86).

The unknown parameters of the approximation d_n could be calculated by various methods of the best approximation of the $K(u)$ by using expression $K_V(u)$ (128) [18]. In particular, the approximation parameters d_n, $n=0,1,2,\ldots,N$ could be found from conditions of coincidence of $K(u)$ and $K_V(u)$ together with their derivatives up to N-th order inclusive at the point $u=0$ of the complex plane $u=\sigma+i\tau$:

$$K^{(m)}(0)=K_V^{(m)}(0) \quad m=0,1,2,\ldots,N \tag{134}$$

where the upper index (m) signs m-th derivative.

The fulfillment of conditions (134) leads to solution of set of the linear algebraic equations relatively unknown parameters d_n:

$$\sum_{j=0}^{N} a_{ij}d_j=b_i,\; i=0,1,2,\ldots,N \quad a_{ij}=\sum_{k=1}^{4}\omega_j^{(i)}(iu,\gamma_k)\bigg|_{u=0} \tag{135}$$

$$q^{\alpha}(u)=\left(\frac{\omega_*^2(u,\gamma_1)}{\omega_*^2(u,\gamma_4)}\right)^{\alpha/2},\quad b_i=\ln\frac{K(u)}{q(u)q^{\alpha}(u)}\bigg|_{u=0}^{(i)}$$

where $q(u)$ is found from (128), the upper index (i) designs i-th derivative and lower index $u=0$ defines point in which it calculates. Fulfillment of conditions (134) in form (135) guarantees existence and uniqueness of the approximation $K_V(u)$ (128) for the function $K(u)$ in some circular $|u|<M$ $(M>0)$ of the complex plane $u=\sigma+i\tau$. By this, the asymptotic representation of $K_V(u)$ at infinity $(|u|\to\infty)$ coincides with asymptotic of $K(u)$ (18) – (20). By increasing number N of the approximation terms in (131), an accuracy of the approximation $K_V(u)$ increases into circular $|u|<M$ or extends the circular $|u|<M$ at the approximation with given error. So, the approximation $K_V(u)$ of the function $K(u)$ in the complex plane $u=\sigma+i\tau$ could be found with any given accuracy. In particular, the unknown factor of the approximation d_0 at $N=0$, defined from zero $(m=0)$ condition (134), is calculated as

$$d_0 = \ln\left[\frac{\eta_R^2\Delta_1}{\beta_2^2\left(1-\beta_2^2\eta_R^2\right)}\left(\frac{1+\beta_1}{1-\beta_1}\right)^{\alpha}\right] \tag{136}$$

where Δ_1 is determined from (16).

The factorization (38) of the approximation of $K(u)$ with form (128) in the considered velocity range II is reached by using elementary means. The functions $K_{\pm}^{+}(u)$ from (38) are defined as

$$\begin{aligned} &K_{\pm}^{+}(u) = q_{\pm}^{+}(u)r_{\pm}^{+}(iu) \\ &r_{+}^{+}(u) = \exp\left(M_{44}^{\alpha}(u) + M_{24}(u)\right) \\ &r_{-}^{+}(u) = \exp\left(M_{11}^{\alpha}(u) + M_{11}(u)\right) \end{aligned} \tag{137}$$

where functions $q_{\pm}^{+}(u)$ are defined by (53), (54). The factorization of approximation of $K(-u)$ with form $K_V(-u)$ is also conducted by elementary means, and functions $K_{\pm}^{-}(u)$ have the forms:

$$\begin{aligned} &K_{\pm}^{-}(u) = q_{\pm}^{-}(u)r_{\pm}^{-}(iu) \\ &r_{+}^{-}(u) = \exp\left(M_{11}^{\alpha}(-u) + M_{11}(-u)\right) \\ &r_{-}^{-}(u) = \exp\left(M_{44}^{\alpha}(-u) + M_{24}(-u)\right), \end{aligned} \tag{138}$$

where functions $q_{\pm}^{-}(u)$ are defined by (55), (56). It should be noted, that the asymptotic representations of $K_{\pm}^{\pm}(u)$ from (48), (53) – (56), obtained at exact factorization of $K(u)$ at great u, coincide with the asymptotic representations of $K_{\pm}^{\pm}(u)$ from (137), (138), obtained at factorization of the approximation of $K_V(u)$ (128) for account of expressions for $M_{kk}^{\alpha}(u)$ $k = 1{,}4$, chosen on the base of the asymptotic representations of $H_{\pm}^{\pm}(u)$ from (85).

As approximation of $K(u)$ in the velocity ranges III, IV, it is assumed the expression $K_V(u)$ with form (128), where

$$r(u) = \exp\left(M_{14}(u)\right). \tag{139}$$

The expression $M_{14}(u)$ is defined by formula (131), in which

$$\omega_n(u,\gamma_k)=\begin{cases}\left(\dfrac{\sqrt{\gamma_k}+\sqrt{\gamma_k-u}}{\sqrt{\gamma_k}}\right)^{-2(n+1)} & k=3,4\\ \left(\dfrac{\sqrt{-\gamma_k}+\sqrt{-\gamma_k+u}}{\sqrt{-\gamma_k}}\right)^{-2(n+1)} & k=1,2\end{cases} \tag{140}$$

and γ_k ($k=1,2,\ldots,6$) are calculated by (21), (22), and their signs in the considered velocity ranges III, IV are found in (25), (26), respectively. The constants d_n, similarly to the case of approximation of the form (128) – (133), are parameters of the approximation, which could be calculated from conditions (134). It is reduced to solution of set of linear algebraic equations (135), in which $q^{\alpha}(u)$ is necessary to equate 1 ($q^{\alpha}(u)=1$) in expression for b_i. Similarly to the velocity range II, the approximation of $K(u)$ in the form of $K_V(u)$ in the complex plane $u=\sigma+i\tau$ for the considered velocity ranges III, IV could be chosen with any given accuracy. In particular case of the approximation (128), (139), (140) at $N=0$, the unique unknown factor of the approximation d_0 is calculated from zero ($m=0$) condition (134) as

$$d_0=\ln\left[\frac{\eta_R^2-4\sqrt{1-\beta_1^2}\sqrt{1-\beta_2^2}}{\beta_2^2\left(1-\eta_R^2\beta_2^2\right)}\right]. \tag{141}$$

The factorization (38) of the approximation of $K(u)$ with form (128), (139), (140) for the considered velocity ranges III, IV is also conducted by elementary means. The functions $K_{\pm}^{+}(u)$ in the considered ranges are defined by formulae (137), in which

$$r_{+}^{+}(u)=\exp\left(M_{34}(u)\right),\ r_{-}^{+}(u)=\exp\left(M_{12}(u)\right) \tag{142}$$

and functions $q_{\pm}^{+}(u)$ are taken from (57), (58) for the range III and from (61), (62) for the range IV, $M_{12}(u)$, $M_{34}(u)$ are calculated by (131), (140). The factorization of the approximation of $K(-u)$ by the expression $K_V(-u)$ is found by formulae (138), in which

$$r_{+}^{-}(u)=\exp\left(M_{12}(-u)\right),\ r_{-}^{-}(u)=\exp\left(M_{34}(-u)\right) \tag{143}$$

where $q_{\pm}^{-}(u)$ are calculated for the range III in (59), (60), and for the range IV in (63), (64), $M_{12}(u)$, $M_{34}(u)$ are computed by (131), (140). It should be noted that at exact factorization of $K(u)$ (48), similar to the factorization of the approximation by function $K_V(u)$ (128), (139), (140), the functions $r_{\pm}^{\pm}(u)$ in these velocity ranges at great values of $|u|$ have the asymptotic representations:

$$r_{\pm}^{\pm}(u) = 1 + O\left(\frac{1}{iu}\right) \text{ at } u \to \infty \quad \eta_- < \text{Im}(u) < \eta_+ \tag{144}$$

8. Effective Solutions of the Problem

The approximation $K_V(u)$ of function $K(u)$, proposed in Section 7 for the velocity rages II – IV, allows one to obtain the problem solutions $\varphi^{\pm}(x,t)$, $\upsilon_{\mp}(x,t)$ without singular quadratures (53) – (64). As above, to define normal stresses $\varphi(x,t)$ under base of moving punch on elastic half-plane, it is used in the velocity ranges II – IV the representation in superposition form (32) at $0 < t < 2a/c_1$, $|x| \le a$, in which $\varphi^{\pm}(x,t)$ are calculated by using formulae (39), taking into account that the functions $K_{\pm}^{\pm}(u)$, included into them, are defined by formulae (137), (138), (142), (143). The function $\varphi^{\infty}(x,t)$ is calculated from formula (42). By calculating quadratures in (39), (40) with account of formulae (137), (138) in the range II, we obtain the next formulae for $\varphi^{\pm}(x,t)$, $\upsilon_{\mp}(x,t)$.

$$\text{II. } \varphi^{+}(x,t) = \varphi^{\infty}(x,t)H(t) + x^{-3/2+\alpha}\sum_{k=1}^{3}\Phi_k^{+}(x,t)H(t-\tau_{k+1}x) \quad t>0,\ x>0 \tag{145}$$

$$\Phi_1^{+}(x,t) = -\frac{\mu}{\beta}\frac{K_{+}^{+}(0)}{\pi}\frac{\partial}{\partial t}\int_0^{t-\tau_2 x}\varepsilon(\tau)G_1^{+}(x,c_2(t-\tau))d\tau$$

$$\Phi_k^{+}(x,t) = -\frac{\mu}{\beta}\frac{K_{+}^{+}(0)}{\pi}\frac{\partial}{\partial t}\int_{t-\tau_k x}^{t-\tau_{k+1} x}\varepsilon(\tau)G_k^{+}(x,c_2(t-\tau))d\tau \quad k=2,3$$

$$G_1^{+}(x,t) = q(x,t)t^{-\alpha}r_{24}^{\alpha}\left(\frac{t}{x}\right),\quad G_2^{+}(x,t) = q(x,t)\left(\sqrt{x\gamma_2}+\sqrt{t-x\gamma_2}\right)^{-2\alpha}r_{24}\left(\frac{t}{x}\right)$$

$$q(x,t) = \frac{(-x\gamma_5+t)(x\gamma_6-t)}{t\sqrt{t-x\gamma_4}},$$

$$r_{24}^{\alpha}(\xi) = \exp(-\text{Re}\, M_{24}(\xi))\cos(2\alpha\theta(\xi) + \text{Im}\, M_{24}(\xi)),\quad \theta(\xi) = \text{arctg}\left(\frac{\sqrt{\xi-\gamma_2}}{\sqrt{\gamma_2}}\right)$$

$$G_3^{+}(x,t) = G_2^{+}(x,t)$$

where $\sqrt{\gamma_k - \xi} = i\sqrt{\xi - \gamma_k}$ in the expression for $M_{24}(\xi)$, if $\gamma_k - \xi \le 0$ $(k = 2,3,4)$, $\varphi^{\infty}(x,t)$ is given in (42).

Asymptotic representation of formula (145) at $x \to 0+0$ coincides totally with (87), if take into account that $\lim_{\xi\to\infty} r_{24}^{\alpha}(\xi) = \cos\pi\alpha$.

$$\upsilon_-(x,t) = (-x)^{-1/2+\alpha} V_0^-(x,t) H(t-\tau_1 x) \qquad t>0\,,\ x>0 \tag{146}$$

$$V_0^-(x,t) = -\frac{c_2}{\pi K_-^+(0)} \int_0^{t-\tau_1 x} \varepsilon(\tau) G_-^+(x, c_2(t-\tau)) d\tau$$

$$G_-^+(x,t) = q(x,t) t^{-\alpha} \cos 2\alpha\theta\left(\frac{t}{-x}\right) r_{11}\left(\frac{t}{x}\right)$$

$$q(x,t) = \frac{\sqrt{-x\gamma_1 + t}}{t}\,,\ \theta(\xi) = \operatorname{arctg}\sqrt{\frac{\gamma_1+\xi}{-\gamma_1}}$$

$$r_{11}(\xi) = \exp\left(-\operatorname{Re} M_{11}(-\xi)\right)\cos\left(\operatorname{Im} M_{11}(-\xi)\right)$$

where $\sqrt{-\gamma_k+\xi} = i\sqrt{\gamma_k - \xi}$ in the expression for $M_{11}(\xi)$, if $-\gamma_k + \xi \le 0$ $(k=1)$.

Asymptotic representation of formula (146) at $x \to 0-0$ coincides totally with (88), obtained before from (70).

$$\varphi^-(x,t) = \varphi^\infty(x,t) H(t) + x^{1/2-\alpha} \Phi_0^-(x,t) H(t+\tau_1 x) \qquad t>0\,,\ x>0 \tag{147}$$

$$\Phi_0^-(x,t) = \frac{\mu}{\beta} \frac{K_+^-(0)}{\pi} \frac{\partial}{\partial t} \int_0^{t+\tau_1 x} \varepsilon(\tau) G_+^-(x, c_2(t-\tau)) d\tau$$

$$G_+^-(x,t) = q(x,t) t^{\alpha} \cos 2\alpha\theta\left(\frac{t}{x}\right) r_{11}\left(\frac{t}{x}\right),$$

$$q(x,t) = \frac{1}{t\sqrt{x\gamma_1 + t}}\,,\ r_{11}(\xi) = \exp\left(-\operatorname{Re} M_{11}(-\xi)\right)\cos\left(\operatorname{Im} M_{11}(-\xi)\right)$$

and function $\theta(\xi)$ is defined by (146).

Asymptotic representation of formula (147) at $x \to 0+0$ coincides totally with (89), obtained before from (71).

$$\upsilon_+(x,t) = (-x)^{3/2-\alpha} \sum_{k=1}^{2} V_k^-(x,t) H(t + x\tau_{2k}) \qquad t>0\,,\ x<0 \tag{148}$$

$$V_1^-(x,t) = -\frac{c_2}{\pi K_-^-(0)} \int_0^{t+\tau_2 x} \varepsilon(\tau) G_1^-(x, c_2(t-\tau)) d\tau$$

$$V_2^-(x,t) = -\frac{c_2}{\pi K_-^-(0)} \int_0^{t+\tau_4 x} \varepsilon(\tau) G_2^-(x, c_2(t-\tau)) d\tau$$

$$G_1^-(x,t) = q(x,t) t^\alpha \cos 2\alpha\theta\left(\frac{t}{-x}\right) r_{24}\left(\frac{t}{x}\right)$$

$$G_2^-(x,t) = q(x,t)\left(\sqrt{-x\gamma_2} + \sqrt{-x\gamma_2 - t}\right)^{2\alpha} r_{24}\left(\frac{t}{x}\right)$$

$$q(x,t) = -\frac{\sqrt{t + x\gamma_4}}{t(x\gamma_5 + t)(x\gamma_6 + t)}, \quad r_{24}(\xi) = \exp(\operatorname{Re}(M_{24}(\xi)))\cos(\operatorname{Im} M_{24}(\xi))$$

and function $\theta(\xi)$ is defined by (145).

The integral in $V_2(x,t)$ is understood in the sense of the Cauchy principal value, because the integrand shows jumps at two points $\tau = t + \tau_{5,6} x$, and vertical displacements of free surface (148) have logarithmic singularities at fronts of the Rayleigh waves behind back edge of punch. By this, the near Rayleigh front propagates to the punch with velocity $V - c_R$ (relatively the punch), and the far front moves from the punch with velocity $V + c_R$. Asymptotic representation of formula (148) at $x \to 0-0$ coincides with (90), obtained from (72).

III. The problem solution $\varphi^\pm(x,t)$, $\upsilon_\mp(x,t)$ in this velocity range, obtained with the help of approximation of $K(u)$ by expression $K_V(u)$ (128), whose factorization pointed by formulae (142), (143), is defined by the corresponding formulae of previous velocity range, in which, it is required to replace $M_{11}(u)$ by $M_{12}(u)$, $M_{24}(u)$ by $M_{34}(u)$ and to suppose $\alpha = 0$.

$$\text{IV. } \varphi^+(x,t) = \varphi^\infty(x,t)H(t) + x^{-1/2}\Phi_0^+(x,t)H(t - \tau_4 x) \tag{149}$$

$$\Phi_0^+(x,t) = -\frac{\mu}{\beta}\frac{K_+^+(0)}{\pi}\frac{\partial}{\partial t}\int_0^{t-\tau_4 x} \varepsilon(\tau) G_+^+(x, c_2(t-\tau)) d\tau$$

$$G_+^+(x,t) = \frac{q(x,t)}{t} r_{12}\left(-\frac{t}{x}\right), \quad q(x,t) = \frac{x\gamma_5 + t}{\sqrt{x\gamma_1 + t}},$$

$$r(\xi) = \exp(-\operatorname{Re} M_{12}(\xi))\cos(\operatorname{Im} M_{12}(\xi)),$$

where $\sqrt{-\gamma_k + \xi} = i\sqrt{\gamma_k - \xi}$ in the expression for $M_{12}(\xi)$, if $-\gamma_k + \xi \le 0$ $(k = 1,2)$.

Asymptotic representation of formula (149) coincides with (91), obtained from (77).

$$\upsilon_-(x,t) = (-x)^{1/2} V_0^-(x,t) H(t - \tau_1 x) \qquad t > 0,\ x < 0 \tag{150}$$

$$V_0^-(x,t) = \frac{c_2}{\pi K_-^+(0)} \int_0^{t-\tau_1 x} \varepsilon(\tau) G_-^+(x, c_2(t-\tau)) d\tau$$

$$G_-^+(x,t) = \frac{q(x,t)}{t} r_{12}\left(-\frac{t}{x}\right), \; q(x,t) = \frac{\sqrt{-x\gamma_1 + t}}{x\gamma_5 + t},$$

$$r_{12}(\xi) = \exp\left(\operatorname{Re} M_{12}(-\xi)\right)\cos\left(\operatorname{Im} M_{12}(-\xi)\right)$$

The integral in (150) is understood in the sense of the Cauchy principal value [15], because integrand function $q(x,t)$ has jump singularity at $\tau = t - \tau_5 x$ on front of the Rayleigh wave. Therefore, the vertical displacements of a free surface before edge of the moving punch (150) show logarithmic singularities at front of the Rayleigh wave, propagating relatively punch with velocity $c_R - V$ along direction to the punch movement. Asymptotic representation of formula (150) at $x \to 0-0$ coincides with (92), obtained from (78).

$$\varphi^-(x,t) = \varphi^\infty(x,t)H(t) + x^{-1/2}\Phi_0^-(x,t)H(t+\tau_1 x) \qquad t>0,\; x>0 \qquad (151)$$

$$\Phi_0^-(x,t) = -\frac{\mu}{\beta}\frac{K_+^-(0)}{\pi}\frac{\partial}{\partial t} \int_0^{t+\tau_1 x} \varepsilon(\tau) G_+^-(x, c_2(t-\tau)) d\tau$$

$$G_+^-(x,t) = \frac{q(x,t)}{t} r_{12}\left(-\frac{t}{x}\right), \; q(x,t) = \frac{x\gamma_5 + t}{\sqrt{x\gamma_1 + t}},$$

$$r_{12}(\xi) = \exp\left(-\operatorname{Re}(M_{12}(\xi))\right)\cos\left(\operatorname{Im}(M_{12}(\xi))\right)$$

Asymptotic representation of formula (151) at $x \to 0+0$ coincides with (93), obtained from (79).

$$\upsilon_+(x,t) = (-x)^{1/2} V_0^+(x,t) H(t+\tau_4 x) \qquad t>0,\; x<0 \qquad (152)$$

$$V_0^-(x,t) = \frac{c_2}{\pi K_-^-(0)} \int_0^{t+\tau_4 x} \varepsilon(\tau) G_-^-(x, c_2(t-\tau)) d\tau$$

$$G_-^-(x,t) = \frac{q(x,t)}{t} r_{34}\left(-\frac{t}{x}\right), \; q(x,t) = \frac{\sqrt{t + x\gamma_4}}{x\gamma_6 + t},$$

$$r_{34}(\xi) = \exp\left(\operatorname{Re}(M_{34}(\xi))\right)\cos\left(\operatorname{Im}(M_{34}(\xi))\right)$$

The integral in (152) is understood in the sense of the Cauchy principal value, because integrand function $q(x,t)$ has jump singularity at $\tau = t + \tau_6 x$ on front of the Rayleigh wave. Due to this, the vertical displacements of a free surface behind back edge of punch (152) show logarithmic singularities at front of the Rayleigh wave [17], propagating relatively to the punch with velocity $V+c_R$ in opposite direction to the punch movement. Asymptotic representation of formula (152) at $x \to 0-0$ coincides with (94), obtained from (80).

The force $P(t)$ is defined in each of the ranges by formulae (111), (112), in which $\varphi^{\pm}(x,t)$ are taken in the form of double inversion in (39). It is more comfortable for obtainment of the formulae $P^{\pm}(t)$ or in evident form for each of ranges. Note, that as before $P^{\infty}(t)$, $M^{\infty}(t)$ are calculated from (115) in all ranges.

$$\text{II. } P^{+}(t) = P_{+}^{\infty}(t)H(t) + \sum_{k=1}^{2} P_k^{+}(t)H(t-\gamma_2 t_2) \qquad t>0 \qquad (153)$$

$$P_{+}^{\infty}(t) = -\frac{\mu}{\beta}\left(t_2\dot{\varepsilon}(t) - p_0^{+}\varepsilon(t)\right)$$

$$P_1^{+}(t) = -\frac{\mu}{\beta}\frac{K_{+}^{+}(0)}{\pi t_2}\int_0^{t-\gamma_2 t_2} \varepsilon(\tau)G_1^{+}\left(\frac{t-\tau}{t_2}\right)d\tau$$

$$P_2^{+}(t) = -\frac{\mu}{\beta}\frac{K_{+}^{+}(0)}{\pi t_2}\int_{t-\gamma_2 t_2}^{t-\gamma_4 t_2} \varepsilon(\tau)G_2^{+}\left(\frac{t-\tau}{t_2}\right)d\tau$$

$$G_1^{+}(\xi) = \frac{q(\xi)}{\xi^2}\xi^{\alpha}\cos 2\alpha\theta(\xi) r_{24}(\xi)$$

$$G_2^{+}(\xi) = \frac{q(\xi)}{\xi^2}\left(\sqrt{\gamma_2} + \sqrt{\gamma_2 - \xi}\right)^{-2\alpha} r_{24}(\xi), \quad q(\xi) = \frac{(-\gamma_5 + \xi)(\gamma_6 - \xi)}{\sqrt{\xi - \gamma_4}},$$

$$p_0^{+} = \frac{K_{+}^{+}(0)}{\pi}\left[\int_{\gamma_2}^{\infty} G_1^{+}(\xi)d\xi + \int_{\gamma_4}^{\gamma_2} G_2^{+}(\xi)d\xi\right],$$

$$r_{24}(\xi) = \exp\left(-\operatorname{Re}(M_{24}(\xi))\right)\cos\left(\operatorname{Im}(M_{24}(\xi))\right)$$

and function $\theta(\xi)$ is given in (145).

The improper integral in p_0^{+} diverges and is understood in the sense of finite part.

$$P^{-}(t) = P_{-}^{\infty}(t)H(t) + P_0^{-}(t)H(t+\gamma_1 t_2) \qquad t>0 \qquad (154)$$

$$P_{-}^{\infty}(t) = -\frac{\mu}{\beta}\left(t_2\dot{\varepsilon}(t) - p_0^{-}\varepsilon(t)\right)$$

$$P_0^{-}(t) = -\frac{\mu}{\beta}\frac{K_{+}^{-}(0)}{\pi t_2}\int_0^{t+\gamma_1 t_2} \varepsilon(\tau)G_1^{-}\left(\frac{t-\tau}{t_2}\right)d\tau$$

$$G_+^-(\xi)=\frac{q(\xi)}{\xi^2}\xi^\alpha\cos2\alpha\theta(\xi)r_{11}(\xi),\ p_0^-=\frac{K_+^-(0)}{\pi}\int_{-\gamma_1}^{\infty}G_+^-(\xi)d\xi$$

$$q(\xi)=\frac{1}{\sqrt{\gamma_1+\xi}},\ r_{11}(\xi)=\exp(-\operatorname{Re}(M_{11}(\xi)))\cos(\operatorname{Im}(M_{11}(\xi)))$$

and function $\theta(\xi)$ is given in (146).

III. Formulae for $P^\pm(t)$ in this velocity range, obtained by using approximation of $K(u)$, are defined by the corresponding formulae of the velocity range II, in which, it is necessary to replace $r_{11}(u)$ by $r_{12}(u)$, $r_{24}(u)$ by $r_{34}(u)$ and to suppose that $\alpha=0$.

IV. $P^+(t)=P_+^\infty(t)H(t)+P_0^+(t)H(t-\gamma_4t_2)\qquad t>0\qquad(155)$

$$P_0^+(t)=-\frac{\mu}{\beta}\frac{K_+^+(0)}{\pi t_2}\int_0^{t-\gamma_4t_2}\varepsilon(\tau)G_+^+\left(\frac{t-\tau}{t_2}\right)d\tau$$

$$G_+^+(\xi)=\frac{q(\xi)}{\xi^2}r_{34}(\xi),\ q(\xi)=\frac{\gamma_6-\xi}{\sqrt{\xi-\gamma_4}},\ p_0^+=\frac{K_+^+(0)}{\pi}\int_{\gamma_4}^{\infty}G_+^+(\xi)d\xi$$

$$r_{34}(\xi)=\exp(-\operatorname{Re}(M_{34}(\xi)))\cos(\operatorname{Im}(M_{34}(\xi)))$$

$$P^-(t)=P_-^\infty(t)H(t)+P_0^-(t)H(t+\gamma_1t_2)\qquad t>0\qquad(156)$$

$$P_0^-(t)=\frac{\mu}{\beta}\frac{K_+^-(0)}{\pi t_2}\int_0^{t+\gamma_1t_2}\varepsilon(\tau)G_+^-\left(\frac{t-\tau}{t_2}\right)d\tau$$

$$G_+^-(\xi)=\frac{q(\xi)}{\xi^2}r_{12}(\xi),\ q(\xi)=\frac{\gamma_5+\xi}{\sqrt{\gamma_1+\xi}},$$

$$r_{12}(\xi)=\exp(-\operatorname{Re}(M_{12}(\xi)))\cos(\operatorname{Im}(M_{12}(\xi))).$$

The formula for computation of p_0^- is present in (154), where $G_+^-(\xi)$ is defined from (156), $P_+^\infty(t)$ given in (153), and $P_-^\infty(t)$ pointed in (154).

The values of moment $M(t)$ in each of velocity ranges are calculated by formulae (113), (114), in which $\varphi^\pm(x,t)$ is present in the form of double quadratures (39) or in evident form for each of the ranges, discussed above in this section.

II. $M^+(t)=M_+^\infty(t)H(t)+\sum_{k=1}^{4}M_k^+(t)H(t-\gamma_2t_2)\qquad t>0\qquad(157)$

$$M_{+}^{\infty}(t) = -\frac{\mu}{\beta}\left(ap_0^{+}\varepsilon(t) - c_2 m_0^{+}\tilde{\varepsilon}(t)\right)$$

$$M_1^{+}(t) = -\frac{\mu}{\beta}\frac{K_{+}^{+}(0)}{\pi t_2} a \int_0^{t-\gamma_2 t_2} \varepsilon(\tau) G_1^{+}\left(\frac{t-\tau}{t_2}\right) d\tau$$

$$M_2^{+}(t) = -\frac{\mu}{\beta}\frac{K_{+}^{+}(0)}{\pi t_2} c_2 \int_{t-\gamma_2 t_2}^{t-\gamma_4 t_2} \varepsilon(\tau) G_2^{+}\left(\frac{t-\tau}{t_2}\right) d\tau$$

$$M_3^{+}(t) = -\frac{\mu}{\beta}\frac{K_{+}^{+}(0)}{\pi t_2} a \int_0^{t-\gamma_2 t_2} \tilde{\varepsilon}(\tau) \tilde{G}_1^{+}\left(\frac{t-\tau}{t_2}\right) d\tau$$

$$M_4^{+}(t) = -\frac{\mu}{\beta}\frac{K_{+}^{+}(0)}{\pi t_2} c_2 \int_{t-\gamma_2 t_2}^{t-\gamma_4 t_2} \tilde{\varepsilon}(\tau) \tilde{G}_2^{+}\left(\frac{t-\tau}{t_2}\right) d\tau$$

$$m_0^{+} = \frac{K_{+}^{+}(0)}{\pi}\left(\int_{\gamma_2}^{\infty} \tilde{G}_1^{+}(\xi) d\xi + \int_{\gamma_4}^{\gamma_2} \tilde{G}_2^{+}(\xi) d\xi\right)$$

$$\tilde{G}_k^{+}(\xi) = \xi^{-1} G_k^{+}(\xi t) \quad k = 1,2, \quad \tilde{\varepsilon}(t) = \int_0^t \varepsilon(\tau) d\tau$$

The functions $G_k^{+}(t)$ $k=1,2$ and also p_0^{+} are given in (153).

$$M^{-}(t) = M_{-}^{\infty}(t) H(t) + \sum_{k=1}^{2} M_k^{-}(t) H(t + \gamma_1 t_2) \qquad t > 0 \tag{158}$$

$$M_{-}^{\infty}(t) = \frac{\mu}{\beta}\left(ap_0^{-}\varepsilon(t) - c_2 m_0^{-}\tilde{\varepsilon}(t)\right)$$

$$M_1^{-}(t) = \frac{\mu}{\beta}\frac{K_{+}^{-}(0)}{\pi t_2} a \int_0^{t+\gamma_1 t_2} \varepsilon(\tau) G_{+}^{-}\left(\frac{t-\tau}{t_2}\right) d\tau$$

$$M_2^{-}(t) = \frac{\mu}{\beta}\frac{K_{+}^{-}(0)}{\pi t_2} c_2 \int_0^{t+\gamma_1 t_2} \tilde{\varepsilon}(\tau) \tilde{G}_{+}^{-}\left(\frac{t-\tau}{t_2}\right) d\tau$$

$$\tilde{G}_{+}^{-}(\xi) = \xi^{-1} G_{+}^{-}(\xi), \quad m_0^{-} = \frac{K_{+}^{-}(0)}{\pi} \int_{-\gamma_1}^{\infty} \tilde{G}_{+}^{-}(\xi) d\xi,$$

where $G_{+}^{-}(\xi)$, $\tilde{\varepsilon}(t)$ and p_0^{-} are calculated by (154), (157), and (154), respectively.

III. Formulae $M^{\pm}(t)$ for this range could be obtained from the formulae of previous range similarly to as in calculation of $P^{\pm}(t)$ by using corresponding formulae of the range II.

$$\text{IV. } M^{+}(t) = M_{+}^{\infty}(t)H(t) + \sum_{k=1}^{2} M_{k}^{+}(t)H(t+\gamma_1 t_2) \qquad t>0 \tag{159}$$

$$M_{+}^{\infty}(t) = -\frac{\mu}{\beta}\left[ap_0^{+}\varepsilon(t) + c_2 m_0^{+}\tilde{\varepsilon}(t)\right]$$

$$M_1^{+}(t) = -\frac{\mu}{\beta}\frac{K_{+}^{+}(0)}{\pi t_2} a \int_0^{t+\gamma_1 t_2} \varepsilon(\tau) G_{+}^{-}\left(\frac{t-\tau}{t_2}\right) d\tau$$

$$M_2^{+}(t) = -\frac{\mu}{\beta}\frac{K_{+}^{+}(0)}{\pi t_2} c_2 \int_0^{t+\gamma_1 t_2} \tilde{\varepsilon}(\tau) \tilde{G}_{+}^{-}\left(\frac{t-\tau}{t_2}\right) d\tau$$

$$\tilde{G}_{+}^{+}(\xi) = \xi^{-1} G_{+}^{+}(\xi),\ m_0^{+} = \frac{K_{+}^{+}(0)}{\pi} \int_{\gamma_4}^{\infty} \tilde{G}_{+}^{+}(\xi) d\xi,$$

where $G_{+}^{+}(\xi)$ and formula for calculation of p_0^{+} are given in (155), and $\tilde{\varepsilon}(t)$ is computed from (157).

$$M^{-}(t) = M_{-}^{\infty}(t)H(t) + \sum_{k=1}^{2} M_{k}^{-}(t)H(t+\gamma_1 t_2) \qquad t>0 \tag{160}$$

$$M_{-}^{\infty}(t) = -\frac{\mu}{\beta}\left(ap_0^{-}\varepsilon(t) - c_2 m_0^{-}\tilde{\varepsilon}(t)\right)$$

$$M_1^{-}(t) = -\frac{\mu}{\beta}\frac{K_{+}^{-}(0)}{\pi t_2} a \int_0^{t+\gamma_1 t_2} \varepsilon(\tau) G_{+}^{-}\left(\frac{t-\tau}{t_2}\right) d\tau$$

$$M_2^{-}(t) = -\frac{\mu}{\beta}\frac{K_{+}^{-}(0)}{\pi t_2} c_2 \int_0^{t+\gamma_1 t_2} \tilde{\varepsilon}(\tau) \tilde{G}_{+}^{-}\left(\frac{t-\tau}{t_2}\right) d\tau$$

$$\tilde{G}_{+}^{-}(\xi) = \xi^{-1} G_{+}^{-}(\xi),\ m_0^{-} = \frac{K_{+}^{-}(0)}{\pi} \int_{-\gamma_1}^{\infty} \tilde{G}_{+}^{-}(\xi) d\xi,$$

where $G_{+}^{-}(\xi)$ and p_0^{-} are determined from (156), and $\tilde{\varepsilon}(t)$ is defined by (157).

ACKNOWLEDGMENTS

This research was supported by the Russian Foundation for Basic Research (grants Nos. 14-07-00271-a, 14-07-00705-a, 14-08-91166-GFEN_a).

REFERENCES

[1] Galin, L. A., *Contact Problems in the Theory of Elasticity*, Gos. Izd. Tekh. Teor. Lit., Moscow, 1953 (In Russian).

[2] Radok, J. R. M., *Quart. J. Appl. Math.*, 1956, *vol. 14,* 289-298.

[3] Apikian, J. G., *Mechanics. Proc. National Academy of Sciences of Armenia,* 1970, *vol. 23,* 31-36 (In Russian).

[4] Gerstle, E. P.; Pearsale, G. W., *Frans. ASME. J. Appl. Mech.*, 1974, *vol. E41*, 1036-1040.

[5] Artan, R., *Acta. Mech.*, 1994, *vol. 104,* 231-239.

[6] Aleksandrov, V. M.; Kovalenko, E. V., *Problems of Mechanics of Continuous Media with Mixed Boundary Conditions*, Nauka, Moscow, 1986 (In Russian).

[7] Belokon, A. V.; Nasedkin, A. V., In: *Mechanics of Contact Interactions*, Fizmatlit, Moscow, 2001, 331-348 (in Russian).

[8] Zelentsov, V. B.; Baturina, N. Yu., *Ecological Bulletin of Research Centers of BSEC,* 2011, *Issue 3,* 40–48 (In Russian).

[9] Lourie, A. I., *Theory of Elasticity*, Springer, Berlin, 2005.

[10] Sneddon, I. N., *Fourier Transforms,* McGraw-Hill, New York, 1951.

[11] Flitman, L. M., *J. Appl. Math. Mech.,* 1959, *vol. 23,* 997 – 1008.

[12] Vorovich, I. I.; Aleksandrov, V. M.; Babeshko, V. A., *Nonclassical Mixed Problems of the Theory of Elasticity,* Nauka, Moscow, 1974 (In Russian).

[13] Babeshko, V. A., *Generalized Factorization Method in Spatial Dynamical Mixed Problems of Elasticity Theory,* Nauka, Moscow, 1984 (In Russian).

[14] Zelentsov, V. B., *J. Appl. Math. Mech.*, 2006, *vol. 70,* 150-161.

[15] Noble, B. *Method Based on the Wiener-Hopf Technique for the Solution of Partial Differential Equations,* Pergamon, London, 1958.

[16] Vilenkin, N. Ya.; Flaherty, R. E., *Functional Analysis,* Wolters-Noordhoff B.V., 1972.

[17] Poruchikov, V. B., *Methods of the Dynamic Theory of Elasticity*, Nauka, Moscow, 1986, (In Russian).

[18] Zelentsov, V. B., *Izvestiya Vuzov. Severo-Kavkazskii Region. Natural Sciences,* 2011, *No. 1,* 27 (in Russian).

[19] Gaier, D., *Lectures on Complex Approximation,* Birkhäuser, 1987.

In: Advanced Materials - Studies and Applications
Editors: I.A. Parinov, S.-H. Chang, S.Theerakulpisut
ISBN: 978-1-63463-749-7

Chapter 17

BEM Analysis of 3-D Equilibrium Problems of Anisotropic Elasticity with Coupled Fields

L. A. Igumnov [*]***, A. A. Ipatov, S. Yu. Litvinchuk and I. P. Markov***
Research Institute for Mechanics, Lobachevsky State University of Nizhni Novgorod,
Nizhny Novgorod, Russia

Abstract

Schemes for analyzing boundary-value problems of linear 3-D anisotropic elasticity with coupled non-mechanical fields are considered. The interaction model corresponds to the theory of piezoelectricity. Special attention is drawn to the construction of static Green's matrix. With this aim, three approaches to the mathematical construction and computer modeling of such matrices are described. Visualizations of elastostatic Green's functions in the form of surfaces are presented. 3-D boundary-value problems are analyzed using BIE method. The BIE are treated using BEM. The BE scheme is based on regularized BIE and a matched approximation. Results of the numerical experiments are presented in comparison with the results published elsewhere. The following static problems are considered: (i) uniform horizontal load acting on a part of the lateral surface of the elastic cube, (ii) uniform load acting on a circular area at the surface of the piezoelectric half-space.

1. Introduction

3-D static problems of anisotropic elasticity with coupled fields can be solved by the boundary element method (BEM). The availability of fundamental solutions (often referred to as "Green's functions"), that are the solutions of the displacements at an observation point due to a unit point load at a source point in an infinite medium, is a key requirement in the direct formulation of the BEM. While Green's functions in closed form are available for some

[*] igumnov@mech.unn.ru

special cases, e.g., transverse isotropy, general anisotropic elastic or piezoelectric materials three-dimensional fundamental solutions must be evaluated numerically. Wang and Achenbach [1] obtained integral expressions of the Green's functions employing Radon transform. Later, semianalytical forms for anisotropic elastic or piezoelectric fundamental solutions were derived using the residue theorem [2, 3]. However, direct implementation of all these approaches into the BEM code becomes unreasonably time-consuming since a whole process of numerical evaluation must be performed for each pair of the load and integration points. Wilson and Cruse [4] proposed to use interpolation scheme for calculating the fundamental solutions. Since such calculations need to be performed only once for each material, computation time required to evaluate Green's functions at the interpolation nodes becomes virtually insignificant. In this paper, we present 3-D static boundary element method for the general anisotropic elasticity and piezoelectricity based on matched approximation for boundary surface and field variables, which is different from conventional BEM approach with isoparametric elements. Static elastic or piezoelectric anisotropic fundamental solutions in the integration points are evaluated by the linear Lagrange interpolation. We find such concept reasonable due to the reduction in overall computation time while maintaining high accuracy of boundary element solutions.

2. Mathematical Models

First we consider a linear 3-D elastic medium. Assuming zero body forces, the equations of equilibrium are given as

$$\sigma_{ij,i} = 0, \quad i, j = \overline{1,3},$$

where σ is the symmetric stress tensor.

The stress-strain relationship (generalized Hooke's law) is:

$$\sigma_{ij} = C_{ijkl}\varepsilon_{kl}, \quad i, j, k, l = \overline{1,3},$$

where C_{ijkl} is the anisotropic material stiffness tensor. The linear strain tensor ε_{kl} is defined by:

$$\varepsilon_{kl} = \frac{1}{2}\left(u_{k,l} + u_{l,k}\right),$$

where $\mathbf{u}$ is the displacement vector.

Combining these equations we derive equations of equilibrium for linear anisotropic elastic solid occupying the domain $\Omega \subset R^3$ with boundary $\Gamma = \partial\Omega$:

$$C_{ijkl}u_{k,il} = 0, \quad \Omega \subset R^3, \quad i, j, k, l = \overline{1,3}.$$

To complete formulation of the boundary-value problem we need to define the boundary conditions:

$$u_i = \widetilde{u}_i \text{ on } \Gamma^u \text{; } t_i = \widetilde{t}_i \text{ on } \Gamma^t,$$

with $\Gamma = \Gamma^u \bigcup \Gamma^t$ and **t** being the traction vector: $t_i = \sigma_{ij} n_j$, where **n** is the normal outward vector to the boundary Γ.

With zero body forces and zero electric charge equilibrium equations, kinematic relations and constitutive equations for a linear 3-D piezoelectric medium are given by

$$\sigma_{ij,j} = 0,\ D_{i,i} = 0,$$

$$\varepsilon_{ij} = \frac{1}{2}\left(u_{i,j} + u_{j,i}\right),\ E_i = -\phi_{,i},$$

$$\sigma_{ij} = C^E_{ijkl}\varepsilon_{kl} - e_{ijk}E_k,\ D_i = e_{kli}\varepsilon_{kl} + \widetilde{\varepsilon}_{ik}E_k,\ i,j,k,l = \overline{1,3},$$

where D_i denotes the electric displacements, E_i is the electric field strength, ϕ is the electric potential; C^E_{ijkl} is the stiffness tensor, e_{ijk} is the piezoelectric tensor and $\widetilde{\varepsilon}_{ik}$ is the permittivity tensor.

By simple reorganization we obtain the field equations of static piezoelectricity:

$$\begin{aligned} C^E_{ijkl}u_{k,li} + e_{lij}\phi_{,li} &= 0, \\ e_{ikl}u_{k,li} - \widetilde{\varepsilon}_{il}\phi_{,li} &= 0. \end{aligned}$$

We now introduce the contracted notations in order to combine elastic and electric variables into generalized vectors and tensors [5]:

$$U_k = \begin{cases} u_k, & k = \overline{1,3}; \\ \phi, & k = 4, \end{cases} \quad T_j = \begin{cases} t_j, & j = \overline{1,3}; \\ q, & j = 4, \end{cases} \quad C_{ijkl} = \begin{cases} C^E_{ijkl}, & i,j,k,l = \overline{1,3}; \\ e_{lij}, & i,l,j = \overline{1,3}, k = 4; \\ e_{ikl}, & i,l,k = \overline{1,3}, j = 4; \\ -\widetilde{\varepsilon}_{il} & i,l = \overline{1,3}, k = j = 4, \end{cases}$$

where $q = D_j n_j$ is the surface charge density.

With this contracted notations the coupled piezoelectric field equations take the following form:

$$C_{ijkl}U_{k,il} = 0,\quad i,l = \overline{1,3},\ j,k = \overline{1,4},$$

which is essentially the same as in the elastic case. This allows us to formulate the generalized boundary-value problem for linear anisotropic elastic or piezoelectric solid occupying the domain $\Omega \subset R^3$ with boundary $\Gamma = \partial\Omega$:

$$C_{ijkl}U_{k,il} = 0, \quad \Omega \subset R^3, \quad i,l = \overline{1,3}, \; j,k = \overline{1,N},$$

with boundary conditions:

$$U_k = \tilde{U}_k \text{ on } \Gamma^U; \; T_k = \tilde{T}_k \text{ on } \Gamma^T,$$

where N denotes the dimension of the problem; $N = 3$ for elasticity and $N = 4$ for piezoelectricity.

To deal with this generalized boundary-value problem we use the boundary integral equation (BIE) method. Based on the Betti's reciprocal work theorem and Somigliana's identity the direct approach BIE is:

$$c_{jk}(\mathbf{y})U_k(\mathbf{y}) = \int_\Gamma g^s_{jk}(\mathbf{x},\mathbf{y})T_k(\mathbf{x})d\Gamma(\mathbf{x}) - \text{p.v.}\int_\Gamma h^s_{jk}(\mathbf{x},\mathbf{y})U_k(\mathbf{x})d\Gamma(\mathbf{x}), \; \mathbf{y} \in \Gamma, \quad (1)$$

where p.v. denote the principal-value integral; c_{jk} are the free term constants; $g^s_{ij}(\mathbf{x},\mathbf{y})$ is the static Green's matrix.

The traction fundamental solution is determined by

$$h^s_{jm}(\mathbf{x},\mathbf{y}) = C_{ijkl}\, g^s_{km,l}\, n_i,$$

where n_i are the components of unit outward vector normal to the boundary Γ at the point $\mathbf{x}$.

2. Boundary Element Method Formulation and Static Green's Functions

For numerical implementation of Equation (1) we employ matched boundary element approach based on regularized representation of BIE (1). Eight-node biquadratic quadrilateral elements are used to discretize the boundary surface, the field variables (generalized displacements and tractions) are approximated inside the boundary elements by linear and constant shape functions, respectively. To successfully apply the proposed BEM formulation to static anisotropic elastic or piezoelectric problems, we need corresponding fundamental solutions. Thus, evaluation of g^s_{ij} and h^s_{ij} is the main issue of the BEM.

Following the article by Wang and Achenbach [1] one can obtain a line integral expression for static anisotropic elastic Green's matrix:

$$g_{jk}^{S}(\mathbf{x},\mathbf{y}) = g_{jk}^{S}(\mathbf{r}) = \frac{1}{8\pi^2}\int_{|\mathbf{n}|=1}\Gamma_{jk}^{-1}(\mathbf{n})\delta(\mathbf{n}\cdot\mathbf{r})\mathrm{d}S(\mathbf{n}) = \frac{1}{8\pi^2 r}\int_{|\mathbf{d}|=1}\Gamma_{jk}^{-1}(\mathbf{d})\mathrm{d}\Omega(\mathbf{d}), \quad (2)$$

$$\mathbf{r} = \mathbf{x}-\mathbf{y},\ \Gamma_{jk}(\mathbf{d}) = C_{ijkl}d_i d_l,\ \mathbf{d}(\varphi) = \mathbf{p}\cos(\varphi)+\mathbf{q}\sin(\varphi),\ \mathrm{d}\Omega(\mathbf{d}) = \{0\le\varphi\le 2\pi\},$$

where $\mathbf{p}$ and $\mathbf{q}$ are a set of orthonormal vectors in the plane normal to $\mathbf{e} = \mathbf{r}/|\mathbf{r}|$. This integral can be numerically evaluated by means of any appropriate quadrature.

To derive an alternative expression for $g_{jk}^{S}(\mathbf{r})$ we rewrite Equation (2) as

$$g_{jk}^{S}(\mathbf{r}) = \frac{1}{8\pi^2}\int_{\Omega}\Gamma_{jk}^{-1}(\mathbf{n})\delta(\mathbf{n}\cdot\mathbf{r})\mathrm{d}\Omega(\mathbf{n}) = \frac{1}{8\pi^2}\int_{\Omega}\frac{A_{jk}(\mathbf{n})}{D(\mathbf{n})}\delta(\mathbf{n}\cdot\mathbf{r})\mathrm{d}\Omega(\mathbf{n}),$$

where $A_{jk}(\mathbf{n})$ and $D(\mathbf{n})$ denote the adjoint matrix and determinant of $\Gamma_{jk}(\mathbf{n})$, respectively.

Applying the residue calculus, static fundamental solutions can be expressed in the following form [2]:

$$g_{jk}^{S}(\mathbf{r}) = -\frac{\mathrm{Im}}{2\pi r}\sum_{m=1}^{3}\frac{A_{jk}(\mathbf{p}+\zeta_m\mathbf{q})}{a_7(\zeta_m-\zeta_m^*)\prod_{k=1,k\ne m}^{3}(\zeta_m-\zeta_k)(\zeta_m-\zeta_k^*)}, \quad (3)$$

where $\mathrm{Im}\,\zeta_m > 0,\ m=\overline{1,3}$ are roots of the polynomial equation: $D(\mathbf{p}+\zeta\mathbf{q}) = 0$, ζ_m^* is the complex conjugate of ζ_m.

Expressions (2) and (3) were originally derived for anisotropic elasticity, but utilizing the contracted notation introduced in Section 1, they as well hold exactly for the case of linear piezoelectricity [3, 5].

Calculation of the static Green's functions should be efficient for successful implementation in BEM software. In spherical coordinates (r,θ,ϕ), the static fundamental solutions can be expressed as follows [6]:

$$g_{ij}^{S}(r,\theta,\phi) = \frac{H_{jk}^{S}(\theta,\phi)}{4\pi r}, \quad i,j=\overline{1,N};$$

where N denotes the dimension of the problem; $N=3$ for elasticity and $N=4$ for piezoelectricity.

All possible values of θ and ϕ are located inside the rectangular area $(-\pi\le\theta\le\pi,\ 0\le\phi\le\pi)$. Thus, value of g_{ij}^{S} at arbitrary θ, ϕ can be obtained by any appropriate interpolation scheme for calculation of H_{jk}^{S} [5, 4]. In the present BEM formulation we employ linear Lagrange interpolation.

With the values of g_{ij}^{S} evaluated by either of the two approaches described above, values of H_{jk}^{S} in the interpolation nodes can be calculated, obviously, as

$$H_{jk}^{S}(\theta,\phi) = 4\pi r g_{ij}^{S}(r,\theta,\phi), \quad i,j=\overline{1,N}.$$

In order to demonstrate the influence of anisotropy on accuracy of the fundamental solution evaluation, Green's function g_{11}^{S} visualizations in form of surfaces in case of observation points located on a unit sphere $(r=1,-\pi\le\theta<\pi, 0\le\phi<\pi)$ are given in Figure 1. Four materials with various degree of anisotropy are employed: isotropic steel, transverse isotropic PZT, orthotropic spruce and monoclinic graphite-epoxy [5]. Their respective stiffness tensors in Voigt notation are given in Table 1.

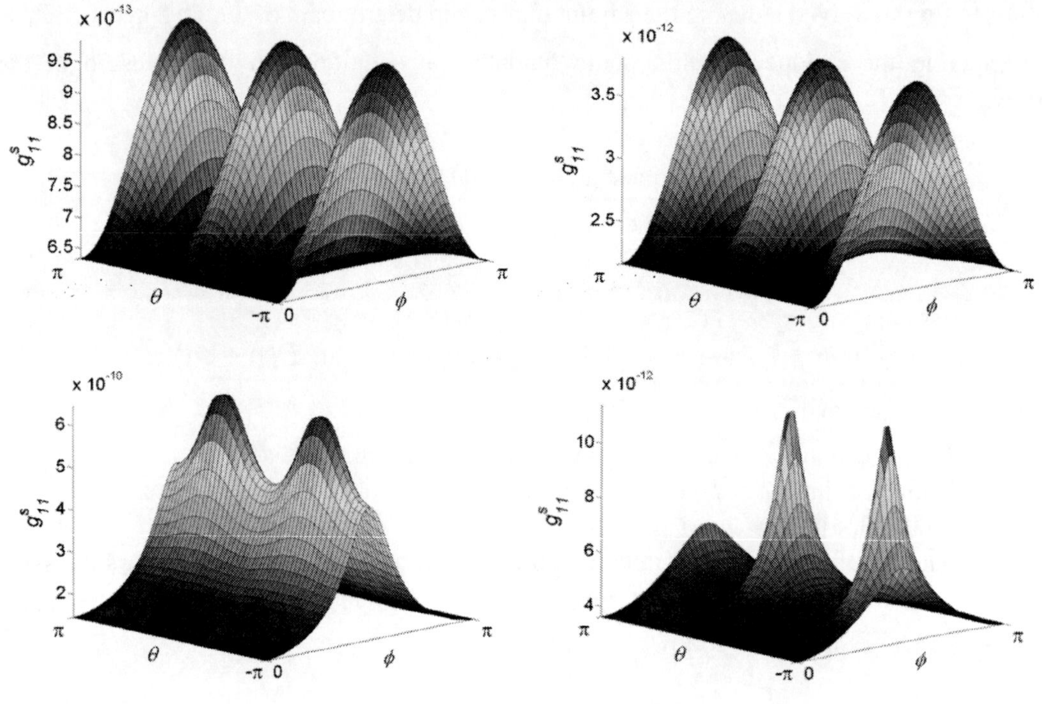

Figure 1. Elastostatic Green's function g_{11}^{S} for: steel, PZT, spruce and graphite-epoxy.

Table 1. Stiffness tensors C in GPa

Steel

282.7	121.2	121.2	0	0	0
121.2	282.7	121.2	0	0	0
121.2	121.2	282.7	0	0	0
0	0	0	80.8	0	0
0	0	0	0	80.8	0
0	0	0	0	0	80.8

PZT

107.6	63.1	63.9	0	0	0
63.1	282.7	63.9	0	0	0
63.9	63.9	100.4	0	0	0
0	0	0	19.6	0	0
0	0	0	0	19.6	0
0	0	0	0	0	22.2

Spruce

$$\begin{bmatrix} 0.44 & 0.32 & 0.19 & 0 & 0 & 0 \\ 0.32 & 16.27 & 0.45 & 0 & 0 & 0 \\ 0.19 & 0.45 & 0.78 & 0 & 0 & 0 \\ 0 & 0 & 0 & 0.61 & 0 & 0 \\ 0 & 0 & 0 & 0 & 0.039 & 0 \\ 0 & 0 & 0 & 0 & 0 & 0.76 \end{bmatrix}$$

Graphite-epoxy

$$\begin{bmatrix} 95.5 & 28.9 & 4.03 & 0 & 0 & 44.7 \\ 28.9 & 25.9 & 4.65 & 0 & 0 & 15.6 \\ 4.03 & 4.65 & 16.3 & 0 & 0 & 0.54 \\ 0 & 0 & 0 & 4.4 & -1.78 & 0 \\ 0 & 0 & 0 & -1.78 & 6.45 & 0 \\ 44.7 & 15.6 & 0.54 & 0 & 0 & 32.7 \end{bmatrix}$$

From these plots it is evident that a finer interpolation grid is required for a higher degree of anisotropy.

3. Numerical Examples

3.1. Anisotropic Elastic Cube under Uniform Surface Pressure

To assay the effect of the Green's functions interpolation grid size on the accuracy of boundary element (BE) solutions, the following anisotropic elastostatic problem is considered [5]. Elastic cube with side length of 0.01 m is clamped at one face and it undergoes to horizontal tractions $\widetilde{t}_1 = t_0$, $t_0 = -1 \cdot 10^8$ Pa applied to one-fifth part of other face (see Figure 2). The anisotropic static Green's functions are evaluated using linear Lagrange interpolation on grids of 100×100, 200×200, 400×400, and 800×800 nodes.

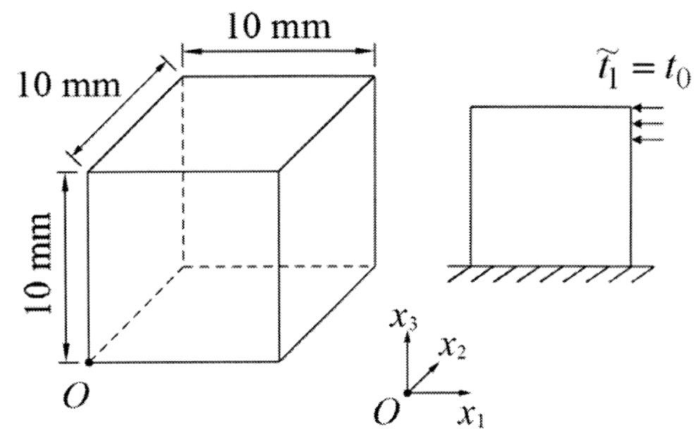

Figure 2. Elastostatic BEM analysis: geometry model and boundary conditions

Table 2. Values of displacements u_1 at the boundary point for steel

Interpolation grid	Displacement $u_1 \cdot 10^{-6}$ m			
	BE-mesh "a"	BE-mesh "b"	BE-mesh "c"	L. Gaul et al. [5]
100×100	-7.6274935	-7.5877104	-7.5546995	-7.489
200×200	-7.6435510	-7.6041117	-7.5711012	-7.506
400×400	-7.6476213	-7.6080357	-7.5752412	-7.511
Kelvin	-7.6463304	-7.6067718	-7.5740006	-7.512

Table 3. Values of displacements u_1 at the boundary point for PZT

Interpolation grid	Displacement $u_1 \cdot 10^{-6}$ m			
	BE-mesh “a”	BE-mesh “b”	BE-mesh “c”	L. Gaul et al. [5]
100×100	-28.963324	-28.699586	-28.532188	-28.73
200×200	-29.033982	-28.770367	-28.602077	-28.81
400×400	-29.051923	-28.787290	-28.619776	-28.83
800×800	-29.056498	-28.791714	-28.624155	-28.83

Table 4. Values of displacements u_1 at the boundary point for spruce

Interpolation grid	Displacement $u_1 \cdot 10^{-4}$ m			
	BE-mesh “a”	BE-mesh “b”	BE-mesh “c”	L. Gaul et al. [5]
100×100	-71.180752	-72.138776	-72.211399	-71.89
200×200	-71.663901	-72.592817	-72.650209	-72.33
400×400	-71.784404	-72.706129	-72.763814	-72.44
800×800	-71.815332	-72.734872	-72.791575	-72.47

Table 5. Values of displacements u_1 at the boundary point for graphite-epoxy

Interpolation grid	Displacement $u_1 \cdot 10^{-6}$ m			
	BE-mesh “a”	BE-mesh “b”	BE-mesh “c”	L. Gaul et al. [5]
100×100	-94.870807	-96.839414	-97.327060	-97.23
200×200	-95.219997	-97.192477	-97.685491	-97.79
400×400	-95.302926	-97.285045	-97.777935	-97.93
800×800	-95.322796	-97.306014	-97.799969	-97.96

Four different elastic materials mentioned in Section 2 are considered. In the case of isotropic material, the well-known closed-form Kelvin's fundamental solution was used instead of 800×800 grid. Three regular boundary element meshes were employed: mesh “a” with 150, mesh “b” with 600 and mesh “c” with 2400 elements. Displacements u_1 at the point with coordinates $(0.01, 0, 0.01)$ m were investigated. In the Tables 2 – 5 the obtained BE-solutions are present in comparison with the corresponding results of [5].

The obtained results indicate once more that a more refined Green's functions interpolation grid is needed when degree of anisotropy greatens.

3.2. Piezoelectric Half-space under Uniform Mechanic or Electric Loading

Let us treat uniform load applied to the surface of piezoelectric half-space [7]. Two loading cases, namely uniform vertical tractions $\tilde{t}_3 = P = 121 \cdot 10^9$ GPa and surface charge $\tilde{q} = Q = 15.8$ C/m^2 within the circular region $r = a = 1$ m, are considered (see Figure 3). Observation points are located inside the half-space, in the case of mechanical loading their coordinates are $(x_1/a \in [0,4],\ x_2/a = 0,\ x_3/a = -1)$, in the case of electric loading their coordinates are $(x_1/a = 0,\ x_2/a = 0,\ -x_3/a \in [0,4])$. Generalized displacements at the points inside the half-space are calculated via Somigliana's identity. Static piezoelectric Green's functions are evaluated using linear Lagrange interpolation on 800×800 grid.

The piezoelectric material of the half-space is transversely isotropic PZT-5A with the following constants [7]:

$$\mathbf{C} = \begin{bmatrix} 121 & 75.4 & 75.2 & 0 & 0 & 0 \\ 75.4 & 121 & 75.2 & 0 & 0 & 0 \\ 75.2 & 75.2 & 111 & 0 & 0 & 0 \\ 0 & 0 & 0 & 21.1 & 0 & 0 \\ 0 & 0 & 0 & 0 & 21.1 & 0 \\ 0 & 0 & 0 & 0 & 0 & 22.8 \end{bmatrix} \text{GPa},$$

$$\mathbf{e} = \begin{bmatrix} 0 & 0 & 0 & 0 & 12.3 & 0 \\ 0 & 0 & 0 & 12.3 & 0 & 0 \\ -5.4 & -5.4 & 15.8 & 0 & 0 & 0 \end{bmatrix} \text{C/m}^2,\ \tilde{\boldsymbol{\varepsilon}} = \begin{bmatrix} 8.11 & 0 & 0 \\ 0 & 8.11 & 0 \\ 0 & 0 & 7.35 \end{bmatrix} \cdot 10^{-9}\ \text{C/V}\cdot\text{m}$$

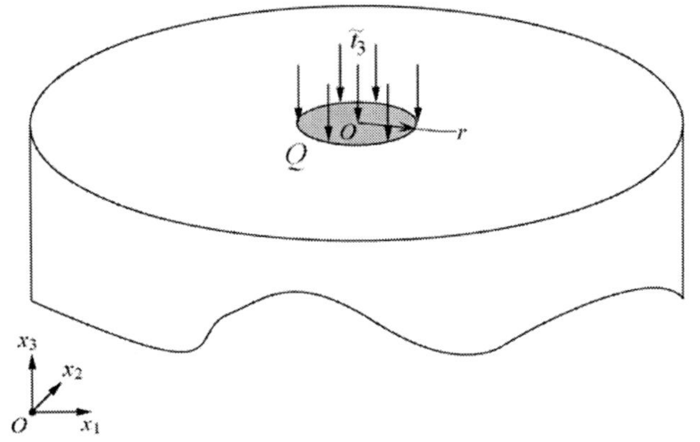

Figure 3. Static piezoelectric BEM analysis: geometry model and boundary conditions.

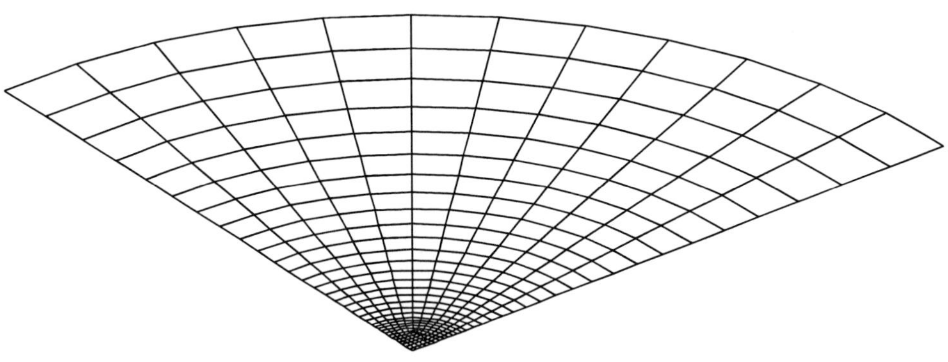

Figure 4. BE-mesh for quarter of the piezoelectric half-space surface.

Due to the symmetry of the problem and the material, only a quarter of the half-space surface was discretized. The employed BE-mesh consisted of 348 elements and 387 nodes (see Figure 4).

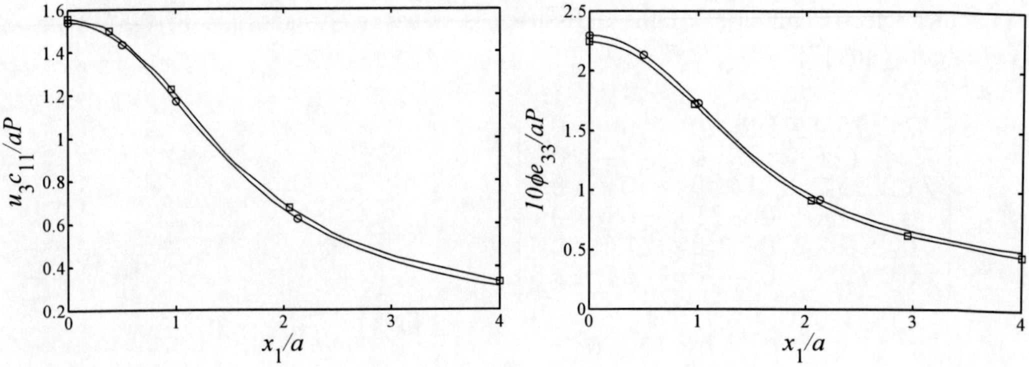

Figure 5. Normalized vertical displacements and electric potential in the case of mechanical load; ○ – obtained BEM results, □ – BE-solutions from [7].

Figures 5, 6 present the obtained BEM results for u_3 and ϕ in comparison with the BEM solutions from [7]. It should be noted that Liew et al. [7] used closed-form fundamental solutions for transversely isotropic piezoelectric solids [8].

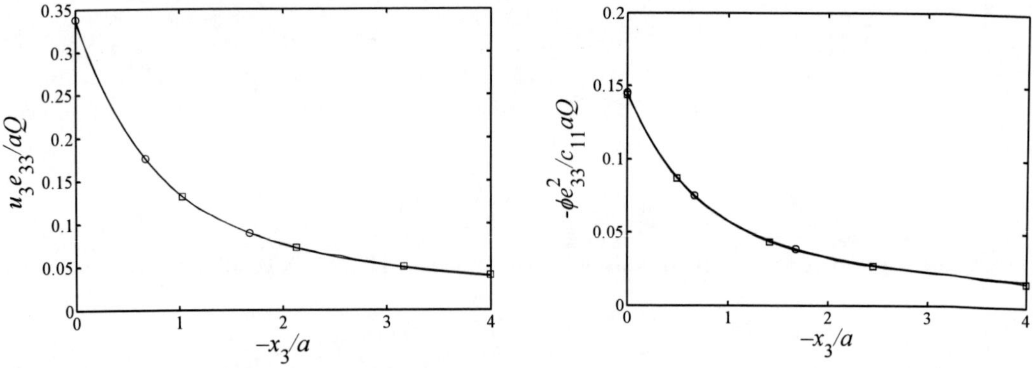

Figure 6. Normalized vertical displacements and electric potential in the case of electrical load; ○ – obtained BEM results, □ – BE-solutions from [7].

It can be observed that the BEM solutions obtained on the base of interpolation scheme for calculation of fundamental solutions are only slightly different from those obtained by other authors who employed closed-form expressions for Green's functions. Thus we can draw a conclusion that the proposed boundary element approach produces highly accurate results.

ACKNOWLEDGMENTS

The work was supported by RFBR under grants Nos.12-08-00984, 12-01-00698, 13-08-00658, 13-08-97091, 14-08-31415, 14-08-31410.

REFERENCES

[1] Wang, C. Y.; Achenbach, J. D., *Proc. R. Soc. Lond. A*, 1995, *vol. 449*, 441.

[2] Tonon, F.; Pan, E.; Amadei, B., *Computers and Structures*, 2001, *vol. 79*, 469.

[3] Pan, E.; Tonon, F., *Int. J. Solids. Struct.*, 2000, *vol. 37*, 943.

[4] Wilson, R. D.; Cruse, T. A., *International Journal for Numerical Methods in Engineering*, 1978, *vol. 12*, 1383.

[5] Gaul, L.; Kogl, M.; Wagner, M., *Boundary Element Methods for Engineers and Scientists,* Springer, Berlin, 2003.

[6] Ting, T. C. T.; Lee, V. G., *Q. J. Mech. Appl. Math*, 1997, *vol. 50*, 407.

[7] Liew, K. M.; Liang, J., *Computational Mechanics*, 2003, *vol. 32*, 29.

[8] Dunn, M. L.; Wienecke, H. A., *Int. J. Solids Structures*, 1996, *vol. 33*, 4571.

In: Advanced Materials - Studies and Applications
Editors: I.A. Parinov, S.-H. Chang, S.Theerakulpisut ISBN: 978-1-63463-749-7

Chapter 18

TREATING COUPLED BOUNDARY VALUE PROBLEMS OF 3D ELASTODYNAMICS BY MEANS OF BOUNDARY INTEGRAL EQUATIONS (BIE) METHOD

L. A. Igumnov**, *Ya. Yu. Rataushko and A. A. Ipatov
Research Institute for Mechanics,
Lobachevsky State University of Nizhni Novgorod,
Nizhny Novgorod, Russia

ABSTRACT

Consider solution schemes for boundary value problems of 3d linear elastodynamics with coupled fields of mechanical and non-mechanical nature. Field interaction models correspond to poroelasticity, thermoelasticity and electro-magneto-elasticity. For porous materials Biot's model is used. Special attention is drawn to the static Green's & Neumann's matrices construction: several approaches are presented. Two approaches to BIE construction – classical and non-classical – are presented. A non-classical approach of boundary integral equations (BIE) method, which allows solving dynamic problems not only in isotropic case, but in anisotropic case too, is considered. Boundary element method (BEM) scheme for BIE is based on regular integral equation of the first kind (integral equation on the plane wave). Numerical results, based on integral equation on the plane wave, for three dimensional dynamic problems are presented. Several numerical experiments are presented: problems of dynamic pressure acting on the surface of unit sphere, unit cube etc. are treated. Results obtained with the help of non-classical approach are given in comparison with the ones of classical approach.

1. INTRODUCTION

Classical formulations for BIE method with their discretized realization and traditional BEM are successful approaches for solving three-dimensional isotropic problems of dynamic

* igumnov@mech.unn.ru.

elasticity. Extension of BIE method and BEM to anisotropic problems and problems, where material models are different from elastic, needs development of special new schemes. Today there are two fundamentally different approaches for constructing such numerical schemes. The first one (classical) is based on using Green-Betti-Somigliana formula and results in the singular BIE. It can rely on dual reciprocity theorem, introduced by Nardini and Brebbia [1]. The second one (non-classical) uses dynamic fundamental solutions and arises from Babeshko paper [2]. It leads to the regular BIE system; for anisotropic models these regular BIE represent exact solution as compared to the approach of double reciprocity theorem application.

2. Problem Formulations

In this section a brief overview of basic theories is given, where non-classical approach can be used.

2.1. Anisotropic Elasticity

The set of linear Lamé equations with respect to second order partial derivatives is as follows:

$$Lu = 0, \quad L = L_0(\partial x_i, \partial x_j) - \partial_t^2, \quad \partial_t = \partial / \partial t, \quad \partial x_i = \partial / \partial x_i, \quad i, j = 1,2,3.$$

Boundary conditions are:

$$(l^0 u)_i \equiv u_i \big|_{\partial\Omega}, (l^1 u)_i \equiv L_0(\partial x_i, \nu), \quad i = 1,2,3,$$

where $\nu \equiv (\nu_1, \nu_2, \nu_3)$ is the outer normal, $\Omega \subset R^3$ is a domain with the smooth boundary $\Gamma = \partial\Omega$.

2.2. Electroelasticity

Differential equations of linear electroelastodynamics (the coupled field is electric potential) take the following form:

$$C_{ijkl} u_{k,li} + e_{lij} \varphi_{,li} = \rho \ddot{u}, \quad e_{ikl} u_{k,li} - \tilde{\varepsilon}_{il} \varphi_{,li} = 0$$

where e_{lij}, $\tilde{\varepsilon}_{il}$ are the tensors of dielectric permittivity and piezoelectric constants, respectively.

2.3. Thermoelasticity

Differential equations of linear thermoelastodynamics (the coupled field is temperature) are as follows:

$$k_{ij}\theta_{,ij} = \theta_0\beta_{ij}\dot{u}_{i,j} + c^{\varepsilon}\dot{\theta} - \tau\dot{q}_{j,j} - \rho R\,;\; C_{ijkl}u_{k,lj} = \rho\ddot{u}_i + \beta_{ij}\theta_{,j} - b_i,$$

where c^{ε} is the specific heat, k_{ij} is the tensor of heat conduction, q is the thermal flux, β_{ij} is the tensor of heat stress, τ is the thermal relaxation time.

2.4. Poroelasticity

Motion equations and formulae of Biot's poroelasticity model have the forms:

$$\rho\ddot{u}_i + \rho_f\ddot{w}_i = \sigma_{ij,j} + \rho f_i;\;\; \rho_f\ddot{u}_i + m_{ij} * \ddot{w}_j = -p_{,i};$$
$$\sigma_{ij} = C_{ijkl}e_{kl} + M^0_{ij}\zeta;\;\; p = M^0_{ij}e_{ij} + M_0\zeta;\;\; \dot{\zeta} = -\text{div}\dot{w} + R,$$

where * denotes convolution, ρ_f is the density of filler, p is the pore pressure deflection, w is the filler displacement relative to the elastic solid skeleton, ζ is the bulk modulus of the filler, M_{ij}, M are the Biot's tensor and modulus, $m_{ij}(t)$ is the linear viscodynamic operator.

3. Non-Classical Approach of BIE Method

The following equations are applicable in dynamic problems [4]:

$$\gamma\hat{A}_\alpha \otimes \hat{A}_\alpha \int_\Gamma \left[l^1 e^{ik_\alpha n(x-y)} l^0\psi(y) - e^{ik_\alpha n(x-y)} l^1\psi(y)\right] d_y\Gamma = 0; \tag{1}$$

$$\gamma\hat{A}_\alpha \otimes \hat{A}_\alpha \int_\Gamma \left[l^1\delta(\tau) * l^0\psi(y,t) - \delta(\tau) * l^1\psi(y,t)\right] d_y\Gamma = 0, \tag{2}$$

$$\det A(k_\alpha) = 0;\; A(k,\omega,n)\hat{A} = 0;\; A^*(k_\alpha \operatorname{sgn}\xi) = trA^*(k_\alpha)\hat{A}_\alpha(\operatorname{sgn}\xi) \otimes \hat{A}_\alpha(\operatorname{sgn}\xi);$$
$$\tau = t - \frac{n(x-y)}{\upsilon_\alpha};$$

$$\pm\gamma c\sum_{\alpha=1}^{N} ig_S(k_\alpha)\int_\Gamma \left[l^1 e^{ik_\alpha n(x-y)} l^0\psi(y) - e^{ik_\alpha n(x-y)} l^1\psi(y)\right] d_y\Gamma = 0; \tag{3}$$

$$\pm \gamma c \sum_{\alpha=1}^{N} i g_S(k_\alpha) \int_\Gamma \left[l^1 \delta(\tau) * \psi(y,t) - \delta(\tau) * l^1 \psi(y,t) \right] d_y \Gamma = 0. \tag{4}$$

Thus, solving a problem with BIE method means that corresponding Equations (1) – (4) are fulfilled for eigenvalues k_α and eigenvectors $\hat{A}_\alpha$.

Green's tensor for linear anisotropic elasticity (as well used in electro-, thermo-, poro-elastic formulations) takes the following form [3]:

$$U(x,\omega) = -c\Delta \int_{S^2} \bar{u}\, dS(n) = -c\Delta \int_{S^2} G(nx,\omega,n) F^{(0)} dS(n);$$

$$G(\xi) = \sum_{\alpha=1}^{N} \frac{e^{ik_\alpha|\xi|}}{2ik_\alpha \lambda_\alpha} A_\alpha(\operatorname{sgn}\xi) \otimes A_\alpha(\operatorname{sgn}\xi), \quad \xi = nx;$$

$$U(x,t) = (2\pi)^{-1} \int_{-\infty}^{+\infty} U(x,\omega) e^{-i\omega t} d\omega, \; \lambda_\alpha = \frac{1}{2} A_\alpha (\upsilon_\alpha^2 M + K) A_\alpha, \quad \upsilon_\alpha = \omega / k_\alpha,$$

where

$K = L_0(n), M = \rho I,\ F^{(0)} = 1;\ g_S = \bar{u}$; ρ is the density and I is the identity matrix, for anisotropic elasticity;

$$F^{(0)} = \rho\left[f^{(0)} - R^{(0)} \mu^{-1} p\right];\ \ K = Q + \mu^{-1} p \otimes p;\ \ M = \rho I;\ \ p_i(n) = e_{kij} n_i n_j;$$

$\mu(n) = \tilde{\varepsilon}_{ij} n_i n_j$, for electroelasticity;

$$F^{(0)} = \left\{\rho f^{(0)}, \rho R^{(0)}/\theta_0\right\}^T; \qquad K = \begin{bmatrix} Q(n) & 0 \\ 0 & k(n) \end{bmatrix}; \qquad M = \begin{bmatrix} \rho I & 0 \\ 0 & \rho c/\theta_0 \end{bmatrix};$$

$k(n) = \left[\theta_0 \left(\tau + (-i\omega)^{-1}\right)\right]^{-1} k_{ij} n_i n_j$, for thermoelasticity;

$$F^{(0)} = \left\{\rho f^{(0)}, R^{(0)}\right\}^T; \qquad K = \begin{bmatrix} \tilde{Q}(n) & 0 \\ 0 & k(n) \end{bmatrix}; \qquad M = \begin{bmatrix} \rho I - \rho_f^2 m^{-1} & 0 \\ 0 & M_0^{-1} \end{bmatrix};$$

$$k(n) = \left(m^{-1}\right)_{ij} n_i n_j; \qquad \tilde{Q}_{ik}(n) = \left(C_{ijkl} - M_0^{-1} M_{ij}^0 M_{kl}^0\right) n_j n_l;$$

$b_i(n) = -\left[M_0^{-1} M_{ij}^0 + \rho_f \left(m^{-1}\right)_{ij}\right] n_j$, for poroelasticity.

Babeshko [2] first introduced this approach to boundary integral equations construction. This method was then applied by Vatulyan [5].

By satisfying the following condition:

$$\frac{iA^*(k_\alpha \operatorname{sgn}\xi)}{\left[\det A(k)\right]'\Big|_{k=k_\alpha}} = \frac{1}{2ik_\alpha\lambda_\alpha} A_\alpha(\operatorname{sgn}\xi)\otimes A_\alpha(\operatorname{sgn}\xi),$$

we obtain for elastic case from (1) BIE coinciding (in case of $x = 0$) with the equation from [2, 5].

For isotropic case of elasticity we have the following equation [6]:

$$\int_S U(x,n)e_u^1(x)dS + \int_S T(x,n)e_u^0(x)dS = 0, \tag{5}$$

$$T = \mu is\cdot\begin{Vmatrix} Am_1+Bn_1 & Am_2+Bn_2 & Am_3+Bn_3 \\ Dn_1-Bn_3 & Dn_2 & Bn_1+Dn_3 \\ Cn_1 & Bn_3+Cn_2 & Cn_3-Bn_2 \end{Vmatrix};\ U = \mathrm{e}\cdot\begin{Vmatrix} n_1 & n_2 & n_3 \\ n_3 & 0 & -n_1 \\ 0 & -n_3 & n_2 \end{Vmatrix};$$

$$s = \operatorname{diag}(-2k_1e_1, k_2e_2, k_2e_2);\ \mathrm{e} = \operatorname{diag}(e_1, e_2, e_2);\ e_j = \exp\left[ik_j nx\right];$$

$$A = \frac{\nu}{1-2\nu},\ B = mn;\ C = n\times i_1 m,\ D = n\times i_2 m.$$

It is possible to build BEM scheme for solving dynamic boundary-value problems with the help of formulae (1) – (4).

4. Green's Matrix Calculation

Mathematically Green's matrix can be defined as follows:

$$\hat{C}_{ijkl}G_{km,li}(x) = -\delta_{jm}\delta(x),\quad i,j,k,l,m = \overline{1,3}.$$

There are several ways to obtain the Green's matrix components.

4.1. Integral Approach

Matrix G_{jk} can be written as follows [7]:

$$G_{jk} = \frac{1}{8\pi^2 r}\int_0^{2\pi}(\mathcal{M}_{jk}^{kk}(z_i(\phi)))^{-1}d\phi$$

Neumann's matrix – first order derivative of the Green's matrix – takes the following form [7]:

$$G_{jk,s} = \frac{1}{8\pi^2 r^2}\int_0^{2\pi}(-r_s^0(\mathcal{M}_{jk}^{kk})^{-1} - z_s F_{jk})d\phi;$$

$$F_{jk} = (\mathcal{M}_{in}^{zz})^{-1}D_{nq}(\mathcal{M}_{qk}^{zz})^{-1};\ D_{nq} = \mathcal{M}_{nq}^{r^0 z} + \mathcal{M}_{nq}^{zr^0};$$

$$\mathcal{M}_{ik}^{zr^0} = C_{ijkl}z_j r_l^0;\ \mathcal{M}_{ik}^{r^0 z} = C_{ijkl}r_j^0 z_l. \tag{6}$$

These integrals can be numerically evaluated with the help of any appropriate quadrature.

4.2. Polynomial Approach

Inverse matrix $\left(M^{zz}(z)\right)^{-1}$ can be represented in terms of adjoint matrix $A_{jk}(z)$ and determinant $D(z)$. Due to the Green's matrix takes the following form [8]:

$$G_{jk} = \frac{1}{8\pi^2}\int_{S^2}\frac{A_{jk}(z_i)}{D(z_i)}\delta(z_i r_i)dS(z_i) = \frac{1}{4\pi^2 r}\int_{-\infty}^{+\infty}\frac{A_{jk}(p+\zeta q)}{D(p+\zeta q)}d\zeta.$$

With the help of the residue theorem, integral (6) can be expressed in the following form [9]:

$$G_{jk}(x) = -\frac{\operatorname{Im}}{2\pi r}\sum_{m=1}^{4}\frac{A_{jk}(p+\zeta_m q)}{a_9(\zeta_m - \zeta_m^*)\prod_{k=1,k\neq m}^{4}(\zeta_m - \zeta_k)(\zeta_m - \zeta_k^*)},$$

where ζ_m are the roots of $D(p+\zeta q) = 0$ being the octic polynomial in terms of ζ; a_9 is the coefficient of ζ^8; $\operatorname{Im}\zeta_m > 0$, $m = \overline{1,4}$; ζ_m^* is the complex conjugate to ζ_m. The Green's matrix derivatives can be obtained with the help of Lagrange interpolation.

For an arbitrary point $x = (x_1, x_2, x_3)$, the Green's tensor G_{JK} derivatives are:

$$\frac{\partial G_{PK}}{\partial x_1} \approx \frac{1}{2h}[G_{PK}(x_1+h, x_2, x_3) - G_{PK}(x_1-h, x_2, x_3)];$$
$$\frac{\partial G_{PK}}{\partial x_2} \approx \frac{1}{2h}[G_{PK}(x_1, x_2+h, x_3) - G_{PK}(x_1, x_2-h, x_3)];$$
$$\frac{\partial G_{PK}}{\partial x_3} \approx \frac{1}{2h}[G_{PK}(x_1, x_2, x_3+h) - G_{PK}(x_1, x_2, x_3-h)].$$

The choice of range h has strong influence on the scheme accuracy. In this paper we choose $h = r \cdot 10^{-3}$.

4.3. Green's Matrices in Spherical Coordinates

Let us consider a spherical coordinate system:

$$x_1 = r\sin\theta_1 \cos\theta_2;\ x_2 = r\sin\theta_1 \sin\theta_2;\ x_3 = r\cos\theta_1.$$

In spherical coordinates (r, θ, ϕ) the static fundamental solutions can be expressed as follows [10]:

$$G_{mk}(r, \theta_1, \theta_2) = \frac{1}{r} G^u_{mk}(\theta_1, \theta_2);\ G_{mk,l} = \frac{1}{r^2} G^u_{mkl}(\theta_1, \theta_2);$$

$$0 \le \theta_1 \le \pi;\ 0 \le \theta_2 \le 2\pi.$$

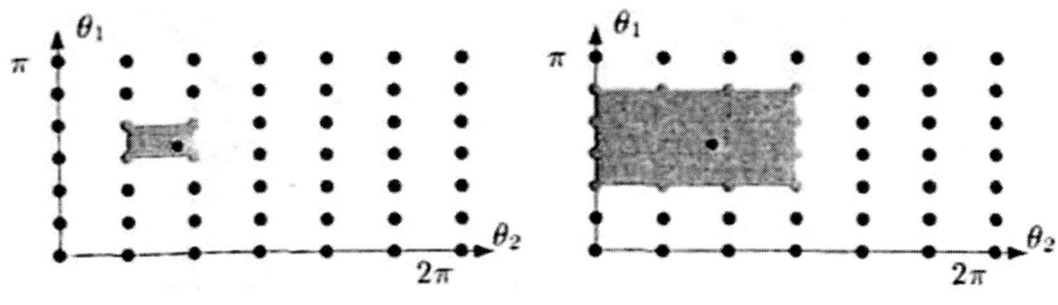

Figure 1. Interpolation grid.

Values of θ_1 and θ_2 are located inside the rectangular area ($0 \le \theta_1 \le \pi, 0 \le \theta_2 \le 2\pi$) (Figure 1). Thus value of G_{mk} at arbitrary θ_1, θ_2 can be obtained by any appropriate interpolation scheme.

4.4. Fundamental Solutions for Anisotropic Elastostatics

In this section, visualizations for some fundamental solution components for different materials are given. Considered materials are: steel, transversely isotropic piezoceramics, orthotropic wood (spruce), monoclinic graphite-epoxy (see Table 1).

Table 1. Anisotropic material properties

Material	Elastic constants, count	Density, kg/m^3
Steel	2	7850
Transversely isotropic piezoceramics	5	7800
Orthotropic wood (spruce)	9	430
Monoclinic graphite-epoxy	13	1600

The stiffness matrices for these materials can be found in [7]. In Figures 2 – 5 some components of Green's and Neumann's matrices are present.

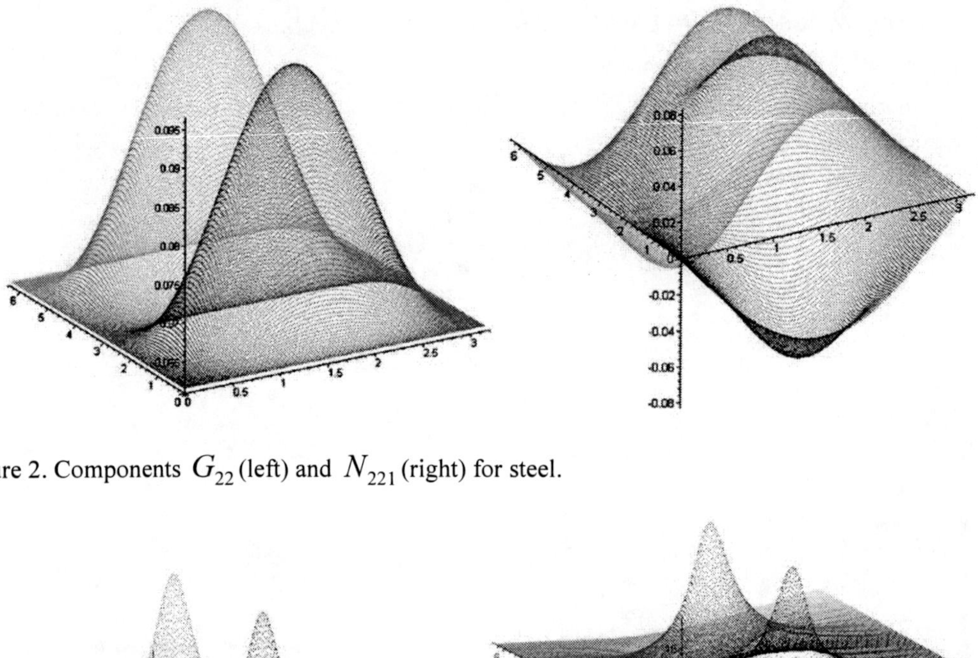

Figure 2. Components G_{22} (left) and N_{221} (right) for steel.

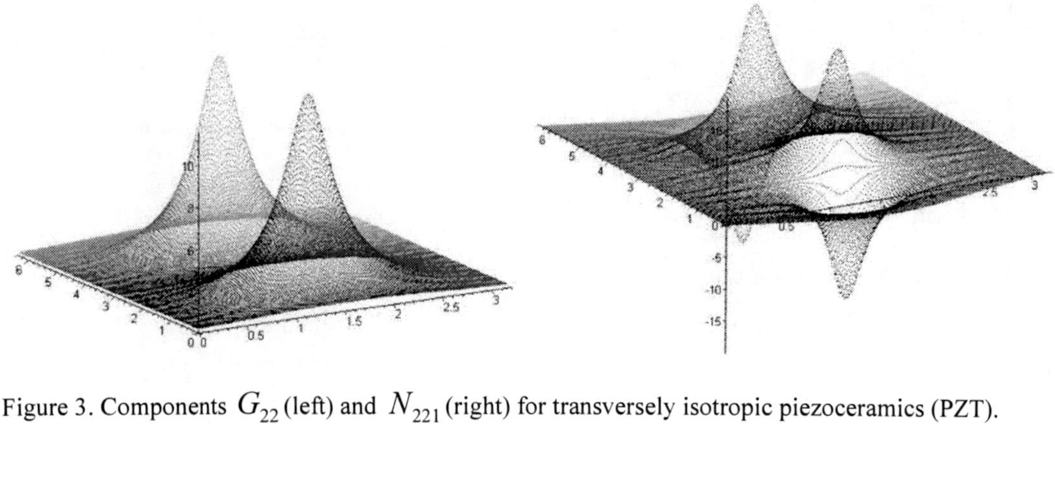

Figure 3. Components G_{22} (left) and N_{221} (right) for transversely isotropic piezoceramics (PZT).

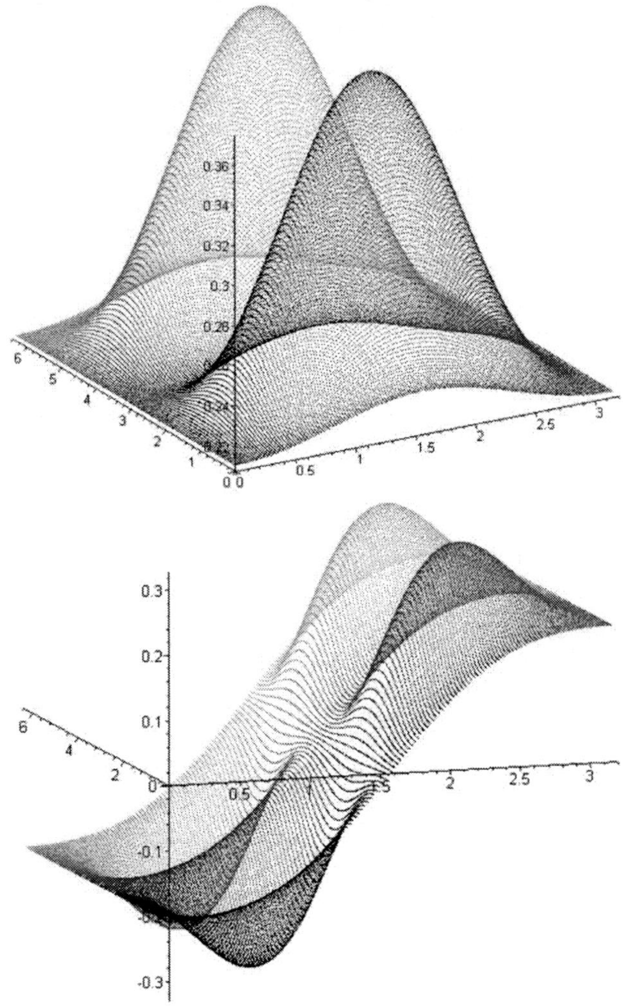

Figure 4. Components G_{22} (upper) and N_{223} (bottom) for orthotropic spruce.

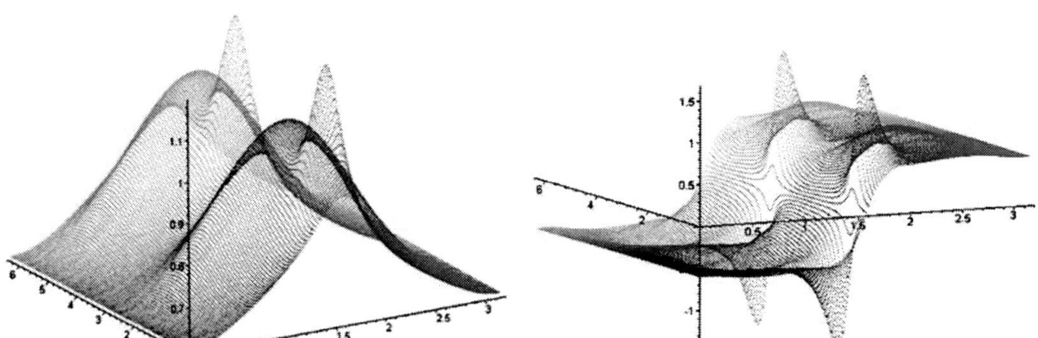

Figure 5. Components G_{22} (left) and N_{223} (right) for monoclinic graphite-epoxy.

5. Numerical Results

First numerical results on the base of Equation (5) were obtained in [6] for the problem of sustained torsional vibrations of isotropic elastic cylindrical body. Numerical results for three-dimensional elastodynamics are still absent.

5.1. Problem of Dynamic Pressure Acting on Surface of Unit Sphere

The problem of dynamic pressure:

$$P(t)=\begin{cases} t,\ 0<t<1; \\ 1,\ t\geq 1, \end{cases} \tag{7}$$

acting on the surface of a unit sphere is considered. Non-classical BIE approach is used. Dimensionless Lamé parameters and density are: $\lambda=6$, $\mu=3$, $\rho=7$. The geometry of the problem and load P(t) visualization are present in Figure 6.

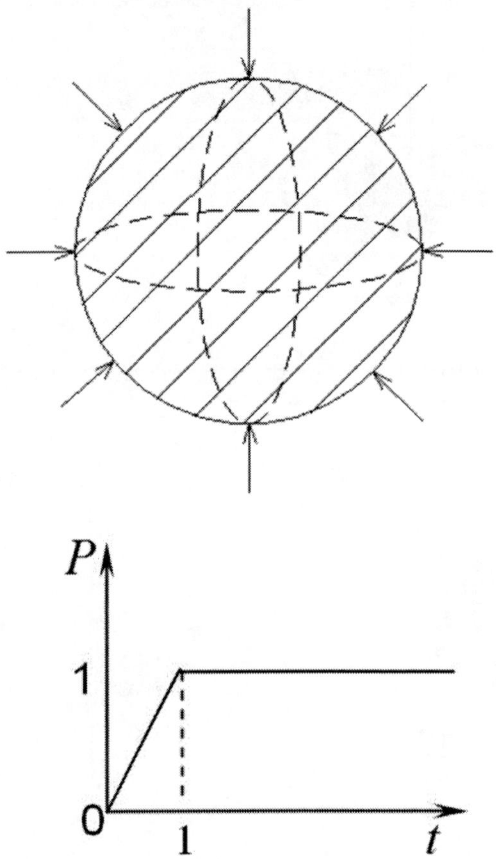

Figure 6. Geometry of the problem and load visualization.

The sphere is covered with 384 quadrilateral boundary elements. A variant of numerical BEM scheme for classical kernels is given in [11]. BEM scheme for solving (5) is based on the scheme from [11], but there are two significant differences: (i) Gaussian quadrature of a fixed order is employed (hierarchic integrating algorithm is not used), (ii) Tikhonov's regularization is used to solve the final system of linear equations.

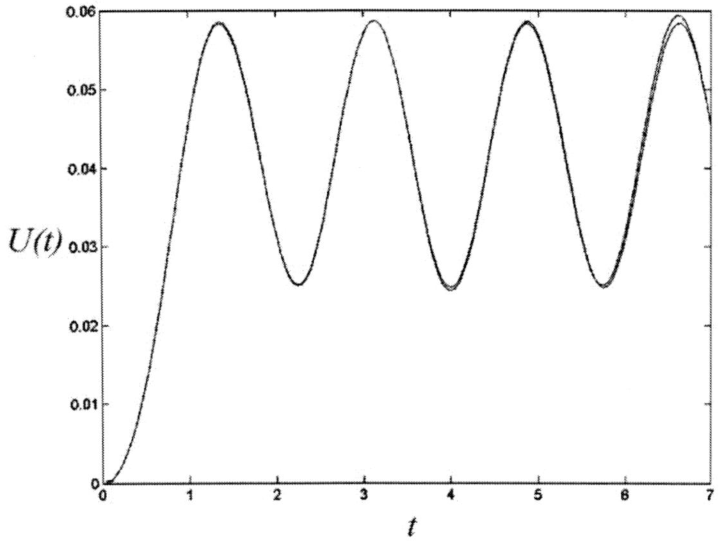

Figure 7. Displacements at point of the sphere surface.

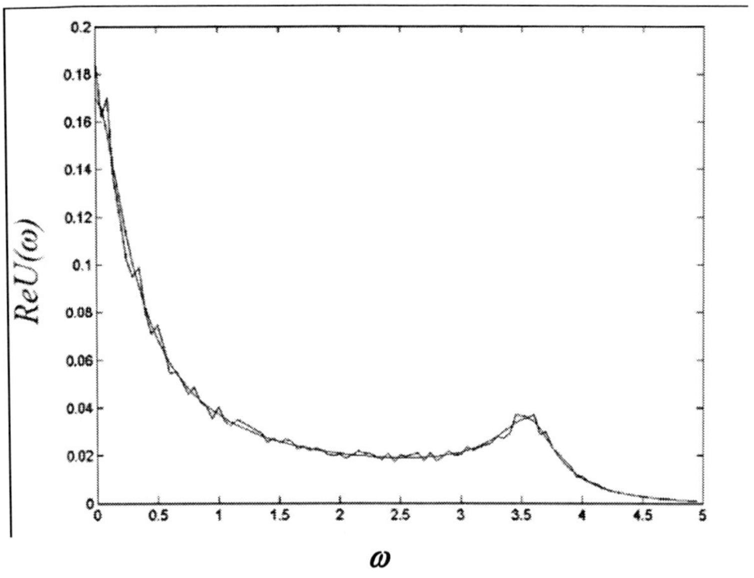

Figure 8. Real part of amplitude-frequency function for displacements.

Displacement response at a point of the sphere surface is shown in Figure 7; the lower line depicts the classical BEM scheme based results. The curves obtained with the help of classical and non-classical BIE show good correspondence. Parameterized by means of the

Tikhonov's regularization the real part of the amplitude-frequency function for displacements is shown in Figure 8, its imaginary part presented in Figure 9, and absolute value of the function demonstrated in Figure 10. Weakly oscillating curves correspond to the solution obtained by means of non-classical approach, smooth ones relate to the solution found by means of classical BEM.

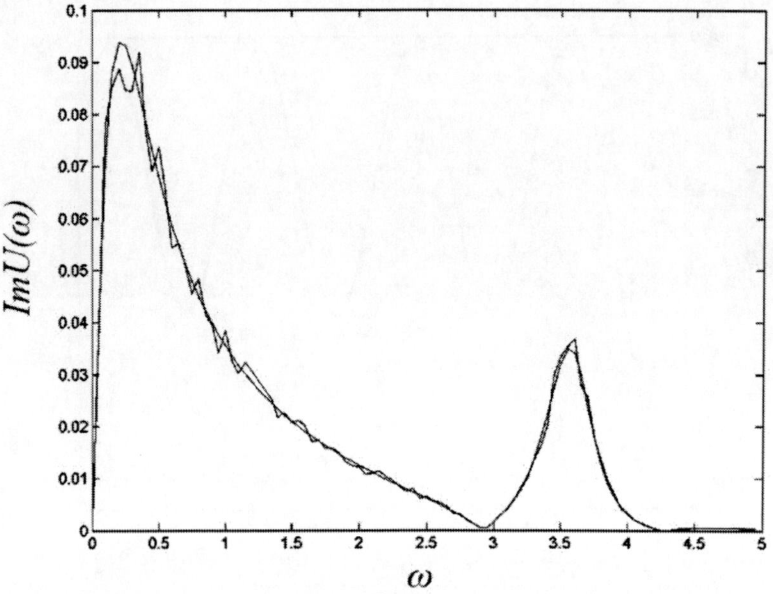

Figure 9. Imaginary part of amplitude-frequency function for displacements.

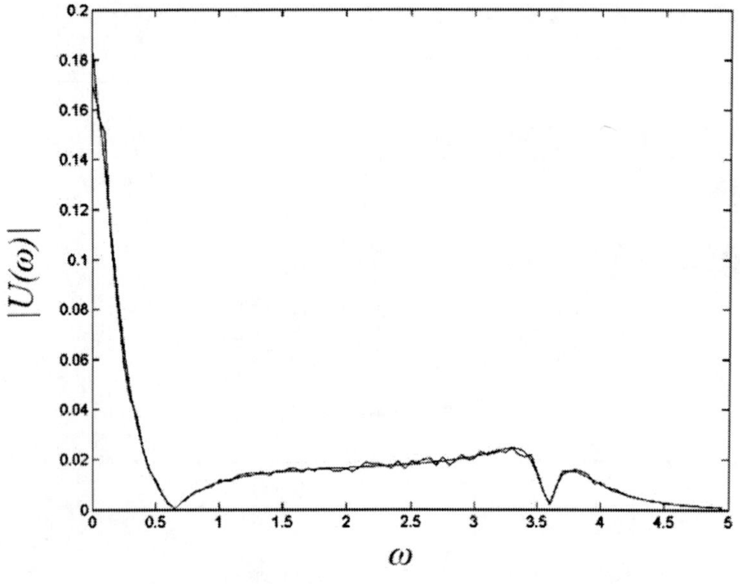

Figure 10. Absolute value of amplitude-frequency function for displacements.

5.2. Problem of Dynamic Pressure Acting on Surface of Unit Cube

The considered cube and its reference points are shown in Figure 11. Dynamic load P(t) from (7) is applied to the surface. Regular boundary element mesh of 216 elements is used. Dimensionless Lamé parameters and density are : $\lambda = 6$, $\mu = 3$, $\rho = 7$.

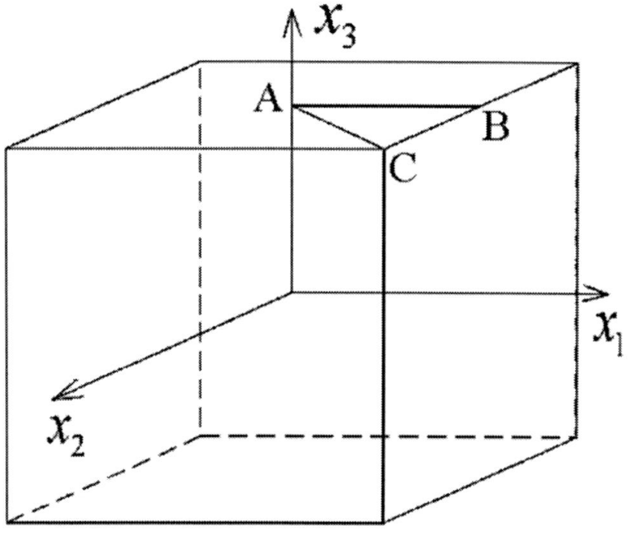

Figure 11. Geometry of the problem and reference points.

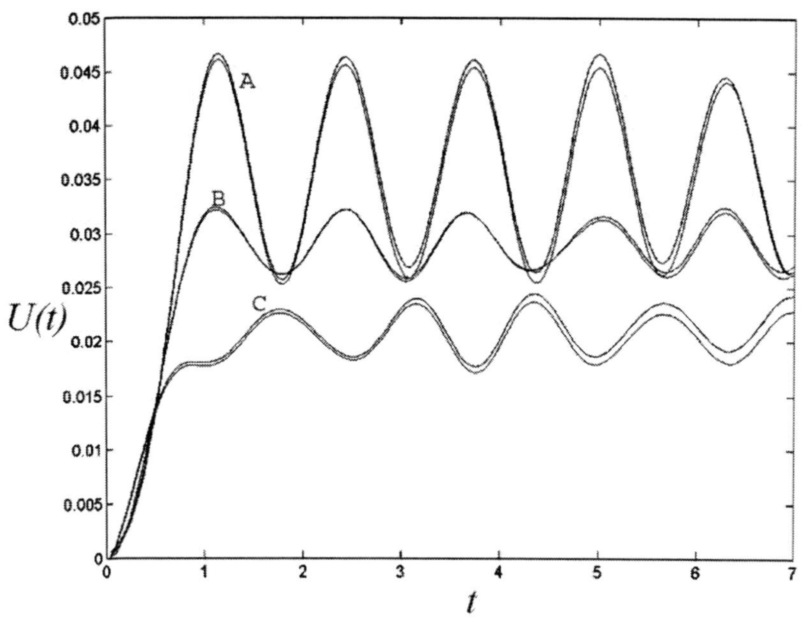

Figure 12. Displacements at reference points.

Displacements at the reference points are demonstrated in Figure 12; the lower curves correspond to the classical BEM scheme based results. The Real, imaginary parts and absolute value of the amplitude-frequency function for displacements at reference points are shown in Figures 13 – 15, respectively. The accuracy achieved with the help of non-classical BIE is no inferior to accuracy of the classical BIEM scheme.

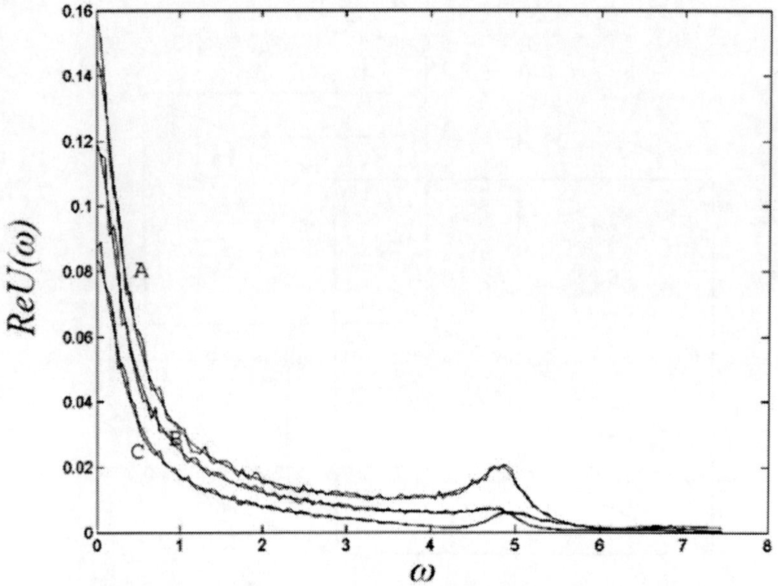

Figure 13. Real part of amplitude-frequency function for displacements.

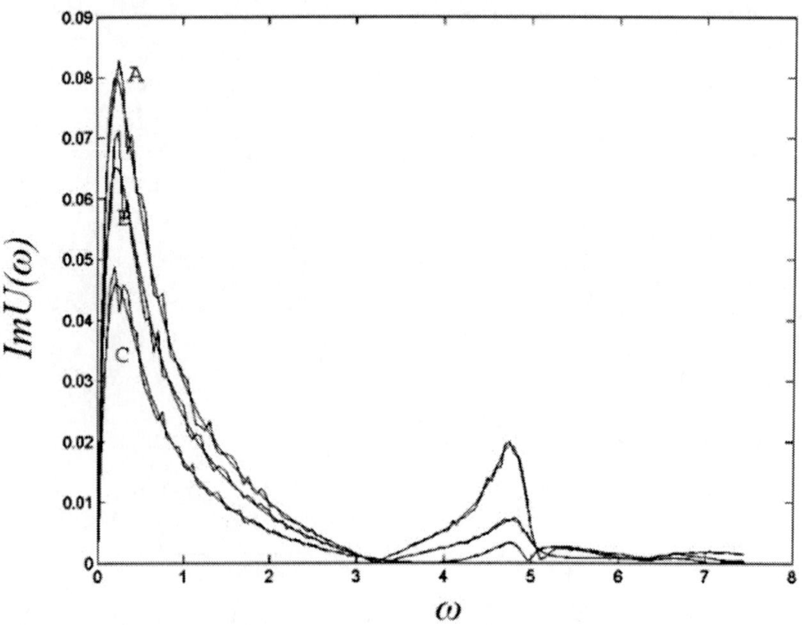

Figure 14. Imaginary part of amplitude-frequency function for displacements.

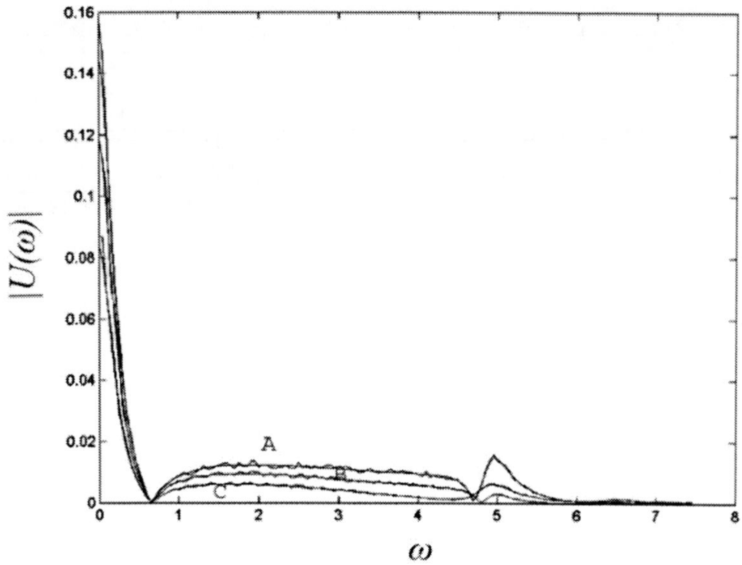

Figure 15. Absolute value of amplitude-frequency function for displacements.

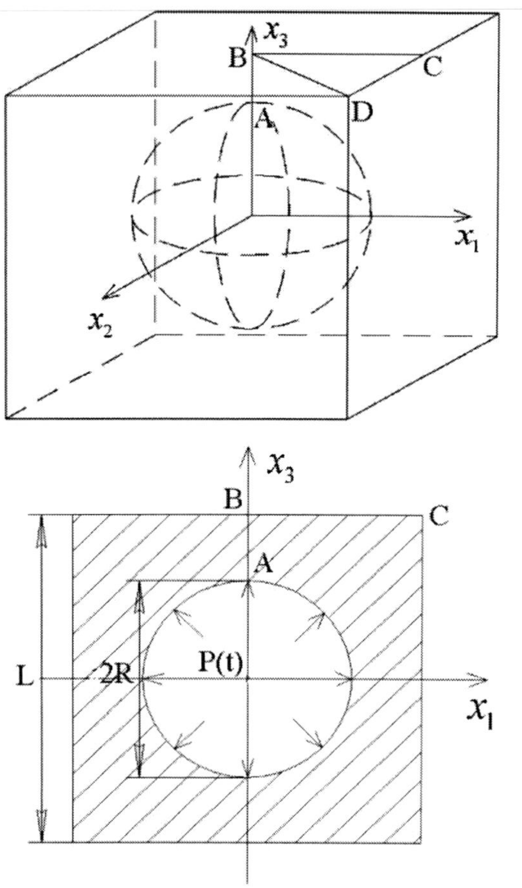

Figure 16. Geometry of the problem and reference points.

5.3. Problem of Dynamic Pressure Acting on Boundary of Spherical Cavity Inside Cube

The considered spherical cavity and cube as well as reference points are present in Figure 16. Dynamic load P(t) from (7) is applied to the cavity boundary. Dimensionless Lamé parameters and density are: $\lambda = 6$, $\mu = 3$, $\rho = 7$. The problem was solved with the geometrical symmetry taken into account. The whole boundary element mesh consisted of 1200 elements.

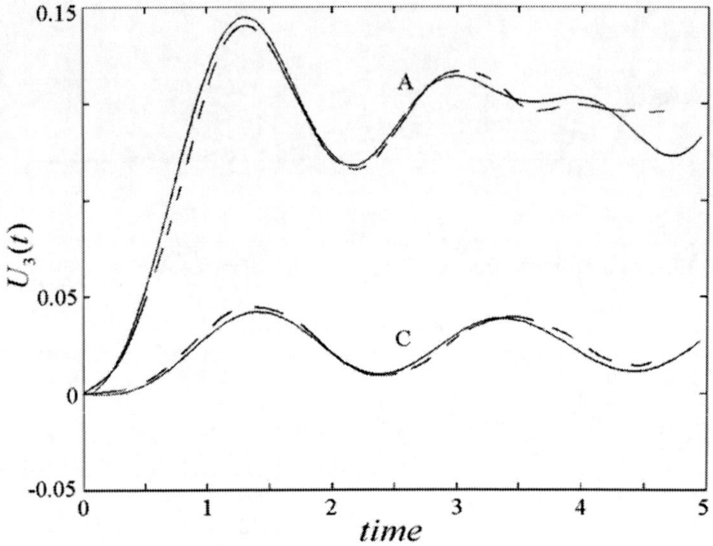

Figure 17. Displacements u_3 at points A and C.

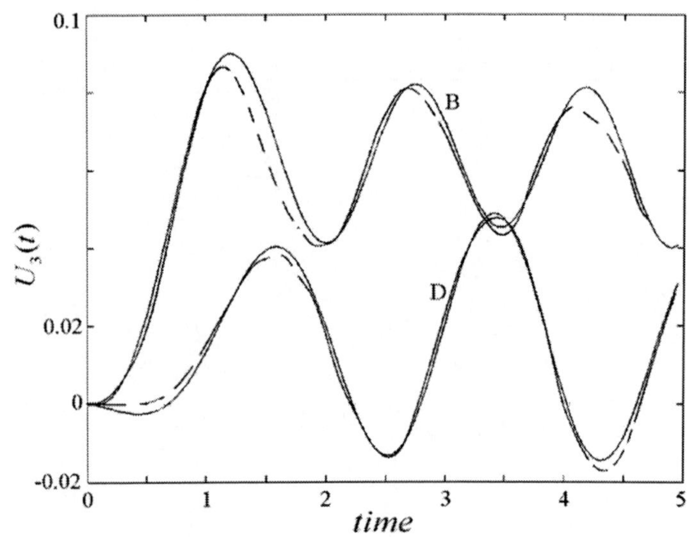

Figure 18. Displacements u_3 at points B and D.

Displacements u_3 at points A and C are present in Figure 17. Displacements u_3 at points B and D are shown in Figure 18.

Central point of a cube face shows maximum displacement values (point B), but maximum amplitude is reached at a corner point of the cube (point D).

The obtained results demonstrate the potential of BIE on single plane wave for solving three-dimensional dynamic problems. The achieved calculation accuracy is no inferior to the accuracy of BEM schemes for classical BIE.

ACKNOWLEDGMENTS

The work was supported by RFBR under grants Nos. 12-08-00984, 12-01-00698, 13-08-00658, 13-08-97091, 14-08-31415, 14-08-31410, 14-08-00811.

REFERENCES

[1] Nardini, D.; Brebbia, C. A., *Boundary Element Methods in Engineering*, C. A. Brebbia (Ed.), Springer, Berlin, 1982, pp. 1 – 312.
[2] Babeshko, V. A., *Proceedings of the USSR Academy of Sciences*, 1985, *vol. 284*, 73.
[3] Norris, A. N., *Proc. R. Soc. Lond. A.*, 1994, *vol. 447*, 175.
[4] Igumnov, L. A., *Proceedings of the USSR Academy of Sciences*, 2006, *vol. 409*, 1.
[5] Vatulyan, A. O., *Proceedings of the USSR Academy of Sciences*, 1993, *vol. 333*, 312.
[6] Vatulyan, A. O., *J. Appl. Math. Mech.*, 1998, *vol. 62*, 492.
[7] Gaul, L.; Kögl, M.; Wagner, M., *Boundary Element Methods for Engineers and Scientists,* Springer, Berlin, 2003.
[8] Wang, C. Y.; Achenbach, J. D., *Proc. R. Soc. Lond. A.*, 1995, *vol. 449*, 441.
[9] Tonon, F.; Pan, E.; Amadei, B., *Computers and Structures*, 2001, *vol. 79*, 469.
[10] Ting, T. C. T.; Lee, V. G., *Q. J. Mech. Appl. Math*, 1997, *vol. 50*, 407.
[11] Igumnov, L. A.; Belov, A. A.; Anufriev, A. A.; Litvinchuk, S. Yu.; Amenitsky, A. V.; Ermolaev, M. D., *Problems of Strength and Plasticity*, 2005, *vol. 67*, 91 (in Russian). http://www.unn.ru/e-library/ppp.html?anum_eng=32

In: Advanced Materials - Studies and Applications
Editors: I.A. Parinov, S.-H. Chang, S.Theerakulpisut
ISBN: 978-1-63463-749-7

Chapter 19

BOUNDARY-ELEMENT MODELING OF THE DYNAMICS OF ELASTIC AND VISCOELASTIC BODIES AND MEDIA

L. A. Igumnov*[*], *A. A. Belov and A. N. Petrov
Research Institute for Mechanics,
Lobachevsky State University of Nizhni Novgorod,
Nizhny Novgorod, Russia

ABSTRACT

Boundary-value problems of three-dimensional (3-D) linear elasticity and viscoelasticity are considered. Viscoelastic material properties are described in the framework of Kelvin-Voigt model and the standard body model, as well as Abel's weakly-singular hereditary kernel. Boundary integral equations (BIE) of the direct approach are used in combination with integral Laplace transform and Durbin's methods or convolution quadratures of its inversion. A concise description of BE-discretization is given. The boundary-element method is presented. Problems of the effect of a pressure jump acting at the end of a compound prismatic viscoelastic body are considered. The problem of vertical force acting on the surface of a half-space with a cavity is analyzed. The form of the cavity affects the response form: in the problem with a spherical cavity, two peaks are observed at the time the surface wave arrives; whereas for a cubic cavity a time range with constant displacements can be found. Displacements were studied in dependence on the distance from the load source. The problem of a deformable prismatic body on a half-space was analyzed. A case was analyzed where a die and a half-space are described by different viscoelastic models.

[*] igumnov@mech.unn.ru.

1. Formulation of the Problem

A piecewise-homogeneous body Ω in 3-D Euclidian space R^3 with Cartesian coordinate system $Ox_1x_2x_3$ is considered. The boundary of the body is designated by Γ, the boundaries of its homogeneous parts Ω_k $(k=1,...,K)$ by Γ_k. Ω_k are assumed to be isotropic viscoelastic bodies [1 – 18]. The following notations for the material parameters of each of the homogeneous parts (substructures) Ω_k will be used: ρ^k is the material density, $\lambda^k(t)$ and $\mu^k(t)$ are Lame functions of the material. The dynamic state of each part of the body Ω_k is described by a differential equation system in displacements:

$$\mu^k(t)*\Delta u^k(x,t)+(\lambda^k(t)+\mu^k(t))*\operatorname{grad}\operatorname{div}u^k(x,t)=\rho^k\,\ddot{u}^k(x,t), \tag{1}$$

where symbol "*" signifies Stieltjes convolution over time *t*. In equations (1), $u^k(x,t)$ is displacement vector of point $x=(x_1,x_2,x_3)$ at time *t*. The physical and geometrical correlations have the following form:

$$\sigma_{ij}=\delta_{ij}\lambda*\varepsilon_{kk}+2\mu*\varepsilon_{ij},\quad \varepsilon_{ij}=\frac{1}{2}\left(\frac{\partial u_j}{\partial u_i}+\frac{\partial u_i}{\partial u_j}\right),$$

where δ_{ij} is Kronecker symbol, $\sigma_{ij},\varepsilon_{ij},i=\overline{1,3},j=\overline{1,3}$ are stress and strain tensors.

Let the displacement vector and Lame functions satisfy the following conditions:

$$u^k(x,0)=\dot{u}^k(x,0)=0,$$
$$\mu(t-\tau)=0,\quad \lambda(t-\tau)=0,\quad \text{where } t<\tau,$$
$$\lim_{t\to0}\dot{\mu}(t)=0,\quad \lim_{t\to0}\dot{\lambda}(t)=0.$$

The particular form of functions $\mu(t)$ and $\lambda(t)$ is determined by the viscoelastic model of the material. The case of proportional memory functions will be considered, then it is sufficient to describe physical relations, for instance, for the case of $i\neq j$:

$$\sigma_{ij}=2\mu*\varepsilon_{ij}=2\int_0^t G(t-\tau)d\varepsilon_{ij}(t),$$

$$\frac{dG(t)}{dt}=-R(t),\quad R(t)\equiv K(t)-\int_0^t K(t-\tau)K(\tau)d\tau+...,\quad \frac{dJ(t)}{dt}=K(t),$$

where $G(t)$ is the memory function of the material, $R(t), K(t)$ are the relaxation and creep kernels of the material. Memory functions of classical viscoelastic models have the forms:

$G(t) = G(0)e^{-\gamma t}$ (Maxwell model),

$J(t) = J(0)(1 - e^{-\beta t})$ (Kelvin-Voigt model),

$G(t) = G(\infty) + (G(0) - G(\infty))e^{-\gamma t}$ (standard model of viscoelastic body),

where γ and β are the values inverse to characteristic times of relaxation and creep, respectively.

Moreover, let the ratio of the value of the modulus over infinity to the value of the modulus at the initial time (for regular models) be determined by parameter $\omega = \dfrac{G(\infty)}{G(0)}$.

A modified theory of viscoelasticity presupposes using the following type of kernels:

$$K(t) = \varphi(t) + \frac{1}{4}\varphi(t) * \varphi(t).$$

Two cases of such kernels will be considered here:

$$\varphi(t) = k\ln\frac{1}{t},\ \varphi(t) = k\frac{t^{-\alpha}}{\Gamma(1-\alpha)},\quad 0 < \alpha < 1.$$

Stress vector $t_n(x,t)$ is introduced at point x on an elementary site with unit normal $n(x)$ as

$$t_n(x,t) \equiv T_{n(x)}u(x,t) = n(x)\lambda(t) * \operatorname{div} u(x,t) + 2\mu(t) * \frac{\partial u(x,t)}{\partial n(x)} + \mu(t) * [n(x) \times \operatorname{rot} u(x,t)]. \tag{2}$$

If $x \in \Gamma_k$, then $n(x)$ will signify a unit vector of the outer (in respect to Ω_k) normal to boundary Γ_k. Stress vector $t_n(x,t) = (t_1, t_2, t_3)$ on boundary Γ_k, corresponding to this normal, will be designated by $t^k(x,t)$.

The following types of boundary conditions for Ω_k will be considered:

$$u_l^k(x,t) = f_l^k(x,t),\quad x \in \Gamma^u \cap \Gamma_k,\quad l = \overline{1,3} \tag{3}$$

$$t_l^k(x,t) = g_l^k(x,t),\quad x \in \Gamma^\sigma \cap \Gamma_k; \tag{4}$$

$$u_l^k(x,t) = u_l^s(x,t),\;\; t_l^k(x,t) = -t_l^s(x,t),\;\; x \in \Gamma'_{ks}\,; \tag{5}$$

$$t_l^k(x,t) = -t_l^s(x,t) = \alpha_{lj}^{k,s}(x)[u_j^k(x,t) - u_j^s(x,t)],\;\; x \in \Gamma''_{ks}\,. \tag{6}$$

Here, Γ^u and Γ^σ are the parts of boundary Γ of body Ω, along which displacements and surface tractions are assigned, respectively; Γ'_{ks} is the boundary of rigid contact of parts Ω_k and Ω_s; Γ''_{ks} is the boundary along which parts Ω_k and Ω_s have elastic connection. Functions $f_l^k(x,t)$ and $g_l^k(x,t)$ are the functions of coordinates and time, coefficients $\alpha_{lj}^{k,s}$ characterize the properties of elastic connection of parts Ω_k and Ω_s of the body.

2. Integral Formulation

Integral Laplace transform is applied to the initial equations:

$$f(p) = \int_0^\infty f(t)e^{-ipt}dt\;,$$

where p is the parameter of Laplace transform.

To solve the equations, the BIE method [1] will be used, based on reducing the boundary-value problem for the differential equation of motion to an integral equation for boundary functions. The displacement vector in internal points of the region is connected with boundary values of displacements and tractions in the following way:

$$u_l^k(x,p) = \int_{\Gamma_k} U_{lj}(x,y,p)t_j^k(y,p)d_yS - \int_{\Gamma_k} T_{lj}(x,y,p)u_j^k(y,p)d_yS\,, \\ l = 1,2,3,\;\; x \in \Omega_k\,. \tag{7}$$

where U_{lj} and T_{lj} are the components of tensors of fundamental and singular solutionsof Equation (1) , respectively. Tensor *T* is written from (2) through tensor *U*, using stress operator T_n as $T(x,y,p) = [T_{n(x)}U(x,y,p)]^T$, where upper symbol "*T*" means transposition. Tensor components *U* and *T* have the following forms:

$$U_{lj}(x-y,p) = \sum_{m=1}^{2}\left[U_{lj}^m(x-y) - \frac{1}{ipr^2}U_{lj}^3(x-y)\left(-\frac{1}{ip}+\frac{ir}{c_m}\right)(-1)^{m+1}\right]e^{-pr/c_m}\;,$$

$$T_{lj}(x,y,p)=\sum_{m=1}^{2}\left[pT_{lj}^{1,m}(x,y)+T_{lj}^{m}(x,y)+\frac{1}{(ip\,r)^2}T_{lj}^{3}(-1)^m\left(\frac{p\,r}{c_m}+1\right)\right]e^{-pr/c_m},$$

where U_{lj}^{m}, U_{lj}^{3}, $T_{lj}^{1,m}$, T_{lj}^{m}, T_{lj}^{3} are the related coordinate functions [1].

For viscoelastic problems, values of $c_1=c_1^k(p)$ and $c_2=c_2^k(p)$ are the operators of longitudinal and transverse wave velocities, respectively.

Formula (7) results in the following BIE:

$$\mathrm{c}_{lj}(x)u_j^k(x,p)+\int_{\Gamma_k}T_{lj}(x,y,p)u_j^k(y,p)d_yS=\int_{\Gamma_k}U_{lj}(x-y,p)t_j^k(y,p)d_yS, \quad (8)$$
$$l=1,2,3,\quad x\in\Gamma_k.$$

The integral in the left part of (8) is singular, that is, understood in the sense of the Cauchy main value, whereas the coefficient at the non-integral member is given by the following formula:

$$c_{lj}(x)=\frac{1-\alpha_\Omega}{2}\delta_{lj}-\int_{\Gamma}T_{lj}^{0}(x,y)d_yS,$$

where

$$T_{lj}^{0}(x,y)=T_{lj}^{1}(x,y)+T_{lj}^{2}(x,y)+\frac{1}{2}\left(\frac{1}{c_2^2}-\frac{1}{c_1^2}\right)*T_{lj}^{3}(x,y)$$

and α_Ω is equal to 1, if region Ω is finite, and -1 for an infinite region Ω_k. If at point x the surface has a unique tangent plane, then $c_{lj}(x)=\delta_{lj}/2$. BIE (8) makes it possible to develop an effective numerical methodology for determining unknown amplitudes of boundary displacements and surface tractions. The solution of the initial boundary-value problem is vector-function $u(x,t)$ obtained by applying inverse Laplace transform to solution (7), (8):

$$f(t)=\frac{1}{2\pi i}\int_{\alpha-i\infty}^{\alpha+i\infty}f(p)e^{ipt}dp. \quad (9)$$

To numerically invert (9), the algorithm introduced by Durbin will be used, as well as the method based on the original integration theorem.

The Durbin's method with a variable integration step [19] uses in the work:

$$F_k=\mathrm{Re}\left[\overline{f}(\alpha+i\omega_k)\right],\quad G_k=\mathrm{Im}\left[\overline{f}(\alpha+i\omega_k)\right],\quad \Delta_k=\omega_{k+1}-\omega_k,$$

$$f(0)=\sum_{k=1}^{\infty}\frac{(F_k+F_{k+1})\Delta_k}{2\pi},$$

$$f(t)\approx\frac{e^{\alpha t}}{\pi t^2}\sum_{k=1}^{\infty}\left[\frac{F_{k+1}-F_k}{\Delta_k}(\cos(\omega_{k+1}t)-\cos(\omega_k t))-\frac{G_{k+1}-G_k}{\Delta_k}(\sin(\omega_{k+1}t)-\sin(\omega_k t))\right].$$

A special quadrature rule of the convolution quadrature method, applied to integrals of the form $y(t)=f(t)*g(t)=\int_0^t f(t-\tau)g(\tau)d\tau$, was introduced in [126, 127]. The convolution integral is substituted for by a quadrature sum, whose weighting factors are determined using Laplace form of function f and a linear multi-step method.

The convolution integral is computed using the following correlation:

$$y(n\Delta t)=\sum_{k=0}^{n}\omega_{n-k}(\Delta t)g(k\Delta t),\qquad n=0,1,\ldots N,$$

where

$$\omega_n(\Delta t)=\frac{R^{-n}}{2\pi}\int_0^{2\pi}\bar{f}\left(\frac{\gamma(Re^{i\varphi})}{\Delta t}\right)e^{-in\varphi}d\varphi$$

The only approximation used in writing the formula was the use of the linear multi-step method for solving Cauchy problem of the first-order differential equation. The rest of the computations were direct. The multi-step method must have the order of accuracy $p\geq 1$, being strictly zero-stable or A-stable, whereas function $\bar{f}(s)$ is limited in the right-hand half-plane relative to straight line $(c-i\infty,c+i\infty)$. Among the examples of multi-step methods there are also reverse differentiation methods of the order $p\leq 6$: for A-stable reverse differentiation method of the second order $(\alpha=90^o)$ the characteristic function can be written as $\gamma(z)=3/2-2z+z^2/2$.

One more approximation that is used in the method approximates the numerical integration, when determining the values of $\omega_n(\Delta t)$ by using L even steps $2\pi/L$:

$$\omega_n(\Delta t)=\frac{R^{-n}}{L}\sum_{l=0}^{L-1}\bar{f}(s)se^{-inl\frac{2\pi}{L}},\quad s=\frac{\gamma(Re^{il\frac{2\pi}{L}})}{\Delta t}.$$

If we assume that function $\bar{f}(s)$ in the equation is computed with a certain inaccuracy ε, the choice $L=2N$ and $R^N=\sqrt{\varepsilon}$ allows for an inaccuracy of order $O(\sqrt{\varepsilon})$ in computing the value of ω_n.

3. NUMERICAL METHODOLOGY

The boundaries of the homogeneous parts of body Ω are approximated by a set of quadrangular and triangular 8-node bi-quadratic elements $S_k\,(k=1,\ldots,N)$, the triangular elements are considered as singular quadrangular elements. The Cartesian coordinates of arbitrary point $y=(y_1,y_2,y_3)$ of element S_k are expressed through the coordinates of the nodal points of this element, using form function $N^l\ (l=1,\ldots,8)$ of the local coordinates $\xi=(\xi_1,\xi_2)$:

$$y_i(\xi)=\sum_{l=1}^{8}N^l(\xi)y_i^{\chi(k,l)},\qquad i=1,2,3, \tag{10}$$

where $\chi(k,l)$ is the global number of the node having in k-th element local number l. Relation (10) makes it possible to find the required characteristics of the element (the normal, the area).

To approximate boundary displacements, bilinear elements will be used, and constant elements will be used for approximating surface tractions. For the computed value of parameter p, we will have the following expressions of the boundary displacements and surface tractions inside element S_k:

$$u_i(y)=\sum_{l=1}^{4}R^l(\xi)u_i^{\chi(k,l)},\qquad i=1,2,3;\ y\in S_k,$$

$$t_i(y)=t_i^{\chi(k,1)},\qquad i=1,2,3;\ y\in S_k.$$

For the elastic connection between h-th and j-th substructures, relation (6) is considered:

$$t_i^h(y)=-t_i^j(y)=-\gamma_i^{hj}(y)\left(u_i^h(y)-u_i^j(y)\right),i=1,2,3;\ y\in S_k,\,S_k\subset\Gamma_{hj}=\Gamma_h\cap\Gamma_j.$$

Here $R^l(\xi)$ are the form functions for a linear quadrangular element.

To obtain a discrete analogue of the BIE, the collocation method will be used. Nodes of the approximation of the initial boundary functions will be chosen as collocation nodes y^m. As a result, sets of linear algebraic equations are formed for each substructure:

$$\frac{1-\alpha_\Omega}{2}u_i^m+\sum_{k=1}^{N}\sum_{l=1}^{4}A_{ij}^{m,k,l}u_j^{\chi(k,l)}=\sum_{k=1}^{N_1}B_{ij}^{m,k}t_j^k-\sum_{k=1}^{N_2}\sum_{l=1}^{4}D_{ij}^{m,k,l}(u_j^{\chi(k,l)}-u_j^{\bar{\chi}(\bar{k},l)}), \tag{11}$$

$$N=N_1+N_2,$$

$$\frac{1-\alpha_\Omega}{8}u_i^m+\sum_{k=1}^{N}\sum_{l=1}^{4}A_{ij}^{m,k,l}u_j^{\chi(k,l)}=\sum_{k=1}^{N_1}B_{ij}^{m,k}t_j^k-\sum_{k=1}^{N_2}\sum_{l=1}^{4}D_{ij}^{m,k,l}(u_j^{\chi(k,l)}-u_j^{\bar{\chi}(\bar{k},l)}), \tag{12}$$

where $\bar{\chi}(\bar{k},l)=\overline{m}$ is the global number of node of certain (adjoint) substructure, which is elastically connected with the substructure along the k-th element, $\bar{k}$ is the number of the element of the adjoint substructure, where elastic connection is assigned; N_1 is the number of boundary elements without elastic connections; N_2 is the number of elements with elastic connections.

Equations (11) are written in the displacement approximation nodes, Equations (12) are written in traction approximation nodes:

$$A_{ij}^{m,k,l}=\int_{-1}^{1}\int_{-1}^{1}\left[R^l(\xi)T_{ij}(x^m,y^k(\xi),p)-\delta_{\chi(k,l),M}T_{ij}^0(x^m,y^k(\xi))\right]J^k(\xi)d\xi_1 d\xi_2,$$

$$B_{ij}^{m,k}=\int_{-1}^{1}\int_{-1}^{1}U_{ij}(x^m,y^k(\xi),p)J_k(\xi)d\xi_1 d\xi_2,$$

$$D_{ij}^{m,k,l}=\int_{-1}^{1}\int_{-1}^{1}R^l(\xi)U_{ij}(x^m,y^k(\xi),p)J_k(\xi)\alpha_j^{\chi\bar{\chi}}(y^k(\xi))d\xi_1 d\xi_2.$$

Note that the coefficients of discrete analogues defined by functions $U_{ij}^1,U_{ij}^2,U_{ij}^3,T_{ij}^{1,1},T_{ij}^{1,2}$ have a singularity of the $\frac{1}{r}$- type, whereas the coefficients of discrete analogues, defined by functions $T_{ij}^1,T_{ij}^2,T_{ij}^3$, have the $\frac{1}{r^2}$- type singularity. That is what determines the special features of the computational process. By computing integrals over the surface, two cases are considered. In the first one, the point x^m does not belong to the element. Integration over the element is reduced to reintegration along local coordinates ξ_1 and ξ_2. Along each of the coordinates, Gauss quadratures are used. In the second one, the point x^m belongs to the element being integrated over. Then a scheme for removing the singularities is used.

4. Numerical Results

Let us consider the problem of an impact on edge with the force $P(t)=1$ N/m^2 of composite prismatic body with rigidly fixed end (Figure 1). This problem has analytical solutions both for elastic and viscoelastic models. To compare the numerical results with analytical ones, consider subregions with identical material parameters: E = $2.11\cdot 10^{11}$ N/m^2; $\nu=0.0$; $\rho=7850$kg/m^3.

The problem is solved in dimensionless quantities: $\bar{x}=x/a$, $\bar{u}=Eu/(Eu)$, $\bar{p}=p/p_0$, $\bar{t}=c_2 t/a$, $\bar{\sigma}=\sigma/p_0$, where a=10, p_0=1 N/m^2.

The following dimensionless parameters result: $\overline{E}=1$; $\overline{\nu}=0$; $\overline{\rho}=0.5$.

Each of the subregions includes 72 elements and 88 points over the quarter-grid; thus, a whole geometrical model includes 576 elements. Figures 2 – 5 compare the analytical elastic solutions (dashed lines) with the numerical BE-results obtained using the convolution quadrature method (curve 1) and using Durbin's method (curve 2). Figures 2 and 3 show stress curves in the region of fixation and contact of the subregions, respectively. Figures 4 and 5 depict displacement curves in the region of contact of the subregions and at the free end of the console, respectively. A homogeneous body cut into two parts was considered.

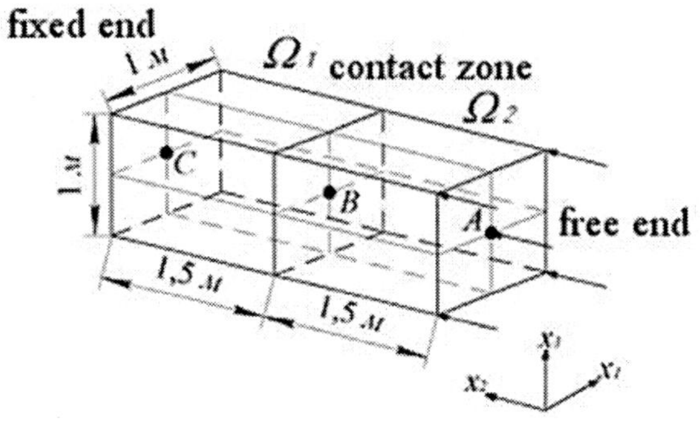

Figure 1. Considered composite prismatic body.

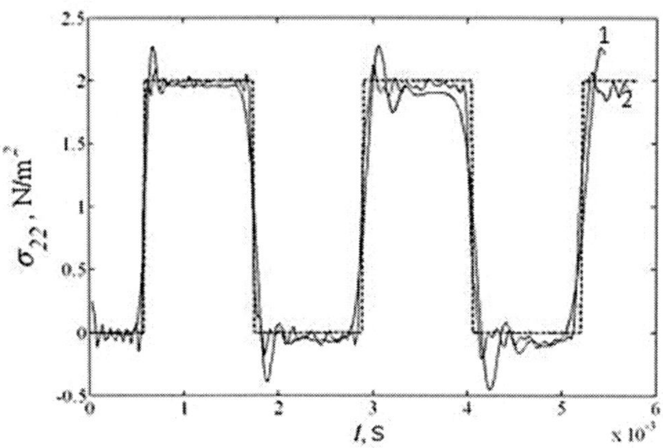

Figure 2. Stresses in the fixation region calculated by using BE-solution: 1 – convolution quadrature method, 2 – Durbin's method, and analytical solution shown by dotted line.

Then we treat the problem on vertical force acting on surface of a half-space with a cavity (see Figure 6). Two geometries of the cavity are considered, namely spherical and cubic ones. The vertical force $P(t)=P_0 f(t)$ acting on the area of $S=1\,\mathrm{m}^2$ of the day surface of a half-space is taken as an acting force. A spherical cavity with the radius of R = 5m is situated inside the half-space at a depth of $h=7.5$ m (the centre of the cavity). Displacements across the surface of the half-space at 15 m from the boundary of the action of the force are analyzed.

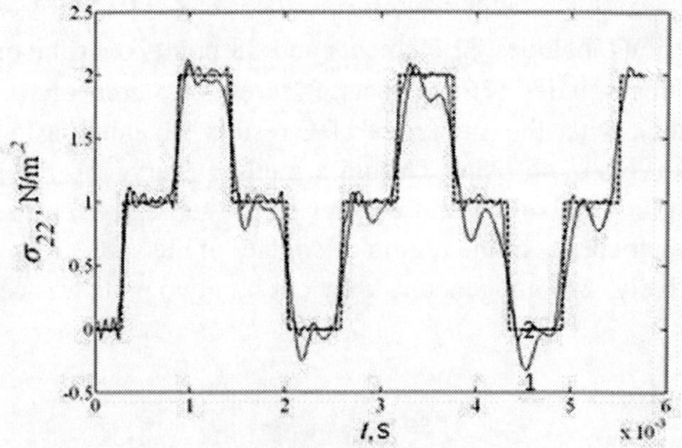

Figure 3. Stresses in the contact region calculated by using BE-solution: 1 – convolution quadrature method, 2 – Durbin's method, and analytical solution shown by dotted line.

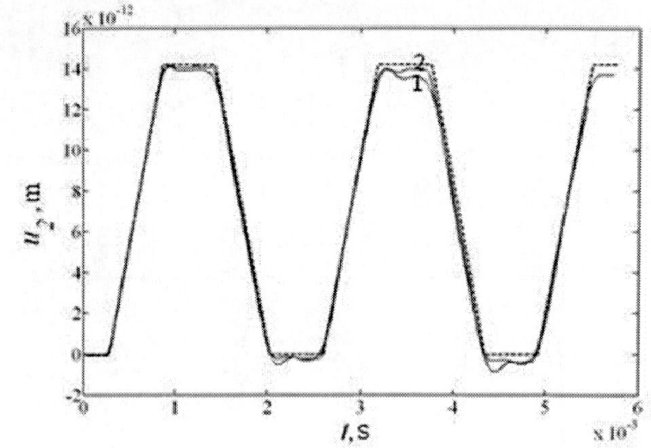

Figure 4. Displacements in the contact region calculated by using BE-solution: 1 – convolution quadrature method, 2 – Durbin's method, and analytical solution shown by dotted line.

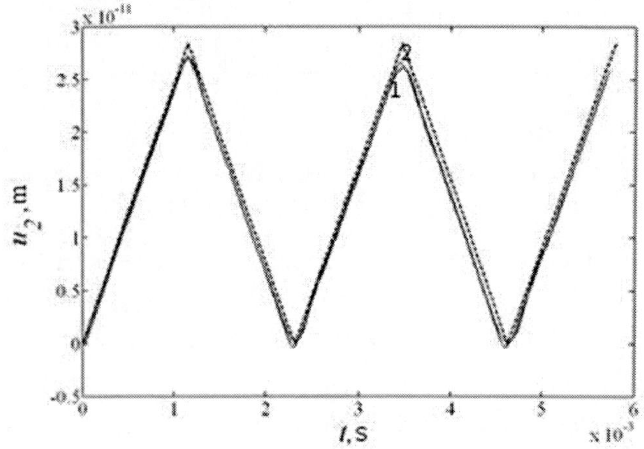

Figure 5. Displacements at the free end calculated by using BE-solution: 1 – convolution quadrature method, 2 – Durbin's method, and analytical solution shown by dotted line.

The BE-grid is constructed, taking advantage of double symmetry in planes. A quarter-grid includes 864 elements and 913 nodes for the half-space, and 150 elements and 171 nodes for the cavity (see Figure 7).

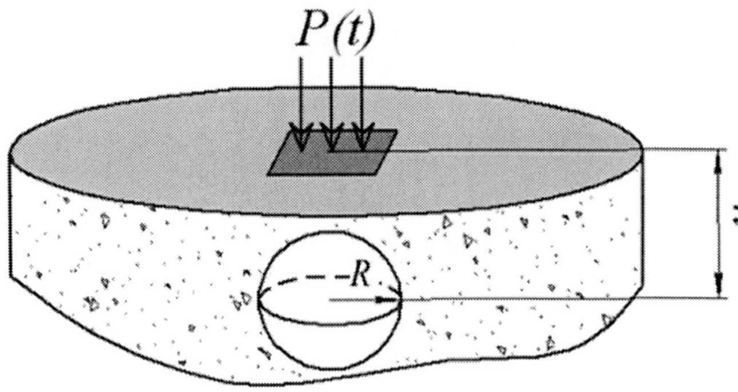

Figure 6. Problem on vertical force acting on surface of half-space with spherical cavity.

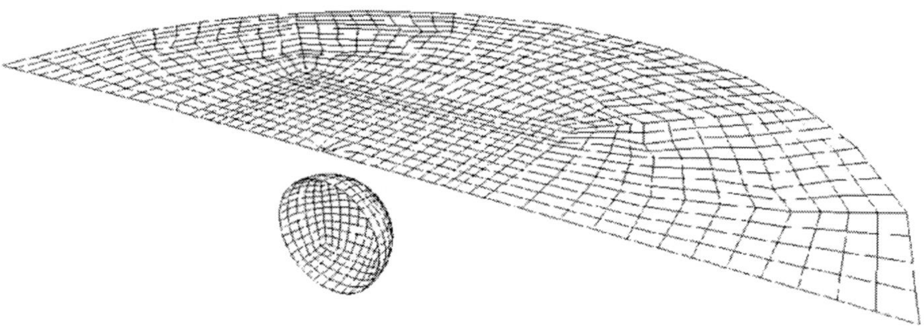

Figure 7. Quarter-grid for the problem on vertical force acting on surface of half-space with spherical cavity.

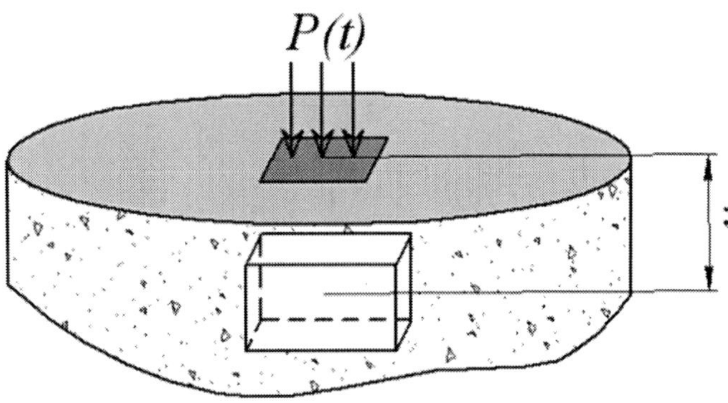

Figure 8. Problem on vertical force acting on surface of half-space with cubic cavity.

We consider also the problem on cubic cavity with the rib length of 10 m situated at a depth of $h = 7.5$ m (the centre of the cube) (see Figures 8, 9).

The material parameters were chosen as follows: $E = 2.5 \cdot 10^8$ N/m^2; $\nu = 0.298$; $\rho = 1884$ kg/m^2 ($c_1 = 425$ m/s; $c_2 = 228$ m/s). Figure 10 presents time histories of displacements obtained using Durbin's method for loads $P(t) = P_0 f(t)$, $f(t) = H(t)$, for $P_0 = 1$ N/m^2.

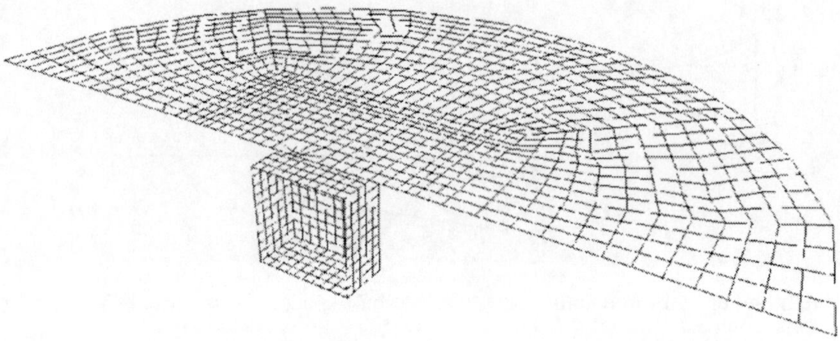

Figure 9. Quarter-grid for the problem on vertical force acting on surface of half-space with cubic cavity.

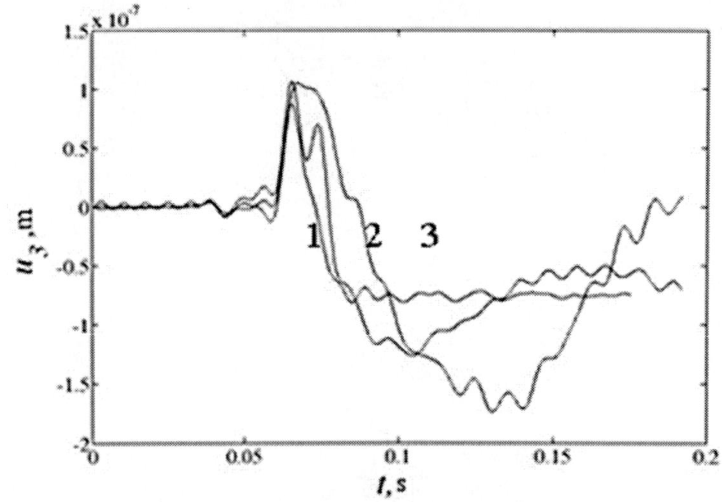

Figure 10. Time histories of displacements obtained using Durbin's method for the cases: 1 – no cavity, 2 – spherical cavity and 3 – cubic cavity.

Figure 11 shows the analyses of displacements for the half-space with a spherical cavity as a function of distance from the load source.

Figure 12 depicts the results of the analysis of displacements for a half-space with a cubic cavity as a function of distance from the load source. It follows from the plots that the response forms for the spherical and cubic cavities changed in comparison with the problem of a half-space. The form of the cavity affects the form of the response: for the problem with a spherical cavity two peaks are observed at the time of the arrival of the surface wave, whereas for the case of a cubic cavity there appeared a time range of constant displacements.

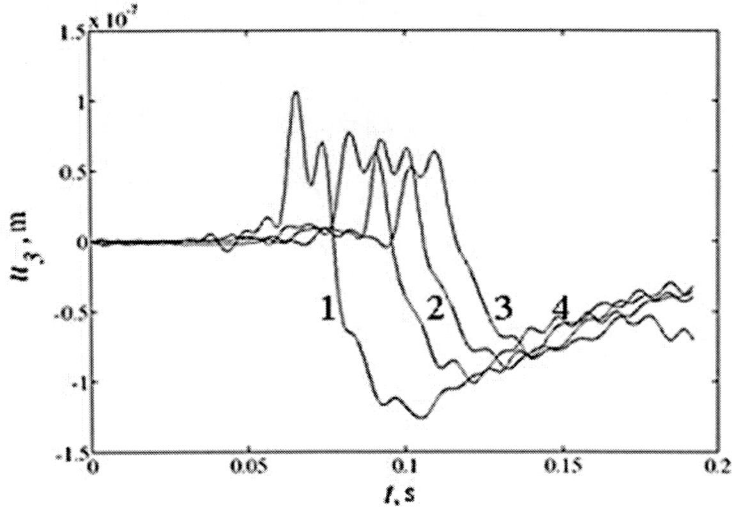

Figure 11. Displacements for half-space with spherical cavity in dependence on distance from the load source: 1 – 15 m, 2 – 19 m, 3 – 21.25 m, 4 – 23.5 m.

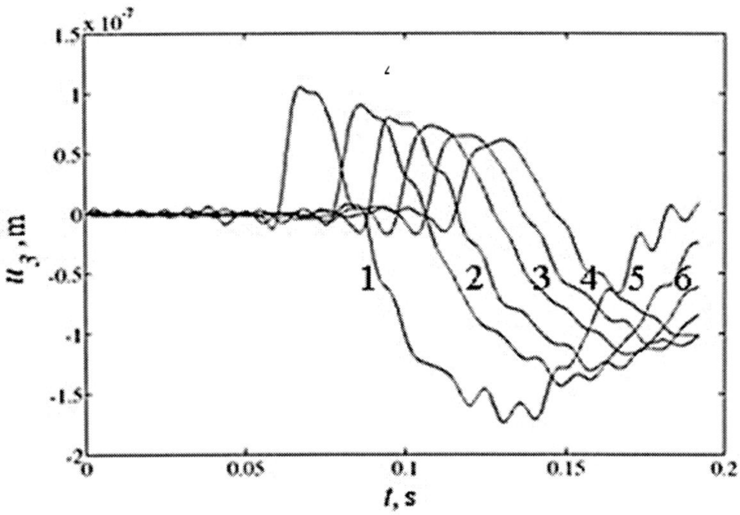

Figure 12. Displacements for half-space with cubic cavity in dependence on distance from the load source: 1 – 15 m, 2 – 19 m, 3 – 21.25 m, 4 – 23.5 m, 5 – 25.5 m, 6 – 27.5 m.

Let us consider the problem of an impact with the force $P(t)$ on the surface of a viscoelastic die $2\,\text{m} \times 2\,\text{m} \times 1\,\text{m}$ situated on a viscoelastic half-space (see Figure 13). One quarter of the even grid of the die includes 96 BE's, and a quarter of the BE-grid of the day surface of the half-space includes 432 quadrangular elements.

$$P(t) = \begin{cases} 1\ \text{N/m}^2, & 0 < t < 0.0085; \\ 0, & t > 0.0085. \end{cases}$$

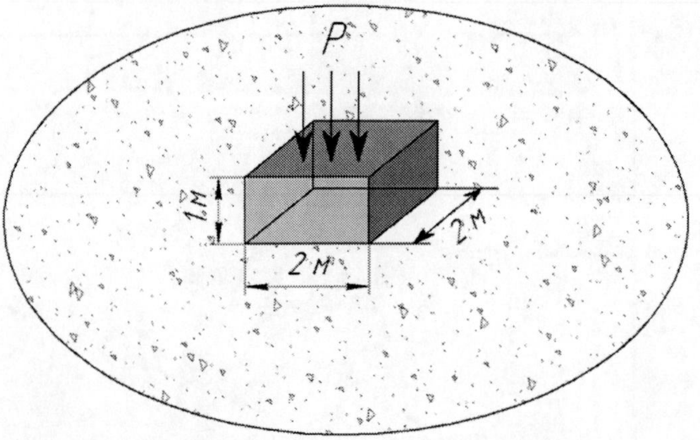

Figure 13. Problem of impact with force $P(t)$ on surface of viscoelastic die situated on viscoelastic half-space.

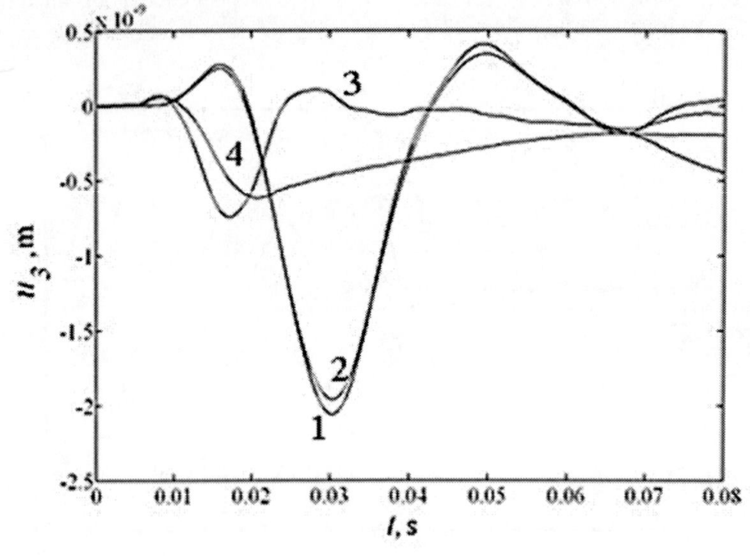

Figure 14. Curves for model of standard viscoelastic body: 1 – elastic solution, 2 – $\gamma = 100\ \text{c}^{-1}$, 3 – $\gamma = 1\ \text{c}^{-1}$, 4 – $\gamma = 0.01\ \text{c}^{-1}$.

The following material parameters were used: (i) for die $E = 3 \cdot 10^8$ N/m^2, $\nu = 0.2$, $\rho =$ 2000kg/m^3; (ii) for half-space $E = 1.38 \cdot 10^8$ N/m^2, $\nu = 0.35$, $\rho = 1966$ kg/m^3.

The values (2.33, 2.33, 0) were chosen as coordinates of the point analyzed. The center of the contact rib of the die was chosen as origin of the coordinates. Figures 14 and 15 present the computational results for the case when both the die and the half-space parameters are computed using the same viscoelastic model. In Figure 14 the curves for the model of standard viscoelastic body are shown. Figure 15 depicts the curves for the power model.

Figure 16 presents the results for elastic die situated on half-space described by the model of standard viscoelastic body.

The effect of viscoelastic properties of materials on the response of the surface displacement wave has been studied. Use of Maxwell and Kelvin-Voigt models can considerably change the response of surface waves. However, there exist certain ranges of characteristic times of creep and relaxation when the response show changes only in decreasing the amplitude. Using the model of standard viscoelastic body and weakly-singular kernel results in the change of the displacement response velocity and the amplitude.

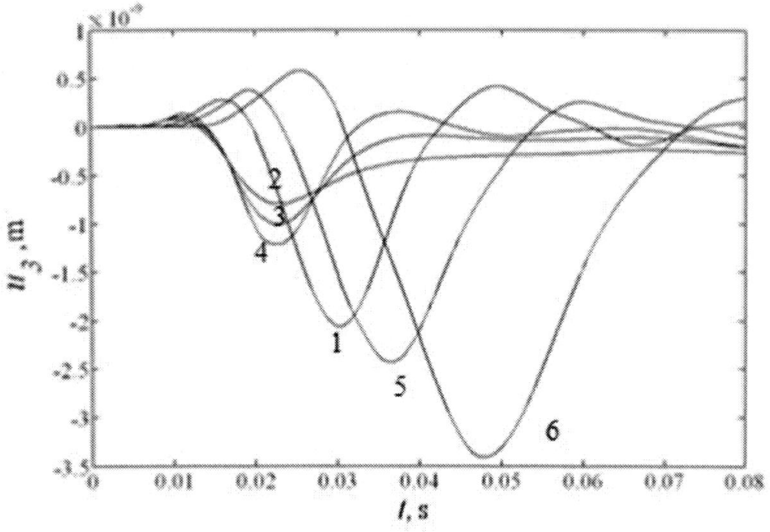

Figure 15. Curves for power model: 1 –elastic solution; 2 – $k=1, \alpha=0.5$; 3 – $k=1, \alpha=0.75$; 4 – $k=1, \alpha=0.95$; 5 – $k=5, \alpha=0.95$; 6 – $k=10, \alpha=0.95$.

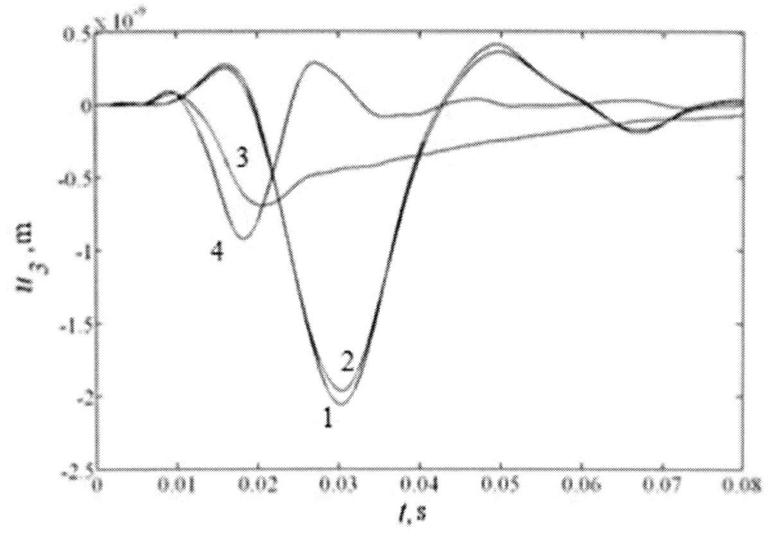

Figure 16. Curves for elastic die situated on half-space described by the model of standard viscoelastic body: 1 – elastic solution, 2 – $\gamma = 100\ c^{-1}$, 3 – $\gamma = 1\ c^{-1}$, 4 – $\gamma = 0.01 c^{-1}$.

ACKNOWLEDGMENTS

The work was supported by RFBR under grants Nos.12-08-00984, 12-01-00698, 13-08-00658, 13-08-97091, 14-08-31415, 14-08-31410.

REFERENCES

[1] Hutoryansky, N. M., *Applied Problems of Strength and Plasticity. Algorothmization and Automation of the Analysis of Problems of Elasticity and Plasticity: All-Rus. Inter-University Collection*, Gorky State University Press, Gorky, 1978, p. 8 (in Russian).

[2] Ugodchikov, A. G.; Hutoryansky, N. M., *Boundary-Element Method in Deformable Solid Mechanics*, Kazan State University Press, Kazan, 1986 (in Russian).

[3] Turilov, V. V., *Analysis of Non-stationary Dynamic Deformation of Three-Dimensional Elastic Structural Elements Using Time-Boundary Elements. PhD thesis*, Gorky, 1986 (in Russian).

[4] Hutoryansky, N. M., *Time-Boundary-Element Method in Special Problems of Non-stationary Dynamics of Elastic and Viscoelastic Solids. DrSc thesis*, Riga, 1988 (in Russian).

[5] Banerjee, P. K.; Ahmad, S., *Proc. ASME Conf. Adv. Boundary Element Anal.*, 1985, *vol. 72*, 65.

[6] Banerjee, P. K.; Ahmad, S.; Manolis, G. D., *Earthquake Eng. Structural Dyn.*, 1986, *vol. 14*, 933.

[7] Ahmad, S.; Banerjee, P. K., *Int. J. Num. Meth. Eng.*, 1988, *vol. 26*, 1709.

[8] Hirose, S.; *Boundary Elem. Proc. 12th Int. Conf.*, 1990, *vol. 1*, 555.

[9] Roytfarb, I. Z.; Kyong, T. V., *Strength of Materials and Theory of Construction*, Budivelnik, Kiev, 1976, p. 32 (in Russian).

[10] Sladek, V.; Sladek, J., *Appl. Math. Model.*, 1984, *vol. 8*, 2.

[11] Beskos, D. E., *Appl. Mech. Reviews.*, 1997, *vol. 50*, 149.

[12] Sutradhar, A.; Paulino, G. H.; Gray, L. J., *Eng. Analysis with Boundary Elements*, 2002, *vol. 26(2)*, 119.

[13] Mesquita, A. D.; Coda, H. B., *Eng. Analysis with Boundary Elements*, 2003, *vol. 27(9)*, 885.

[14] Marrero, M.; Dominguez, J., *Eng. Analysis with Boundary Elements*, 2003, *vol. 27(1)*, 39.

[15] *Developing the Boundary-Element Method and Software Complex for Analyzing Vibrations of Three-Dimensional Structures Interacting with Liquid*: Report on Scientific Research Work No. SR X35838 (Principal Investigator: Hutoryansky N. M.), Research Institute of Mechanics, Nizhny. Novgorod, 1991 (in Russian).

[16] Igumnov, L. A., *Analyzing Axisymmetric Problems of Non-stationary Dynamic Deformation of Viscoelastic Structural Elements Using the Time-Boundary Integral Equation Method. PhD thesis*, Gorky, 1987 (in Russian).

[17] Hutoryansky, N. M.; Turilov, V. V.; Igumnov, L. A., *AIAA Pap.*, 1995, *vol. 2639*, 186 (in Russian).

[18] Lokshin, A. A.; Suvorova, Yu. V., *Mathematical Theory of Wave Propagation in Media with the Memory Effect*, Moscow State University Press, Moscow, 1982 (in Russian).

[19] Lubich, C., *Numer. Math.*, 1988, *vol. 52*, 129.

[20] Lubich, C., *Numer. Math.*, 1988, *vol. 52*, 413.

In: Advanced Materials - Studies and Applications
Editors: I.A. Parinov, S.-H. Chang, S.Theerakulpisut
ISBN: 978-1-63463-749-7

Chapter 20

ANALYSIS OF 3D POROELASTODYNAMICS WITH THE HELP OF BOUNDARY ELEMENT METHOD

L. A. Igumnov*, Ya. Yu. Rataushko, S. Yu. Litvinchuk and A. A. Ipatov
Research Institute for Mechanics, Lobachevsky State University of Nizhni Novgorod, Nizhny Novgorod, Russia

ABSTRACT

3D poroelastic problems are considered in full and strict mathematical formulation. Regularized Boundary Integral Equations (BIE), matched element-by-element approximation and adaptive numerical integration in combination with the singularity decreasing algorithm are used. Boundary element schemes are based on the traditional time-step method and Runge-Kutta relying methods for numerical inversion of Laplace transform. The effectiveness of the approach is demonstrated using the boundary element analysis of typical 3D dynamic problems of poroelastic solids deformation.

Appearance of the third wave kind in a 3D poroelastic body is numerically proved. Boundary element modeling of surface waves on homogeneous and compound poroelastic bodies is given. The effect of poroelastic material properties on the wave processes is demonstrated. Boundary element analysis of homogeneous and layered poroelastic half-space is presented. The effect of weakening half-space with cavity on the wave propagation is shown.

1. INTRODUCTION

The paper is dedicated to the development of 3D poroelastic dynamic problems numerical modeling technique based on using the Boundary Element Method (BEM) in Laplace domain and time-stepping schemes for Laplace transform numerical inversion.

* Corresponding author: L. A. Igumnov. Research Institute of Mechanics, Lobachevsky State University of Nizhny Novgorod, 603950, Nizhny Novgorod, Russia. E-mail: igumnov@mech.unn.ru.

Specially designed software allows analyzing wave processes in homogeneous and piecewise homogeneous solids. Finite and semi-finite solids are considered, including ones weakened by cavities.

Study of wave propagation processes in saturated porous continua began from the works of Y. I. Frenkel (1944) and M. Biot (1956). L. Y. Kosachevsky in 1959 showed that Frenkel's and Biot's models rely on the same stress-strain relations but the latter is of a greater generality. After that waves in saturated porous continua have been studied by C. Zwikker and C. W. Kosten, J. Geertsma and D. C. Smith, P. P. Zolotaryov, V. N. Nikolayevsky, V. P. Stepanov, L. M. Doroginitskaya, C. McCann and D.M. McCann, S. H. Yakubov, A. A. Gubaydullin etc., but two classical Biot's works are considered the most influential. Common state of the art can be found in works of R. de Boer (2000), M. Schanz (2001, 2009), V. N. Nikolayevsky (2005). Works of T. I. Belyankova, V. V. Kalinchuk, S. Diebels, W. Ehlers contain an overview of Biot's theory as well as other approaches to porous continua modeling.

Biot's model is based on the description of how two phases – porous elastic skeleton and gas or liquid filler – interact. Historically Biot's theory was the first model to predict all three possible types of waves in porous continuum: fast shear wave, fast and slow compression waves. By their nature both poroelastic fast waves are close to the corresponding ones of elastic continuum. The principal difference between elastic and poroelastic continua is the presence of slow compression wave in poroelastic case. This wave is caused by transfer of pore filler particles with respect to the skeleton. Ignoring the slow wave leads to serious inaccuracies in estimation of fast waves damping.

There are two major approaches to dynamic processes modeled by means of BEM: solving BIE system directly in time domain [1] or in Laplace or Fourier domain followed by the respective transform inversion [2]. Traditional stepping schemes with spline approximation in scope of the first approach require fundamental solutions in time domain what severely restricts their usage. Often it is only possible to construct fundamental solution matrices in Laplace or Fourier domain.

Thus first BEM formulations for Biot's poroelastodynamics, published by G. D. Manolis and D. E. Beskos, employ Laplace transform.

The formulation in time domain has been developed in [3] based on analytical inversion of Laplace transform for fundamental solutions. The main shortcoming of the approach is again the demand of fundamental solutions in time domain. They exist for quasi-static problems of poroviscoelasticity, but are quite cumbersome; damping effects cannot be accounted at all. As well the methodology is characterized by significant computational costs and low stability rate. In 1988, C. Lubich introduced brand new Convolution Quadrature Method (CQM) [4, 5]. Now it is widely applied in construction of time-step boundary element schemes on the basis of fundamental solutions in Laplace domain [6, 7]. CQM helps to get rid of fundamental solutions in time domain requirement and shows better stability but is still costly. Work [8] addresses the issue of Runge-Kutta-based CQM usage. Various combinations of this CQM modification with BEM stepping schemes were developed in [9, 10]. An overview of different boundary element approaches based on CQM is present in [11]. Works [9, 12] estimate the accuracy and precision of Runge-Kutta-based CQM. In this paper a similar boundary element scheme based on the stepping method for Laplace transform numerical inversion is considered. A modification of the scheme with varied integration step relying on highly oscillatory quadrature principles is employed.

2. PROBLEM FORMULATION

Biot's saturated poroelastic material model is used for description of porous materials [6]. Saturated poroelastic continuum is assumed to be a two-phase continuum consisting of an elastic skeleton and compressible fluid or gas filler. Saturation means $V = V^f + V^s$ where V is the overall volume of the considered body, V^f is the summary pore (filler) volume, V^s is the volume of the elastic skeleton. The filler can seep through the pores freely; closed pores are assumed to be a part of the skeleton.

There are several Biot's theory formulations distinguished by the number of unknowns: three skeleton displacements and pore pressure or three filler displacements along with the skeleton ones [13, 14]. However in 1987, G. Bonnet showed that four components are enough for saturated porous continuum description [14]. Boundary-value problem for full Biot's model of linear saturated poroelastic continuum in Laplace domain in terms of four unknowns – skeleton displacements $\bar{u}_i$ and pore pressure $\bar{p}$ – takes the following form [6]:

$$
\begin{aligned}
&G\bar{u}_{i,jj} + \left(K + \frac{G}{3}\right)\bar{u}_{j,ij} - (\alpha - \beta)\bar{p}_{,i} - s^2(\rho - \beta\rho_f)\bar{u}_i = -\bar{F}_i, \\
&\qquad \frac{\beta}{s\rho_f}\bar{p}_{,ii} - \frac{\phi^2 s}{R}\bar{p} - (\alpha - \beta)s\bar{u}_{i,i} = -\bar{a}, \quad x \in \Omega, \\
&\qquad \bar{u}'(x,s) = \tilde{u}', \quad x \in \Gamma^u, \quad \bar{u}' = (\bar{u}_1, \bar{u}_2, \bar{u}_3, \bar{p}), \\
&\qquad \bar{t}_n'(x,s) = \tilde{t}_n', \quad x \in \Gamma^\sigma, \quad \bar{t}' = (\bar{t}_1, \bar{t}_2, \bar{t}_3, \bar{q}),
\end{aligned}
\tag{1}
$$

where Γ^u denotes Dirichlet boundary and Γ^σ denotes Neumann boundary; G, K are the elastic moduli; $\phi = V^f / V$ is the porosity; $\bar{F}_i, \bar{a}$ are the bulk body forces; $\beta = \dfrac{\kappa\rho_f\phi^2 s}{\phi^2 + s\kappa(\rho_a + \phi\rho_f)}$, $\alpha = 1 - \dfrac{K}{K_s}$ and $R = \dfrac{\phi^2 K_f K_s^2}{K_f(K_s - K) + \phi K_s(K_s - K_f)}$ are the constants reflecting interaction between the skeleton and filler; κ is the permeability; ρ, ρ_a, ρ_f are the material density, apparent mass density and filler density, respectively; K_s, K_f are the elastic bulk moduli of the skeleton and filler, respectively; $\rho_a = C\phi\rho_f$ is the apparent mass density presenting an imaginary mass density introduced in [13] to describe dynamic interaction between the skeleton and filler phases; coefficient C depends on the pores geometry and excitation frequency. G. Bonnet and J.-L. Auriault measured $C = 0.66$ for low-frequency excitation and a system of spherical pores distributed right up to each other [15].

BIE system in terms of elastic skeleton displacements and pore pressure can be represented in time domain only in conditions of $\kappa \to \infty$. This model corresponds to the filler passing through pores without friction. Sometimes incompressible Biot's model implying $\alpha = 1$ and $R \to \infty$ is applied; in this case pore pressure is completely defined by elastic skeleton deformations and makes no degree of freedom anymore.

3. BEM APPLICATION

Boundary-value problem (1) can be reduced to the following BIE [6,16,17]:

$$\alpha_\Omega \bar{u}_k(x,s) + \int_\Gamma \left(\widetilde{T}_{ik}(x,y,s)\bar{u}_i(y,s) - \widetilde{T}^0_{ik}(x,y,s)\bar{u}_i(x,s) - \widetilde{U}_{ik}(x,y,s)\bar{t}_i(y,s) \right) d\Gamma = 0,$$

$$(x \in \Gamma),\quad \bar{t} = (\bar{t}_1, \bar{t}_2, \bar{t}_3, \bar{q})^T,\quad \bar{u} = (\bar{u}_1, \bar{u}_2, \bar{u}_3, \bar{p})^T,$$

where $\widetilde{U}(x,s)$, $\widetilde{T}(x,s)$ are fundamental and singular solutions, respectively; $\widetilde{T}^0(x,s)$ contains the isolated singularities.

To approximate boundary of a piecewise homogeneous body, a set of quadrangular and triangular elements is used, where triangular elements are considered singular quadrangular. Quadrangular elements help to obtain a higher precision as compared to fully triangular approximation. Every boundary element is mapped to a reference one (canonical square $\xi = (\xi_1, \xi_2) \in [-1,1]^2$ or triangle $0 \le \xi_1 + \xi_2 \le 1,\ \xi_1 \ge 0,\ \xi_2 \ge 0$) with the help of the formula:

$$y_i(\xi) = \sum_{l=1}^{8} N^l(\xi) y_i^{\beta(k,l)}, \qquad i = 1,2,3,$$

where $\beta(k,l)$ is the global node number, l is the local node number in element k, $N^l(\xi)$ are the shape functions. Boundary functions are interpolated by a subset of geometrical grid nodes. Local approximation of generalized boundary displacements and tractions follows the Goldshteyn's displacement-stress matched model [18]. This means that generalized displacements are approximated by bilinear element and constant tractions.

After applying collocation method on the set of approximation nodes of the boundary functions, the following system of linear equations is obtained for each homogeneous region of the whole compound body ($N = N_1 + N_2$):

$$\frac{1-\alpha_\Omega}{2} u_i^m + \sum_{k=1}^{N} \sum_{l=1}^{4} A_{ij}^{m,k,l} u_j^{\chi(k,l)} = \sum_{k=1}^{N_1} B_{ij}^{m,k} t_j^k - \sum_{k=1}^{N_2} \sum_{l=1}^{4} D_{ij}^{m,k,l} (u_j^{\chi(k,l)} - u_j^{\bar{\chi}(\bar{k},l)}),$$

$$\frac{1-\alpha_\Omega}{8} u_i^m + \sum_{k=1}^{N} \sum_{l=1}^{4} A_{ij}^{m,k,l} u_j^{\chi(k,l)} = \sum_{k=1}^{N_1} B_{ij}^{m,k} t_j^k - \sum_{k=1}^{N_2} \sum_{l=1}^{4} D_{ij}^{m,k,l} (u_j^{\chi(k,l)} - u_j^{\bar{\chi}(\bar{k},l)}).$$

here, the first equation is written in the approximation nodes of generalized displacements, the second one defined in the approximation nodes of generalized tractions; the first right-side term in both equations is related to boundary elements with no contact, the second one satisfies the conditions of contact between elements k and $\bar{k}$ belonging to different regions; coefficients A, B and D are computed with the help of the following formulae:

$$A_{ij}^{m,k,l} = \int_{-1}^{1}\int_{-1}^{1}\left[R^l(\xi)T_{ij}(x^m, y^k(\xi), s) - \delta_{\chi(k,l),M} T_{ij}^0(x^m, y^k(\xi))\right]J^k(\xi)d\xi_1 d\xi_2,$$

$$B_{ij}^{m,k} = \int_{-1}^{1}\int_{-1}^{1} U_{ij}(x^m, y^k(\xi), s)J_k(\xi)d\xi_1 d\xi_2,$$

$$D_{ij}^{m,k,l} = \int_{-1}^{1}\int_{-1}^{1} R^l(\xi)U_{ij}(x^m, y^k(\xi), s)J_k(\xi)\alpha_j^{\chi\overline{\chi}}(y^k(\xi))d\xi_1 d\xi_2,$$

where s is the Laplace transform parameter. The integrals are calculated using Gaussian quadrature on the elements containing no singularities. Once an element contains singularity, the singularity decreasing or eliminating algorithm is used [17]. An adaptive integration algorithm is used: the order of Gaussian quadrature in an element is chosen by the satisfaction of the prescribed precision if possible, otherwise the element is recursively subdivided to smaller elements for integration.

4. Laplace Transform Inversion

CQM is generally a method to calculate a convolution integral:

$$y(t) = f(t) * g(t) = \int_0^t f(t-\tau)g(\tau)d\tau$$

from one of the convoluted functions ($f(t)$) known in Laplace domain and some discrete values of other one ($g(t)$) [4–6]. The idea consists in rewriting the convolution integral in the following form with the help of the inverse Laplace transform formula:

$$\int_0^t f(t-\tau)g(\tau)d\tau = \frac{1}{2\pi i}\lim_{R\to\infty}\int_{c-iR}^{c+iR}\bar{f}(s)\underbrace{\int_0^t e^{s(t-\tau)}g(\tau)d\tau}_{x(t,s)} ds, \qquad (2)$$

where $\bar{f}(s)$ denotes the function in Laplace domain.

Calculation of $x(t,s)$ numerically is the only approximation to be used in scope of CQM. $x(t,s)$ is obviously the solution of the following Cauchy problem:

$$\frac{d}{dt}x(t,s) = sx(t,s) + g(t), \quad x(0,s) = 0.$$

A scheme to solve this problem defines the particular case of CQM. Employing a linear multistep method for solution of $x(t,s)$, evaluating the outer integral in (2) with the help of residue theory and expanding $\bar{f}(\gamma(z)/\Delta t)$ in power series by z, we get the convolution integral in the following form [6]:

$$y(n\Delta t) = \sum_{k=0}^{n} \omega_{n-k}(\Delta t) g(k\Delta t), \qquad n = 0,1,...N. \tag{3}$$

Coefficients $\omega_n(\Delta t)$ can be computed by means of integral Cauchy formula:

$$\omega_n(\Delta t) = \frac{1}{2\pi i} \int_{|z|=R} \bar{f}\left(\frac{\gamma(z)}{\Delta t}\right) z^{-n-1} dz = \frac{R^{-n}}{2\pi} \int_0^{2\pi} \bar{f}\left(\frac{\gamma(Re^{i\varphi})}{\Delta t}\right) e^{-in\varphi} d\varphi \tag{4}$$

where R denotes radius of the analyticity region for $\bar{f}(\gamma(z)/\Delta t)$, and $\gamma(z)$ is the characteristic function for the applied linear multistep method. Backward differentiation formula (BDF) based methods of order ≤ 6 are applicable in scope of this solution scheme. For BDF-2 (traditional method, or 2nd order Euler one) we have $\gamma(z) = 3/2 - 2z + z^2/2$ [6].

Time-step method for Laplace transform numerical inversion is similar to CQM, but whereas CQM is based on the convolution theorem, it is based on the integration theorem, hence dedicated to the calculation of the original function integral. In order to get as a result the original function $f(t)$ such as $f(0) = 0$ from $\bar{f}(s)$ we need to apply the theorem to its derivative: $f(t) = \int_0^t f'(\tau) d\tau$.

In this case (3), (4) will be rewritten as:

$$f(0) = 0, \quad f(n\Delta t) = \sum_{k=1}^{n} \omega_k(\Delta t), \quad n = 1,...N,$$

$$\omega_n(\Delta t) = \frac{R^{-n}}{L} \sum_{l=0}^{L-1} \bar{f}(s) s e^{-in\varphi}, \quad s = \frac{\gamma(z)}{\Delta t}, \quad z = Re^{i\varphi}, \quad \varphi = 2\pi \tfrac{l}{L}.$$

Several modifications of the method can be derived from different ways to calculate $\omega_n(\Delta t)$. For varied step by φ and linear approximation of the integrand function we obtain

$$\omega_n(\Delta t) = \frac{R^{-n}}{2\pi} \sum_{k=0}^{L-1} \left[\bar{f}(s_k) s_k e^{-in\varphi_k} + \bar{f}(s_{k+1}) s_{k+1} e^{-in\varphi_{k+1}}\right] \frac{(\varphi_{k+1} - \varphi_k)}{2}, \quad s_k = \frac{\gamma(Re^{i\varphi_k})}{\Delta t}$$

or, basing on the highly oscillatory quadrature [19]:

$$\omega_n(\Delta t) = \frac{R^{-n}}{2\pi}\sum_{k=0}^{L-1}\frac{\varphi_{k+1}-\varphi_k}{2}e^{-in\frac{\varphi_k+\varphi_{k+1}}{2}}\left[D_1(w)\overline{f}(s_k)s_k + D_2(w)\overline{f}(s_{k+1})s_{k+1}\right]\quad s_k = \frac{\gamma(Re^{i\varphi_k})}{\Delta t}$$

$$w = -n\frac{\varphi_{k+1}-\varphi_k}{2},\quad D_{1,2}(w) = \begin{cases} \dfrac{\sin w}{w} \pm \dfrac{w\cos w - \sin w}{w^2}i, & \text{where } |w| > w_*; \\ e^{\mp wi}, & \text{where } |w| \le w_*, \end{cases}$$

here w_* is specific key value, and the overall scheme using this formula is informally called 'scheme with a key'.

Substituting basic linear multistep method for Runge-Kutta method referred to as Butcher tableau $\frac{c|A^T}{b^T}$, $A \in R^{m\times m}$, $b, c \in R^m$, we obtain the following representation [10,11]:

$$f(0) = 0,\quad f(n\Delta t) = b^T A^{-1}\sum_{k=1}^{n}\omega_k(\Delta t),\quad n = 1,\ldots N.$$

Formulae for coefficients $\omega_n(\Delta t)$ can still be used here taking into account that now they contain matrices instead of scalars. Characteristic function $\gamma(z)$ takes the following form [10,11]: $\gamma(z) = A^{-1} - zA^{-1}[1]b^T A^{-1}$, where $[1] = (1,\ldots,1)^T$.

Radau (Radau IIA) scheme [10] is chosen to further represent Runge-Kutta family.

5. Slow Compression Wave in 1D Case

An effect of slow compression wave appearance in poroelastic 1d column of length l = 9m under axial load $F = 1$ N/m^2 is considered (Figure 1). Material properties of the column are K=4.8·10^9 N/m^2, G=7.2·10^9 N/m^2, ρ=2458 kg/m^3, ϕ=0.19, K_s=3.6·10^{10} N/m^2, ρ_f=1000 kg/m^3, K_f=3.3·10^9 N/m^2. The problem has analytical solution in Laplace domain [6] which just transforms back to time domain with the help of the time-step method for Laplace transform numerical inversion. N time steps and L nodes by argument φ are used in scope of the method.

Hereafter L = 2L is considered rather than L for two-stage Runge-Kutta-based methods so that their costs could be easily compared to the ones of traditional Euler-based method. L nodes are distributed percentage wise 95% / 5% between ranges $[0, \pi/2]$ and $[\pi/2, \pi]$, and regularly inside these ranges.

The results for pore pressure and flux responses at point A (l_1=1.5 m) in time domain are shown in Figures 2 – 5.

Figures 2 – 5 demonstrate the effect of slow wave appearance in porous continuum for large permeability coefficient value (Figures 2, 4 at approximately of 0.00155 sec).

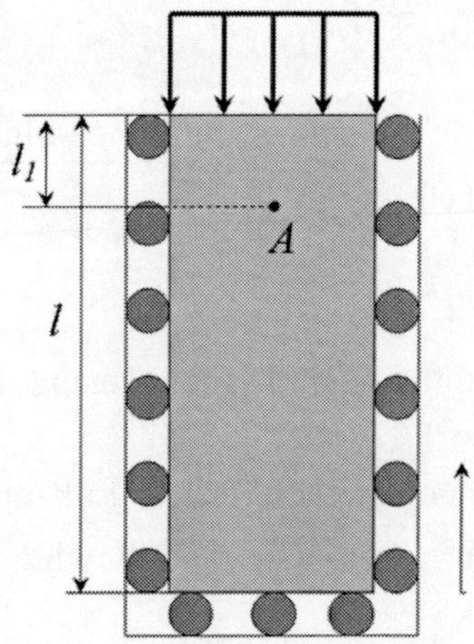

Figure 1. Poroelastic 1d column.

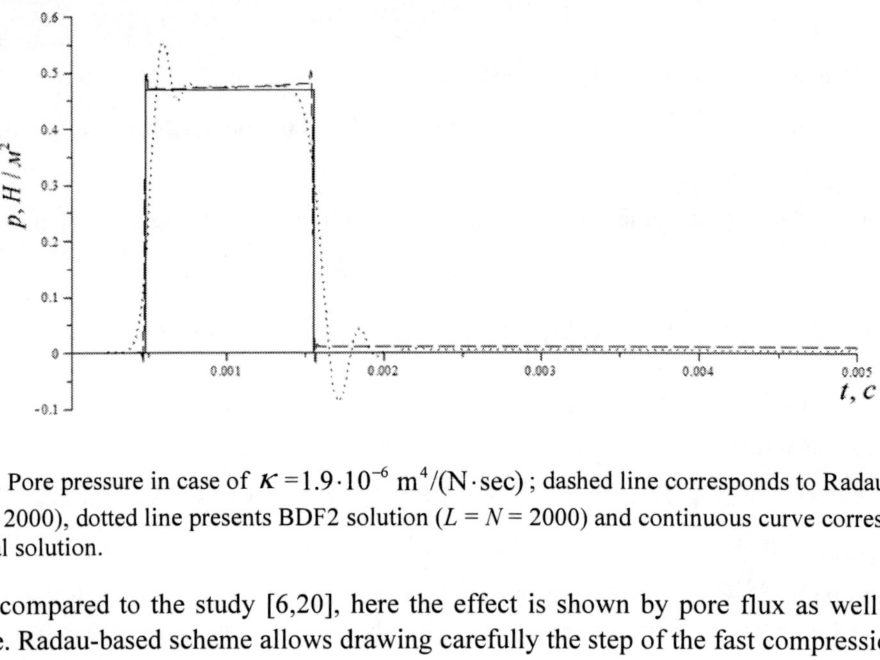

Figure 2. Pore pressure in case of $\kappa = 1.9 \cdot 10^{-6}\ \mathrm{m^4/(N \cdot sec)}$; dashed line corresponds to Radau scheme ($Ł = N = 2000$), dotted line presents BDF2 solution ($L = N = 2000$) and continuous curve corresponds to analytical solution.

As compared to the study [6,20], here the effect is shown by pore flux as well as pore pressure. Radau-based scheme allows drawing carefully the step of the fast compression wave before the slow wave arrives (0 – 0.0015 sec).

6. Slow Compression Wave in 3D Case

Let us consider the problem of frontal load $f(t) = 1\ \mathrm{N/m^2}$ to 3D poroelastic column 3 × 1 × 1 m^3 with other fixed front (Figure 6). The lateral faces are traction-free but as well transverse displacements u_1, u_2 are prohibited. Pore flux vanishes everywhere except the loaded front where zero pore pressure is assumed.

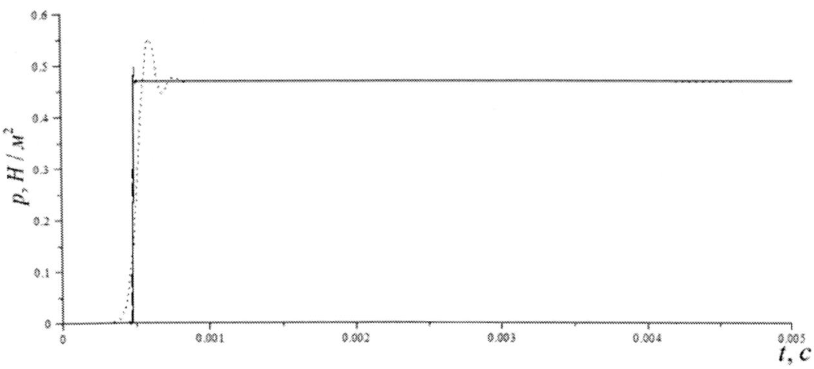

Figure 3. Pore pressure in case of $\kappa = 1.9 \cdot 10^{-10}$ m^4/(N·sec); dashed line corresponds to Radau scheme ($Ł = N = 2000$), dotted line presents BDF2 solution ($L = N = 2000$) and continuous curve corresponds to analytical solution.

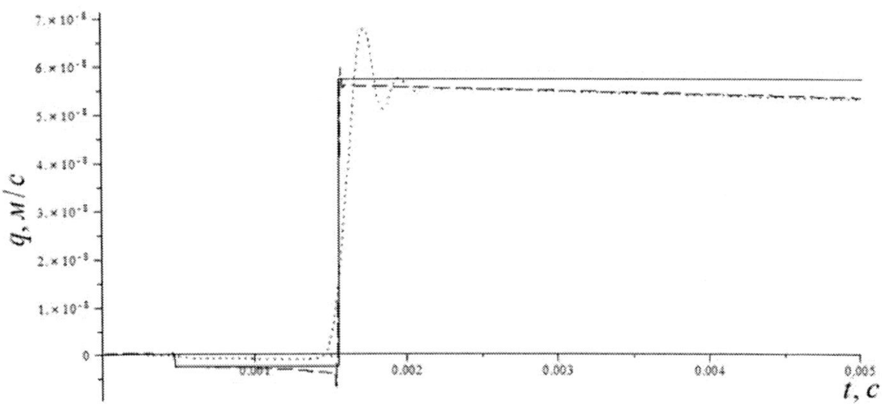

Figure 4. Pore flux in case of $\kappa = 1.9 \cdot 10^{-6}$ m^4/(N·sec); dashed line corresponds to Radau scheme ($Ł = N = 2000$), dotted line presents BDF2 solution ($L = N = 2000$) and continuous curve corresponds to analytical solution.

Axial displacement u_3 and pore pressure p responses are observed at points A and B, and with the help of imaginary boundary at the center of the column.

The problem is analogous to one-dimensional, so 1d analytical solutions are applicable here. Material properties are the same as for the previous problem. The problem is solved with the help of the Radau-based scheme with $Ł = 4N = 1000$; a mesh of 504 boundary elements is chosen. Figures 7 – 12 show the results of slow compression wave appearance modeling depending on the permeability coefficient.

An influence of the porous permeability on the slow compression wave appearance ($t = 0$ and $t \approx 0.0062$ sec for point A, $t \approx 0.0031$ sec and $t \approx 0.0093$ sec for point B, and $t \approx 0.00155$ sec, $t \approx 0.00465$ sec and $t \approx 0.00775$ sec for the central point) is shown subsequently to [6, 20]. Whereas works [6,20] only account for 1d analytical solution for displacements and pore pressure, here pore pressure, flux and skeleton displacement responses are numerically obtained for 3D case. Corresponding 1d numerical-analytical solutions are as well added for comparison.

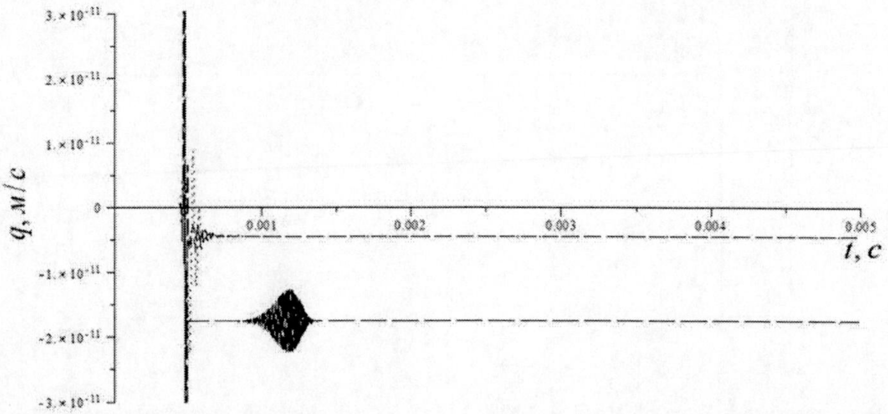

Figure 5. Pore flux in case of $\kappa = 1.9 \cdot 10^{-10}$ $m^4/(N \cdot sec)$; dashed line corresponds to Radau scheme ($L = N = 2000$), dotted line presents BDF2 solution ($L = N = 2000$) and continuous curve corresponds to analytical solution.

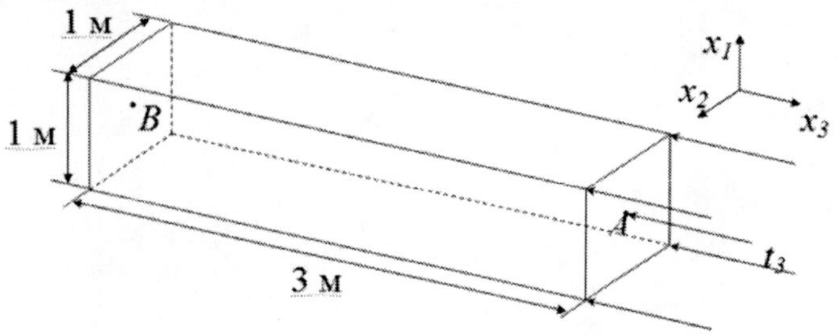

Figure 6. Poroelastic 3D column.

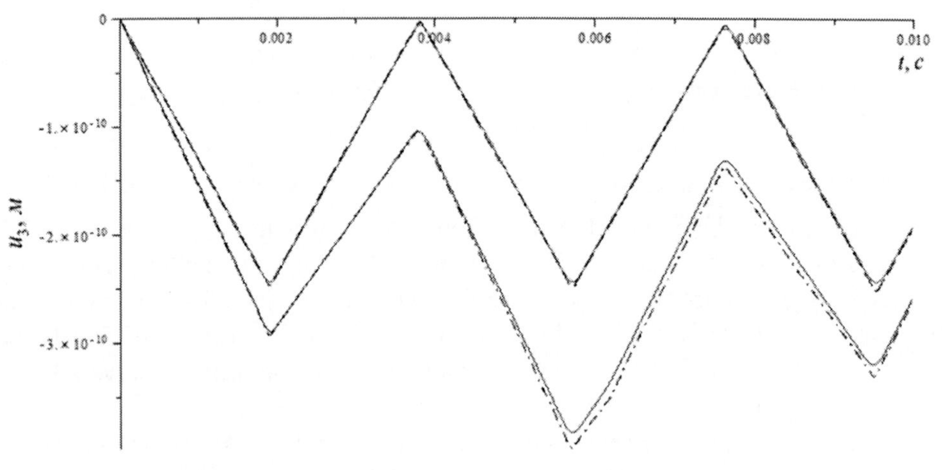

	BEM solution	Numerical-analytical solution
$\kappa = 1.9 \cdot 10^{-10}$ $m^4/(N \cdot sec)$	———	— — —
$\kappa = 1.9 \cdot 10^{-6}$ $m^4/(N \cdot sec)$	———	—·—·—

Figure 7. Displacements at point A.

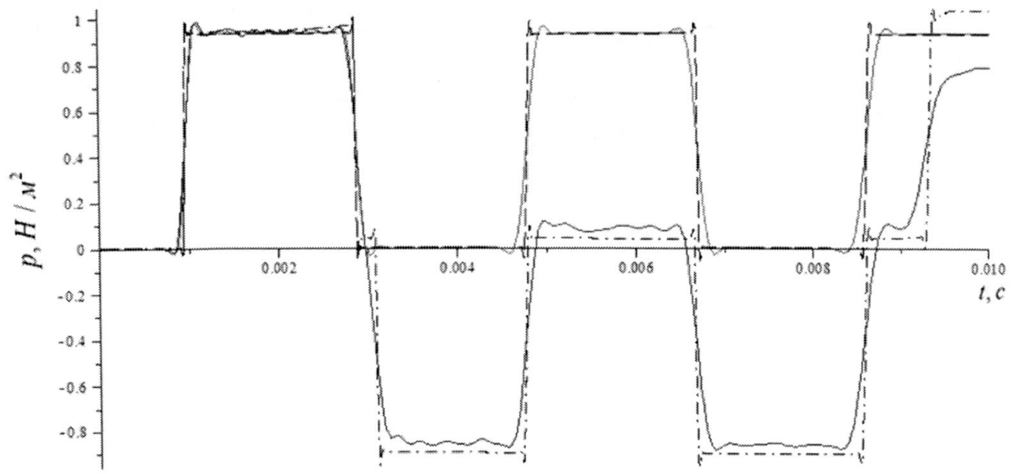

	BEM solution	Numerical-analytical solution
$\kappa = 1.9 \cdot 10^{-10}$ m^4/(N·sec)	———	– – –
$\kappa = 1.9 \cdot 10^{-6}$ m^4/(N·sec)	———	–·–·–

Figure 8. Pore pressure at point B.

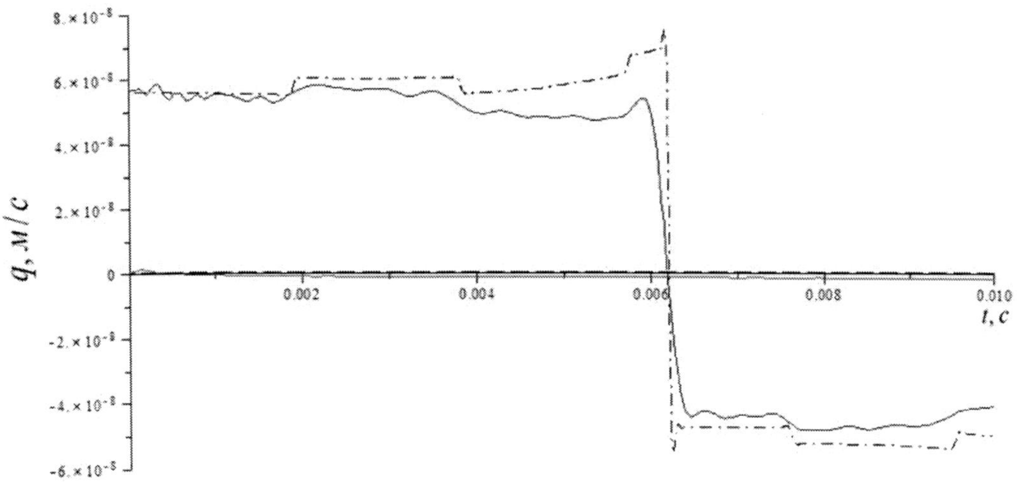

	BEM solution	Numerical-analytical solution
$\kappa = 1.9 \cdot 10^{-10}$ m^4/(N·sec)	———	– – –
$\kappa = 1.9 \cdot 10^{-6}$ m^4/(N·sec)	———	–·–·–

Figure 9. Pore flux at point A.

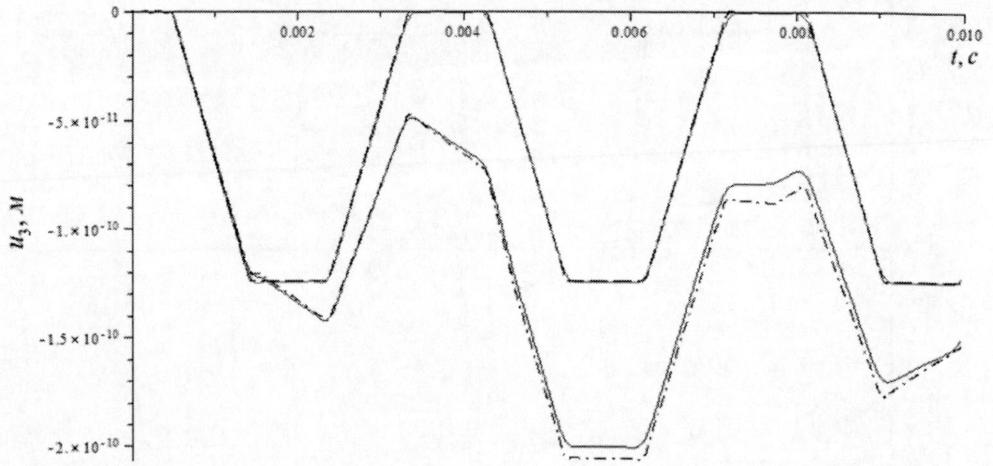

	BEM solution	Numerical-analytical solution
$\kappa = 1.9 \cdot 10^{-10}$ m^4/(N·sec)	———	— — —
$\kappa = 1.9 \cdot 10^{-6}$ m^4/(N·sec)	———	—·—·—

Figure 10. Displacements at the central point.

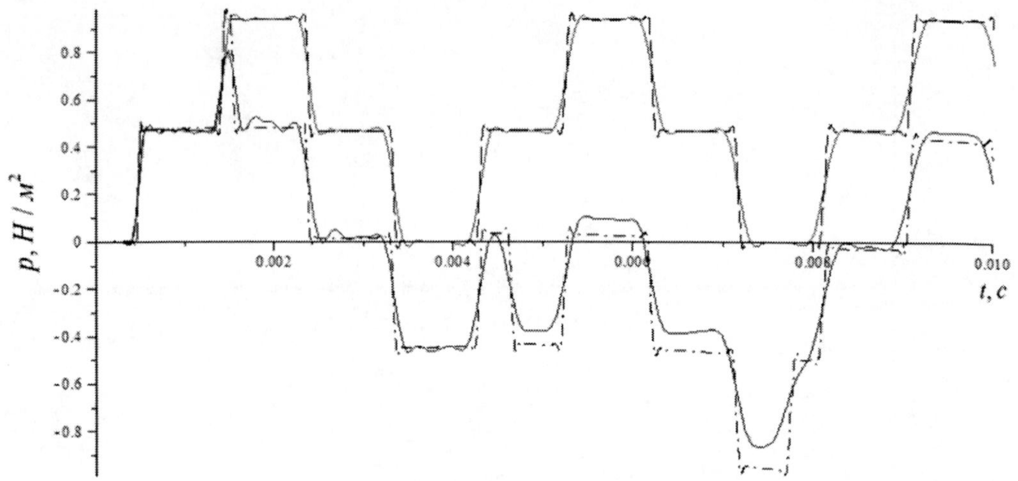

	BEM solution	Numerical-analytical solution
$\kappa = 1.9 \cdot 10^{-10}$ m^4/(N·sec)	———	— — —
$\kappa = 1.9 \cdot 10^{-6}$ m^4/(N·sec)	———	—·—·—

Figure 11. Pore pressure at the central point.

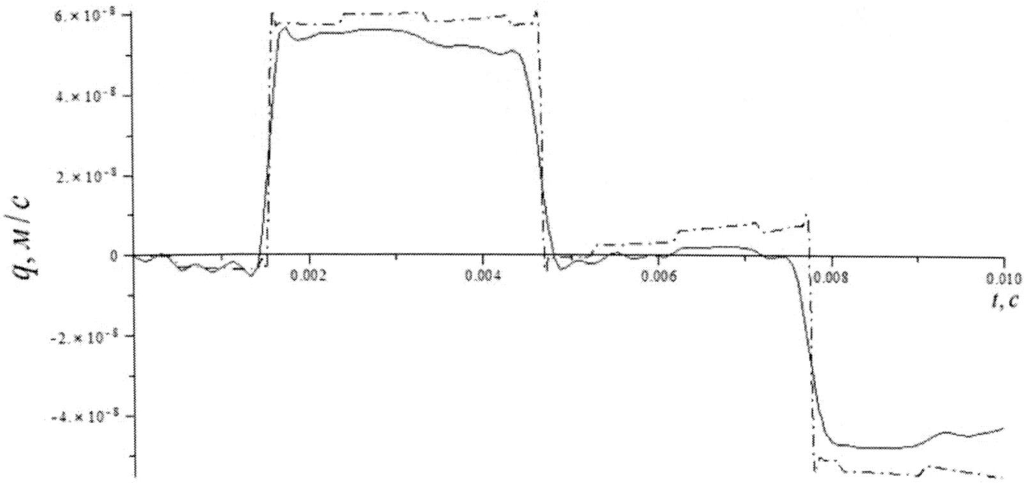

	BEM solution	Numerical-analytical solution
$\kappa = 1.9 \cdot 10^{-10}$ m^4/(N·sec)	———	– – –
$\kappa = 1.9 \cdot 10^{-6}$ m^4/(N·sec)	———	–·–·–

Figure 12. Pore flux at the central point.

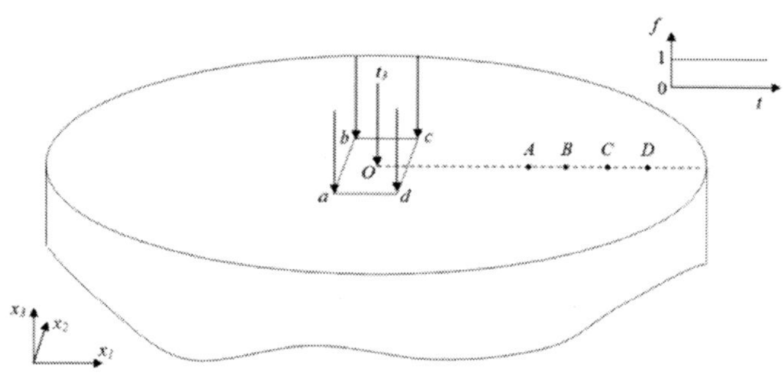

Figure 13. Homogeneous half-space.

7. Homogeneous Half-Space

A problem of vertical load $t_3 = t^0 f(t)$, $t^0 = -1000\,\text{N/m}^2$ acting on a surface element $abcd$ of homogeneous poroelastic half-space is considered, where $f(t)$ is Heaviside function (Figure 13). The area of $abcd$ is equal to $1\,\text{m}^2$. The rest of the surface is free and permeable: $p = 0$. The poroelastic material is rock with the properties: $K = 8 \cdot 10^9$ N/m^2, $G = 6 \cdot 10^9$ N/m^2, $\rho = 2458$ kg/m^3, $\phi = 0.19$, $K_s = 3.6 \cdot 10^{10}$ N/m^2, $\rho_f = 1000$ kg/m^3, $K_f = 3.3 \cdot 10^9$ N/m^2, $\kappa = 1.9 \cdot 10^{-10}$ m^4/(N·sec). The problem is treated on several boundary element meshes.

Results obtained for point D at a distance 15 m from the excitation area with help of the 1980 elements mesh are shown in Figures 14, 15. One of the studied aspects was modeling slow compression wave in terms of the half-space. For this purpose point E at 3 m depth under the excitation area is considered (see Figure 16). To calculate generalized displacement and traction responses at this point a fictitious boundary at a depth of 3 m is introduced in parallel with the surface. In this formulation the problem is quite close to the problem of layered half-space. The Radau scheme used with N =125 and $Ł$ = 1000 on $[0,\pi]$, BDF2 demanded of N = 1000. The results are shown in Figures 17 – 20 in comparison with other solutions where available. The study validates and improves the results of [7]. As compared to [7], here all basic functions including displacements, tractions, pore pressure and flux are drawn; a mesh of 1176 boundary elements is used whereas [7] uses 342 triangular elements.

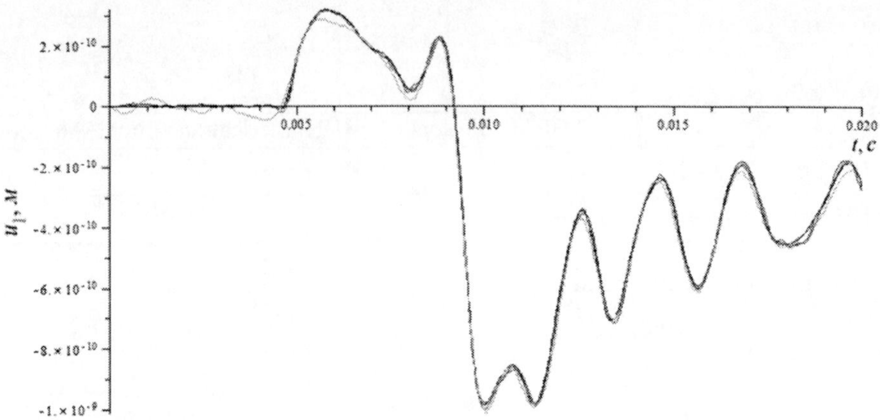

——— BDF2, $L = N = 1000$ ——— Radau, $Ł = 2N = 250$.
——— Radau, $Ł = 2N = 500$ ——— Radau, $Ł = 2N = 1000$.

Figure 14. Horizontal displacements at point D.

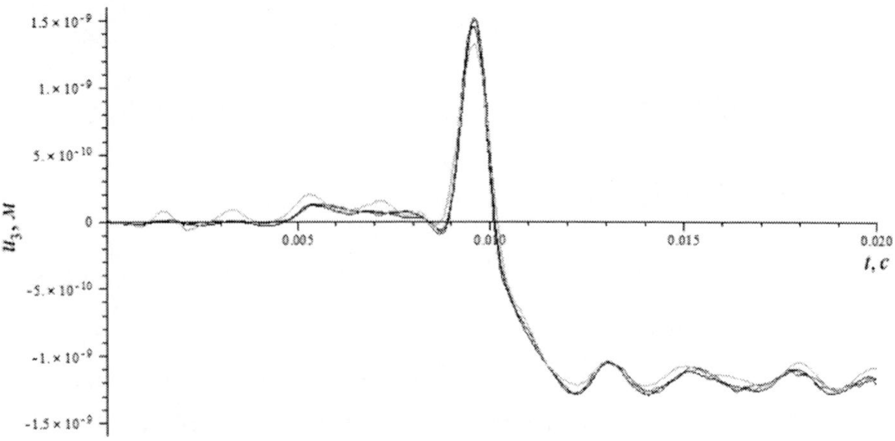

——— BDF2, $L = N = 1000$ ——— Radau, $Ł = 2N = 250$.
——— Radau, $Ł = 2N = 500$ ——— Radau, $Ł = 2N = 1000$.

Figure 15. Vertical displacements at point D.

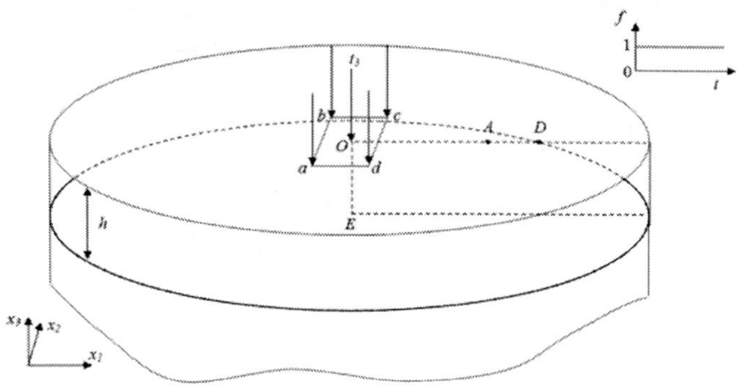

Figure 16. Two-layered half-space.

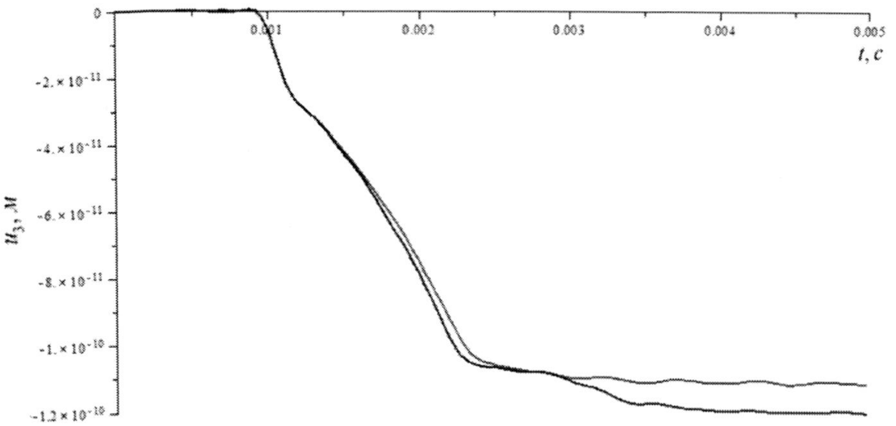

Figure 17. Vertical displacements at point E; red line presents case of $\kappa = 1.9 \cdot 10^{-10}$ m^4/(N·sec) and blue line corresponds to case of $\kappa = 1.9 \cdot 10^{-6}$ m^4/(N·sec).

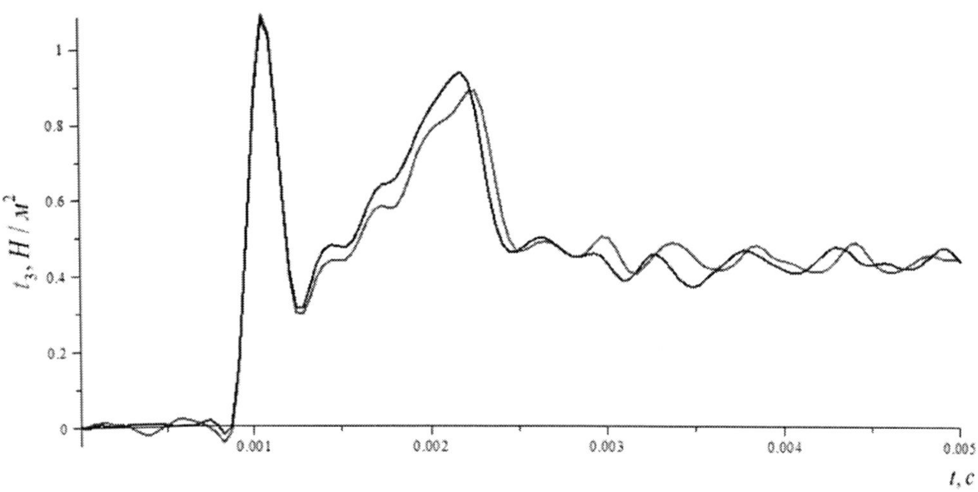

Figure 18. Vertical tractions at point E; red line presents case of $\kappa = 1.9 \cdot 10^{-10}$ m^4/(N·sec) and blue line corresponds to case of $\kappa = 1.9 \cdot 10^{-6}$ m^4/(N·sec).

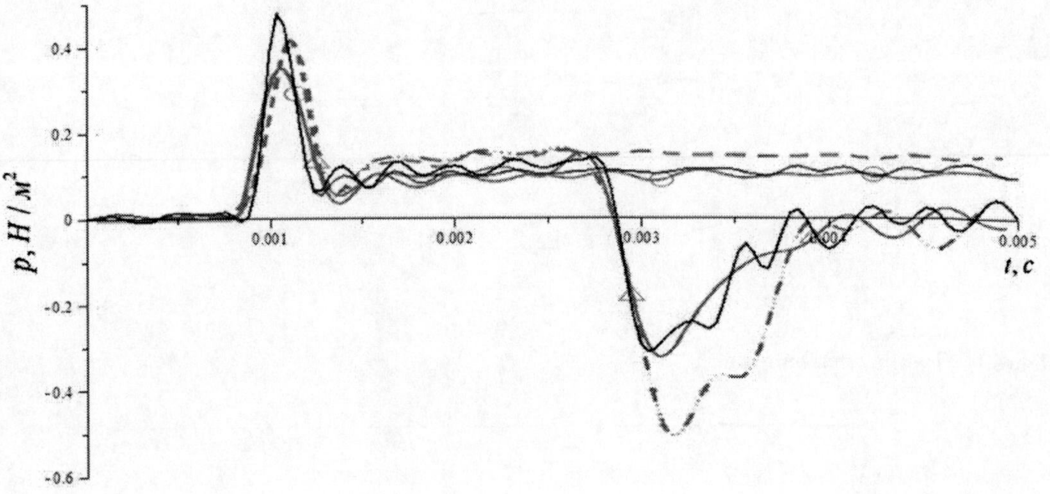

	Radau	BDF2	Solution from [7]
$\kappa = 1.9 \cdot 10^{-10}$ m^4/(N·sec)	———	—○—	– – – –
$\kappa = 1.9 \cdot 10^{-6}$ m^4/(N·sec)	———	—Δ—	—··—··—

Figure 19. Pore pressure at point E.

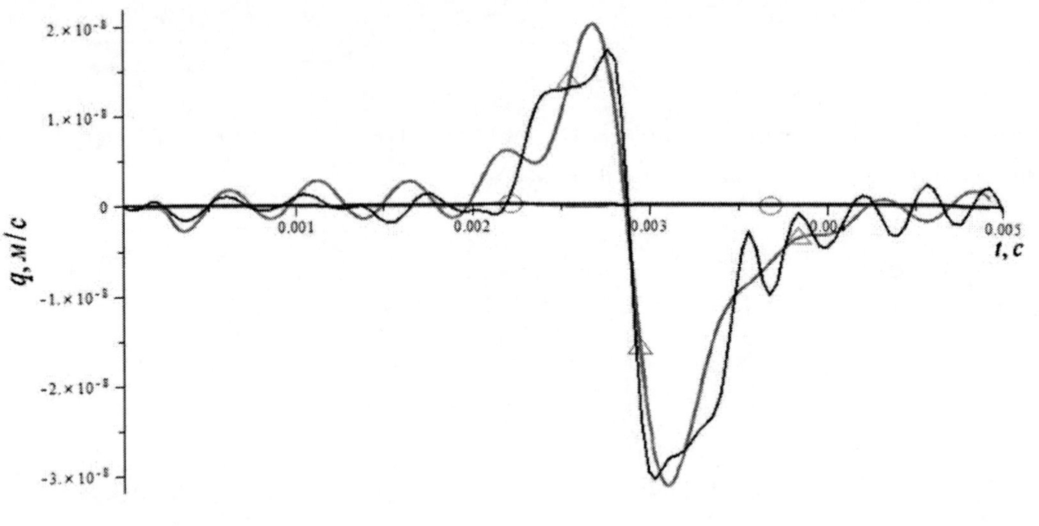

	Radau	BDF2
$\kappa = 1.9 \cdot 10^{-10}$ m^4/(N·sec)	———	—○—
$\kappa = 1.9 \cdot 10^{-6}$ m^4/(N·sec)	———	—Δ—

Figure 20. Pore flux at point E.

8. Layered Half-Space

A problem of two-layer poroelastic half-space subjected to a Heaviside-type load is considered (see Figure 16). All the parameters are the same as for the problem of homogeneous half-space except that here the layers are of different material properties: a layer of sandstone is located over the rock half-space. The sandstone properties are: $K = 2.1 \cdot 10^8$ N/m^2, $G = 9.8 \cdot 10^7$ N/m^2, $\rho = 1884$ kg/m^3, $\phi = 0.48$, $K_s = 1.1 \cdot 10^{10}$ N/m^2, $\rho_f = 1000$ kg/m^3, $K_f = 3.3 \cdot 10^9$ N/m^2, $\kappa = 3.55 \cdot 10^{-9}$ m^4/(N$\cdot$sec). Figures 21, 22 represent vertical displacements at point A in 10 m from the excitation area for sandstorm layer of thickness $h = 5$ m and $h = 10$ m, respectively.

Rayleigh wave appearance ($t \approx 0.043$ sec) can be observed after the fast compression wave arrival ($t \approx 0.01$ sec). Then ($t > 0.05$ sec), some perturbance can as well be seen, that argues for the arrival of compression waves, reflected by the sandstone/rock boundary. The chosen sandstone layer thickness still allows distinguishing the effect of Rayleigh wave due to the fact that reflected compression waves arrive later. The results here as compared to ones in [7] rely on the matched displacement-stress model whose usage is better motivated than of isoparametric one [18].

9. Half-Space with Cavity

A problem of vertical load applied to a surface element of poroelastic rock half-space with spherical or cubic cavity is considered (Figure 23).

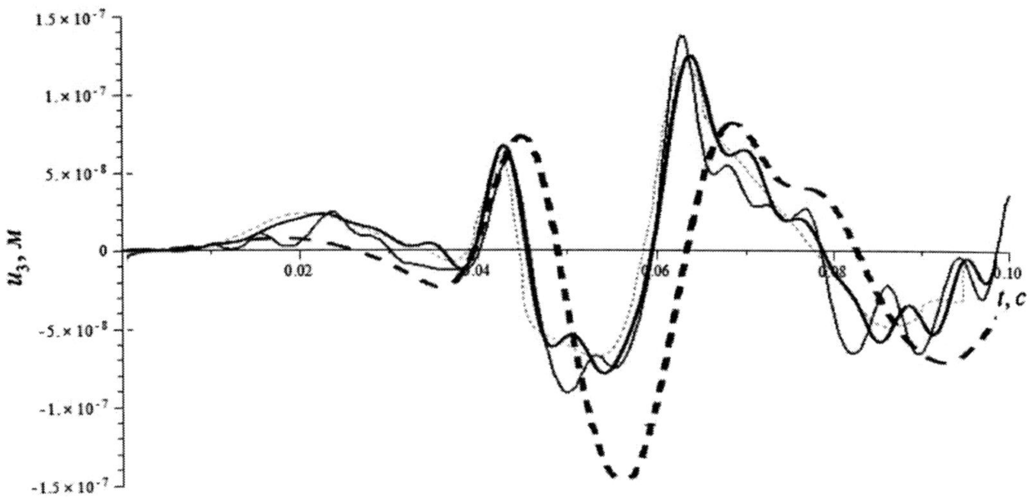

——— Radau, $L = 1000$, $N = 270$ · ········ BDF2, L = N = 1000.
——— Durbin's method ($\alpha = 5$) - - - - solution of [7].

Figure 21. Vertical displacements at point A, $h = 5$ m.

The load $P(t) = P_0 f(t)$, $P_0 = 1000\ \text{N/m}^2$ acts on the area of $1\ \text{m}^2$; the cube side/sphere diameter is 10 m; the center of cavity is located at a depth of $h = 7.5\ \text{m}$ under the excitation area.

The half-space surface is impermeable: $q = 0$. The problem is treated with help of the Radau-based scheme with $N = 250$ and $Ł = 1000$ on $[0, \pi]$. Boundary element mesh takes into account 2 symmetry planes; quarter of the mesh contains 420 elements on the surface and 324 elements on the cavity boundary. Responses of displacements and pore pressure at point 15 m from the excitation area are shown in Figures 24 – 26 in comparison with the case of poroelastic half-space without cavity.

It is seen from the plots that shape of the cavity affects the forms of responses qualitatively. Both displacements and pore pressure for the case without cavity are closer to the spherical cavity case than to the cubic cavity case.

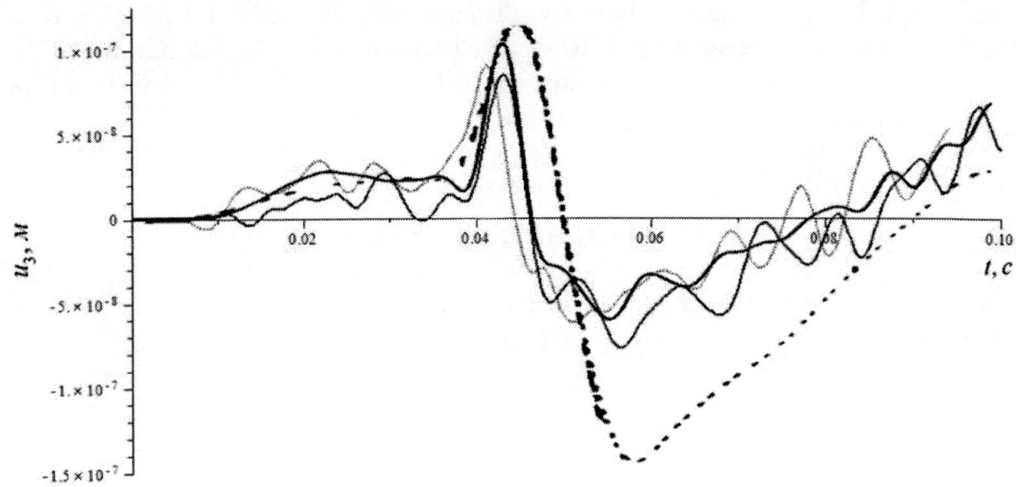

——— Radau, $Ł = 1000$, $N = 270$ ········ BDF2, $L = N = 1000$.
——— Durbin's method ($\alpha = 5$) - - - - solution of [7].

Figure 22. Vertical displacements at point A, $h = 10$ m.

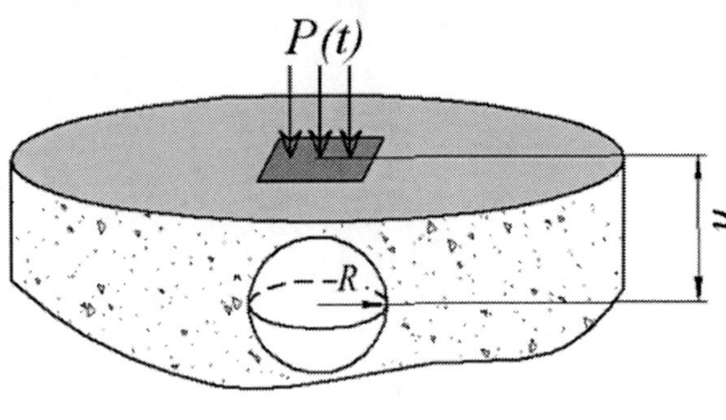

Figure 23. (Continued).

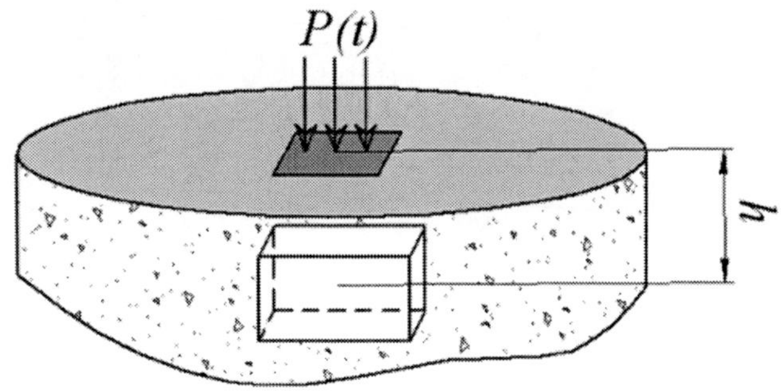

Figure 23. Half-space with cavity.

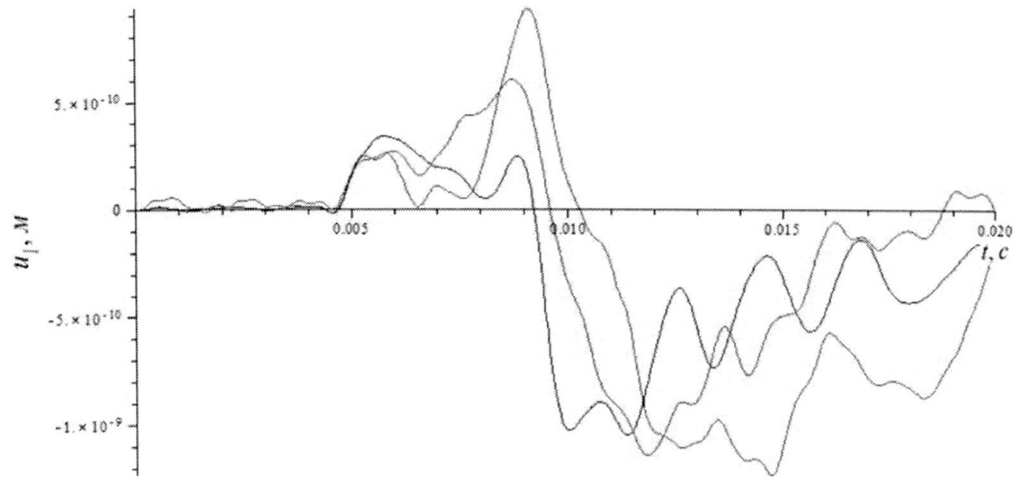

Figure 24. Horizontal displacements; red line presents case of spherical cavity, green curve corresponds to case of cubic cavity and blue line demonstrates case without cavity.

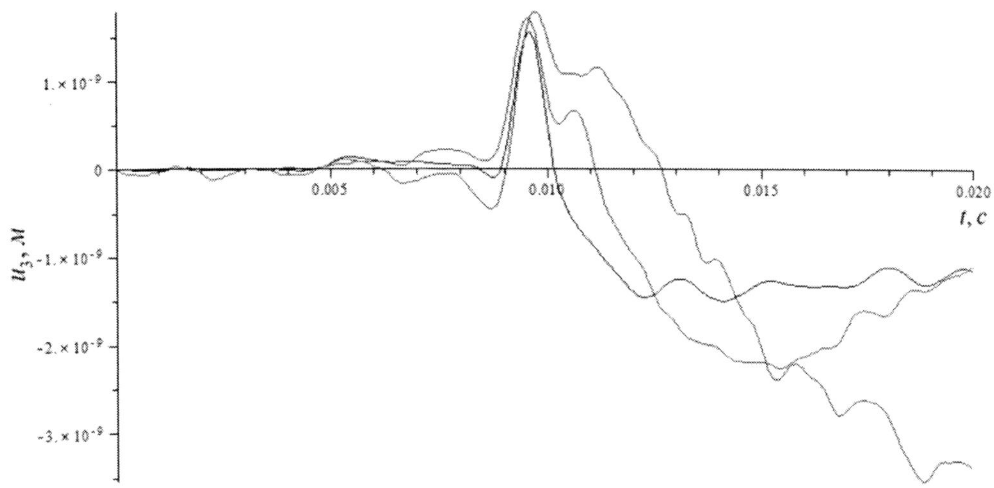

Figure 25. Vertical displacements; red line presents case of spherical cavity, green curve corresponds to case of cubic cavity and blue line demonstrates case without cavity.

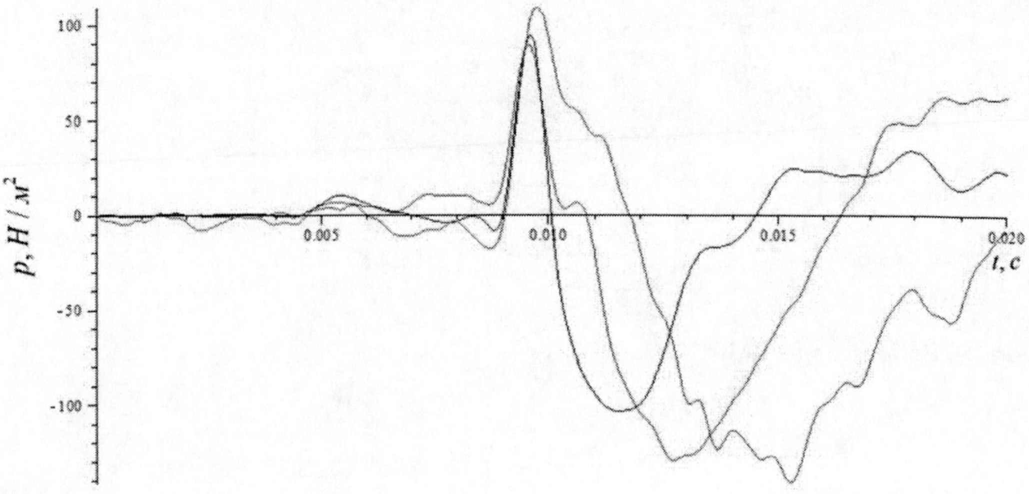

Figure 26. Pore pressure; red line presents case of spherical cavity, green curve corresponds to case of cubic cavity and blue line demonstrates case without cavity.

However there is still certain difference. For the cavity case, vertical displacements gain one more peak ($t \approx 0.011$ sec) following the arrival of Rayleigh wave in the central part of the plot ($t \approx 0.0095$ sec). In the plot of horizontal displacements the peak right before the Rayleigh wave arrival becomes much stronger ($t \approx 0.009$ sec).

ACKNOWLEDGMENTS

The work was done with the financial support of RFBR (Grants Nos. 12-08-00984, 12-01-00698, 13-08-00658, 13-08-97091, 14-08-31415, 14-08-31410).

REFERENCES

[1] Mansur, W. J., Time-Stepping Technique to Solve Wave Propagation Problems Using the Boundary Element Method, *PhD thesis*, University of Southampton, 1983.

[2] Cruze, T. A., Rizzo, F. J., *J. Math. Anal. Appl.*, 1968, vol. 22(1), 244.

[3] Chen, J., Dargush, G. F., *Int. J. Solids Structures*, 1995, vol. 32(15), 2257.

[4] Lubich, C., *Numer. Math.*, 1988, vol. 52, 129.

[5] Lubich, C., *Numer. Math.*, 1988, vol. 52, 413.

[6] Schanz, M., *Wave Propagation in Viscoelastic and Poroelastic Continua*, Springer, Berlin, 2001.

[7] Schanz, M., Antes, H., *Porous Media: Theory, Experiments and Numerical Applications,* W. Ehlers, J. Bluhm (Eds.), Springer, Berlin, 2002, pp. 1-383.

[8] Lubich, C., Ostermann, A., *Math. Comput.*, 1993, vol. 60, 105.

[9] Calvo, M. P., Cuesta, E., Palencia, C., *Numer. Math.*, 2007, vol. 107, 589.

[10] Banjai, L., Messner, M., Schanz, M., *Comput. Methods Appl. Mech. Engrg.*, 2012, vol. 245-246, 90.

[11] Banjai, L., Schanz, M., *Fast Boundary Element Methods in Engineering and Industrial Applications*, U. Langer (Ed.), Springer, Berlin, 2012, pp. 145.

[12] Banjai, L., Lubich, C., Melenk, J. M., *Numer. Math.*, 2011, vol. 119, 1.

[13] Biot, M. A., *J. Acoust. Soc. Am.*, 1956, vol. 28(2), 168.

[14] Bonnet, G., *J. Acoust. Soc. Am.*, 1987, vol. 82(5), 1758.

[15] Bonnet, G., Auriault, J.-L., *Physics of Finely Divided Matter.*, N. Boccara, M. Daoud (Eds.), Springer, Berlin, 1985, pp. 306.

[16] Amenitsky, A. V., Belov, A. A., Igumnov, L. A., Karelin, I. S., *Problems of Strength and Plasticity,* 2009, vol. 71, 164 (in Russian). http://www.unn.ru/e-library/ppp.html?anum_eng=113.

[17] Bazhenov, V. G., Igumnov, L. A., Boundary Integral Equations and Boundary Element Methods in Treating the Problems of 3D Elastodynamics with Coupled Fields, PhysMathLit, Moscow, 2008 (in Russian).

[18] Goldshteyn, R. V., *Boundary Integral Equations Method: Numerical Aspects and Application in Mechanics* (Mechanics: New in Foreign Science), Mir, Moscow, 1978, pp. 183 (in Russian).

[19] Bazhenov, V. G., Belov, A. A., Igumnov, L. A., *Boundary Element Modeling the Dynamics of Piecewise Homogeneous Media and Constructions*, Lobachevsky State University of Nizhny Novgorod, Nizhny Novgorod, 2009 (in Russian).

[20] Schanz, M., *Appl. Mech. Rev.*, 2009, vol. 62(3), 030803.

In: Advanced Materials - Studies and Applications
Editors: I.A. Parinov, S.-H. Chang, S.Theerakulpisut
ISBN: 978-1-63463-749-7

Chapter 21

SEMI-INVERSE METHOD IN THE PROBLEM OF PLASTIC DEFORMATION OF CYLINDRICAL SHELL

A. S. Yudin[1,*], S. A. Yudin[1] and T. V. Sigaeva[1,2]
[1]Institute of Mathematics, Mechanics and Computer Science, Southern Federal University, Rostov-on-Don, Russia
[2]University of Alberta, Edmonton, Canada

ABSTRACT

Plastic deformations of thin cylindrical shells under internal pressure are considered. Particular emphasis is placed on analytical approximations for the shells deformation. Large displacements and rotations as well as deformations with significant geometrical changes are included into the mathematical model. Physical nonlinearity of the material is simulated by using relations of the Davis-Nadai type.

1. INTRODUCTION

Many difficulties arise in solving large deformation problems for shells when using direct numerical methods like the shooting method [1]. Firstly, properties of the material depend on its deformable state and it is impossible to resolve the set of equations with respect to the high order derivatives explicitly. Thus, it is necessary to use the continuation method (with respect to the parameter of loading) coupled with an iterative process. Secondly, choosing initial parameters approximations for Cauchy problems, needed for application of shooting method, is also a difficult task. This difficulty can be overcome if the problem is solved in steps, starting with a small value of the loading parameter and using particular cases of systems in elastic zones. Another problem arises in simulating fixed-clamped boundary conditions. During plastic deformation, abrupt bending with the formation of a plastic joint appears on the support contour. A simple way to model this process is to establish a fixed hinge condition, while a more adequate approach is based on the introduction of a coefficient of

* Corresponding author: yudin@math.sfedu.ru.

resistance to relate the bending moment and angle of rotation. This coefficient can be determined experimentally or theoretically (using three-dimensional theory). To solve such a problem one may use the semi-inverse method, in which a deformed shape of the shell meridian is assigned *a priori*. This approach was successfully used in a number of problems for the extraction of dome-shaped shells [2–4].

The purpose of this paper is to analyze opportunities to use analytical approximations to aid in the solution of the problem of inflating cylindrical shells which are fixed on their end faces. An analytical form of the meridian for deformed shells was previously validated in [5], while this paper deals with the wider range of test functions describing the form of meridian.

2. Equations and Method

To solve this problem, we first should assume that the strains e_k are not negligible and $e_3 = \varepsilon_3$ (compression of the material in a normal direction) does not depend on the lateral coordinate ζ. Kinematic equations of the Reissner type generalized for large strains are employed [6]:

$$e_1 = (\varepsilon_1 + \zeta\kappa_1),\ \varepsilon_1 = (w'\sin\Phi + u'\cos\Phi)/\alpha_o + \cos(\Phi - \Phi_o) - 1,$$
$$\kappa_1 = \Phi_o'/\alpha_o - (1+\varepsilon_3)K_1,$$
$$e_2 = (\varepsilon_2 + \zeta\kappa_2),\ \varepsilon_2 = u/r_o,\ \kappa_2 = (\sin\Phi_o)/r_o - (1+\varepsilon_3)K_2,$$

$$K_1 = \Phi'/\alpha_o,\ K_2 = (\sin\Phi)/r_o; \tag{1}$$

$$\gamma_1 = \gamma_o/(1+\varepsilon_1),\ \gamma_o = (w'\cos\Phi - u'\sin\Phi)/\alpha_o - \sin(\Phi - \Phi_o); \tag{2}$$

$$u' = \alpha_o[\varepsilon_1\cos\Phi - \gamma_o\sin\Phi + (\cos\Phi - \cos\Phi_o)],$$
$$w' = \alpha_o[\varepsilon_1\sin\Phi + \gamma_o\cos\Phi + (\sin\Phi - \sin\Phi_o)]. \tag{3}$$

Here Φ_o and Φ are, respectively, the angles between the normal and rotation axis before and after deformation; κ_1 and κ_2 describe changes of the main curvatures. Positive direction of the normal is assumed to be directed inwards, into the shell.

All variable quantities represent functions of the longitudinal Lagrangian curvilinear coordinates. In particular. for a cylindrical shell, this coordinate is given by the axial coordinate z, while the slope angle is $\Phi_o = \pi/2$, $\alpha_o(z) = 1$, $r_o = r_c$, where r_c is the radius of the cylindrical shell's middle surface.

To reduce the three-dimensional problem to the shell problem, we introduce integral characteristics of the stress state in the virtual work of internal forces ($\zeta \in [-h_o/2,\ h_o/2]$):

$$N_1^o = (1+\varepsilon_3)\int_{(h_o)}\sigma_1(1+e_2)d\zeta,\quad N_2^o = (1+\varepsilon_3)\int_{(h_o)}\sigma_2(1+e_1)d\zeta,$$

$$Q^o = (1+\varepsilon_3) \int\limits_{(h_o)} [\sigma_{13} + \gamma\sigma_1(1+e_2)]d\zeta, \quad M_1^o = (1+\varepsilon_3)^2 \int\limits_{(h_o)} \sigma_1(1+e_2)\zeta d\zeta,$$

$$M_2^o = (1+\varepsilon_3)^2 \int\limits_{(h_o)} \sigma_2(1+e_1)\zeta d\zeta. \tag{4}$$

The principle of virtual displacements leads to the following equilibrium equations:

$$(r_o\overline{V}^o)' + \alpha_o r_o p_w^o = 0, \ (r_o\overline{H}^o)' - \alpha_o \overline{N}_2^o + \alpha_o r_o p_u^o = 0,$$

$$(r_o M_1^o)' - \alpha_o M_2^o \cos\Phi + \alpha_o r_o[\gamma_o \overline{N}_1^o - (1+\varepsilon_1)Q^o] = 0, \tag{5}$$

where

$$\overline{V}^o = \overline{N}_1^o \sin\Phi + Q^o \cos\Phi, \ \overline{H}^o = \overline{N}_1^o \cos\Phi - Q^o \sin\Phi,$$

$$p_w^o = \delta_1\delta_2 p_w, \ p_u^o = \delta_1\delta_2 p_u, \ \delta_1 = 1+\varepsilon_1, \ \delta_2 = 1+\varepsilon_2.$$

Symbols $\overline{V}^o$ and $\overline{H}^o$ denote the internal forces, oriented, respectively, along the symmetry axis and radius in the cylindrical coordinate system; p_w, p_u are the components of an external follower force acting on the surface. Additionally, the angle of transverse shear γ_o is claimed to be small and negligible in the following process of deriving constitutive equations. Furthermore, we assume that the material is incompressible (valid for metals and alloys) and this gives us

$$\overline{N}_j^o = N_j^o + (K_1 M_1^o + K_2 M_2^o)/\delta_j, \ j = 1, 2.$$

The loading diagram demonstrating the material properties can be approximated by the power function:

$$\sigma = C\bar{e}^{\eta} = \Lambda(\bar{e})\bar{e}, \ \bar{e} = (2/\sqrt{3})\sqrt{\bar{e}_1^2 + \bar{e}_1\bar{e}_2 + \bar{e}_2^2},$$

where $\bar{e}$ is the logarithmic strain intensity of the incompressible material; $\bar{e}_k = \ln(1+e_k)$; C, η are the material constants; $\Lambda(\bar{e}) = C\bar{e}^{\eta-1}$ is the secant modulus.

The Davis–Nadai constitutive equations for an incompressible material are used to relate the stresses and logarithmic strains:

$$\sigma_1 = (4/3)\Lambda(\bar{e})(\bar{e}_1 + \nu\bar{e}_2), \ \sigma_2 = (4/3)\Lambda(\bar{e})(\bar{e}_2 + \nu\bar{e}_1). \tag{6}$$

After substituting (6) into (4), the constitutive equations for generalized internal forces can be derived with the help of the method, presented in work [7]. We approximate function $\Lambda(\bar{e})$ by the first terms of the Taylor expansion of ζ in the vicinity of zero (three terms is usually enough). After integration we neglect the terms with h_o of the order higher than three. Therefore, we deal with the nonlinear equations, relating forces and moments with the logarithmic strains $\overline{\varepsilon}_k = \overline{e}_k |_{\zeta=0}$, $k = 1, 2, 3$, and the curvatures κ_1, κ_2.

To simplify the constitutive equations, it can be assumed that $\Lambda(\overline{e}) = \Lambda(\overline{\varepsilon})$, i.e., the secant modulus is replaced by its own value on the middle surface. Basically, this means that the material remains homogeneous along the thickness after deformation, what is valid for the problems with significant drawing out. Moreover, let us assume there are no zones where curvature changes dramatically, i.e., $\max\{\zeta\overline{\kappa}_1, \zeta\overline{\kappa}_2\} << 1$. In addition to this, bending of the edge is not allowed. As a result, we have the following nonlinear constitutive equations:

$$N_1^o = \overline{B}_1(\overline{\varepsilon}_1 + 0.5\overline{\varepsilon}_2),\ N_2^o = \overline{B}_2(\overline{\varepsilon}_2 + 0.5\overline{\varepsilon}_1),$$

$$M_1^o = \overline{D}_1(\overline{\kappa}_1 + 0.5\overline{\kappa}_2),\ M_2^o = \overline{D}_2(\overline{\kappa}_2 + 0.5\overline{\kappa}_1), \tag{7}$$

where

$$\overline{\kappa}_1 = \kappa_1/\delta_1,\ \overline{\kappa}_2 = \kappa_2/\delta_2,\ \overline{B}_1 = \overline{B}/\delta_1,$$
$$\overline{B}_2 = \overline{B}/\delta_2,\ \overline{D}_1 = \delta_3^2\delta_2\overline{D},\ \overline{D}_2 = \delta_3^2\delta_1\overline{D}. \tag{8}$$

Initially, the cylindrical shell has a constant thickness h_o, a diameter d_c, a radius $r_o = r_c = d_c/2$, a length $L_c = 0.2$ and a half-length $L_{05} = L_c/2$. Let us set the symmetry axis z of the circular cylinder horizontally and put the origin of the system into the middle. Then the coordinates of the end faces are: $z = -L_{05}$ and $z = L_{05}$. Because of symmetry, it is convenient to consider half of the cylinder. Assume that the second coordinate r lies along the radius. Let the end faces of the cylinder be fixed. Because of internal hydraulic pressure the shell plastically deforms and changes its form to a barrel. An ordinate of the deformed shell meridian is determined by the function $r_1(\zeta)$. Let us specify the difference between the coordinates z and ζ along the rotation axis of the shell (see Figure 1).

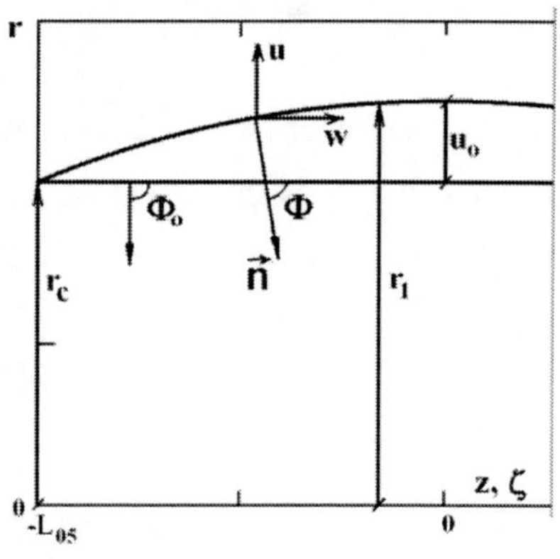

Figure 1. The coordinate system.

The Lagrangian coordinate z denotes the position of a material point on the shell meridian. It is also an independent variable in the equations of the mathematical model. ζ is the Eulerian coordinate which denotes a point in the space on the intersection of the deformed shell meridian and radial ray taking its origin in the point with a coordinate z. The coordinates z and ζ differ by a value of the axial displacement so that that $\zeta = z + w$. When w(z) is determined, it is possible to come back to the Lagrangian coordinates using the above relation.

For large deformations, the initial thickness is no longer constant along the axial coordinate due to lateral contraction (compression along the normal). Consideration of this factor has a significant impact on the strains in tangential directions.

Maximum thickness can be observed in the points of binding; minimum thickness is located in the point of the largest displacements u_0. In the drawing out process of the dome, with an average height, from a circular plate, the form of the dome's meridian resembles a circular arc (weakly elliptical), while the thickness distribution depends quadratically on the slope angle of the normal (or on the arc length) [2– 4]. It is validated theoretically as well as experimentally.

Based on this result and similarity of the meridian forms the thickness can be assumed to be a quadratic form of the coordinate ζ:

$$h(\zeta) = h_o[1-\delta(1-\zeta^2)] \cdot \tag{9}$$

The coefficient δ can be determined from relations between shell volumes V_0 and V_1 in the initial and deformed states, respectively

$$V_0 = 2\pi r_c h_o L_{05}, \quad V_1 = 2\pi \int_{-L_{05}}^{0} r_1(\zeta)\alpha_o(\zeta)h(\zeta)d\zeta,$$

$$2\pi \int_{-L_{05}}^{0} r_1(\zeta)\alpha_o(\zeta)h(\zeta)d\zeta = 2\pi r_c h_o L_{05} . \tag{10}$$

On substituting (9) into (10) we obtain

$$\int_{-L_{05}}^{0} r_1(\zeta)\alpha_o(\zeta)[1-\delta(1-\zeta^2)]d\zeta = r_c L_{05} . \tag{11}$$

And after integration, from (11)

$$\delta = (J_1 - r_c L_{05})/(J_1 - J_2), \tag{12}$$

where

$$J_1 = \int_{-L_{05}}^{0} r_1(\zeta)\alpha_o(\zeta)d\zeta, \quad J_2 = \int_{-L_{05}}^{0} r_1(\zeta)\alpha_o(\zeta)\zeta^2 d\zeta \cdot \tag{13}$$

For the thickness to be adjusted, it should be given by a fourth-order power function, e.g.

$$h(\zeta) = h_o[1 - \delta(1-\zeta^2) - \delta_1(1-\zeta^4)]. \tag{14}$$

The coefficients δ and δ_1 can be related with the help of the relation between V_0 and V_1:

$$\delta_1 = [J_1 + \delta(J_4 - J_1) - r_c L_{05}]/(J_1 - J_4), \tag{15}$$

where

$$J_4 = \int_{-L_{05}}^{0} r_1(\zeta)\alpha_o(\zeta)\zeta^4\, d\zeta,$$

and J_1 is the same as in the expressions (12) and (13). Here δ is the primary parameter and δ_1 depends on it.

It should be noted that numerical experiments do not demonstrate any advantages of the dependence (14) over (9). Therefore, in this work we claim that the quadratic approximation is good enough for our purposes.

To determine the axial displacement w(z) we use the incompressibility equation, but this time the volumes of the initial and deformed states depend on the coordinate z: $V_z(\zeta(z)) - V_{0z}(z) = 0$, or simply

$$\int_{-L_{05}}^{\zeta(z)} [r_1(\xi)\alpha_o(\xi)h(\xi)]d\xi - h_o(z + L_{05}) = 0. \tag{16}$$

Let us add and subtract a constant h_o in (16), integrate it and then substitute $\zeta(z) = z + w(z)$:

$$h_o[z + w(z) + L_{05}] = h_o(z + L_{05}) + \int_{-L_{05}}^{z+w(z)} [h_o - r_1(\zeta)\alpha_o(\zeta)h(\zeta)]d\zeta. \tag{17}$$

From (17) we obtain:

$$w(z) = \frac{1}{h_o}\int_{-L_{05}}^{z+w(z)} [h_o - r_1(\zeta)\alpha_o(\zeta)h(\zeta)]d\zeta. \tag{18}$$

Using (18), we can build a simple iterative process on the basis of contraction mapping principle [8]

$$w_k(z) = \frac{1}{h_o}\int_{-L_{05}}^{z+w_{k-1}(z)} [h_o - r_1(\zeta)\alpha_o(\zeta)h(\zeta)]d\zeta. \tag{19}$$

The function w(z) is equal to zero at $z = -L_{05}$ и $z = 0$, negative inside this range and has a minimum there. The following function meets these conditions

$$w_0(z) = W_0 z(1 + z / L_{05}) \tag{20}$$

and can be used as a zero-order approximation. The iterative process (19) converges rapidly after one or two iterations only. Moreover, the second iteration is used to verify the first one. If W_0 in (20) is iteratively chosen so that (20) coincides with the minimum value of w(z), then (20) is already a good approximation and can be used in the following derivations.

After w(z) has been determined, all parameters of the shell in the deformed state can be represented in the Lagrangian coordinates. Local application of the incompressibility property to a point on the middle surface allows for determining displacement and strains from the relation

$$[1 + \varepsilon_1(z)] \cdot [1 + \varepsilon_2(z)] \cdot [1 + \varepsilon_3(z)] = 1 \cdot \tag{21}$$

Thus, the radial displacement can be expressed as $u(z) = r_1(\zeta(z)) - r_c$; the elongation in the radial direction is $\varepsilon_2(z) = u(z) / r_c$; the normal strain along the thickness is

$$\varepsilon_3(z) = [h(\zeta(z)) - h_o] / h_o] < 0.$$

The strain along the meridian is found from (21):

$$\varepsilon_1(z) = [\delta_2(z) \cdot \delta_3(z)]^{-1} - 1,$$

where $\delta_j(z) = 1 + \varepsilon_j(z)$, $j = 1, 2, 3$. The thickness of the deformed shell is $h(z) = h_o \cdot \delta_3(z)$. The intensity of the logarithmic strains of the middle surface is given by

$$\bar{\varepsilon}(z) = (2 / \sqrt{3}) \sqrt{\bar{\varepsilon}_1(z)^2 + \bar{\varepsilon}_1(z) \cdot \bar{\varepsilon}_2(z) + \bar{\varepsilon}_2(z)^2},$$

where

$$\bar{\varepsilon}_j(z) = \ln(1 + \varepsilon_j(z)) = \ln(\delta_j(z)).$$

Let the following functions to denote the slope angles of the tangent and normal:

$$f_1(\zeta) = -\mathrm{arctg}(r_1(\zeta)), \; f(\zeta) = f_1(\zeta) + \pi / 2,$$
$$\Phi(z) = f(\zeta(z)), \; \Phi_1(z) = \pi / 2 - \Phi(z).$$

By integrating the first two equilibrium equations from (5) and on substituting the obtained relations into (3), we find the uniform pressure

$$p = G_1(z)/G_2(z) \approx \text{const}, \tag{22}$$

where

$$G_1(z) = [-(r_o M_1^o)' + M_2^o(z)\cos\Phi(z)]\delta_1^{-1}(z) + I_{21}(z)\sin\Phi(z) + C_H \sin\Phi(z) - C_V \cos\Phi(z),$$

$$G_2(z) = I_{11}(z)\cos\Phi(z) + I_{22}(z)\sin\Phi(z),$$

$$I_{11}(z) = \int_{-L_{05}}^{z} \delta_1(\xi)\delta_2(\xi)\sin\Phi_1(\xi)d\xi,\ I_{21}(z) = \int_{-L_{05}}^{z} \overline{N}_2^o(\xi)d\xi,$$

$$I_{22}(z) = \int_{-L_{05}}^{z} \delta_1(\xi)\delta_2(\xi)\cos\Phi_1(\xi)d\xi.$$

In fact, the pressure should be constant. Therefore, in the semi-inverse method we look for parameters of the deformed shell satisfying this condition. The equation (22) includes constants of integration C_H and C_V. They denote, respectively, the axial and radial reactions corresponding to the binding of the end faces. Moreover, they represent controlled parameters from the right-hand side of the equation (22).

3. CALCULATIONS

Let us consider a cylindrical shell that has constant initial thickness $h_o = 3.8 \cdot 10^{-3}$ m; diameter $d_c = 0.2$ m, radius $r_o = r_c = d_c / 2 = 0.1$ m, length $L_c = 0.2$ m, half-length $L_{05} = L_c / 2 = 0.1$ m. The side view of the shell can be investigated with the help of 0.2m x 0.2 m square.

Noncorrosive steel 12X18H10T can be used as a material for the cylinder because of its significant plasticity properties. Major points of the plasticity diagram are: the proportional limit is $\sigma_{02} = 360$ MPa, $\varepsilon_{02} = 0.001714$; the fracture point is $\sigma_B = 720$ MPa, $\varepsilon_B = 0.615$. With the help of these limits, we can determine parameters to approximate the diagram displaying the nonlinear material properties as follows

$$\sigma(\overline{\varepsilon}) = C\overline{\varepsilon}^{\eta},$$

where

$$\eta = [\ln(\sigma_B) - \ln(\sigma_{02})]/[\ln(\overline{\varepsilon}_B) - \ln(\varepsilon_{02})] = 0.1178,\ C = \sigma_B / \overline{\varepsilon}^{\eta} = 762.45.$$

The secant modulus is $\Lambda(\overline{\varepsilon}) = C\overline{\varepsilon}^{\eta-1}$.

In the dimensionless analysis we use values of σ_B and the initial shell radius to normalize stresses and displacements, respectively.

For instance, let us set the maximum displacement $u_0 = 0.035\,\text{m}$, what is 35% of the initial radius. Then the maximum radius of the shell in the deformed state will be $r_b = 0.135$ m. Therefore, the dimensionless parameters are $\tilde{r}_c = 1.0$, $\tilde{u}_b = 0.35$, $\tilde{r}_b = 1.35$, $\tilde{L}_{05} = 1.0$.

Now, let us use different smooth curves passing through the edge points $(\mp\tilde{L}_{05}, \tilde{r}_c)$ and the maximum displacement point $(0, \tilde{r}_b)$ as test functions to describe a form of the meridian in the deformed state. To ensure the smoothness of the curvature, these functions specifying dependence of the meridian ordinate on the axial variable $\tilde{z}$ must be easily differentiated. Examples of such functions are the following:

1) Power function $\tilde{f}_p(\tilde{z}) = \tilde{r}_b - \tilde{u}_b|\tilde{z}|^s$,
   where s is varied and can be real; when s=2 the function describes a parabola.
2) Trigonometric powers of cosine: $\tilde{f}_c(\tilde{z}) = \tilde{A}_c[\cos(N_c\tilde{z})]^s + \tilde{B}_c$,
   where $\tilde{A}_c = -\tilde{u}_b / \{[\cos(N_c\tilde{L}/2)]^s - 1\}$, $\tilde{B}_c = \tilde{r}_b - \tilde{A}_c$; N_c and s are varied.
3) Hyperbolic powers of cosine: $\tilde{f}_h(\tilde{z}) = \tilde{A}_h[\cos(N_h\tilde{z})]^s + \tilde{B}_h$,
   where $\tilde{A}_h = -\tilde{u}_b / \{[\mathrm{ch}(N_h\tilde{L}/2)]^s - 1\}$, $\tilde{B}_h = \tilde{r}_b - \tilde{A}_c$; N_h and s are varied.
4) Circular arc: $\tilde{f}_o(\tilde{z}) = \sqrt{R_o - \tilde{z}^2} - \tilde{B}_o$,
   where $\tilde{R}_o = \tilde{r}_b + \tilde{B}_o$ is arc radius, $\tilde{B}_o = [(\tilde{L}/2)^2 - 2\tilde{r}_c\tilde{u}_b - \tilde{u}_b^2] / (2\tilde{u}_b)$.
   This function, in turn, is uniquely determined since the arc can be specified using three points only.
5) Elliptical arc: $\tilde{f}_e(\tilde{z}) = k_e\sqrt{a_e^2 - \tilde{z}^2} - \tilde{B}_e$,
   where a_e, b_e are the semi-axes of an ellipse, $k_e = b_e / a_e$,

$$\tilde{B}_e = [(2a_e/\tilde{L})^2 - 1]\tilde{u}_b - \tilde{r}_c + (2a_e/\tilde{L})\tilde{u}_b\sqrt{[1 - (2a_e/\tilde{L})^2};$$

   while a_e, k_e are the parameters to control a form.

6) Combinations of the function with weight coefficients.

Option 1 is the mixture of the functions 2) and 5):

$$\tilde{f}_1(\tilde{z}) = m_1\tilde{f}_c(\tilde{z}) + (1 - m_1)\tilde{f}_o(\tilde{z});$$

Option 2 is the mixture of the functions 2) and 1):

$$\tilde{f}_2(\tilde{z}) = m_2\tilde{f}_c(\tilde{z}) + (1 - m_2)\tilde{f}_p(\tilde{z});$$

Option 3 is the mixture of the functions 2), 1) and 5):

$$\tilde{f}_3(\tilde{z}) = m_1\tilde{f}_c(\tilde{z}) + m_2\tilde{f}_p(\tilde{z}) + m_3\tilde{f}_o(\tilde{z}),$$

where $m_1 + m_2 + m_3 = 1$.

Since any function of 1) –5) passes through the certain points, their combinations also pass through them. Some other functions, e.g., of the logarithmic type, were considered and thoroughly analyzed, but not given therein.

The algorithm of matching for acceptable approximations can be described as follows. Firstly, it is necessary to choose parameters for every approximation 1) – 5), which will make the function from (22) as close to the constant as possible. Then two or three of the best functions are selected to be combined in mixtures.

Let us consider an example of the shell dilatation, in which the displacements in the middle will be $\tilde{u}_b = 0.35$, thus, $\tilde{r}_b = 1.35$. The matching process for the test functions gives following values of the unknown parameters:

1) $s = 1.5 - 1.8$; 2) $s = 1.5$, $N_c = 1.5$, $\tilde{A}_c = 0.3567$, $\tilde{B}_c = 0.9933$;

3) $s = 1.2$, $N_h = 0.5$, $\tilde{B}_h = 2.3627$, $\tilde{A}_h = -1.0127$; 4) $\tilde{B}_o = 0.2536$, $\tilde{R}_o = 1.6036$.

Comparison of these results demonstrates that the best mixture is composed of the cosine, power function and circular arc with the weight coefficients: $m_1 = 0.4$, $m_2 = 0.25$, $m_3 = 0.35$. In the central part of the profile $\tilde{z} \in [-0.95,\ 0.1]$ the pressure has only 4 – 6% deviation from its mean value, $p_{mid} \approx 1.7$ MPa. This error is much higher on the edges (~14%). However, the solution of the Cauchy problem for the system of equations (1)–(3), (5)–(7), subject to initial conditions on the left edge which are given by the semi-inverse method, agrees well with the selected geometry. To improve accuracy, it is necessary to develop algorithms to solve this boundary-value problem iteratively with initial approximations given by semi-inverse method solutions.

ACKNOWLEDGMENTS

The chapter was prepared under theme No. 213.01-11/2014-29 as a basic part of the government research financed by the Russian Department of Education and Science.

REFERENCES

[1] Yudin, SA. In: *Proceedings of Post-graduate Students and Competitors of Rostov State* University, TerraPrint, Rostov-on-Don, 2006, vol. 12, 38–40 (In Russian).

[2] Yudin, AS; Yudin, SA. In: *Proceedings of the 10th International Conference on Modern Problems of Mechanics of Solid Medium*, CVVR, Rostov-on-Don, 2006, vol. 1, 290–294 (In Russian).

[3] Yudin, AS; Yudin, SA. In: *Proceedings of the 11th International Conference on Modern Problems of Mechanics of Solid Medium*, CVVR, Rostov-on-Don, 2007, vol. 1, 255–259 (In Russian).

[4] Yudin, AS; Yudin, SA. In: *Proceedings of the International Conference on Models and Algorithms for Simulation of Physical-Chemical Processes*, TSPI Press, Taganrog, 2008, 86–94 (In Russian).

[5] Yudin, SA; Yudin, AS. In: *Proceedings of the13th International Conference on Modern Problems of Mechanics of Solid Medium*, Southern Federal University Press, Rostov-on-Don, 2009, vol. 1, 213–217 (In Russian).

[6] Yudin, AS. Izvestiya of North-Caucasus Scientific Center of High School. *Nat. Sci.*, 1978, No. 3, 34–37 (In Russian).

[7] Yudin, AS; Yudin, SA. In: *Proceedings of the 12th International Conference on Modern Problems of Mechanics of Solid Medium*, CVVR, Rostov-on-Don, 2008, vol. 1, 223 – 227 (In Russian).

[8] Lebedev, LP; Vorovich, II. Functional Analysis in Mechanics, Springer, Berlin, 2002, 1–238.

In: Advanced Materials - Studies and Applications
Editors: I.A. Parinov, S.-H. Chang, S.Theerakulpisut
ISBN: 978-1-63463-749-7

Chapter 22

DEFINITION OF STRESS-STRAIN STATE OF LAYERED ANISOTROPIC ELLIPTIC CONSTRUCTION UNDER ACTION OF IMPULSE LOAD

***I. P. Miroshnichenko*[1,*] *and V. P. Sizov*[2]**
[1]Don State Technical University, Rostov-on-Don, Russia
[2]Southern Federal University, Rostov-on-Don, Russia

ABSTRACT

New method is present to define dynamic elastic fields of displacements, stresses and strains in layered elliptic constructions with transverse-isotropic layers under action impulse load, based on tensor relations of generalized method of scalarization of the dynamic elastic fields in the transverse-isotropic media.

1. CONSTITUTIVE EQUATIONS AND MAIN CONCEPTS

Let us consider a multilayer construction consisting of N layers, fixed rigidly one to others (see Figure 1). The layers are numerated by numbers from 1 to N beginning with initial layer, internal boundaries of each n-th layer are designated as S_{n-1}, and internal ones as S_n ($n = 1,\ldots, N$).

The layers can have arbitrary thickness and be fabricated from various isotropic and transverse-isotropic construction materials with their specific physical and mechanical characteristics and also demonstrate properties of liquid and gaseous matters.

Some areas (V_1, V_2, V_3,...) of the construction are subject to external loading depending on time and coordinates. The influence could be of impulse kind (single pulse or several pulses of pressure). Restrictions on a form of the given pulses and areas of their applications are absent.

* Corresponding author: I. P. Miroshnichenko, e-mail: ipmir@rambler.ru.

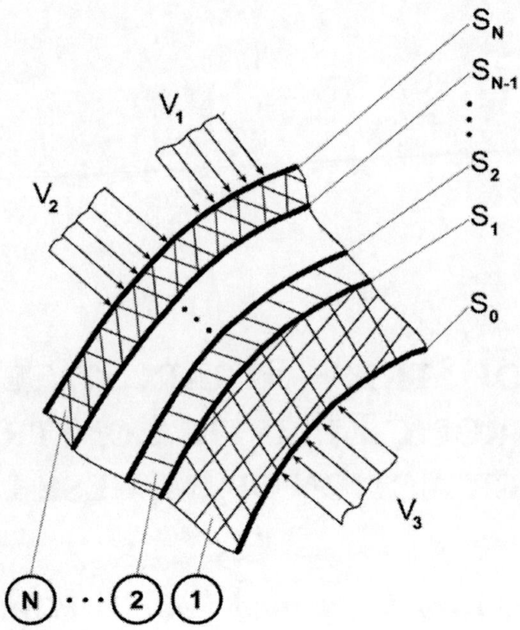

Figure 1. Considered problem.

Under these conditions, complex stress-strain state arises in the construction, formed not only initial load but perturbations, caused by numerous re-reflections of waves from heterogeneities (boundaries of the layers), interference phenomena and diffraction.

It is required to define this stress-strain state in the construction. We must define fields of displacements, stresses and strains in arbitrary given areas of the construction at any moment of time as during loading process as after it.

The unknown displacements, stresses and strains in elastic statement of the problem are described in each of the layers without sources by using the following constitutive relations written in invariant form [1,2]:

(i) motion equations in stresses for unit element of continuous medium:

$$\rho \frac{\partial^2 U^i}{\partial t^2} = \nabla_j \sigma^{ij}; \tag{1}$$

(ii) Hooks constitutive relations:

$$\sigma^{ij} = C^{ijrs} \varepsilon_{rs}; \tag{2}$$

(iii) Cauchy relations (geometrical relations):

$$\varepsilon_{rs} = \frac{1}{2}\left(\nabla_r U_s + \nabla_s U_r\right). \tag{3}$$

We use in Equation (1) and relations (2), (3) the following designations: ρ is the density of construction material, U^i are the contravariant components of displacement vector, ∇_j is the covariant derivative, σ^{ij} are the contravariant components of stress tensor, C^{ijrs} are the contravariant components of elastic moduli of construction material layer, ε_{rs} are the covariant components of small strains tensor, U_i are the covariant components of displacement vector, i, j, r, s are the tensor indices.

By substituting (3) into (2) and then into (1), we obtain the set of motion equations in displacements:

$$\rho \frac{\partial^2 U^i}{\partial t^2} = \nabla_j C^{ijrs} \nabla_{(r} U_{s)}, \tag{4}$$

in which we conduct a symmetrization on indices *r* and *s* in brackets.

To define displacements from (4), it is necessary to form boundary and initial conditions to which a stress-strain state should satisfy in the treated layered construction.

The solution of the set of motion equations (4) satisfying the given boundary and initial conditions describes totally the stress-strain state in treated layer of the construction.

A definition of the solution in common case of arbitrary geometry and anisotropy presents significant difficulties, connected with vector character of the equations. In this case to define a field, one numerical value only is insufficient (as in the case, for instance, of liquid medium, where vector of the liquid velocity is determined by gradient of one parameter, namely scalar potential) [4].

In the considered case, it is necessary to calculate three numerical values in each point of space. It is very important to choose rightly three scalar functions defining a vector field. For example, if we select as these functions the Cartesian components of vector U_x, U_y, U_z, then a satisfaction of the obtained solutions to the boundary conditions will be difficult, because the Cartesian components are connected with a curve-linear (cylindrical or elliptic) boundary surface by a complex way.

If we define vector $\overline{U}$ via its components in the coordinate system corresponding boundary, then other difficulty arises, namely scalar equations do not divide into three independent equations each of which including only one component. So we obtain a set of three equations each of which contains all three functions, that leads to necessity to solve a differential equation of sixth order [5].

Conventional solution method of the vector boundary-value problems of the dynamic elasticity theory in isotropic media, when Equation (4) transforms to Lame equation of motion, consists in division of vector field into two parts, namely longitudinal and transverse ones. In this case, a simplification of the problem is possible only by using the orthogonal coordinate systems in which the division of variables is possible and which satisfy Bromwich conditions [5].

As a result, three obtained scalar wave equations describe longitudinal and transverse waves, respectively of two mutually perpendicular polarizations. The difficulty, connected with mutual transformation of these waves at boundaries, is solved in those cases, when construction surface coincides with any coordinate surface. In the paper [6], this method has been generalized to the case of transverse-isotropic media and coordinate systems with admitted transforms of the kind:

$$\tilde{x}^J = x^J;\ \tilde{x}^K = f^K\left(x^K, x^N\right);\ \tilde{x}^N = f^N\left(x^K, x^N\right), \tag{5}$$

where the indices pointed by letters J, K, N are fixed and correspond to local affine basis $\bar{e}_J, \bar{e}_K, \bar{e}_N$, but $f^K\left(x^K, x^N\right)$ and $f^N\left(x^K, x^N\right)$ are the continuously differentiated unique functions; the mark of tilde corresponds to "new" coordinates, the "old" coordinates have not the mark of tilde (they could be the Cartesian coordinates).

The group of transforms (5) includes also orthogonal transforms being their subgroup relatively of the "multiplication" operation.

At coincidence of main axis of symmetry for the treated medium with invariant vector $\bar{e}_J$, Equation (4) may be present in the case of monochromatic waves as

$$-\left\{\left[-\left(\frac{C_{11}-C_{12}}{2}\right)+C_{44}\right]\nabla_J\nabla_J+\omega^2\rho\right\}U_i = C_{11}\nabla_i\nabla^s U_s - \left(\frac{C_{11}-C_{12}}{2}\right)\delta_{ijk}^{rsk}\nabla^j\nabla_s U_s +$$

$$+\left[-\left(\frac{C_{11}-C_{12}}{2}\right)+C_{13}+C_{44}\right]\nabla_i\nabla_J U_J + \delta_i^J\left\{\left[-\left(\frac{C_{11}-C_{12}}{2}\right)+C_{13}+C_{44}\right]\nabla_J\nabla^s U_s +\right.$$

$$\left.+\left[-\left(\frac{C_{11}-C_{12}}{2}\right)+C_{44}\right]\nabla^s\nabla_s U_J + \left[C_{13}+C_{33}-2\left(C_{13}+2C_{44}\right)\right]\nabla_J\nabla_J U_J\right\}, \tag{6}$$

where ω is the circular frequency; constants C_{ij} are the elements of the matrix of elastic constants written on convolution index [7], but δ_i^J is Krönecker symbol.

The dependence of wave movement on coordinate x has the form $\exp\left[ihx^J\right]$ and therefore

$$\nabla_J = \nabla^J = \frac{\partial}{\partial x^J} = ih, \tag{7}$$

where h is the absolute value of wave vector projection on direction of $\bar{e}_J$.

As it has been shown in [5], the solution of Equation set (6) reduced to solution of three scalar wave equations in respect to functions φ, ω, v, describing quasilongitudinal, quasitransverse and transverse waves:

$$\nabla \begin{pmatrix} \phi \\ \omega \\ v \end{pmatrix} + g^2 \begin{pmatrix} \varphi \\ \omega \\ v \end{pmatrix} = 0. \tag{8}$$

In order to find wave numbers $g = \omega / v_\Phi$, where v_Φ is the phase velocity of propagation of the corresponding waves, we use the following dispersion equations [6]:

$$2v_\Phi^2 \rho = (C_{11} - C_{33})\sin^2 Q + (C_{33} + C_{44}) \pm$$

$$\pm \left(\left[(C_{11} + C_{33} - 2C_{44})\sin^2 Q - (C_{33} - C_{44}) \right]^2 + \left[2(C_{13} + C_{44})\sin Q \cos Q \right]^2 \right)^{\frac{1}{2}}, \tag{9}$$

where upper sign before figure brackets relates to the quasilongitudinal waves, but lower sign corresponds to the quasitransverse waves.

To find the wave number of the quasitransverse waves $\chi = \omega / v_t$ (where v_t is the phase velocity of propagation of the transverse waves), we use the dispersion relation in the form:

$$v_1^2 \rho = 1/2 \left(C_{11} - C_{12} \right) \sin^2 Q + C_{44} \cos^2 Q. \tag{10}$$

In relations (9), (10), Q is the angle between wave vector and direction of $\bar{e}_J$.

By using (9), (10), we define the wave numbers as

$$g^2 = A_1 \pm \sqrt{A_1^2 - A_2^2}; \tag{11}$$

$$\chi^2 = 2\left(\omega^2 \rho - C_{44} h^2 \right) / (C_{11} - C_{12}) + h^2, \tag{12}$$

where

$$A_1 = \{ (C_{11} + C_{44})\omega^2 \rho - \left[C_{11} - (C_{33} - 2C_{44}) - C_{13}(C_{13} + 2C_{44}) \right] h^2 \} / (2C_{11} C_{44});$$

$$A_2 = \left[C_{13}\left(C_{13}+C_{44}\right)+C_{44}\left(C_{11}+C_{13}\right)+C_{33}\left(C_{44}-C_{11}\right)\right]h^4 -$$
$$-\left(C_{33}-C_{11}\right)h^2\omega^2\rho+\omega^4\rho^2/\left(C_{11}C_{44}\right).$$

The sign of minus in (11) relates to the wave number $\overset{(L)}{g}$ of the quasilongitudinal waves, but the sign of plus defines the wave number $\overset{(T)}{g}$ of the quasitransverse waves.

The common solution of differential equations (8) at the wave numbers, defined by (11) and (12), is used to determine displacements U_i, stresses σ_{ij} and strains ε_{ij}, which are written [6] as

$$U_i = \left(\overset{(L)}{D_1}\nabla_i + \delta_i^J \overset{(L)}{D_2}\nabla_J\right)\varphi + \left(\overset{(T)}{D_1}\frac{1}{\overset{(T)}{g}}\nabla_i\nabla_J + \overset{(T)}{g}\,\delta_i^J\right)\omega + \sqrt{g}\left(\delta_i^K\nabla^N - \delta_i^N\nabla^K\right)v; \tag{13}$$

$$\sigma_{ij} = \left(\overset{(L)}{d_1}g_{ij} + \overset{(L)}{d_2}\delta_i^J\delta_j^J + \overset{(L)}{d_3}\delta_{(i}^J\nabla_{j)}\nabla_J + \overset{(L)}{d_4}\nabla_i\nabla_j\right)\varphi +$$

$$+\left(\overset{(T)}{d_1}g_{ij}\nabla_J + \overset{(T)}{d_2}\delta_i^J\delta_j^J\nabla_J + \overset{(T)}{d_3}\delta_{(i}\nabla_{j)} + \overset{(T)}{d_4}\nabla_i\nabla_j\nabla_J\right)\omega +$$

$$+2\sqrt{g}\left[a_2\left(\nabla_{(i}\delta_{j)}^K\nabla^N - \nabla_{(i}\delta_{j)}^N\nabla^K\right) + a_4\left(\delta_{(i}^J\delta_{j)}^K\nabla^N - \delta_{(i}^J\delta_{j)}^N\nabla^K\right)\nabla_J\right]v; \tag{14}$$

$$\varepsilon_{ij} = \left(\overset{(L)}{D_1}\nabla_i\nabla_j + \overset{(L)}{D_2}\nabla_{(i}\delta_{j)}^J\nabla_J\right)\varphi + \left(\overset{(T)}{D_1}\frac{1}{\overset{(T)}{g}}\nabla_i\nabla_j\nabla_J + \overset{(T)}{g}\,\nabla_{(i}\delta_{j)}^J\right)\omega +$$

$$+\sqrt{g}\left(\nabla_{(i}\delta_{j)}^K\nabla^N - \nabla_{(i}\delta_{j)}^N\nabla^K\right)v, \tag{15}$$

where

$$\overset{(L)}{d}_1 = -\overset{(L)}{g}{}^2 a_1 - h^2 a_3\left(\overset{(L)}{D}_1 + \overset{(L)}{D}_2\right); \quad \overset{(L)}{d}_2 = -\overset{(L)}{g}{}^2 a_3 - h^2\left(2a_4 \overset{(L)}{D}_2 + a_5 \overset{(L)}{D}_1 + a_5 \overset{(L)}{D}_2\right);$$

$$\overset{(L)}{d}_3 = 2a_2 \overset{(L)}{D}_2 + 2a_4\left(2\overset{(L)}{D}_1 + \overset{(L)}{D}_2\right); \quad \overset{(L)}{d}_4 = 2a_2 \overset{(L)}{D}_1; \tag{16}$$

$$\overset{(T)}{d}_1 = a_1 \overset{(T)}{g}\left(1 - \overset{(T)}{D}_1\right) + a_3\left(\overset{(T)}{g} - \frac{\overset{(T)}{h}{}^2}{\overset{(T)}{g}}\overset{(T)}{D}_1\right); \quad \overset{(T)}{d}_2 = a_3 \overset{(T)}{g}\left(1 - \overset{(T)}{D}_1\right) + 2a_4 \overset{(T)}{g} + a_5\left(\overset{(T)}{g} - \frac{\overset{(T)}{h}{}^2}{\overset{(T)}{g}}\overset{(T)}{D}_1\right);$$

$$\overset{(T)}{d}_3 = 2a_2 \overset{(T)}{g} + 2a_4\left(\overset{(T)}{g} - 2\overset{(T)}{D}_1\frac{\overset{(T)}{h}{}^2}{\overset{(T)}{g}}\right); \quad \overset{(T)}{d}_4 = 2a_2 \overset{(T)}{D}_1\frac{1}{\overset{(T)}{g}}; \tag{17}$$

$$a_1 = C_{12}; \qquad a_2 = (C_{11} - C_{12})/2; \qquad a_3 = C_{13} - C_{12};$$
$$a_4 = C_{44} - (C_{11} - C_{12})/2; \qquad a_5 = C_{11} + C_{33} - 2(C_{13} + 2C_{44}); \tag{18}$$

$$\overset{(L)}{D}_1 = \frac{\overset{(L)}{g}{}^2 b_4}{b_3 - \left(\overset{(L)}{g}{}^2 - \overset{(L)}{h}{}^2\right)(1 - b_4)}; \quad \overset{(L)}{D}_2 = \frac{\overset{(L)}{g}{}^2}{\overset{(L)}{h}{}^2}\,\frac{b_3 - \left(\overset{(L)}{g}{}^2 - \overset{(L)}{h}{}^2\right) - b_4 \overset{(L)}{h}{}^2}{b_3 - \left(\overset{(L)}{g}{}^2 - \overset{(L)}{h}{}^2\right)(1 - b_4)}; \tag{19}$$

$$\overset{(T)}{D}_1 = -\frac{\overset{(T)}{g}{}^2 b_4}{b_3 - \left(\overset{(T)}{g}{}^2 - \overset{(T)}{h}{}^2\right) - b_4 \overset{(T)}{h}{}^2}; \tag{20}$$

$$b_3 = \left(\omega^2 \rho - h^2 C_{44}\right) / C_{11};\; b_4 = \left(C_{13} + C_{44}\right) / C_{11}. \tag{21}$$

The obtained functions satisfy the motion equation (6), the solution of which is reduced to solution of three well-studied scalar equations (8) with eigenvalues (11) and (12).

To solve specific problem, it is necessary to find a particular solution corresponding to the given boundary conditions.

The solutions of the problems for layered plain and cylindrical constructions have been constructed in the works [8] – [10] and [10] – [14], respectively, on the base of relations (13) – (15) of the generalized method of scalarization of the dynamic elastic fields in transverse-isotropic media [6]. These studies proposed a method to define fields of displacements, stresses and strains in above-mentioned constructions, considered features of wave processes at different kinds of external non-stationary influence for different physical and mechanical characteristics of layered materials and opened new possibilities for solution of various applied problems.

2. Definition of Elastic Fields in Problems of Excitation of Elliptic Cylinders

In the case of constructions with boundaries having the elliptic cylinder form, we use the cylindrical elliptic coordinate system ξ, η, z, connected with Cartesian coordinates x, y, z by the transformation formulae:

$$x = f \cdot ch\xi \cdot \cos\eta;\; y = f \cdot sh\xi \cdot \sin\eta;\; z = z, \tag{22}$$

where $0 \le \xi < \infty$; $0 \le \eta \le 2\pi$; $-\infty < z < \infty$; $2f$ is the distance between focuses (see Figures 2, 3).

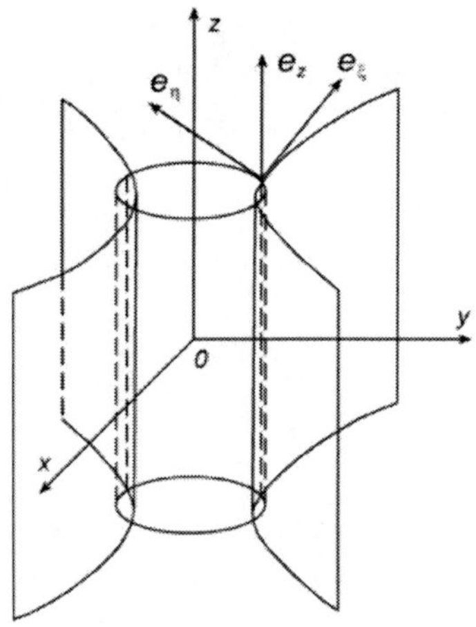

Figure 2. Cylindrical and elliptic coordinates.

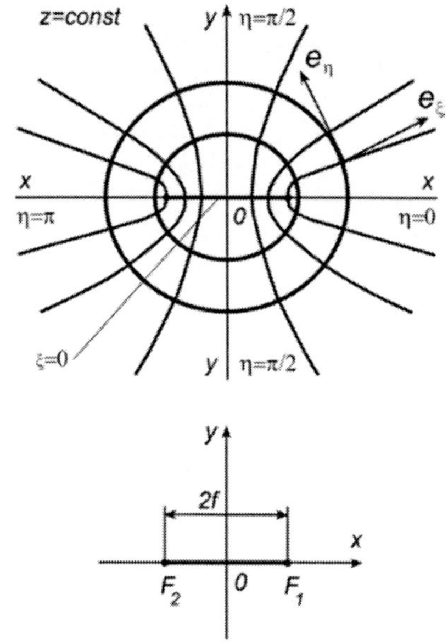

Figure 3. Distance between focuses.

An elliptic rod, space with hollow in the form of elliptic cylinder, thick elliptic tube and layered construction, similar to layered circular cylinder, relate to these constructions.

We use a time law in the form of $e^{-i\omega t}$ and perform transit to other time dependence by using Fourier transform. Let us introduce the next designations in accordance with before introduced basis $\bar{e}_J, \bar{e}_K, \bar{e}_N$:

$$x^J = z;\ x^K = \xi;\ x^N = \eta. \tag{23}$$

We specify for the treated coordinate system the differential operators, presented in formulae for tensor fields of displacements (13) and stresses (14), by using the matrix of metric tensor g_{ij}, corresponding to transform (22), values of Christoffel symbols [2] and covariant differentiation. As a result, we obtain expressions for covariant components of elastic field tensors through scalar functions φ, w, v in coordinates of elliptic cylinder:

$$U_z = ih\left(\overset{(L)}{D}_1 + \overset{(L)}{D}_2\right)\varphi + \left(\overset{(T)}{g} - \overset{(T)}{D}_1 \frac{h^2}{\overset{(T)}{g}}\right)w;\quad U_\xi = \overset{(L)}{D}_1 \frac{\partial \varphi}{\partial \xi} + ih\frac{\overset{(T)}{D}_1}{\overset{(T)}{g}}\frac{\partial w}{\partial \xi} + \frac{\partial v}{\partial \eta};$$

$$U_\eta = \overset{(L)}{D}_1 \frac{\partial \varphi}{\partial \eta} + ih\frac{\overset{(T)}{D}_1}{\overset{(T)}{g}}\frac{\partial w}{\partial \eta} - \frac{\partial v}{\partial \xi}; \tag{24}$$

$$\sigma_{zz} = \left[\overset{(L)}{d}_1 + \overset{(L)}{d}_2 - h^2 \left(\overset{(L)}{d}_3 + \overset{(L)}{d}_4 \right) \right] \varphi + ih \left(\overset{(T)}{d}_1 + \overset{(T)}{d}_2 + \overset{(T)}{d}_3 - h^2 \overset{(T)}{d}_4 \right) w \; ;$$

$$\sigma_{z\xi} = ih \left(\frac{1}{2} \overset{(L)}{d}_3 + \overset{(L)}{d}_4 \right) \frac{\partial \varphi}{\partial \xi} + \left(\frac{1}{2} \overset{(T)}{d}_3 - h^2 \overset{(T)}{d}_4 \right) \frac{\partial w}{\partial \xi} + ih(a_2 + a_4) \frac{\partial v}{\partial \eta} \; ;$$

$$\sigma_{z\eta} = ih \left(\frac{1}{2} \overset{(L)}{d}_3 + \overset{(L)}{d}_4 \right) \frac{\partial \varphi}{\partial \eta} + \left(\frac{1}{2} \overset{(T)}{d}_3 - h^2 \overset{(T)}{d}_4 \right) \frac{\partial w}{\partial \eta} - ih(1 + a_4) \frac{\partial v}{\partial \xi} \; ;$$

$$\sigma_{\xi\xi} = \left[\overset{(L)}{d}_1 f^2 (ch^2\xi - \cos^2\eta) + \overset{(L)}{d}_4 \left(\frac{\partial^2}{\partial \xi^2} - \frac{1}{2} \frac{sh 2\xi}{ch^2\xi - \cos^2\eta} \frac{\partial}{\partial \xi} + \frac{1}{2} \frac{\sin 2\eta}{ch^2\xi - \cos^2\eta} \frac{\partial}{\partial \eta} \right) \right] \varphi +$$

$$+ ih \left[\overset{(T)}{d}_1 f^2 (ch^2\xi - \cos^2\eta) + \overset{(T)}{d}_4 \left(\frac{\partial^2}{\partial \xi^2} - \frac{1}{2} \frac{sh 2\xi}{ch^2\xi - \cos^2\eta} \frac{\partial}{\partial \xi} + \frac{1}{2} \frac{\sin 2\eta}{ch^2\xi - \cos^2\eta} \frac{\partial}{\partial \eta} \right) \right] w +$$

$$+ 2a_2 \left(\frac{\partial^2}{\partial \xi \partial \eta} - \frac{1}{2} \frac{\sin 2\eta}{ch^2\xi - \cos^2\eta} \frac{\partial}{\partial \xi} - \frac{1}{2} \frac{sh 2\xi}{ch^2\xi - \cos^2\eta} \frac{\partial}{\partial \eta} \right) v \; ;$$

$$\sigma_{\xi\eta} = \overset{(L)}{d}_4 \left(\frac{\partial^2}{\partial \xi \partial \eta} - \frac{1}{2} \frac{\sin 2\eta}{ch^2\xi - \cos^2\eta} \frac{\partial}{\partial \xi} - \frac{1}{2} \frac{sh 2\xi}{ch^2\xi - \cos^2\eta} \frac{\partial}{\partial \eta} \right) \varphi +$$

$$+ ih \overset{(T)}{d}_4 \left(\frac{\partial^2}{\partial \xi \partial \eta} - \frac{1}{2} \frac{\sin 2\eta}{ch^2\xi - \cos^2\eta} \frac{\partial}{\partial \xi} - \frac{1}{2} \frac{sh 2\xi}{ch^2\xi - \cos^2\eta} \frac{\partial}{\partial \eta} \right) w +$$

$$+ a_2 \left(\frac{\partial^2}{\partial \eta^2} - \frac{\sin 2\eta}{ch^2\xi - \cos^2\eta} \frac{\partial}{\partial \eta} - \frac{\partial^2}{\partial \xi^2} + \frac{sh 2\xi}{ch^2\xi - \cos^2\eta} \frac{\partial}{\partial \xi} \right) v \; ;$$

$$\sigma_{\eta\eta} = \left[\overset{(L)}{d}_1 f^2 (ch^2\xi - \cos^2\eta) + \overset{(L)}{d}_4 \left(\frac{\partial^2}{\partial \eta^2} + \frac{1}{2} \frac{sh 2\xi}{ch^2\xi - \cos^2\eta} \frac{\partial}{\partial \xi} - \frac{1}{2} \frac{\sin 2\eta}{ch^2\xi - \cos^2\eta} \frac{\partial}{\partial \eta} \right) \right] \varphi +$$

$$+ ih \left[\overset{(T)}{d}_1 f^2 (ch^2\xi - \cos^2\eta) + \overset{(T)}{d}_4 \left(\frac{\partial^2}{\partial \eta^2} + \frac{1}{2} \frac{sh 2\xi}{ch^2\xi - \cos^2\eta} \frac{\partial}{\partial \xi} - \frac{1}{2} \frac{\sin 2\eta}{ch^2\xi - \cos^2\eta} \frac{\partial}{\partial \eta} \right) \right] w -$$

$$- 2 \left(\frac{\partial^2}{\partial \xi \partial \eta} - \frac{1}{2} \frac{sh 2\xi}{ch^2\xi - \cos^2\eta} \frac{\partial}{\partial \eta} + \frac{1}{2} \frac{\sin 2\eta}{ch^2\xi - \cos^2\eta} \frac{\partial}{\partial \xi} \right) v \; ; \qquad (25)$$

The coefficients $a_{i,} \overset{(L)}{d}_i, \overset{(T)}{d}_i, \overset{(L)}{D}_i, \overset{(T)}{D}_i$ can be found from expressions (16) – (20). The physical components of elastic field tensors are determined from formulae (24), (25) by conventional way [15].

The potential functions φ, w, v are the common solutions of Helmholtz equations with corresponding wave numbers:

$$\nabla_m \nabla^m F + g^2 F = 0, \tag{26}$$

where g takes values of $\overset{(L)}{g}$ for the quasilongitudinal waves, $F = \varphi$; $\overset{(T)}{g}$ for the quasitransverse waves, $F = w$; and χ for pure transverse waves, $F = v$.

By following the method of variable division, the common solution of wave Equation (26), to which the potentials φ, w, v satisfy, could be present as a superposition of particular solutions.

Based on [16, 17], unique particular solutions of Equations (26) are present through the wave functions of elliptic cylinder:

$$Ce_m(\xi,q)ce_m(\eta,q)e^{ihz}; \quad a = a_m, \quad m = 0,1,...;$$

$$Se_m(\xi,q)se_m(\eta,q)e^{ihz}; \quad a = b_m, \quad m = 1,2,...;$$

$$Me_m^{(1),(2)}(\xi,q)ce_m(\eta,q)e^{ihz}; \quad a = a_m, \quad m = 0,1,...;$$

$$Ne_m^{(1),(2)}(\xi,q)se_m(\eta,q)e^{ihz}; \quad a = b_m, \quad m = 1,2,..., \tag{27}$$

where

$$Me_m^{(1),(2)}(\xi,q) = Ce_m(\eta,q) \pm iFey_m(\xi,q); \quad a = a_m;$$

$$Ne_m^{(1),(2)}(\xi,q) = Se_m(\eta,q) \pm iGey_m(\xi,q); \quad a = b_m; \tag{28}$$

$ce_m(\xi,q)$; $se_m(\eta,q)$ are even and odd Mathieu functions of first order, respectively [17], corresponding to eigenvalues a_m and b_m; $Ce_m(\xi,q)$, $Se_m(\eta,q)$ are the modified Mathieu functions [17] (first particular solutions of the modified Mathieu equation); they are regular inside of ellipse into plain $z = const$ and similar on their properties to Bessel functions [18]; $Fey_m(\xi,q)$, $Gey_m(\xi,q)$ are the modified Mathieu functions [17] (second particular solutions of the modified Mathieu equation); they are similar to Newman functions [18] and have a singularity at $\xi = 0$; $Me_m^{(1),(2)}(\xi,q)$, $Ne_m^{(1),(2)}(\xi,q)$ are the combined Mathieu functions [17]; they satisfy radiation conditions and are similar on their properties to Hankel functions [18] of first and second kind, respectively.

Then the solution regular into cylinder, which is used in "internal problems", has the form:

$$F(z,\xi,\eta)=\sum_{m=0}^{\infty}\int_{-\infty}^{\infty}\left[A_m Ce_m(\xi,q)ce_m(\eta,q)+B_m Se_m(\xi,q)se_m(\eta,q)\right]e^{ihz}dh \quad , \tag{29}$$

and common solution for "external problems" should satisfy the radiation condition and is written as

$$F(z,\xi,\eta)=\sum_{m=0}^{\infty}\int_{-\infty}^{\infty}\left[C_m Me_m^{(1),(2)}(\xi,q)ce_m(\eta,q)+D_m Ne_m^{(1),(2)}(\xi,q)se_m(\eta,q)\right]e^{ihz}dh \quad . \tag{30}$$

The coefficients A_m, B_m, C_m, D_m correspond to eigenvalues m, h in these series; they are calculated by satisfying boundary conditions of specific problem.

3. Common Case of Excitation of Elliptic Constructions. System of Linear Algebraic Equations in Respect to Elastic Field Amplitudes

The simple method of solution, described in [10,19], cannot be applied to the problem solution, when different types of waves are excited in construction in the form of elliptic cylinders. It is connected with that the angle Mathieu functions, describing the pointed types of waves, are no orthogonal because parameter q is various for quasilongitudinal, quasitransverse and transverse waves. The same conclusion takes place in the case, when we consider even the waves of one kind in layered construction with different physical and mechanical properties of various layers. In these more complex cases, it may be applied the method of division of the angle Mathieu functions into Fourier series on trigonometric functions.

Let us discuss this method on example of definition of the elastic fields in the layer of elliptic cylinder (the waves of w-type), when other types of waves are excited, too.

In this case, the common solution for wave potential functions at any parameter h has the form:

$$\varphi=\sum_{m=0}^{\infty}\Big[\varphi_m^{+1}Me_m^{(1)}(\xi,\overset{(L)}{q})ce_m(\eta,\overset{(L)}{q})+\varphi_m^{+2}Ne_m^{(1)}(\xi,\overset{(L)}{q})se_m(\eta,\overset{(L)}{q})+$$

$$+\varphi_m^{-1}Me_m^{(2)}(\xi,\overset{(L)}{q})ce_m(\eta,\overset{(L)}{q})+\varphi_m^{-2}Ne_m^{(2)}(\xi,\overset{(L)}{q})se_m(\eta,\overset{(L)}{q})\Big]e^{ihz} \quad ;$$

$$w=\sum_{m=0}^{\infty}\left[w_m^{+1}Me_m^{(1)}(\xi,\overset{(T)}{q})ce_m(\eta,\overset{(T)}{q})+w_m^{+2}Ne_m^{(1)}(\xi,\overset{(T)}{q})se_m(\eta,\overset{(T)}{q})+\right.$$

$$\left.+w_m^{-1}Me_m^{(2)}(\xi,\overset{(T)}{q})ce_m(\eta,\overset{(T)}{q})+w_m^{-2}Ne_m^{(2)}(\xi,\overset{(T)}{q})se_m(\eta,\overset{(T)}{q})\right]e^{ihz};$$

$$v=\sum_{m=0}^{\infty}\left[v_m^{+1}Me_m^{(1)}(\xi,\mu)ce_m(\eta,\mu)+v_m^{+2}Ne_m^{(1)}(\xi,\mu)se_m(\eta,\mu)+\right.$$

$$\left.+v_m^{-1}Me_m^{(2)}(\xi,\mu)ce_m(\eta,\mu)+v_m^{-2}Ne_m^{(2)}(\xi,\mu)se_m(\eta,\mu)\right]e^{ihz}. \quad (31)$$

In these formulae $\overset{(L,T)}{q}=\frac{f^2}{4}(\overset{(L,T)}{g}{}^2-h^2);\quad \mu=\frac{f^2}{4}(\chi^2-h^2)$.

Let us write expressions for components of displacements and stresses, which are present in boundary conditions of the problem and after multiplying on $(ch^2\xi-\cos^2\eta)$, we obtain:

$$U_\xi=\overset{(L)}{D_1}\frac{\partial}{\partial\xi}\varphi+ih\frac{\overset{(T)}{D_1}}{\overset{(T)}{g}}\frac{\partial}{\partial\xi}w+\frac{\partial}{\partial\eta}v;\quad U_\eta=\overset{(L)}{D_1}\frac{\partial}{\partial\eta}\varphi+ih\frac{\overset{(T)}{D_1}}{\overset{(T)}{g}}\frac{\partial}{\partial\eta}w-\frac{\partial}{\partial\xi}v;$$

$$U_z=ih\left(\overset{(L)}{D_1}+\overset{(L)}{D_2}\right)\varphi+\left(\overset{(T)}{g}-\overset{(T)}{D_1}\frac{h^2}{\overset{(T)}{g}}\right)w; \quad (32)$$

$$\sigma_{\xi\xi}(ch^2\xi-\cos^2\eta)=$$

$$=\left\{\overset{(L)}{d_1}f^2(ch^2\xi-\cos^2\eta)^2+\overset{(L)}{d_4}\left[(ch^2\xi-\cos^2\eta)\frac{\partial^2}{\partial\xi^2}-\frac{1}{2}sh2\xi\frac{\partial}{\partial\xi}+\frac{1}{2}\sin 2\eta\frac{\partial}{\partial\eta}\right]\right\}\varphi+$$

$$+ih\left\{\overset{(T)}{d_1}f^2(ch^2\xi-\cos^2\eta)^2+\overset{(T)}{d_4}\left[(ch^2\xi-\cos^2\eta)\frac{\partial^2}{\partial\xi^2}-\frac{1}{2}sh2\xi\frac{\partial}{\partial\xi}+\frac{1}{2}\sin 2\eta\frac{\partial}{\partial\eta}\right]\right\}w+$$

$$+2a_2\left[(ch^2\xi-\cos^2\eta)\frac{\partial^2}{\partial\xi\partial\eta}-\frac{1}{2}\sin 2\eta\frac{\partial}{\partial\xi}-\frac{1}{2}sh2\xi\frac{\partial}{\partial\eta}\right]v;$$

$$\sigma_{\xi\eta}(ch^2\xi-\cos^2\eta)=$$
$$=\overset{(L)}{d_4}\left[(ch^2\xi-\cos^2\eta)\frac{\partial^2}{\partial\xi\partial\eta}-\frac{1}{2}\sin 2\eta\frac{\partial}{\partial\xi}-\frac{1}{2}sh2\xi\frac{\partial}{\partial\eta}\right]\varphi+$$

$$+ih\overset{(T)}{d_4}\left[(ch^2\xi-\cos^2\eta)\frac{\partial^2}{\partial\xi\partial\eta}-\frac{1}{2}\sin 2\eta\frac{\partial}{\partial\xi}-\frac{1}{2}sh2\xi\frac{\partial}{\partial\eta}\right]w+$$

$$+a_2\left[(ch^2\xi-\cos^2\eta)\frac{\partial^2}{\partial\eta^2}-\sin 2\eta\frac{\partial}{\partial\eta}-(ch^2\xi-\cos^2\eta)\frac{\partial^2}{\partial\xi^2}+sh2\xi\frac{\partial}{\partial\xi}\right]v\,;$$

$$\sigma_{\xi z}=ih\left(\frac{1}{2}\overset{(L)}{d_3}+\overset{(L)}{d_4}\right)\frac{\partial}{\partial\xi}\varphi+\left(\frac{1}{2}\overset{(T)}{d_3}-h^2\overset{(T)}{d_4}\right)\frac{\partial}{\partial\xi}w+ih(a_2+a_4)\frac{\partial}{\partial\eta}v\,. \tag{33}$$

We substitute the expressions for potential functions (31) in the given relations, preliminary representing the Mathieu functions in the form of division into trigonometric Fourie series:

$$ce_m(\eta,q)=\sum_{r=0}^{\infty}A_r^m(q)\cos r\eta\;;\qquad se_m(\eta,q)=\sum_{r=0}^{\infty}B_r^m(q)\sin r\eta\,. \tag{34}$$

As a result, we obtain

$$U_\xi(\xi,\eta,z)=\sum_m\left[\varphi_m^{+1}\overset{(L)}{P_m^{+1}}(\xi,\eta)+\varphi_m^{+2}\overset{(L)}{P_m^{+2}}(\xi,\eta)+\varphi_m^{-1}\overset{(L)}{P_m^{-1}}(\xi,\eta)+\varphi_m^{-2}\overset{(L)}{P_m^{-2}}(\xi,\eta)+\right.$$

$$+w_m^{+1}\overset{(T)}{P_m^{+1}}(\xi,\eta)+w_m^{+2}\overset{(T)}{P_m^{+2}}(\xi,\eta)+w_m^{-1}\overset{(T)}{P_m^{-1}}(\xi,\eta)+w_m^{-2}\overset{(T)}{P_m^{-2}}(\xi,\eta)+$$

$$\left.+v_m^{+1}\overset{(\mu)}{P_m^{+1}}(\xi,\eta)+v_m^{+2}\overset{(\mu)}{P_m^{+2}}(\xi,\eta)+v_m^{-1}\overset{(\mu)}{P_m^{-1}}(\xi,\eta)+v_m^{-2}\overset{(\mu)}{P_m^{-2}}(\xi,\eta)\right]e^{ihz}\,;$$

$$U_\eta(\xi,\eta,z)=\sum_m\left[\varphi_m^{+1}\overset{(L)}{Q_m^{+1}}(\xi,\eta)+\varphi_m^{+2}\overset{(L)}{Q_m^{+2}}(\xi,\eta)+\varphi_m^{-1}\overset{(L)}{Q_m^{-1}}(\xi,\eta)+\varphi_m^{-2}\overset{(L)}{Q_m^{-2}}(\xi,\eta)+\right.$$

$$+w_m^{+1}\overset{(T)}{Q_m^{+1}}(\xi,\eta)+w_m^{+2}\overset{(T)}{Q_m^{+2}}(\xi,\eta)+w_m^{-1}\overset{(T)}{Q_m^{-1}}(\xi,\eta)+w_m^{-2}\overset{(T)}{Q_m^{-2}}(\xi,\eta)+$$

$$+ v_m^{+1} \overset{(\mu)}{Q_m^{+1}}(\xi,\eta) + v_m^{+2} \overset{(\mu)}{Q_m^{+2}}(\xi,\eta) + v_m^{-1} \overset{(\mu)}{Q_m^{-1}}(\xi,\eta) + v_m^{-2} \overset{(\mu)}{Q_m^{-2}}(\xi,\eta) \Bigg] e^{ihz};$$

$$U_z(\xi,\eta,z) = \sum_m \Bigg[\varphi_m^{+1} \overset{(L)}{R_m^{+1}}(\xi,\eta) + \varphi_m^{+2} \overset{(L)}{R_m^{+2}}(\xi,\eta) + \varphi_m^{-1} \overset{(L)}{R_m^{-1}}(\xi,\eta) + \varphi_m^{-2} \overset{(L)}{R_m^{-2}}(\xi,\eta) +$$

$$+ w_m^{+1} \overset{(T)}{R_m^{+1}}(\xi,\eta) + w_m^{+2} \overset{(T)}{R_m^{+2}}(\xi,\eta) + w_m^{-1} \overset{(T)}{R_m^{-1}}(\xi,\eta) + w_m^{-2} \overset{(T)}{R_m^{-2}}(\xi,\eta) \Bigg] e^{ihz}; \qquad (35)$$

$$\sigma_{\xi\xi}(\xi,\eta,z)(ch^2\xi - \cos^2\eta) =$$

$$= \sum_m \Bigg[\varphi_m^{+1} \overset{(L)}{f_m^{+1}}(\xi,\eta) + \varphi_m^{+2} \overset{(L)}{f_m^{+2}}(\xi,\eta) + \varphi_m^{-1} \overset{(L)}{f_m^{-1}}(\xi,\eta) + \varphi_m^{-2} \overset{(L)}{f_m^{-2}}(\xi,\eta) +$$

$$+ w_m^{+1} \overset{(T)}{f_m^{+1}}(\xi,\eta) + w_m^{+2} \overset{(T)}{f_m^{+2}}(\xi,\eta) + w_m^{-1} \overset{(T)}{f_m^{-1}}(\xi,\eta) + w_m^{-2} \overset{(T)}{f_m^{-2}}(\xi,\eta) +$$

$$+ v_m^{+1} \overset{(\mu)}{f_m^{+1}}(\xi,\eta) + v_m^{+2} \overset{(\mu)}{f_m^{+2}}(\xi,\eta) + v_m^{-1} \overset{(\mu)}{f_m^{-1}}(\xi,\eta) + v_m^{-2} \overset{(\mu)}{f_m^{-2}}(\xi,\eta) \Bigg] e^{ihz};$$

$$\sigma_{\xi\eta}(\xi,\eta,z)(ch^2\xi - \cos^2\eta) =$$

$$= \sum_m \Bigg[\varphi_m^{+1} \overset{(L)}{h_m^{+1}}(\xi,\eta) + \varphi_m^{+2} \overset{(L)}{h_m^{+2}}(\xi,\eta) + \varphi_m^{-1} \overset{(L)}{h_m^{-1}}(\xi,\eta) + \varphi_m^{-2} \overset{(L)}{h_m^{-2}}(\xi,\eta) +$$

$$+ w_m^{+1} \overset{(T)}{h_m^{+1}}(\xi,\eta) + w_m^{+2} \overset{(T)}{h_m^{+2}}(\xi,\eta) + w_m^{-1} \overset{(T)}{h_m^{-1}}(\xi,\eta) + w_m^{-2} \overset{(T)}{h_m^{-2}}(\xi,\eta) +$$

$$+ v_m^{+1} \overset{(\mu)}{h_m^{+1}}(\xi,\eta) + v_m^{+2} \overset{(\mu)}{h_m^{+2}}(\xi,\eta) + v_m^{-1} \overset{(\mu)}{h_m^{-1}}(\xi,\eta) + v_m^{-2} \overset{(\mu)}{h_m^{-2}}(\xi,\eta) \Bigg] e^{ihz};$$

$$\sigma_{\xi z}(\xi,\eta,z) = \sum_m \Bigg[\varphi_m^{+1} \overset{(L)}{z_m^{+1}}(\xi,\eta) + \varphi_m^{+2} \overset{(L)}{z_m^{+2}}(\xi,\eta) + \varphi_m^{-1} \overset{(L)}{z_m^{-1}}(\xi,\eta) + \varphi_m^{-2} \overset{(L)}{z_m^{-2}}(\xi,\eta) +$$

$$+ w_m^{+1} \overset{(T)}{z_m^{+1}}(\xi,\eta) + w_m^{+2} \overset{(T)}{z_m^{+2}}(\xi,\eta) + w_m^{-1} \overset{(T)}{z_m^{-1}}(\xi,\eta) + w_m^{-2} \overset{(T)}{z_m^{-2}}(\xi,\eta) +$$

$$+ \overset{(\mu)}{v_m^{+1}} \overset{(\mu)}{z_m^{+1}}(\xi,\eta) + \overset{(\mu)}{v_m^{+2}} \overset{(\mu)}{z_m^{+2}}(\xi,\eta) + \overset{(\mu)}{v_m^{-1}} \overset{(\mu)}{z_m^{-1}}(\xi,\eta) + \overset{(\mu)}{v_m^{-2}} \overset{(\mu)}{z_m^{-2}}(\xi,\eta) \Bigg] e^{ihz}, \quad (36)$$

where

$$\overset{(L)}{P_m^{+1}}(\xi,\eta) = \overset{(L)}{D_1} \frac{\partial}{\partial \xi} Me_m^{(1)}(\xi, \overset{(L)}{q}) \sum_{r=0}^{\infty} A_r^m(\overset{(L)}{q}) \cos r\eta;$$

$$\overset{(L)}{P_m^{+2}}(\xi,\eta) = \overset{(L)}{D_1} \frac{\partial}{\partial \xi} Ne_m^{(1)}(\xi, \overset{(L)}{q}) \sum_{r=0}^{\infty} B_r^m(\overset{(L)}{q}) \sin r\eta;$$

$$\overset{(T)}{P_m^{+1}}(\xi,\eta) = ih \frac{\overset{(T)}{D_1}}{\overset{(T)}{g}} \frac{\partial}{\partial \xi} Me_m^{(1)}(\xi, \overset{(T)}{q}) \sum_{r=0}^{\infty} A_r^m(\overset{(T)}{q}) \cos r\eta;$$

$$\overset{(T)}{P_m^{+2}}(\xi,\eta) = ih \frac{\overset{(T)}{D_1}}{\overset{(T)}{g}} \frac{\partial}{\partial \xi} Ne_m^{(1)}(\xi, \overset{(T)}{q}) \sum_{r=0}^{\infty} B_r^m(\overset{(T)}{q}) \sin r\eta;$$

$$\overset{(\mu)}{P_m^{+1}}(\xi,\eta) = -Me_m^{(1)}(\xi,\mu) \sum_{r=0}^{\infty} A_r^m(\mu) r \sin r\eta;$$

$$\overset{(\mu)}{P_m^{+2}}(\xi,\eta) = Ne_m^{(1)}(\xi,\mu) \sum_{r=0}^{\infty} B_r^m(\mu) r \cos r\eta;$$

$$\overset{(L)}{Q_m^{+1}}(\xi,\eta) = -\overset{(L)}{D_1} Me_m^{(1)}(\xi, \overset{(L)}{q}) \sum_{r=0}^{\infty} A_r^m(\overset{(L)}{q}) r \sin r\eta;$$

$$\overset{(L)}{Q_m^{+2}}(\xi,\eta) = \overset{(L)}{D_1} Ne_m^{(1)}(\xi, \overset{(L)}{q}) \sum_{r=0}^{\infty} B_r^m(\overset{(L)}{q}) r \cos r\eta;$$

$$\overset{(T)}{Q_m^{+1}}(\xi,\eta) = -ih \frac{\overset{(T)}{D_1}}{\overset{(T)}{g}} Me_m^{(1)}(\xi, \overset{(T)}{q}) \sum_{r=0}^{\infty} A_r^m(\overset{(T)}{q}) r \sin r\eta;$$

$$\overset{(T)}{Q}{}_m^{+2}(\xi,\eta) = ih\frac{\overset{(T)}{D_1}}{\overset{(T)}{g}} Ne_m^{(1)}(\xi,\overset{(T)}{q})\sum_{r=0}^{\infty} B_r^m(\overset{(T)}{q}) r\cos r\eta \quad ;$$

$$\overset{(\mu)}{Q}{}_m^{+1}(\xi,\eta) = -\frac{\partial}{\partial\xi} Me_m^{(1)}(\xi,\mu)\sum_{r=0}^{\infty} A_r^m(\mu)\cos r\eta \quad ;$$

$$\overset{(\mu)}{Q}{}_m^{+2}(\xi,\eta) = -\frac{\partial}{\partial\xi} Ne_m^{(1)}(\xi,\mu)\sum_{r=0}^{\infty} B_r^m(\mu)\sin r\eta \quad ;$$

$$\overset{(L)}{R}{}_m^{+1}(\xi,\eta) = ih\left(\overset{(L)}{D_1} + \overset{(L)}{D_2}\right) Me_m^{(1)}(\xi,\overset{(L)}{q})\sum_{r=0}^{\infty} A_r^m(\overset{(L)}{q})\cos r\eta \quad ;$$

$$\overset{(L)}{R}{}_m^{+2}(\xi,\eta) = ih\left(\overset{(L)}{D_1} + \overset{(L)}{D_2}\right) Ne_m^{(1)}(\xi,\overset{(L)}{q})\sum_{r=0}^{\infty} B_r^m(\overset{(L)}{q})\sin r\eta \quad ;$$

$$\overset{(T)}{R}{}_m^{+1}(\xi,\eta) = \left(\overset{(T)}{g} - \overset{(T)}{D_1}\frac{h^2}{\overset{(T)}{g}}\right) Me_m^{(1)}(\xi,\overset{(T)}{q})\sum_{r=0}^{\infty} A_r^m(\overset{(T)}{q})\cos r\eta \quad ;$$

$$\overset{(T)}{R}{}_m^{+2}(\xi,\eta) = \left(\overset{(T)}{g} - \overset{(T)}{D_1}\frac{h^2}{\overset{(T)}{g}}\right) Ne_m^{(1)}(\xi,\overset{(T)}{q})\sum_{r=0}^{\infty} B_r^m(\overset{(T)}{q})\sin r\eta \quad ; \tag{37}$$

$$\overset{(L)}{f}{}_m^{+1}(\xi,\eta) = \overset{(L)}{d_1} f^2 (ch^2\xi - \cos^2\eta)\, Me_m^{(1)}(\xi,\overset{(L)}{q})\sum_{r=0}^{\infty} A_r^m(\overset{(L)}{q})\cos r\eta +$$

$$+\overset{(L)}{d_4}(ch^2\xi - \cos^2\eta)\frac{\partial^2}{\partial\xi^2} Me_m^{(1)}(\xi,\overset{(L)}{q})\sum_{r=0}^{\infty} A_r^m(\overset{(L)}{q})\cos r\eta -$$

$$-\frac{\overset{(L)}{d_4}}{2} sh2\xi\frac{\partial}{\partial\xi} Me_m^{(1)}(\xi,\overset{(L)}{q})\sum_{r=0}^{\infty} A_r^m(\overset{(L)}{q})\cos r\eta +$$

$$+\frac{\overset{(L)}{d_4}}{2} Me_m^{(1)}(\xi, \overset{(L)}{q}) \sin 2\eta \sum_{r=0}^{\infty} -r A_r^m(\overset{(L)}{q}) \sin r\eta \quad ;$$

$$\overset{(L)}{f_m^{+2}}(\xi,\eta) = \overset{(L)}{d_1} f^2 (ch^2\xi - \cos^2\eta) Ne_m^{(1)}(\xi, \overset{(L)}{q}) \sum_{r=0}^{\infty} B_r^m(\overset{(L)}{q}) \sin r\eta +$$

$$+\overset{(L)}{d_4}(ch^2\xi - \cos^2\eta) \frac{\partial^2}{\partial \xi^2} Ne_m^{(1)}(\xi, \overset{(L)}{q}) \sum_{r=0}^{\infty} B_r^m(\overset{(L)}{q}) \sin r\eta -$$

$$-\frac{\overset{(L)}{d_4}}{2} sh2\xi \frac{\partial}{\partial \xi} Ne_m^{(1)}(\xi, \overset{(L)}{q}) \sum_{r=0}^{\infty} B_r^m(\overset{(L)}{q}) \sin r\eta +$$

$$+\frac{\overset{(L)}{d_4}}{2} Ne_m^{(1)}(\xi, \overset{(L)}{q}) \sin 2\eta \sum_{r=0}^{\infty} r B_r^m(\overset{(L)}{q}) \cos r\eta \quad ;$$

$$\overset{(T)}{f_m^{+1}}(\xi,\eta) = ih \Bigg[\overset{(T)}{d_1} f^2 (ch^2\xi - \cos^2\eta) Me_m^{(1)}(\xi, \overset{(T)}{q}) \sum_{r=0}^{\infty} A_r^m(\overset{(T)}{q}) \cos r\eta +$$

$$+\overset{(T)}{d_4}(ch^2\xi - \cos^2\eta) \frac{\partial^2}{\partial \xi^2} Me_m^{(1)}(\xi, \overset{(T)}{q}) \sum_{r=0}^{\infty} A_r^m(\overset{(T)}{q}) \cos r\eta -$$

$$-\frac{\overset{(T)}{d_4}}{2} sh2\xi \frac{\partial}{\partial \xi} Me_m^{(1)}(\xi, \overset{(T)}{q}) \sum_{r=0}^{\infty} A_r^m(\overset{(T)}{q}) \cos r\eta +$$

$$+\frac{\overset{(T)}{d_4}}{2} Me_m^{(1)}(\xi, \overset{(T)}{q}) \sin 2\eta \sum_{r=0}^{\infty} -r A_r^m(\overset{(T)}{q}) \sin r\eta \Bigg] ;$$

$$\overset{(T)}{f_m^{+2}}(\xi,\eta) = ih \Bigg[\overset{(T)}{d_1} f^2 (ch^2\xi - \cos^2\eta) Ne_m^{(1)}(\xi, \overset{(T)}{q}) \sum_{r=0}^{\infty} B_r^m(\overset{(T)}{q}) \sin r\eta +$$

$$+\overset{(T)}{d_4}(ch^2\xi - \cos^2\eta) \frac{\partial^2}{\partial \xi^2} Ne_m^{(1)}(\xi, \overset{(T)}{q}) \sum_{r=0}^{\infty} B_r^m(\overset{(T)}{q}) \sin r\eta -$$

$$-\frac{\overset{(T)}{d_4}}{2} sh2\xi \frac{\partial}{\partial \xi} Ne_m^{(1)}(\xi, \overset{(T)}{q}) \sum_{r=0}^{\infty} B_r^m(\overset{(T)}{q}) \sin r\eta +$$

$$\left. + \frac{\overset{(T)}{d_4}}{2} Ne_m^{(1)}(\xi, \overset{(T)}{q}) \sin 2\eta \sum_{r=0}^{\infty} rB_r^m(\overset{(T)}{q}) \cos r\eta \right];$$

$$\overset{(\mu)}{f_m^{+1}}(\xi,\eta) = 2a_2(ch^2\xi - \cos^2\eta) \frac{\partial}{\partial \xi} Me_m^{(1)}(\xi, \overset{(\mu)}{q}) \sum_{r=0}^{\infty} -rA_r^m(\overset{(\mu)}{q}) \sin r\eta -$$

$$-a_2 \frac{\partial}{\partial \xi} Me_m^{(1)}(\xi,\mu) \sin 2\eta \sum_{r=0}^{\infty} A_r^m(\mu) \cos r\eta +$$

$$+ a_2 sh2\xi\, Me_m^{(1)}(\xi,\mu) \sum_{r=0}^{\infty} rA_r^m(\mu) \sin r\eta \quad ;$$

$$\overset{(\mu)}{f_m^{+2}}(\xi,\eta) = 2a_2(ch^2\xi - \cos^2\eta) \frac{\partial}{\partial \xi} Ne_m^{(1)}(\xi,\mu) \sum_{r=0}^{\infty} rB_r^m(\mu) \cos r\eta -$$

$$-a_2 \frac{\partial}{\partial \xi} Ne_m^{(1)}(\xi,\mu) \sin 2\eta \sum_{r=0}^{\infty} B_r^m(\mu) \sin r\eta -$$

$$-a_2 sh2\xi\, Ne_m^{(1)}(\xi,\mu) \sum_{r=0}^{\infty} rB_r^m(\mu) \cos r\eta \quad ;$$

$$\overset{(L)}{h_m^{+1}}(\xi,\eta) = \overset{(L)}{d_4}(ch^2\xi - \cos^2\eta) \frac{\partial}{\partial \xi} Me_m^{(1)}(\xi, \overset{(L)}{q}) \sum_{r=0}^{\infty} -rA_r^m(\overset{(L)}{q}) \sin r\eta -$$

$$-\frac{\overset{(L)}{d_4}}{2} \frac{\partial}{\partial \xi} Me_m^{(1)}(\xi, \overset{(L)}{q}) \sin 2\eta \sum_{r=0}^{\infty} A_r^m(\overset{(L)}{q}) \cos r\eta -$$

$$-\frac{\overset{(L)}{d_4}}{2} sh2\xi\, Me_m^{(1)}(\xi, \overset{(L)}{q}) \sum_{r=0}^{\infty} -rA_r^m(\overset{(L)}{q}) \sin r\eta \quad ;$$

$$\overset{(L)}{h}{}_m^{+2}(\xi,\eta)=\overset{(L)}{d}_4(ch^2\xi-\cos^2\eta)\frac{\partial}{\partial\xi}Ne_m^{(1)}(\xi,\overset{(L)}{q})\sum_{r=0}^{\infty}rB_r^m(\overset{(L)}{q})\cos r\eta-$$

$$-\frac{\overset{(L)}{d}_4}{2}\frac{\partial}{\partial\xi}Ne_m^{(1)}(\xi,\overset{(L)}{q})\sin 2\eta\sum_{r=0}^{\infty}B_r^m(\overset{(L)}{q})\sin r\eta-$$

$$-\frac{\overset{(L)}{d}_4}{2}sh2\xi\,Ne_m^{(1)}(\xi,\overset{(L)}{q})\sum_{r=0}^{\infty}rB_r^m(\overset{(L)}{q})\cos r\eta\;;$$

$$\overset{(T)}{h}{}_m^{+1}(\xi,\eta)=ih\left[\overset{(T)}{d}_4(ch^2\xi-\cos^2\eta)\frac{\partial}{\partial\xi}Me_m^{(1)}(\xi,\overset{(T)}{q})\sum_{r=0}^{\infty}-rA_r^m(\overset{(T)}{q})\sin r\eta-\right.$$

$$-\frac{\overset{(T)}{d}_4}{2}\frac{\partial}{\partial\xi}Me_m^{(1)}(\xi,\overset{(T)}{q})\sin 2\eta\sum_{r=0}^{\infty}A_r^m(\overset{(T)}{q})\cos r\eta-$$

$$\left.-\frac{\overset{(T)}{d}_4}{2}sh2\xi\,Me_m^{(1)}(\xi,\overset{(T)}{q})\sum_{r=0}^{\infty}-rA_r^m(\overset{(T)}{q})\sin r\eta\right];$$

$$\overset{(T)}{h}{}_m^{+2}(\xi,\eta)=ih\left[\overset{(T)}{d}_4(ch^2\xi-\cos^2\eta)\frac{\partial}{\partial\xi}Ne_m^{(1)}(\xi,\overset{(T)}{q})\sum_{r=0}^{\infty}rB_r^m(\overset{(T)}{q})\cos r\eta-\right.$$

$$-\frac{\overset{(T)}{d}_4}{2}\frac{\partial}{\partial\xi}Ne_m^{(1)}(\xi,\overset{(T)}{q})\sin 2\eta\sum_{r=0}^{\infty}B_r^m(\overset{(T)}{q})\sin r\eta-$$

$$\left.-\frac{\overset{(T)}{d}_4}{2}sh2\xi\,Ne_m^{(1)}(\xi,\overset{(T)}{q})\sum_{r=0}^{\infty}rB_r^m(\overset{(T)}{q})\cos r\eta\right];$$

$$\overset{(\mu)}{h}{}_m^{+1}(\xi,\eta)=a_2\left[(ch^2\xi-\cos^2\eta)\frac{\partial}{\partial\xi}Me_m^{(1)}(\xi,\mu)\sum_{r=0}^{\infty}-r^2A_r^m(\mu)\cos r\eta-\right.$$

$$-(ch^2\xi-\cos^2\eta)\frac{\partial^2}{\partial\xi^2}Me_m^{(1)}(\xi,\mu)\sum_{r=0}^{\infty}A_r^m(\mu)\cos r\eta-$$

$$-Me_m^{(1)}(\xi,\mu)\sin 2\eta\sum_{r=0}^{\infty}-rA_r^m(\mu)\sin r\eta +$$

$$\left.+sh2\xi\frac{\partial}{\partial\xi}Me_m^{(1)}(\xi,\mu)\sum_{r=0}^{\infty}A_r^m(\mu)\cos r\eta\right];$$

$$\overset{(\mu)}{h_m^{+2}}(\xi,\eta)=a_2\left[(ch^2\xi-\cos^2\eta)Ne_m^{(1)}(\xi,\mu)\sum_{r=0}^{\infty}-r^2B_r^m(\mu)\cos r\eta-\right.$$

$$-(ch^2\xi-\cos^2\eta)\frac{\partial^2}{\partial\xi^2}Ne_m^{(1)}(\xi,\mu)\sum_{r=0}^{\infty}B_r^m(\mu)\cos r\eta-$$

$$-Ne_m^{(1)}(\xi,\mu)\sin 2\eta\sum_{r=0}^{\infty}rB_r^m(\mu)\cos r\eta+$$

$$\left.+sh2\xi\frac{\partial}{\partial\xi}Ne_m^{(1)}(\xi,\mu)\sum_{r=0}^{\infty}B_r^m(\mu)\sin r\eta\right];$$

$$\overset{(L)}{z_m^{+1}}(\xi,\eta)=ih\left(\frac{1}{2}\overset{(L)}{d_3}+\overset{(L)}{d_4}\right)\frac{\partial}{\partial\xi}Me_m^{(1)}(\xi,\overset{(L)}{q})\sum_{r=0}^{\infty}A_r^m(\overset{(L)}{q})\cos r\eta;$$

$$\overset{(L)}{z_m^{+2}}(\xi,\eta)=ih\left(\frac{1}{2}\overset{(L)}{d_3}+\overset{(L)}{d_4}\right)\frac{\partial}{\partial\xi}Ne_m^{(1)}(\xi,\overset{(L)}{q})\sum_{r=0}^{\infty}B_r^m(\overset{(L)}{q})\sin r\eta;$$

$$\overset{(T)}{z_m^{+1}}(\xi,\eta)=\left(\frac{1}{2}\overset{(T)}{d_3}-h^2\overset{(L)}{d_4}\right)\frac{\partial}{\partial\xi}Me_m^{(1)}(\xi,\overset{(T)}{q})\sum_{r=0}^{\infty}A_r^m(\overset{(T)}{q})\cos r\eta;$$

$$\overset{(T)}{z_m^{+2}}(\xi,\eta)=\left(\frac{1}{2}\overset{(T)}{d_3}-h^2\overset{(L)}{d_4}\right)\frac{\partial}{\partial\xi}Ne_m^{(1)}(\xi,\overset{(T)}{q})\sum_{r=0}^{\infty}B_r^m(\overset{(T)}{q})\sin r\eta;$$

$$\overset{(\mu)}{z_m^{+1}}(\xi,\eta)=-ih(a_2+a_4)Me_m^{(1)}(\xi,\mu)\sum_{r=0}^{\infty}A_r^m(\mu)r\sin r\eta;$$

$$\overset{(\mu)}{z}{}_m^{+2}(\xi,\eta) = ih\left(a_2 + a_4\right)Ne_m^{(1)}(\xi,\mu)\sum_{r=0}^{\infty} B_r^m(\mu) r \cos r\eta \quad . \tag{38}$$

The functions, corresponding to the waves, propagating in the back direction (from boundary ξ_1 toward boundary ξ_0, index with minus sign), are defined by the expressions (37), (38) at the replacement of the combined Mathieu functions of first kind by the combined Mathieu functions of second kind, $Me_m^{(1)}(\xi,q) \to Me_m^{(2)}(\xi,q)$, $Ne_m^{(1)}(\xi,q) \to Ne_m^{(2)}(\xi,q)$.

Let us multiply left and right parts of expressions (35), (36) on $\cos k\eta$ and integrate on η into range from 0 to 2π. Then we multiply the expressions (35), (36) on $\sin k\eta$ and integrate on η in the same limits. By using an orthogonality of trigonometric functions and computing integrals on η, we obtain infinite set of linear algebraic equations in respect to amplitudes of wave harmonics which propagate into elliptic layer. We write this set without multiplier e^{ihz} :

$$P_k^1(\xi) = \sum_m \Bigg[\varphi_m^{+1} \overset{(L)}{P}{}_{mk}^{+1}(\xi) + w_m^{+1} \overset{(T)}{P}{}_{mk}^{+1}(\xi) + v_m^{+2} \overset{(\mu)}{P}{}_m^{+2}(\xi) +$$

$$+ \varphi_m^{-1} \overset{(L)}{P}{}_{mk}^{-1}(\xi) + w_m^{-1} \overset{(T)}{P}{}_{mk}^{-1}(\xi) + v_m^{-2} \overset{(\mu)}{P}{}_{mk}^{-2}(\xi) \Bigg];$$

$$Q_k^2(\xi) = \sum_m \Bigg[\varphi_m^{+2} \overset{(L)}{Q}{}_{mk}^{+2}(\xi) + w_m^{+2} \overset{(T)}{Q}{}_{mk}^{+2}(\xi) + v_m^{+1} \overset{(\mu)}{Q}{}_{mk}^{+1}(\xi) +$$

$$+ \varphi_m^{-2} \overset{(L)}{Q}{}_{mk}^{-2}(\xi) + w_m^{-2} \overset{(T)}{Q}{}_{mk}^{-2}(\xi) + v_m^{-1} \overset{(\mu)}{Q}{}_{mk}^{-1}(\xi) \Bigg];$$

$$R_k^1(\xi) = \sum_m \Bigg[\varphi_m^{+1} \overset{(L)}{R}{}_{mk}^{+1}(\xi) + w_m^{+1} \overset{(T)}{R}{}_{mk}^{+1}(\xi) + v_m^{+2} \overset{(\mu)}{R}{}_{mk}^{+2}(\xi) +$$

$$+ \varphi_m^{-1} \overset{(L)}{R}{}_{mk}^{-1}(\xi) + w_m^{-1} \overset{(T)}{R}{}_{mk}^{-1}(\xi) + v_m^{-2} \overset{(\mu)}{R}{}_{mk}^{-2}(\xi) \Bigg]; \tag{39}$$

$$P_k^2(\xi) = \sum_m \Bigg[\varphi_m^{+2} \overset{(L)}{P}{}_{mk}^{+2}(\xi) + w_m^{+2} \overset{(T)}{P}{}_{mk}^{+2}(\xi) + v_m^{+1} \overset{(\mu)}{P}{}_{mk}^{+1}(\xi) +$$

$$
\left. + \varphi_m^{-2} \overset{(L)}{P_{mk}^{-2}}(\xi) + w_m^{-2} \overset{(T)}{P_{mk}^{-2}}(\xi) + v_m^{-1} \overset{(\mu)}{P_{mk}^{-1}}(\xi) \right];
$$

$$
Q_k^1(\xi) = \sum_m \left[\varphi_m^{+1} \overset{(L)}{Q_{mk}^{+1}}(\xi) + w_m^{+1} \overset{(T)}{Q_{mk}^{+1}}(\xi) + v_m^{+2} \overset{(\mu)}{Q_{mk}^{+2}}(\xi) + \right.
$$

$$
\left. + \varphi_m^{-1} \overset{(L)}{Q_{mk}^{-1}}(\xi) + w_m^{-1} \overset{(T)}{Q_{mk}^{-1}}(\xi) + v_m^{-2} \overset{(\mu)}{Q_{mk}^{-2}}(\xi) \right];
$$

$$
R_k^2(\xi) = \sum_m \left[\varphi_m^{+2} \overset{(L)}{R_{mk}^{+2}}(\xi) + w_m^{+2} \overset{(T)}{R_{mk}^{+2}}(\xi) + v_m^{+1} \overset{(\mu)}{R_{mk}^{+1}}(\xi) + \right.
$$

$$
\left. + \varphi_m^{-2} \overset{(L)}{R_{mk}^{-2}}(\xi) + w_m^{-2} \overset{(T)}{R_{mk}^{-2}}(\xi) + v_m^{-1} \overset{(\mu)}{R_{mk}^{-1}}(\xi) \right]; \tag{40}
$$

$$
\Phi_k^1(\xi) = \sum_m \left[\varphi_m^{+1} \overset{(L)}{\Phi_{mk}^{+1}}(\xi) + w_m^{+1} \overset{(T)}{\Phi_{mk}^{+1}}(\xi) + v_m^{+2} \overset{(\mu)}{\Phi_{mk}^{+2}}(\xi) + \right.
$$

$$
\left. + \varphi_m^{-1} \overset{(L)}{\Phi_{mk}^{-1}}(\xi) + w_m^{-1} \overset{(T)}{\Phi_{mk}^{-1}}(\xi) + v_m^{-2} \overset{(\mu)}{\Phi_{mk}^{-2}}(\xi) \right];
$$

$$
H_k^2(\xi) = \sum_m \left[\varphi_m^{+1} \overset{(L)}{H_{mk}^{+1}}(\xi) + w_m^{+1} \overset{(T)}{H_{mk}^{+1}}(\xi) + v_m^{+2} \overset{(\mu)}{H_{mk}^{+2}}(\xi) + \right.
$$

$$
\left. + \varphi_m^{-1} \overset{(L)}{H_{mk}^{-1}}(\xi) + w_m^{-1} \overset{(T)}{H_{mk}^{-1}}(\xi) + v_m^{-2} \overset{(\mu)}{H_{mk}^{-2}}(\xi) \right];
$$

$$
Z_k^1(\xi) = \sum_m \left[\varphi_m^{+1} \overset{(L)}{Z_{mk}^{+1}}(\xi) + w_m^{+1} \overset{(T)}{Z_{mk}^{+1}}(\xi) + v_m^{+2} \overset{(\mu)}{Z_{mk}^{+2}}(\xi) + \right.
$$

$$
\left. + \varphi_m^{-1} \overset{(L)}{Z_{mk}^{-1}}(\xi) + w_m^{-1} \overset{(T)}{Z_{mk}^{-1}}(\xi) + v_m^{-2} \overset{(\mu)}{Z_{mk}^{-2}}(\xi) \right]; \tag{41}
$$

$$\Phi_k^2(\xi) = \sum_m \Bigg[\varphi_m^{+2} \overset{(L)}{\Phi}{}_{mk}^{+2}(\xi) + w_m^{+2} \overset{(T)}{\Phi}{}_{mk}^{+2}(\xi) + v_m^{+1} \overset{(\mu)}{\Phi}{}_{mk}^{+1}(\xi) +$$

$$+ \varphi_m^{-2} \overset{(L)}{\Phi}{}_{mk}^{-2}(\xi) + w_m^{-2} \overset{(T)}{\Phi}{}_{mk}^{-2}(\xi) + v_m^{-1} \overset{(\mu)}{\Phi}{}_{mk}^{-1}(\xi) \Bigg];$$

$$H_k^1(\xi) = \sum_m \Bigg[\varphi_m^{+2} \overset{(L)}{H}{}_{mk}^{+2}(\xi) + w_m^{+2} \overset{(T)}{H}{}_{mk}^{+2}(\xi) + v_m^{+1} \overset{(\mu)}{H}{}_{mk}^{+1}(\xi) +$$

$$+ \varphi_m^{-2} \overset{(L)}{H}{}_{mk}^{-2}(\xi) + w_m^{-2} \overset{(T)}{H}{}_{mk}^{-2}(\xi) + v_m^{-1} \overset{(\mu)}{H}{}_{mk}^{-2}(\xi) \Bigg];$$

$$Z_k^2(\xi) = \sum_m \Bigg[\varphi_m^{+2} \overset{(L)}{Z}{}_{mk}^{+2}(\xi) + w_m^{+2} \overset{(T)}{Z}{}_{mk}^{+2}(\xi) + v_m^{+1} \overset{(\mu)}{Z}{}_{mk}^{+1}(\xi) +$$

$$+ \varphi_m^{-2} \overset{(L)}{Z}{}_{mk}^{-2}(\xi) + w_m^{-2} \overset{(T)}{Z}{}_{mk}^{-2}(\xi) + v_m^{-1} \overset{(\mu)}{Z}{}_{mk}^{-1}(\xi) \Bigg], \qquad (42)$$

where

$$\overset{(L)}{P}{}_{mk}^{+1}(\xi) = \overset{(L)}{D}_1 \frac{\partial}{\partial \xi} Me_m^{(1)}(\xi, \overset{(L)}{q}) A_k^m(\overset{(L)}{q}) \delta_{ok}; \quad \overset{(T)}{P}{}_{mk}^{+1}(\xi) = ih \frac{\overset{(T)}{D}_1}{\overset{(T)}{g}} \frac{\partial}{\partial \xi} Me_m^{(1)}(\xi, \overset{(T)}{q}) A_k^m(\overset{(T)}{q}) \delta_{ok};$$

$$\overset{(\mu)}{P}{}_{mk}^{+2}(\xi) = Ne_m^{(1)}(\xi, \mu) k B_k^m(\mu) \delta_{ok}; \quad \overset{(L)}{Q}{}_{mk}^{+2}(\xi) = \overset{(L)}{D}_1 Ne_m^{(1)}(\xi, \overset{(L)}{q}) k B_k^m(\overset{(L)}{q}) \delta_{ok};$$

$$\overset{(T)}{Q}{}_{mk}^{+2}(\xi) = ih \frac{\overset{(T)}{D}_1}{\overset{(T)}{g}} Ne_m^{(1)}(\xi, \overset{(T)}{q}) k B_k^m(\overset{(T)}{q}) \delta_{ok}; \quad \overset{(\mu)}{Q}{}_{mk}^{+1}(\xi) = -\frac{\partial}{\partial \xi} Me_m^{(1)}(\xi, \mu) A_k^m(\mu) \delta_{ok};$$

$$\overset{(L)}{R}{}_{mk}^{+1}(\xi, \eta) = ih \left(\overset{(L)}{D}_1 + \overset{(L)}{D}_2 \right) Me_m^{(1)}(\xi, \overset{(L)}{q}) A_k^m(\overset{(L)}{q}) \delta_{ok};$$

$$\overset{(T)}{R}{}^{+1}_{mk}(\xi) = \left(\overset{(T)}{g} - \overset{(T)}{D}_1 \frac{h^2}{\overset{(T)}{g}} \right) Me^{(1)}_m(\xi, \overset{(T)}{q}) A^m_k(\overset{(T)}{q}) \delta_{ok}\,; \quad \overset{(\mu)}{R}{}^{+1}_{mk}(\xi) = 0\,; \tag{43}$$

$$\overset{(L)}{P}{}^{+2}_{mk}(\xi) = \overset{(L)}{D}_1 \frac{\partial}{\partial \xi} Ne^{(1)}_m(\xi, \overset{(L)}{q}) B^m_k(\overset{(L)}{q})\,; \quad \overset{(T)}{P}{}^{+2}_{mk}(\xi) = ih \frac{\overset{(T)}{D}_1}{\overset{(T)}{g}} \frac{\partial}{\partial \xi} Ne^{(1)}_m(\xi, \overset{(T)}{q}) B^m_k(\overset{(T)}{q})\,;$$

$$\overset{(\mu)}{P}{}^{+1}_{mk}(\xi) = -Me^{(1)}_m(\xi, \mu) k A^m_k(\mu)\,; \quad \overset{(L)}{Q}{}^{+1}_{mk}(\xi) = -\overset{(L)}{D}_1 Me^{(1)}_m(\xi, \overset{(L)}{q}) k A^m_k(\overset{(L)}{q})\,;$$

$$\overset{(T)}{Q}{}^{+1}_{mk}(\xi) = -ih \frac{\overset{(T)}{D}_1}{\overset{(T)}{g}} Me^{(1)}_m(\xi, \overset{(T)}{q}) k A^m_k(\overset{(T)}{q})\,; \quad \overset{(\mu)}{Q}{}^{+2}_{mk}(\xi) = -\frac{\partial}{\partial \xi} Ne^{(1)}_m(\xi, \mu) B^m_k(\mu)\,;$$

$$\overset{(L)}{R}{}^{+2}_{mk}(\xi) = ih \left(\overset{(L)}{D}_1 + \overset{(L)}{D}_2 \right) Ne^{(1)}_m(\xi, \overset{(L)}{q}) B^m_r(\overset{(L)}{q})\,;$$

$$\overset{(T)}{R}{}^{+2}_{mk}(\xi) = \left(\overset{(T)}{g} - \overset{(T)}{D}_1 \frac{h^2}{\overset{(T)}{g}} \right) Ne^{(1)}_m(\xi, \overset{(T)}{q}) B^m_r(\overset{(T)}{q})\,; \quad \overset{(\mu)}{R}{}^{+2}_{mk}(\xi) = 0\,; \tag{44}$$

$$\overset{(L)}{\Phi}{}^{+1}_{mk}(\xi) = \left[\overset{(L)}{d}_1 f^2 ch^4\xi\, Me^{(1)}_m(\xi, \overset{(L)}{q}) + \overset{(L)}{d}_4 ch^2\xi \frac{\partial^2}{\partial \xi^2} Me^{(1)}_m(\xi, \overset{(L)}{q}) - \right.$$

$$\left. - \frac{\overset{(L)}{d}_4}{2} sh2\xi \frac{\partial}{\partial \xi} Me^{(1)}_m(\xi, \overset{(L)}{q}) \right] A^m_k - \left[2\overset{(L)}{d}_1 f^2 ch^2\xi\, Me^{(1)}_m(\xi, \overset{(L)}{q}) + \overset{(L)}{d}_4 \frac{\partial^2}{\partial \xi^2} Me^{(1)}_m(\xi, \overset{(L)}{q}) \right] \times$$

$$\times \left[\frac{1}{4}\left(A^m_{k-2} + A^m_{2-k} + A^m_{k+2} \right) + \frac{1}{2} A^m_k \right] +$$

$$+ \overset{(L)}{d}_1 f^2 Me^{(1)}_m(\xi, \overset{(L)}{q}) \left[\frac{1}{16}\left(A^m_{k-4} + A^m_{4-k} + A^m_{k+4} \right) + \frac{1}{4}\left(A^m_{k-2} + A^m_{2-k} + A^m_{k+2} \right) + \frac{3}{8} A^m_k \right] -$$

$$-\frac{\overset{(L)}{d_4}}{4} Me_m^{(1)}(\xi, \overset{(L)}{q})\left[(2-k)A_{k-2}^m + (2+k)A_{k+2}^m + (2-k)A_{2-k}^m\right].$$

here we take parameter $\overset{(L)}{q}$ at calculation of factors A_i^j .

$$\overset{(T)}{\Phi}{}_{mk}^{+1}(\xi) = ih\left[\overset{(T)}{d_1} f^2 ch^4\xi\, Me_m^{(1)}(\xi, \overset{(T)}{q}) + \overset{(T)}{d_4} ch^2\xi \frac{\partial^2}{\partial \xi^2} Me_m^{(1)}(\xi, \overset{(T)}{q}) - \right.$$

$$\left. -\frac{\overset{(L)}{d_4}}{2} sh2\xi \frac{\partial}{\partial \xi} Me_m^{(1)}(\xi, \overset{(T)}{q})\right] A_k^m \delta_{ok} - ih\left[2\overset{(T)}{d_1} f^2 ch^2\xi\, Me_m^{(1)}(\xi, \overset{(T)}{q}) + \overset{(T)}{d_4} \frac{\partial^2}{\partial \xi^2} Me_m^{(1)}(\xi, \overset{(T)}{q})\right]\times$$

$$\times\left[\frac{1}{4}\left(A_{k-2}^m + A_{2-k}^m + A_{k+2}^m\right) + \frac{1}{2}A_k^m\right]\delta_{ok} +$$

$$+ ih\overset{(T)}{d_1} f^2\, Me_m^{(1)}(\xi, \overset{(T)}{q})\left[\frac{1}{16}\left(A_{k-4}^m + A_{4-k}^m + A_{k+4}^m\right) + \frac{1}{4}\left(A_{k-2}^m + A_{2-k}^m + A_{k+2}^m\right) + \frac{3}{8}A_k^m\right]\delta_{ok} -$$

$$- ih\frac{\overset{(T)}{d_4}}{4} Me_m^{(1)}(\xi, \overset{(T)}{q})\left[(2-k)A_{k-2}^m + (2+k)A_{2+k}^m + (2-k)A_{2-k}^m\right]\delta_{ok}.$$

here we take parameter $\overset{(T)}{q}$ at calculation of factors A_i^j .

$$\overset{(\mu)}{\Phi}{}_{mk}^{+2}(\xi) = \left[2a_2 ch^2\xi \frac{\partial}{\partial \xi} Ne_m^{(1)}(\xi, \mu) - a_2 sh2\xi\, Ne_m^{(1)}(\xi, \mu)\right] kB_k^m \delta_{ok} -$$

$$- 2a_2 \frac{\partial}{\partial \xi} Ne_m^{(1)}(\xi, \mu)\left[\frac{1}{4}(k-2)B_{k-2}^m + \frac{1}{4}(2-k)B_{2-k}^m + \frac{1}{4}(k+2)B_{k+2}^m + \frac{1}{2}kB_k^m\right]\delta_{ok} -$$

$$- a_2 \frac{\partial}{\partial \xi} Ne_m^{(1)}(\xi, \mu)\frac{1}{2}\left(-B_{k-2}^m + B_{k+2}^m + B_{2-k}^m\right)\delta_{ok}.$$

here we take parameter μ at calculation of factors B_i^j .

$$\overset{(L)}{H}{}^{+1}_{mk}(\xi)=\left[-\overset{(L)}{d_4}ch^2\xi\frac{\partial}{\partial\xi}Me^{(1)}_m(\xi,\overset{(L)}{q})+\frac{1}{2}\overset{(L)}{d_4}sh2\xi\,Me^{(1)}_m(\xi,\overset{(L)}{q})\right]kA^m_k\delta_{ok}+$$

$$+\overset{(L)}{d_4}\frac{\partial}{\partial\xi}Me^{(1)}_m(\xi,\overset{(L)}{q})\left[-\frac{1}{4}(2-k)A^m_{2-k}+\frac{1}{4}(k+2)A^m_{k+2}+\right.$$

$$\left.+\frac{1}{4}(k-2)A^m_{k-2}+\frac{1}{2}kA^m_k\right]\delta_{kn}-\frac{1}{4}\overset{(L)}{d_4}\frac{\partial}{\partial\xi}Me^{(1)}_m(\xi,\overset{(L)}{q})\left(-A^m_{k+2}+A^m_{2-k}+A^m_{k-2}\right)\delta_{ok}.$$

here we take parameter $\overset{(L)}{q}$ at calculation of factors A^j_i.

$$\overset{(T)}{H}{}^{+1}_{mk}(\xi)=ih\left\{\left[-\overset{(T)}{d_1}ch^2\xi\frac{\partial}{\partial\xi}Me^{(1)}_m(\xi,\overset{(T)}{q})+\frac{1}{2}\overset{(T)}{d_4}sh2\xi\,Me^{(1)}_m(\xi,\overset{(T)}{q})\right]kA^m_k\delta_{ok}+\right.$$

$$+\overset{(T)}{d_4}\frac{\partial}{\partial\xi}Me^{(1)}_m(\xi,\overset{(T)}{q})\left[-\frac{1}{4}(2-k)A^m_{2-k}+\frac{1}{4}(k+2)A^m_{k+2}+\frac{1}{4}(k-2)A^m_{k-2}+\frac{1}{2}kA^m_k\right]\delta_{ok}-$$

$$\left.-\frac{\overset{(T)}{d_4}}{4}\frac{\partial}{\partial\xi}Me^{(1)}_m(\xi,\overset{(T)}{q})\left[-A^m_{k+2}+A^m_{2-k}+A^m_{k-2}\right]\delta_{ok}\right\},$$

here we take parameter $\overset{(T)}{q}$ at calculation of factors A^j_i.

$$\overset{(\mu)}{H}{}^{+2}_{mk}(\xi)=a_2\left\{-ch^2\xi\,Ne^{(1)}_m(\xi,\mu)k^2B^m_k+\right.$$

$$+\left[ch^2\xi\frac{\partial^2}{\partial\xi^2}Ne^{(1)}_m(\xi,\mu)+sh2\xi\frac{\partial}{\partial\xi}Ne^{(1)}_m(\xi,\mu)\right]B^m_k+$$

$$+Ne^{(1)}_m(\xi,\mu)\left[-\frac{1}{4}(2-k)^2B^m_{2-k}+\frac{1}{4}(k+2)^2B^m_{k+2}+\frac{1}{4}(k-2)^2B^m_{k-2}+\frac{1}{2}k^2B^m_k\right]+$$

$$+\frac{\partial^2}{\partial\xi^2}Ne^{(1)}_m(\xi,\mu)\left[\frac{1}{4}(B^m_{2-k}+B^m_{k+2}+B^m_{k-2})+\frac{1}{2}B^m_k\right]-$$

$$- Ne_m^{(1)}(\xi,\mu)\frac{1}{2}\left[-(k+2)B_{k+2}^m + (2-k)B_{2-k}^m + (k-2)B_{k-2}^m\right]\Bigg\}\delta_{ok}.$$

here we take parameter μ at calculation of factors B_i^j.

$$\overset{(L)}{Z}{}_{mk}^{+1}(\xi) = ih\left(\frac{1}{2}\overset{(L)}{d}_3 + \overset{(L)}{d}_4\right)\frac{\partial}{\partial\xi}Me_m^{(1)}(\xi,\overset{(L)}{q})A_k^m(\overset{(L)}{q})\delta_{ok};$$

$$\overset{(T)}{Z}{}_{mk}^{+1}(\xi) = \left(\frac{1}{2}\overset{(T)}{d}_3 - h^2\overset{(L)}{d}_4\right)\frac{\partial}{\partial\xi}Me_m^{(1)}(\xi,\overset{(T)}{q})A_k^m(\overset{(T)}{q})\delta_{ok};$$

$$\overset{(\mu)}{Z}{}_{mk}^{+2}(\xi) = ih(a_2 + a_4)Ne_m^{(1)}(\xi,\mu)B_k^m(\mu)\delta_{ok};$$

$$\overset{(L)}{\Phi}{}_{mk}^{+2}(\xi) = \Bigg[\overset{(L)}{d}_1 f^2 ch^4\xi\, Ne_m^{(1)}(\xi,\overset{(L)}{q}) + \overset{(L)}{d}_4 ch^2\xi\frac{\partial^2}{\partial\xi^2}Ne_m^{(1)}(\xi,\overset{(L)}{q}) -$$

$$-\frac{\overset{(L)}{d}_4}{2}sh2\xi\frac{\partial}{\partial\xi}Me_m^{(1)}(\xi,\overset{(L)}{q})\Bigg]B_k^m\delta_{ok} -$$

$$-\left[\overset{(L)}{d}_1 f^2 ch^2\xi 2\, Ne_m^{(1)}(\xi,\overset{(L)}{q}) + \overset{(L)}{d}_4\frac{\partial^2}{\partial\xi^2}Ne_m^{(1)}(\xi,\overset{(L)}{q})\right]\times$$

$$\times\left[\frac{1}{4}(-B_{2-k}^m + B_{k+2}^m + B_{k+2}^m) + \frac{1}{2}B_k^m\right]\delta_{ok} +$$

$$+\overset{(L)}{d}_1 f^2 Ne_m^{(1)}(\xi,\overset{(L)}{q})\left[\frac{1}{16}\left(-B_{4-k}^m + B_{k-4}^m + B_{k+4}^m\right) + \frac{1}{4}\left(-B_{2-k}^m + B_{k+2}^m + B_{k-2}^m\right) + \frac{3}{8}B_k^m\right]\delta_{ok} +$$

$$+\frac{1}{4}\overset{(L)}{d}_4 Ne_m^{(1)}(\xi,\overset{(L)}{q})\left[-(k+2)B_{k+2}^m + (2-k)B_{2-k}^m + (k-2)B_{k-2}^m\right]\delta_{ok}.$$

here we take parameter $\overset{(L)}{q}$ at calculation of factors B_i^j.

$$\overset{(T)}{\Phi}{}^{+2}_{mk}(\xi) = ih\left[\overset{(T)}{d_1} f^2 ch^4\xi \, Ne^{(1)}_m(\xi, \overset{(T)}{q}) + \overset{(T)}{d_4} ch^2\xi \frac{\partial^2}{\partial \xi^2} Ne^{(1)}_m(\xi, \overset{(T)}{q}) - \right.$$

$$\left. - \frac{\overset{(T)}{d_4}}{2} sh2\xi \frac{\partial}{\partial \xi} Ne^{(1)}_m(\xi, \overset{(T)}{q})\right] B^m_k \delta_{ok} - ih\left[\overset{(T)}{d_1} f^2 ch^2\xi 2\, Ne^{(1)}_m(\xi, \overset{(T)}{q}) + \overset{(T)}{d_4} \frac{\partial^2}{\partial \xi^2} Ne^{(1)}_m(\xi, \overset{(T)}{q})\right] \times$$

$$\times \left[\frac{1}{4}(-B^m_{2-k} + B^m_{k+2} + B^m_{k-2}) + \frac{1}{2} B^m_k\right]\delta_{ok} +$$

$$+ ih\overset{(T)}{d_1} f^2 \, Ne^{(1)}_m(\xi, \overset{(T)}{q})\left[\frac{1}{16}\left(-B^m_{4-k} + B^m_{k-4} + B^m_{k+4}\right) + \frac{1}{4}\left(-B^m_{2-k} + B^m_{k+2} + B^m_{k-2}\right) + \frac{3}{8} B^m_k\right]\delta_{ok} +$$

$$+ ih\frac{\overset{(T)}{d_4}}{4} Ne^{(1)}_m(\xi, \overset{(T)}{q})\left[-(k+2)B^m_{k+2} + (2-k)B^m_{2-k} + (k-2)B^m_{k-2}\right]\delta_{ok}$$

.

here we take parameter $\overset{(T)}{q}$ at calculation of factors B^j_i.

$$\overset{(\mu)}{\Phi}{}^{+1}_{mk}(\xi) = -a_2\left[2ch^2\xi \frac{\partial}{\partial \xi} Me^{(1)}_m(\xi, \mu) - sh2\xi \, Me^{(1)}_m(\xi, \mu)\right] kA^m_k \delta_{ok} +$$

$$+ 2a_2 \frac{\partial}{\partial \xi} Me^{(1)}_m(\xi, \mu)\left[\frac{1}{4}((2-k)A^m_{2-k} + (k+2)A^m_{k+2} + (k-2)A^m_{k-2}) + \frac{1}{2} kA^m_k\right]\delta_{ok} -$$

$$- a_2 \frac{\partial}{\partial \xi} Me^{(1)}_m(\xi, \mu)\frac{1}{2}\left(-A^m_{k+2} + A^m_{2-k} + A^m_{k-2}\right)\delta_{ok}$$

.

here we take parameter μ at calculation of factors A^j_i.

$$\overset{(L)}{H}{}^{+2}_{mk}(\xi) = \left[\overset{(L)}{d_4} ch^2\xi \frac{\partial}{\partial \xi} Ne^{(1)}_m(\xi, \overset{(L)}{q}) - \frac{1}{2}\overset{(L)}{d_4} sh2\xi \, Ne^{(1)}_m(\xi, \overset{(L)}{q})\right] B^m_k \delta_{ok} -$$

$$- \overset{(L)}{d_4} \frac{\partial}{\partial \xi} Ne^{(1)}_m(\xi, \overset{(L)}{q})\left[\frac{1}{4}(k-2)B^m_{k-2} + \frac{1}{4}(2-k)B^m_{2-k} + \right.$$

$$\left. + \frac{1}{4}(k+2)B^m_{k+2} + \frac{1}{2} kB^m_k\right]\delta_{ok} - \frac{1}{4}\overset{(L)}{d_4} \frac{\partial}{\partial \xi} Ne^{(1)}_m(\xi, \overset{(L)}{q})\left(-B^m_{k-2} + B^m_{k+2} + B^m_{2-k}\right)\delta_{ok}$$

.

here we take parameter $\overset{(L)}{q}$ at calculation of factors B_i^j.

$$\overset{(T)}{H}{}^{+2}_{mk}(\xi) = ih\left[\overset{(T)}{d_1} ch^2\xi \frac{\partial}{\partial\xi} Ne_m^{(1)}(\xi, \overset{(T)}{q}) - \frac{1}{2}\overset{(T)}{d_4} sh2\xi \frac{\partial^2}{\partial\xi^2} Ne_m^{(1)}(\xi, \overset{(T)}{q})\right] B_k^m \delta_{ok} -$$

$$- ih\overset{(T)}{d_4}\frac{\partial}{\partial\xi} Ne_m^{(1)}(\xi, \overset{(T)}{q})\left[-\frac{1}{4}(k-2)B_{k-2}^m + \frac{1}{4}(2-k)B_{2-k}^m + \right.$$

$$\left. + \frac{1}{4}(k+2)B_{k+2}^m + \frac{1}{2}kB_k^m\right]\delta_{ok} + ih\frac{\overset{(T)}{d_4}}{4}\frac{\partial}{\partial\xi} Ne_m^{(1)}(\xi, \overset{(T)}{q})\left[-B_{k-2}^m + B_{k+2}^m + B_{2-k}^m\right]\delta_{ok}$$

here we take parameter $\overset{(T)}{q}$ at calculation of factors B_i^j.

$$\overset{(\mu)}{H}{}^{+1}_{mk}(\xi) = -ch^2\xi\, Me_m^{(1)}(\xi,\mu)k^2 A_k^m \delta_{ok} +$$

$$+ \left[-ch^2\xi \frac{\partial^2}{\partial\xi^2} Me_m^{(1)}(\xi,\mu) + sh2\xi \frac{\partial}{\partial\xi} Me_m^{(1)}(\xi,\mu)\right] A_k^m \delta_{ok} +$$

$$+ \frac{\partial^2}{\partial\xi^2} Me_m^{(1)}(\xi,\mu)\left[\frac{1}{4}(A_{k-2}^m + A_{2-k}^m + A_{k+2}^m) + \frac{1}{2}A_k^m\right]\delta_{ok} +$$

$$+ Me_m^{(1)}(\xi,\mu)\left[\frac{1}{4}(k-2)^2 A_{k-2}^m + \frac{1}{4}(2-k)^2 A_{2-k}^m + \frac{1}{4}(k+2)^2 A_{k+2}^m + \frac{1}{2}k^2 A_k^m\right]\delta_{ok} +$$

$$+ Me_m^{(1)}(\xi,\mu)\frac{1}{2}\left[-(k-2)A_{k-2}^m + (k+2)A_{k+2}^m + (2-k)A_{2-k}^m\right]\delta_{ok}$$

here we take parameter μ at calculation of factors A_i^j.

$$\overset{(L)}{Z}{}^{+2}_{mk}(\xi) = ih\left(\frac{1}{2}\overset{(L)}{d_3} + \overset{(L)}{d_4}\right)\frac{\partial}{\partial\xi} Ne_m^{(1)}(\xi, \overset{(L)}{q})B_k^m(\overset{(L)}{q});$$

$$\overset{(T)}{Z}{}^{+2}_{mk}(\xi) = \frac{1}{2}\left(\overset{(T)}{d_3} - h^2\overset{(T)}{d_4}\right)\frac{\partial}{\partial\xi} Ne_m^{(1)}(\xi, \overset{(T)}{q})B_k^m(\overset{(T)}{q});$$

$$\overset{(\mu)}{Z}{}^{+1}_{mk}(\xi) = -ih(a_2 + a_4)Me_m^{(1)}(\xi,\mu)kA_k^m(\mu); \quad \delta_{ok} = \begin{cases} 2 & \text{at} \quad k = m = 0; \\ 1 & \text{at} \quad \text{other } k, m. \end{cases} \tag{45}$$

The functions of ξ, present in (39) – (42) with minus index, are calculated by using formulae (43), (44) at replacement of the combined Mathieu functions of first kind by the corresponding Mathieu functions of second kind.

Left sides of the expressions (39) – (42) have the following forms:

$$P_k^1(\xi) = \frac{1}{\pi}\int_0^{2\pi} U_\xi(\xi,\eta)\cos k\eta d\eta; \quad Q_k^2(\xi) = \frac{1}{\pi}\int_0^{2\pi} U_\eta(\xi,\eta)\cos k\eta d\eta;$$

$$R_k^1(\xi) = \frac{1}{\pi}\int_0^{2\pi} U_z(\xi,\eta)\cos k\eta d\eta; \quad P_k^2(\xi) = \frac{1}{\pi}\int_0^{2\pi} U_\xi(\xi,\eta)\sin k\eta d\eta;$$

$$Q_k^1(\xi) = \frac{1}{\pi}\int_0^{2\pi} U_\eta(\xi,\eta)\sin k\eta d\eta; \quad R_k^2(\xi) = \frac{1}{\pi}\int_0^{2\pi} U_z(\xi,\eta)\sin k\eta d\eta; \tag{46}$$

$$\Phi_k^1(\xi) = \frac{1}{\pi}\int_0^{2\pi} \sigma_{\xi\xi}(\xi,\eta)(ch^2\xi - \cos^2\eta)\cos k\eta d\eta;$$

$$H_k^2(\xi) = \frac{1}{\pi}\int_0^{2\pi} \sigma_{\xi\eta}(\xi,\eta)(ch^2\xi - \cos^2\eta)\sin k\eta d\eta;$$

$$Z_k^1(\xi) = \frac{1}{\pi}\int_0^{2\pi} \sigma_{\xi z}(\xi,\eta)\cos k\eta d\eta; \quad \Phi_k^2(\xi) = \frac{1}{\pi}\int_0^{2\pi} \sigma_{\xi\xi}(\xi,\eta)(ch^2\xi - \cos^2\eta)\sin k\eta d\eta;$$

$$H_k^1(\xi) = \frac{1}{\pi}\int_0^{2\pi} \sigma_{\xi\eta}(\xi,\eta)(ch^2\xi - \cos^2\eta)\cos k\eta d\eta;$$

$$Z_k^2(\xi) = \frac{1}{\pi}\int_0^{2\pi} \sigma_{\xi z}(\xi,\eta)\sin k\eta d\eta. \tag{47}$$

By changing number k from 0 to ∞ in (39) – (42), we obtain infinite set of equations relatively of wave amplitudes, propagating in elliptic layer. As it is seen, this set of equations consists of two independent subsets. The first subset (39), (41) is combined relatively of even (index 1) harmonics of quasilongitudinal $\varphi_m^{\pm 1}$, quasitransverse $w_m^{\pm 1}$ waves and odd (index 2) harmonics of pure transverse waves $v_m^{\pm 2}$. Odd harmonics of quasilongitudinal $\varphi_m^{\pm 2}$, quasitransverse $w_m^{\pm 2}$ waves and even harmonics of pure transverse waves $v_m^{\pm 1}$ are the unknown variables in Equation subset (40), (42).

So, if we know distributions of displacements $U_\xi(\xi',\eta), U_\eta(\xi',\eta), U_z(\xi',\eta)$ and stresses $\sigma_{\xi\xi}(\xi',\eta), \sigma_{\xi\eta}(\xi',\eta), \sigma_{\xi z}(\xi',\eta)$ on coordinate η on some surface $\xi = \xi'$ (for example, on boundary), then, if to solve Equation set (39) – (42) with heterogeneous terms (45), (46) at $\xi = \xi'$, we can find amplitudes of harmonics of the elastic waves and determine potential functions from (31). Then by knowing scalar functions $\varphi(\xi,\eta,z)$, $w(\xi,\eta,z)$, $v(\xi,\eta,z)$, we can calculate tensors of elastic fields by using formulae (24), (25) in any area of the construction.

4. Wave Matrix of Transform for Elliptic Layer

To simplify the next formulae, we write Equations (39), (41) relatively of unknown parameters $\varphi_m^{\pm 1}, w_m^{\pm 1}, v_m^{\pm 1}$ in matrix form as

$$\bar{B}_k = \bar{C}_k \bar{\psi} \,. \tag{48}$$

here vector-column $\bar{B}_k$ for each value of *k* consists of two blocks with three elements in every block. These blocks correspond to the displacement *U* and stress *σ*, given on surface *ξ*:

$$\bar{B}_k = \begin{pmatrix} U_k \\ \sigma_k \end{pmatrix}, \text{ where } U_k = (P_k^1, Q_k^2, R_k^1)^T \,;\ \sigma_k = (\Phi_k^1, H_k^2, Z_k^1)^T \,. \tag{49}$$

The vector elements $\bar{B}_k$ are defined by formulae (45), (46).

The rectangular matrix $\bar{C}_k$ has six rows and is defined as $\bar{C}_k = (C_{0k} C_{1k} C_{2k} \ldots C_{mk} \ldots)$, where

$$C_{mk} = \begin{pmatrix} \overset{(L)}{P}{}^{+1}_{mk} & \overset{(T)}{P}{}^{+1}_{mk} & \overset{(\mu)}{P}{}^{+1}_{mk} & \overset{(L)}{P}{}^{-1}_{mk} & \overset{(T)}{P}{}^{-1}_{mk} & \overset{(\mu)}{P}{}^{-1}_{mk} \\ \overset{(L)}{Q}{}^{+1}_{mk} & \overset{(T)}{Q}{}^{+1}_{mk} & \overset{(\mu)}{Q}{}^{+1}_{mk} & \overset{(L)}{Q}{}^{-1}_{mk} & \overset{(T)}{Q}{}^{-1}_{mk} & \overset{(\mu)}{Q}{}^{-1}_{mk} \\ \overset{(L)}{R}{}^{+1}_{mk} & \overset{(T)}{R}{}^{+1}_{mk} & \overset{(\mu)}{R}{}^{+1}_{mk} & \overset{(L)}{R}{}^{-1}_{mk} & \overset{(T)}{R}{}^{-1}_{mk} & \overset{(\mu)}{R}{}^{-1}_{mk} \\ \overset{(L)}{\Phi}{}^{+1}_{mk} & \overset{(T)}{\Phi}{}^{+1}_{mk} & \overset{(\mu)}{\Phi}{}^{+1}_{mk} & \overset{(L)}{\Phi}{}^{-1}_{mk} & \overset{(T)}{\Phi}{}^{-1}_{mk} & \overset{(\mu)}{\Phi}{}^{-1}_{mk} \\ \overset{(L)}{H}{}^{+1}_{mk} & \overset{(T)}{H}{}^{+1}_{mk} & \overset{(\mu)}{H}{}^{+1}_{mk} & \overset{(L)}{H}{}^{-1}_{mk} & \overset{(T)}{H}{}^{-1}_{mk} & \overset{(\mu)}{H}{}^{-1}_{mk} \\ \overset{(L)}{Z}{}^{+1}_{mk} & \overset{(T)}{Z}{}^{+1}_{mk} & \overset{(\mu)}{Z}{}^{+1}_{mk} & \overset{(L)}{Z}{}^{-1}_{mk} & \overset{(T)}{Z}{}^{-1}_{mk} & \overset{(\mu)}{Z}{}^{-1}_{mk} \end{pmatrix} = \begin{pmatrix} C^1_{mk} & C^2_{mk} \\ C^3_{mk} & C^4_{mk} \end{pmatrix} \quad (50)$$

The blocks C^i_{mk} of matrix C_{mk} have sizes 3×3, but their elements are calculated from formulae (43), (44).

The vector-column $\bar{\psi}$ corresponds to amplitudes of harmonics of the different wave types by the following way: $\bar{\psi} = (\psi^+_0\, \psi^-_0 \psi^+_1 \psi^-_1 \psi^+_2 \psi^-_2 ... \psi^+_m \psi^-_m ...)^T$, where

$$\psi^+_m = (\varphi^{+1}_m w^{+1}_m v^{+2}_m)^T \,;\, \psi^-_m = (\varphi^{-1}_m w^{-1}_m v^{-2}_m)^T \,. \quad (51)$$

Let us restrict the number of harmonics taken into account by value of M. Then index m changes from 0 to $(M - 1)$. By changing index k in the same limits, we obtain a set of $6M$ linear algebraic equations with $6M$ unknown variables, which is present in the matrix form:

$$\bar{B} = \bar{C}\bar{\psi} \,, \quad (52)$$

where

$$\bar{B} = (B_0\, B_1\, B_2 ... B_k ... B_{M-1})^T \,. \quad (53)$$

The elements of vector $\bar{B}$ are found by expressions (49). The quadratic matrix $\bar{C}$ with sizes of $6M\times6M$ consists of blocks C_{mk} (50), number of which in the matrix is equal to $M\times M$. The vector-column $\bar{\psi}$ is defined by relations (51), where last index has number $m = M - 1$.

The equations, similar to the above-pointed in this section, are obtained also for amplitudes $\varphi^{\pm2}_m, w^{\pm2}_m, v^{\pm1}_m$, which satisfy Equations (40), (42). All expressions in these equations coincide with those of (48) – (52) at mutual replacement of indices 1 and 2.

To solve the problem of definition of field in elliptic cylinder, it is useful to regroup rows in (52) and (53) by the next way:

$$B = (U_0 U_1 U_2 ... U_k ... U_{M-1} \sigma_0 \sigma_1 \sigma_2 ... \sigma_k ... \sigma_{M-1})^{(T)} = (U\sigma)^{(T)}, \tag{54}$$

where vector U corresponds to the Fourier-harmonics of displacement and vector σ corresponds to the Fourier-harmonics of stress; U_k and σ_k are found by expressions (48);

$$\psi = (\psi_0^+ \psi_1^+ \psi_2^+ ... \psi_m^+ ... \psi_{M-1}^+ \psi_0^- \psi_1^- \psi_2^- ... \psi_m^- ... \psi_{M-1}^-)^{(T)} = (\psi^+ \psi^-)^{(T)}, \tag{55}$$

where vector ψ^+ corresponds to the amplitudes of wave harmonics propagating in direct direction, but vector ψ^- corresponds to the waves propagating back, ψ_m^+ and ψ_m^- are defined by relations (51).

In accordance with this regrouping, the matrix $\bar{C}$ is defined by using (51) as

$$\bar{C} = \begin{pmatrix} C^1 & C^2 \\ C^3 & C^4 \end{pmatrix}, \tag{56}$$

where i-th block $(i = 1,2,3,4)$ is written as

$$C^i = \begin{pmatrix} C_{00}^i & C_{10}^i ... C_{m0}^i ... C_{M-1,0}^i \\ C_{10}^i & C_{11}^i ... C_{m1}^i ... C_{M-1,1}^i \\ \cdots\cdots\cdots & \\ C_{0k}^i & C_{1k}^i ... C_{mk}^i ... C_{M-1,k}^i \\ \cdots\cdots\cdots & \\ C_{0,M-1}^i & C_{1,M-1}^i ... C_{m,M-1}^i ... C_{M-1,M-1}^i \end{pmatrix}. \tag{57}$$

The elements of matrix C^i are the blocks C_{mk}^i, whose expressions are present in (50).

By using these transformed matrices instead of relation (52), we obtain

$$B = C\psi, \tag{58}$$

where the vector-column B, defined by formula (54) similarly to the case of circular cylinder, could be named vector of states. It characterizes the boundary conditions of the problem at $\xi = \xi_0$ and $\xi = \xi_1$.

Matrix C with elements, defined finally by formulae (39) and (41), is the wave transfer matrix for elliptic layer. It is similar to matrix C, introduced for circular cylinder in [10], and plays main role in the following solution.

4. STRESS-STRAIN STATE OF CONSTRUCTION IN FORM OF ELLIPTIC CYLINDER

Let us use the expressions (54) – (58) for solution of the stated problem on definition of the elastic field tensors in elliptic cylinder, whose internal surface is subject to external loading.

The boundary conditions for the considered construction in general case of existence of all types of waves are written by using the matrix designations in the form:

$$\sigma=\sigma_0 \text{ at } \xi=\xi_0 ;\ \sigma=0 \text{ at } \xi=\xi_1 . \tag{59}$$

In order to satisfy the boundary conditions, we substitute (59) into expression (58) at $\xi=\xi_0$ and $\xi=\xi_1$, respectively and obtain two relations:

$$B^{\downarrow}=C^{\downarrow}\psi \text{ or } \begin{pmatrix} U^{\downarrow} \\ \sigma_0 \end{pmatrix}=\begin{pmatrix} C^1 & C^2 \\ C^3 & C^4 \end{pmatrix}^{\downarrow}\begin{pmatrix} \psi^+ \\ \psi^- \end{pmatrix}; \tag{60}$$

$$B^{\uparrow}=C^{\uparrow}\psi \text{ or } \begin{pmatrix} U^{\uparrow} \\ 0 \end{pmatrix}=\begin{pmatrix} C^1 & C^2 \\ C^3 & C^4 \end{pmatrix}^{\uparrow}\begin{pmatrix} \psi^+ \\ \psi^- \end{pmatrix}. \tag{61}$$

here as before, arrow $\downarrow$ designs internal boundary ($\xi=\xi_0$), but arrow $\uparrow$ defines external boundary ($\xi=\xi_1$). We define vector ψ from these expressions. With this aim by assuming that $\det C\neq 0$, we write

$$\psi=C^{\downarrow-1}B^{\downarrow}=C^{\uparrow-1}B^{\uparrow}. \tag{62}$$

This formula connects components of vector of states on the construction boundary by the following way:

$$B^{\downarrow}=C^{\downarrow}C^{\uparrow-1}B^{\uparrow} \text{ or } B^{\uparrow}=C^{\uparrow}C^{\downarrow-1}B^{\downarrow}. \tag{63}$$

We design

$$A=C^{\downarrow}C^{\uparrow-1}=\begin{pmatrix} A_1 & A_2 \\ A_3 & A_4 \end{pmatrix};\quad \bar{A}=C^{\uparrow}C^{\downarrow-1}=A^{-1}=\begin{pmatrix} \bar{A}_1 & \bar{A}_2 \\ \bar{A}_3 & \bar{A}_4 \end{pmatrix} \tag{64}$$

and rewrite expressions (63) in the form:

$$\begin{pmatrix} U^{\downarrow} \\ \sigma_0 \end{pmatrix} = \begin{pmatrix} A_1 & A_2 \\ A_3 & A_4 \end{pmatrix} \begin{pmatrix} U^{\uparrow} \\ 0 \end{pmatrix} \text{ or } \begin{pmatrix} U^{\uparrow} \\ 0 \end{pmatrix} = \begin{pmatrix} \bar{A}_1 & \bar{A}_2 \\ \bar{A}_3 & \bar{A}_4 \end{pmatrix} \begin{pmatrix} U^{\downarrow} \\ \sigma_0 \end{pmatrix}. \tag{65}$$

By solving Equations (64) in respect to $U^{\uparrow}$ and $U^{\downarrow}$, we obtain two the same sets of formulae for the vector of states on the construction boundaries in the form:

$$B^{\downarrow} = \begin{pmatrix} A_1 A_3^{-1} \sigma_0 \\ \sigma_0 \end{pmatrix}; \quad B^{\uparrow} = \begin{pmatrix} A_3^{-1} \sigma_0 \\ 0 \end{pmatrix}; \tag{66}$$

$$B^{\downarrow} = \begin{pmatrix} -\bar{A}_3^{-1} \bar{A}_4 \sigma_0 \\ \sigma_0 \end{pmatrix}; \quad B^{\uparrow} = \begin{pmatrix} -\bar{A}_1 \bar{A}_3^{-1} \bar{A}_4 \sigma_0 + \bar{A}_2 \sigma_0 \\ 0 \end{pmatrix}. \tag{67}$$

By substituting (66) and (67) into formula (62), we obtain the following equivalent relations for calculation of amplitudes of harmonics in elliptic layer:

$$\begin{pmatrix} \psi^+ \\ \psi^- \end{pmatrix} = \begin{pmatrix} C^1 & C^2 \\ C^3 & C^4 \end{pmatrix}^{\downarrow -1} \begin{pmatrix} A_1 A_3^{-1} \sigma_0 \\ \sigma_0 \end{pmatrix} = \begin{pmatrix} C^1 & C^2 \\ C^3 & C^4 \end{pmatrix}^{\uparrow -1} \begin{pmatrix} A_3^{-1} \sigma_0 \\ 0 \end{pmatrix}; \tag{68}$$

$$\begin{pmatrix} \psi^+ \\ \psi^- \end{pmatrix} = \begin{pmatrix} C^1 & C^2 \\ C^3 & C^4 \end{pmatrix}^{\downarrow -1} \begin{pmatrix} -\bar{A}_3^{-1} \bar{A}_4 \sigma_0 \\ \sigma_0 \end{pmatrix} = \begin{pmatrix} C^1 & C^2 \\ C_3 & C_4 \end{pmatrix}^{\uparrow -1} \begin{pmatrix} -\bar{A}_1 \bar{A}_3^{-1} \bar{A}_4 \sigma_0 + \bar{A}_2 \sigma_0 \\ 0 \end{pmatrix}. \tag{69}$$

By using any of these relations, we can calculate wave amplitudes in elliptic layer and define scalar functions φ, w, v from (31). Then by using these functions and formulae (32), (33), we can determine all components of displacements, stresses and strains in any point of the considered construction.

Note that the obtained solutions present the angle-frequency spectrum of searched field, that is $f(h,\omega)$. To obtain a result in common case, when load depends on time and coordinate z on arbitrary given law, it is necessary to apply inverse Fourier transform as it is described for the case of circular cylinder [10].

Thus the problem on definition of stress-strain state in construction in the form of elliptic cylinder layer loaded according to given law is solved, in principle. This solution is the base of definition of elastic fields in layered elliptic construction.

5. ELASTIC FIELDS IN LAYERED ELLIPTIC CONSTRUCTION

Let us treat a problem is similar to the above-formulated one for the layered construction in the form of circular cylinder [10] in the case, when layers are elliptic. We expand the problem statement and main concepts, presented in [10], on the considered case of elliptic construction.

We assume that the boundary conditions of continuity of the displacements and normal stresses (rigid junction) are also carried out for their Fourier components, i.e., for elements of the state vector B. Then we can write

$$B_n^{\uparrow} = B_{n+1}^{\downarrow}, \tag{60}$$

where index n relates to layer number.

Vectors $B_n^{\uparrow}$ and $B_{n+1}^{\downarrow}$ are defined by expressions for B at $\xi = \xi_n$. By using these expressions, we obtain

$$C_n^{\uparrow}\psi_n = C_{n+1}^{\downarrow}\psi_{n+1}. \tag{61}$$

here $C_n^{\uparrow}$ and $C_{n+1}^{\downarrow}$ are calculated by a formula, in which it is necessary to state $\xi = \xi_k$ and take parameters corresponding to physical and mechanical characteristics of n-th and (n+1)-th layer, respectively.

By assuming that the matrices have not singularities, we obtain the formula connecting the wave amplitudes in adjacent layers:

$$\psi_n = C_n^{\uparrow -1} C_{n+1}^{\downarrow}\psi_{n+1}. \tag{62}$$

This formula has the same form as in [10] for circular cylinders. The formal representation of recurrent relations [10] between fields in different layers of the construction, defining amplitudes of fields in (n+1)-th layer through amplitudes of fields in n-th layer, performs also in the considered case of elliptic cylinders. However, the sense and matrix relations, presented in these formulae, have been found above in this chapter and differed significantly from the case of the layered circular cylinders. Nevertheless, the formal solution of the problems on definition of field in the layered elliptic construction does not differ from above case of circular cylinders, if to use the matrices, described in this chapter.

So the formal algorithm for solution of the problem for layered construction, consisting of elliptic cylinders, coincides with the solution algorithm of similar problems for circular cylinders.

ACKNOWLEDGMENTS

The work was supported in part by the Russian Foundation for Basic Research under grant No.13-08-00754.

References

[1] Novatski, V., *Theory of Elasticity*, Moscow, Mir, 1975, pp. 1-872 (in Russian).

[2] McConnell, A. J., *Application of Tensor Analysis*, Dover, New York, 1947, pp. 1-411.

[3] Christensen, R. M., *Mechanics of Composite Materials*, J. Wiley, 2nd ed. 1979, pp. 1-334.

[4] Grinchenko, V. T., Meleshko, V. V. *Harmonic Oscillations and Waves in Elastic Bodies*. Naukova Dumka, Kiev, 1981, pp. 1-284 (in Russian).

[5] Morse, P. M., Feshbach, H., *Methods of Theoretical Physics*, Pt. 2, McGraw-Hill, New York, 1953, pp. 1-896.

[6] Sizov, V. P. *Izvestiya of USSR Academy of Science. Mechanics of Solids*, 1988, No 5, 55-58.

[7] Fedorov, F. N. *Theory of Elastic Waves in Crystals*. Plenum, New York, 1968, pp. 1-386.

[8] Nasledyshev, A. G., Sizov, V. P., Chernov, A. A. In: *Waves and Diffraction-90*, Physical Society, Moscow, 1990, vol. 1, 359-362 (in Russian).

[9] Yevtushenko, N. A., Lobzenko, P. V., Sizov, V. P. *Izvestiya of Russian Academy of Science. Mechanics of Solids*, 1996, No 6, 53-59.

[10] Sizov, V. P., Miroshnichenko, I. P. *Elastic Waves in Layered Anisotropic Structures*, LAP LAMBERT Academic Publishing, Saarbrucken (Germany), 2012, pp. 1-270.

[11] Miroshnichenko, I. P., Sizov, V. P. *Izvestia of Higher School. North-Caucases Region. Natural Sciences*, 1999, No 2, 19-22 (in Russian).

[12] Miroshnichenko, I. P., Sizov, V. P., *Izvestia of Higher School. North-Caucases Region. Natural Sciences*, 1999, No 3, 40-42 (in Russian).

[13] Petrov, A. M., Sizov, V. P., *Izvestiya of Russian Academy of Science. Mechanics of Solids*, 2000, No 3, 34-39.

[14] Miroshnichenko, I. P., Sizov, V. P., *Izvestiya of Russian Academy of Science. Mechanics of Solids*, 2000, No 1, 97-104.

[15] Rashevsky, P. K. *Riemannian Geometry and Tensor Analysis*. Nauka, Moscow, 1967, 1-664 (in Russian).

[16] Ivanov, E. A. *Diffraction of Electromagnetic Waves on Two Bodies*. Nauka i Tekhnika, Minsk, 1968, 1-583 (in Russian).

[17] McLachlan, N. W., Theory and Application of Mathieu Functions, Oxford University Press, 1947, 1-475.

[18] Korn, G. A., Korn, T. M., Mathematical Handbook for Scientists and Engineers. McGraw-Hill, New York, 1968, 1-720.

[19] Miroshnichenko, I. P., In: *Advanced Materials. Physics, Mechanics and Applications*, S.-H. Chang, I. A. Parinov, V. Yu. Topolov (Eds.), Springer Proceedings in Physics, vol. 152, Chapter 14. Springer, Heidelberg, New York, Dordrecht, London, 2014, pp. 163-179.

PART IV. APPLICATIONS OF ADVANCED MATERIALS

In: Advanced Materials - Studies and Applications
Editors: I.A. Parinov, S.-H. Chang, S.Theerakulpisut
ISBN: 978-1-63463-749-7

Chapter 23

THEORETICAL MODELING AND EXPERIMENTAL STUDY OF HIGH INTENSITY FOCUSED ULTRASOUND TRANSDUCERS

A. N. Rybyanets[1,2,*], V. A. Khokhlova[3], A. Sapozhnikov[3], A. A. Naumenko[2], M. A. Lugovaya[2] and S. A. Shcherbinin[2]
[2]Southern Federal University, Rostov-on-Don, Russia
[3]Moscow State University, Moscow, Russia

ABSTRACT

Recent advances in the fields of physical acoustics, imaging technologies, piezoelectric materials, and ultrasonic transducer design have led to emerging of novel methods and apparatus for ultrasonic diagnostics, therapy, and anesthetics as well as an expansion of traditional application fields. The chapter presents the results of theoretical modeling and experimental study of different High Intensity Focused Ultrasound (HIFU) transducers. Exact solutions for acoustic fields of single and multi-element spherically focusing transducers in the linear approximation were calculated using the Rayleigh integral. Numerical solutions of parabolic Khokhlov-Zabolotskaya-Kuznetsov equation were obtained for nonlinear focused fields. HIFU transducers comprising spherical piezoceramic caps immersed in acoustic contact fluids were fabricated and tested. Acoustic field measurements for the transducers were acquired using a calibrated fiber optic hydrophone and ultrasonic measurement system (UMS). *In-vitro* experiments in tissues (fresh porcine adipose tissue and bovine liver) and *in-vivo* experiments on sonicating blood vessels were carried out using different protocols. The results of theoretical modeling and tissue experiments that prove the efficacy, safety, and selectivity of using the developed HIFU transducers and irradiation methods enhancing tissue lysis and hemostasis were presented.

---

[*] arybyanets@gmail.com.

1. INTRODUCTION

Ultrasound has found usage in all aspects of the medical field, including diagnostic, therapeutic, and surgical applications. The use of ultrasound as a valuable diagnostic and therapeutic tool in several fields of clinical medicine is now so well established that it can be considered essential for good patient care [1]. However, remarkable advances in ultrasound imaging technology over the last decade have permitted us now to envision the combined use of ultrasound both for imaging/diagnostics and for therapy. It does not take much imagination to picture the tremendous promise of ultrasound- or MR-guided High Intensity Focused Ultrasound (HIFU) therapy or surgery. Just as smaller and smaller diagnostic ultrasound systems will soon replace the stethoscope, so eventually will increasingly precise therapeutic ultrasound systems replace the scalpel.

Therapeutic ultrasound has been in use for many years [1-4]. Early applications were those for which tissue heating was the goal, and so it was used to treat soft tissue injuries such as may be incurred during sport activities. More recently, attention has been drawn both to high intensity focused beams that may be used for thermal ablation of selected regions and also to low intensity fields that appear to be able to stimulate physiological processes, such as tissue repair, without biologically significant temperature rises. Ultrasonic tools are used for therapeutic effect in dentistry and are being investigated for use in thrombolysis.

Ultrasound beyond the diagnostic ranges can initiate various kinds of effects when insonated into biological tissue. The ability of ultrasound to interact with tissue to produce biological changes has been known for a long time [3]. Much of the early drive to understand these interactions came from an interest in harnessing ultrasonically induced changes for therapeutic benefit. More recently, the concern has been to understand any possible hazard that may arise from diagnostic ultrasound imaging. It is convenient to divide therapeutic ultrasound into two classes, i.e., applications that utilize ‘low’ ($0.125 - 3 W/cm^2$) and ‘high’ intensities ($\geq 5 W/cm^2$). The intention of the low intensity treatments is to stimulate normal physiological responses to injury or to accelerate some processes such as the transport of drugs across the skin. The purpose of the high intensity treatments is rather to selectively destroy tissue in a controlled fashion. The resulting effects include thermal, mechanical, chemical, and optical reactions. Mechanical effects, more specifically, may consist of acoustic cavitation, radiation force, shear stress, and acoustic streaming/microstreaming. Among them, the thermal effect and acoustic cavitation are the most significant, and their mechanisms of action have been relatively well-understood. [1,2]. It is however often extremely difficult to identify positively the mechanisms involved in producing biological change, and indeed to isolate non-thermal effects from the thermal ones.

The thermal effect is caused by the absorption of ultrasound in biological tissue. Ultrasonic waves cause vibration or rotation of molecules or part of macromolecules in tissue, and this movement results in frictional heat. Depending on the temperature and the duration of contact, the tissue may become more susceptible to chemotherapy or radiotherapy (> 43 °C, 1 hour) or alternatively, protein denaturation may occur (coagulation necrosis) (56 °C, 1 sec) [1,2,5]). Beneficial effects that arise from ultrasonically induced heating include an increase in extensibility of collagenous structures such as tendons and scar tissue, a decrease in joint stiffness, pain relief, changes in blood flow, decrease in muscle spasm, and, at high intensities, selective tissue ablation as achieved in focused ultrasound surgery [4].

Excluding the effects of thermal transfer, the temperature elevation in biological tissue induced by ultrasound (plane wave) absorption is theoretically linearly proportional to acoustic intensity in the following manner ($\partial T/\partial t = 2\alpha I/\rho C_P = 0.014\ I$; T is the temperature [°C], t is the time [sec], α is the absorption coefficient [≈ 0.03 Np/cm in tissue-like medium at 1 MHz], I is the acoustic intensity, ρ is the density [≈ 1 g/cm^3 in tissue-like medium], C_P = specific heat [≈ 4.2 J/g°C in tissue-like medium]) [6, 7]. Because of this linearity and predictability, a thermal effect was traditionally preferred to a mechanical effect in the medical applications of unfocused and focused ultrasound.

Non-thermal mechanisms that can produce beneficial (therapeutic) changes in tissue may be cyclic or non-cyclic in nature. The early literature refers to "micro-massage" [8]. This is presumably thought to be an effect due to the periodic nature of the sound pressure field. The one of non-cyclic effects thought to be involved in ultrasound therapy is acoustic streaming. This may be due to stable, oscillating cavities, or to radiation forces in intra- or extracellular fluids. Streaming may act to modify the local environment of a cell leading, for example, to alter concentration gradients in the vicinity of an extracellular membrane. The concentration gradient affects the diffusion of ions and molecules across a membrane, and thus streaming may account for the reported changes in the potassium and calcium content of cells following ultrasonic exposure [1].

Acoustic cavitation, defined as the formation and activity of a gas- or vapor-filled cavity (bubble) in a medium exposed to ultrasound, plays a major role in the mechanical effects and minor roles in the chemical and the optical effects of ultrasound medical technology. If biological tissue is insonated by an ultrasound wave, more intense than a specific threshold, negative pressure representing the rarefaction of the wave, may be large enough to draw gas out of the tissue solution to form a bubble. It is easy to understand the underlying mechanism if this is compared to the numerous bubbles formed by vigorous rotation of a motorboat screw. This bubble either repeats radial oscillations being in a resonant size with the ultrasound frequency (stable cavitation; non-inertial cavitation) or oscillates in a similar manner expanding gradually above its resonant size due to net influxes of gas and vapor into the bubble (rectified diffusion), and finally disintegrates by a violent and asymmetrical collapse (unstable cavitation, inertial cavitation) [1,9]. Acoustic cavitation, particularly inertial cavitation, can cause a significant degree of mechanical and thermal effects as well as chemical and optical effects. The thermal effect caused by acoustic cavitation is larger than that caused by ultrasound absorption alone. Mechanical and thermal effects by acoustic cavitation are generally known to be complex, unpredictable, and, sometimes, detrimental. The threshold of acoustic cavitation depends on the (negative) pressure amplitude, ultrasound frequency, and the tissue where cavitation occurs [1,8,10].

Radiation force is a force exerted at an interface between two media or inhomogeneity in a medium due to the passage of ultrasound waves. An acoustic field in fluid may set up acoustic streaming; the transfer of momentum to liquid, by the absorption of energy from an acoustic field, causes acoustic streaming. The fluid velocity caused by acoustic streaming is spatially non-uniform thereby generating a velocity gradient in the field. This gradient causes shear stress. Acoustic streaming caused by an oscillating bubble in a sound field immediately surrounding the bubbles is specifically referred to as acoustic microstreaming. Shear stress formed by microstreaming is an important mechanism underlying many biological reactions [1,10].

Traditional therapeutic applications of ultrasound include the treatment of soft tissue and bone injuries, wound healing, hyperthermic cancer treatment, focused ultrasound surgery of Parkinson's disease, glaucoma and retinal detachment and for sealing traumatic capsular tears, benign prostatic hyperplasia, the liver, the kidney, prostate and bladder tumours, vascular occlusion therapy, and tool surgery [1–5]. Most physiotherapy units offer spatial average intensities up to 3 W/cm^2 and offer one or more transducers operating at discrete frequencies in the range of 0.75 – 5 MHz. The choice of transducer depends on the depth of the target to be treated deeper targets require lower frequencies because of the frequency dependence of ultrasonic attenuation. Devices offer either discrete intensity settings or continuously variable controls. The output may be continuous or pulsed. Pulsed exposures are often chosen when thermal effects are to be kept to a minimum. Commonly available pulsing regimes are 2:2 and 2:8 ms [1].

Therapeutic transducers are usually made of low loss lead zirconate-titanate (PZT) or recently from 1-3 connectivity type piezocomposites [11–13]. They are mounted in a light-weight, hand-held waterproof housing, and are typically air-backed. For physiotherapy applications, the sound is usually coupled directly into the patient through a thin layer of coupling medium (e.g., aqueous gel or mineral oil), although for awkward geometries the limb to be treated may immersed in a small water bath. Despite the fact that medical ultrasound is a relatively older field compared to Magnetic Resonance Imaging (MRI) or Positron Emission Tomography (PET), ongoing advances have allowed to continuously expand the field in its numerous applications. In the past decade, with the advent of faster processing, specialized contrast agents, understanding of nonlinear wave propagation, novel real-time signal and image processing as well as new piezoelectric materials, processing technologies and ultrasound transducer designs and manufacturing, ultrasound imaging and therapy have enjoyed a multitude of new features and clinical applications [1,14–21].

In this chapter, we review some significant advances in ultrasound technology as well as new applications of medical ultrasound developed during the past decade. The chapter is divided, somewhat arbitrarily, into two informative sections. Section 2 describes *theoretical calculations and numerical modeling of HIFU fields.* The importance of nonlinear effects for tissue heating at high output levels of ultrasound has been considered. Section 3 presents a new multi-frequency and harmonics methods for dynamical focusing of HIFU, describes the original *in-vitro* experiments in tissues and discusses their applications for the body aesthetics therapy. This section explores also the application of HIFU for hemostasis and describes HIFU transducer design, results of non-linear acoustic field calculations and original *in-vivo* experiments on sonicating blood vessels.

2. Theoretical Calculations and Numerical Modeling of HIFU

The characterization of medical acoustic devices that operate at high output levels has been a research topic and an issue of practical concern for several decades [22]. The importance of nonlinear effects has been considered and addressed even at diagnostic levels of ultrasound [23]. In lithotripsy and HIFU, these effects are critical because acoustic pressures up to 100 MPa or higher can be reached; such pressures are two or even three orders higher in magnitude than diagnostic ultrasound.

Numerical modeling has been used to predict high amplitude acoustic fields from medical devices. One advantage of modeling is that it can be used to determine the acoustic field in both water and tissue. Numerical algorithms, most commonly based on the nonlinear parabolic Khokhlov-Zabolotskaya-Kuznetsov (KZK) equation, have been developed and applied to the nonlinear fields of lithotripters, unfocused ultrasonic piston sources, diagnostic ultrasonic transducers operating in tissue harmonic imaging mode [24], focused ultrasound sources [25], and HIFU sources [26,27]. For highly focused sources, the Rayleigh integral can be used to capture diffraction effects more accurately in the model. More comprehensive models based on full-wave nonlinear equations have also been developed, but they are much more computationally intensive.

2.1. Theoretical Calculation of HIFU Fields

As is known, the system of hydrodynamic equations in the general case cannot be solved analytically, and the basic approach is to use numerical simulation. It is possible to solve numerically the complete system of hydrodynamic equations, but in practice it is usually unrealizable because of limitations on available computer memory and its computational speed even using the modern computing clusters. However, one can find reasonably accurate solution to the practical problem of interest with some simplifying assumptions.

It is often assumed that the amplitude of acoustic waves is sufficiently low so that the ultrasound field can be described in approximation of linear acoustic. In the linear case we can reduce the problem to considering harmonic (sinusoidal) waves: $p(\mathbf{r},t) = P(\mathbf{r})\exp(-i\omega t)/2 + P^*(\mathbf{r})\exp(i\omega t)/2$, were p is the acoustic pressure, P is its complex amplitude, ω is the angular frequency. The corresponding equation for harmonic waves in homogeneous medium is the Helmholtz equation: $\Delta P + \left(\omega^2/c_0^2\right)P = 0$, where c_0 is the sound speed in the medium. The solution of this equation for one-way propagation from ultrasound transducer can be written as [28]:

$$P(\mathbf{r}) = -i\rho_0 c_0 \frac{k}{2\pi}\int_S \frac{V(\mathbf{r}')\,\mathrm{e}^{ik|\mathbf{r}-\mathbf{r}'|}}{|\mathbf{r}-\mathbf{r}'|}dS'. \tag{1}$$

This expression is called the Rayleigh integral.

In the paper [28] O'Neill used Rayleigh integral for the calculation of acoustic field generated by the source in the shape of a spherical cap under the assumption that the focusing angle is small and the radius of curvature of the radiating surface is much larger than the wavelength. Then locally the surface elements radiate as a flat source, and it can be assumed that the resulting field given by the equation (1) is sufficiently accurate. With a uniform velocity distribution on the cap one can obtain a simple analytical expression for the amplitude of the pressure on the axis of the transducer:

$$P = P_0 \frac{e^{ikz} - e^{ikR_{max}}}{1 - z/F}, \tag{2}$$

where z is the axial coordinate, R_{max} is the distance from the observation point to the edge of the transducer, P is the complex amplitude of the acoustic pressure on the axis of the transducer (at the point with coordinate z, measured from the center of the bowl), F is the radius of curvature of the bowl, $P_0 = \rho_0 c_0 V$ is the characteristic acoustic pressure on the surface of the transducer, V is the amplitude of the normal velocity component of the radiating surface. The distance $R_{max} = R_{max}(z)$ can be written as follows:

$$R_{max} = F\sqrt{1 + \left(1 - \frac{z}{F}\right)^2 - 2\left(1 - \frac{z}{F}\right)\cos\alpha}, \tag{3}$$

where α is the half-opening angle of the cap, i.e., the angle between the rays directed from the geometrical focus ($z = F$) to the center of the bowl and to its edge.

The above result for the sources in the form of a spherical cap can be simply extended to the case of a source in the form of an annular spherical segment. Indeed, since the problem is linear, we can embed in a spherical cap another coaxial one (with the same curvature, but different diameter). Then it is possible to obtain oscillation only of an annular spherical segment if excite the surface of the above mentioned cap in antiphase. Consequently, due to the principle of superposition, the field of the ring can be obtained by subtracting fields of two caps of different diameters.

Therefore, if the center of the transducer has a hole, then the solution will be:

$$p(z,t) = \frac{p_0 e^{-i\omega t}}{1 - z/F}\left(e^{ikR\min} - e^{ikR\max}\right),\ A(z) = \frac{2p_0}{1 - z/F}\left|\sin\left(k\frac{R_{\min} - R_{\max}}{2}\right)\right|, \tag{4}$$

where $R_{\min} = F\sqrt{1 + (1 - z/F)^2 - 2(1 - z/F)\cos\beta_1}$, $\sin\beta_1 = a_1/F$.

If the beam propagates in absorbing medium, then attenuation for the pressure amplitude $A(z)$ of the source with a hole can be taken into account as follows:

$$A(z) = \left|\frac{2p_0}{1 - z/F}\sin\left(k\frac{R_{\min} - R_{\max}}{2}\right)\right|\exp(-\alpha z), \tag{5}$$

where α is the attenuation at the frequency of the radiation.

For the calculation of the medium heating one can use a standard formula: $Q = 2\alpha I$.

The frequency dependence of the attenuation α, obtained experimentally for a number of liquids and tissues, can be found in [29].

Since the regime of high intensities is required for stopping the bleeding, nonlinear propagation of acoustic waves occurs, and theoretical description requires more complex models than for the case of low intensity.

For strongly focused fields non-linear models such as Westervelt equation can be used, which is a generalization of the classical wave equation to the nonlinear case in the approximation of the absence of back propagating waves [30]. Even more complex models based on the solution of the full nonlinear wave equation have been developed [31]. However, these approaches require large computing power and time-consuming calculations (up to several days) on supercomputers, i.e., practically inapplicable to practical problems. This difficulty can be significantly reduced by using the evolution equation for the quasi-plane wave. The corresponding equation in nonlinear acoustics equation is known as the KZK equation [26,32].

2.2. Numerical Modeling

Numerical modeling of experimental conditions was performed using a KZK-type nonlinear parabolic equation generalized for the frequency-dependent absorption properties of the propagation medium:

$$\frac{\partial}{\partial\tau}\left[\frac{\partial p}{\partial z}-\frac{\beta}{\rho_0 c_0^3}p\frac{\partial p}{\partial\tau}-L_{abs}(p)\right]=\frac{c_0}{2}\Delta_\perp p, \tag{6}$$

where p is the acoustic pressure, z is the propagation coordinate along the axis of the beam, $\tau = t - z/c_0$ is the retarded time, c_0 is the ambient sound speed, ρ_0 is the ambient density of the medium (is the coefficient of nonlinearity), $\Delta_\perp$ is the Laplacian with respect to the transverse coordinate r, and L_{abs} is the linear operator that accounts for the absorption and dispersion of the medium.

For simulations in water, thermoviscous absorption was included as

$$L_{abs}=\frac{b}{2\rho_0 c_0^3}\frac{\partial^2 p}{\partial\tau}, \tag{7}$$

where b is the dissipative parameter of water. For simulations in gel, the propagation path for ultrasound comprised a two-layer medium consisting of water followed by tissue-mimicking gel phantom. The frequency-dependent absorption of ultrasound in the gel was included in the model according to a nearly linear power law combined with weak thermoviscous absorption, as in the water:

$$\alpha(f)=\frac{2\pi^2 f^2 b}{\rho_0 c_0^3}+\alpha_0\left(\frac{f}{f_0}\right)^\eta \tag{8}$$

Here α_0 is the absorption parameter of the gel at the fundamental frequency f_0, and variation of the sound speed with frequency was calculated for the power law term (η) in Equation (8) using the local dispersion relations [26,32].

The boundary condition for Equation (6) was set by translating the pressure amplitude, p_0, uniformly distributed over the curved surface of the source to the plane $z = 0$. The translation of the amplitude was performed using a geometrical acoustics approximation following the spherical convergence of the field. The focusing phase shift along the radial coordinate was introduced in the parabolic approximation as

$$p(z=0,r,\tau) = \frac{p_0}{\sqrt{1+r_0^2/F^2}} \sin\left(2\pi f_0 \left(\tau + \frac{r^2}{2c_0 F}\right)\right), \tag{9}$$

if $r < r_0/\sqrt{(1+r_0^2/F^2)}$ and $p(z=0,r,\tau)=0$, elsewhere. Here, $2r_0$ is the aperture of the source and F is its radius of curvature.

Equation (6) was solved numerically in the frequency domain using a previously developed finite difference algorithm [33]. The acoustic pressure waveform was represented as a Fourier series expansion as

$$p(z,r,\tau) = \sum_{n=1}^{\infty} c_n\,(z,r) e^{in2\pi f_0 \tau}, \tag{10}$$

where c_n is the complex amplitude of the n-th harmonic. A set of nonlinear coupled differential equations for the amplitudes of the harmonics was derived and integrated numerically using the method of fractional steps and an operator-splitting procedure. The simulations were performed assuming radial symmetry of the HIFU source.

To characterize the HIFU output level in the focal zone in water or in gel, two values of the spatial peak intensity were introduced. These values will be referred to in the paper as focal intensities. The first value, I_N, was calculated from the numerically modeled nonlinear waveform as a combination of the focal intensities of all harmonic components:

$$I_N = f_0 \int_0^{1/f_0} \frac{p^2}{\rho_0 c_0} d\tau = \frac{2}{\rho_0 c_0} \sum_{n=1}^{\infty} |c_n|^2. \tag{11}$$

The second value, I_L, was calculated based on the results of the linear acoustic propagation modeling as

$$I_L = \frac{p_F^2}{2\rho_0 c_0}, \tag{12}$$

where p_F is the focal (i.e., spatial peak) pressure amplitude *in situ.* The linear focusing gain of the source was defined as a ratio of the focal and source pressure amplitude obtained from linear modeling in water:

$$G = p_F/p_0. \tag{13}$$

The medium heating was calculated as follows:

$$Q = \sum_{n=1}^{\infty} 2\alpha_n I_n, \tag{14}$$

where $\alpha_n = \alpha(n\omega_0)$ is the attenuation of n-th harmonic and $I_n = |C_n|^2 / (2\rho_0 c_0)$ is the intensity of this harmonic. That is total heat source in the non-harmonic (in particular, in non-linear) wave is the sum of the heat sources of the individual harmonics, and each harmonic absorption may be considered in the plane wave approximation.

3. New Methods for Therapeutic Applications of HIFU

High-intensity focused ultrasound (HIFU) continues to be a very attractive option for minimally invasive procedures. Using well-established principles, this ablative therapy can be used to treat a number of benign and malignant diseases with few side effects. During the last 15 years, there has been an enormous amount of work, both laboratory based and in the form of clinical trials, aimed at developing devices that can deliver treatments with safe and effective outcomes [34,35]. However, the application of this technology still has many drawbacks. It is expected that current obstacles to implementation will be resolved in the near future. Although most clinical studies on HIFU therapy have dealt only with thermal ablations (focused ultrasound surgery – FUS), the range of its potential applications in medical science appears to be much wider when numerous ongoing investigations on cosmetics, hemostasis, thrombolysis and targeted drug/gene-delivery systems are considered.

Main limitation of HIFU for body shaping application (fat reduction, cellulite therapy) is a small size of focal zones and a small thickness of target tissues along with its spatial extension, leading to long procedure time, pain and injury of bones and vital organs. In the field of body aesthetic applications, uses of new HIFU technologies (dynamical focusing and natural harmonics methods) in combination with other therapeutic and diagnostic modalities such as ultrasonic standing waves therapy, shear waves elasticity imaging and therapy, targeted drug/gene-delivery and combinational treatments (ultrasound-RF-vacuum massage), can be anticipated and has a more revolutionary clinical impact [36–45].

Acoustic hemostasis may provide an effective method in surgery and prehospital settings for treating trauma and elective surgery patients. The main drawback of the hemostasis applications is low ultrasound absorption ability of blood and as a result, low heating and coagulation rate at real blood flow. Recent technological advances are expected to resolve these problems.

In this section, new promising approaches for enhancing tissue lysis and hemostasis based on HIFU transducers designs, new materials, dynamical focusing and multifrequency harmonics methods [46–58] are described.

3.1. Dynamical Focusing Method

The method of direct synthesis of dynamically focused acoustic fields is based on the continuous cyclic generation of different focal patterns at optimal repetition frequency by simultaneous applying of M different frequency signals to N sectors of spherically shaped sectored transducer.

For illustration of basic principles of the method HIFU transducers comprising a spherical piezoceramic caps with back electrodes divided circumferentially into 4 regular shaped sectors were designed and tested. The sectors were simultaneously powered by 4 sinus or burst drive signals with different frequencies from 6 dB bandwidth of the transducer. Calculation and modeling of acoustic field patterns for different transducer configurations and frequencies sets were performed. The results of *ex-vivo* experiments in different tissues confirming the advantages of dynamical focusing method are present.

3.1.1. Calculations and Modeling

Dynamical acoustic fields were calculated and modeled for HIFU transducer comprising a spherical piezoceramic cap (radius of curvature of 50 mm, aperture of 80 mm) with back electrode divided circumferentially into 4 regularly shaped sectors. FEM modeling and calculations were performed using ANSYS and Matlab software packages. Calculations were made for water loading and different pressures on the surface of piezoelement.

A result of summary of ultrasonic beams with different frequencies in the focal area is an equivalent of continuous changes in phase shifts of signal for i-th sector according to the following equation: $\varphi_i = 3600\ \Delta f_i\ (F/C - t)$, where Δf_i frequency shift from f_0, F is the focal distance, C is the sound velocity, t is the time. Frequency excitation protocols: $f_1 = 700$ kHz, $f_2 = 690$ kHz, $f_3 = 700$ kHz, $f_4 = 690$ kHz for 4 regularly shaped sectors. Repetition rate of focal patterns appearance was a common divider of the applied frequencies (10 kHz). The results of FEM calculations of acoustic field structure for regularly sectored dynamically focused transducer are shown in Figure 1.

3.1.2. Acoustic Field Measurements

Acoustic measurements of the dynamically focused sectored transducers were made in an oil bath using calibrated hydrophone and acoustic intensity measurement system (AIMS). The electric power and hydrophone signal were measured using LeCroy WS 24Xs oscilloscope. Transducer sectors were powered using function generator Agilent 33220A and power amplifier AG 1012 LF. For measurements of acoustic field patterns for 4 regularly shaped sectored transducer, the sectors were powered by single frequency bursts at appropriate phase shifts, corresponding to initial and final focal zones.

The results of AIMS 2D Scan in the focal plane of dynamically focused transducer at low power excitation (power = 40W, burst length = 20 cycles) are listed in Table 1. The photos of focal zones patterns of dynamical acoustic fields in mineral oil are shown in Figure 2.

(a)

Figure 1. (Continued).

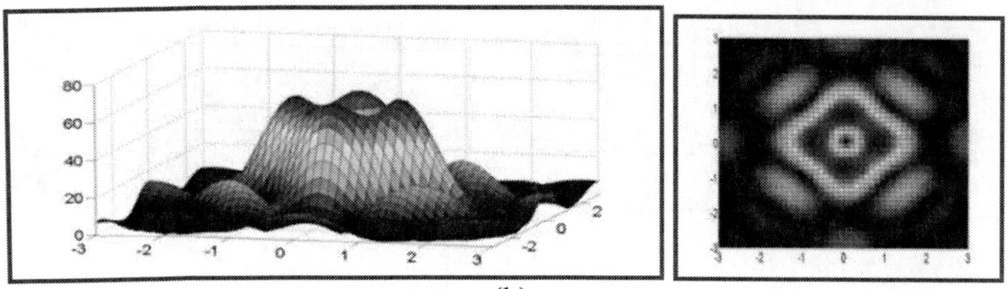

(b)

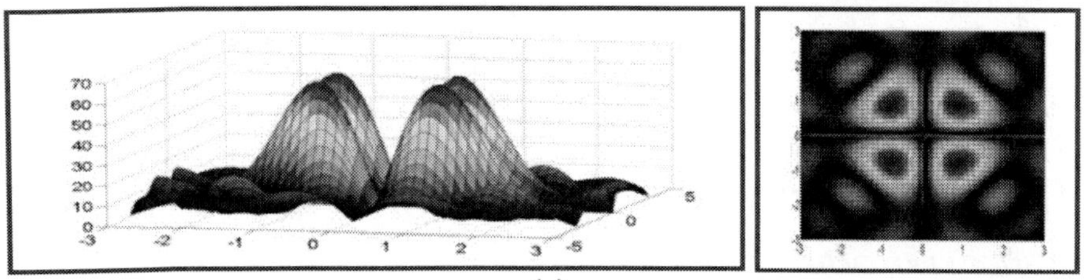

(c)

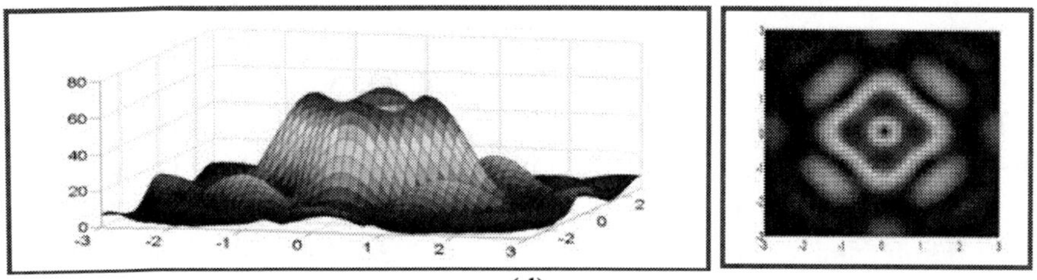

(d)

Figure 1. FEM calculated sequential samples of focal zones for regularly sectored dynamically focused transducer. Corresponding phase shifts for neighbor sectors are: (a) Θ = 0 degrees; (b) Θ = 90 degrees; Θ = 180 degrees; Θ = 270 degrees.

Table 1.

Single Focus f1, f2, f3, f4 = 700 kHz , Θ1, Θ2, Θ3, Θ4 = 0°			Four Foci f1, f2, f3, f4 = 700 kHz, Θ1 , Θ3 = 0˚, Θ2, Θ4 = 180°		
Focal Diameter @ 3 dB [mm2]	Peak Pressure [kPa]	Power @40 dB [W]	Focal Diameter @ 3 dB [mm2] for each of 4 focus	Peak Pressure [kPa] for each of 4 focus	Power @40 dB [W] for each of 4 focus
1.5×1.6	6110.6	28.5	1.4×1.6	3090.3	7.3

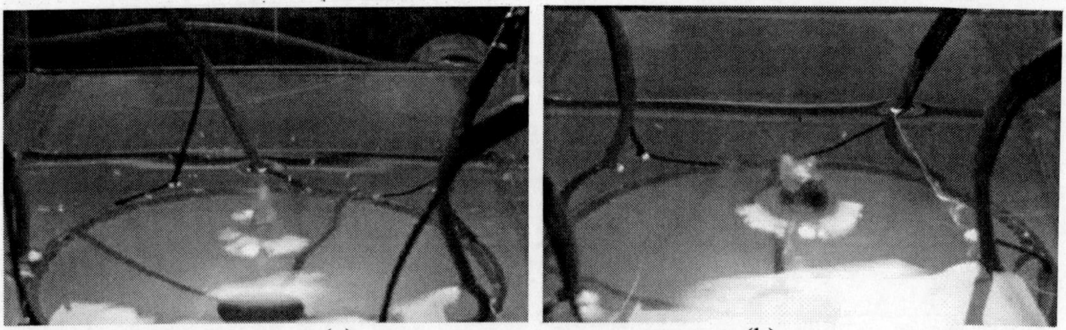

(a) (b)

Figure 2. Visualization of dynamical acoustic fields (end points of dynamical cycle) generated of sectored focusing transducer at 700 kHz in a mineral oil: (a) single focus - Θ1, Θ2, Θ3, Θ4 = 0°; (b) four focus 0 - Θ1 , Θ3 = 0°, Θ2, Θ4 = 180°.

3.1.3. In Vitro Experiments on Tissues

The HIFU dynamical focusing transducers comprising 4 regular sectors were used for tissue treatment (aperture of 80 mm, focal length of 50 mm, resonance frequency of 700 kHz). The sectors were simultaneously powered by sinus bursts drive signals with different frequencies from 6 dB bandwidth of the transducer using two function generator and power amplifier. The acoustic intensity was kept at 750 W/cm^2 (I_{SAL}).

The fresh porcine adipose samples were placed in the oil bath and positioned right under the transducer such that the focal point was placed inside the sample. The samples were irradiated by the dynamical HIFU for 3 – 10 s. After exposure the samples were sectioned along the beam axis respectively to compare the dimensions of the lesions. Preliminary experiments on thin Al foil in oil bath were also made for cavitation detection.

The photographs of cavitational holes in thin Al foil as well as cavitational lesions in the porcine samples induced by HIFU transducer working in the dynamical focusing and single focusing regimes are shown in Figure 3.

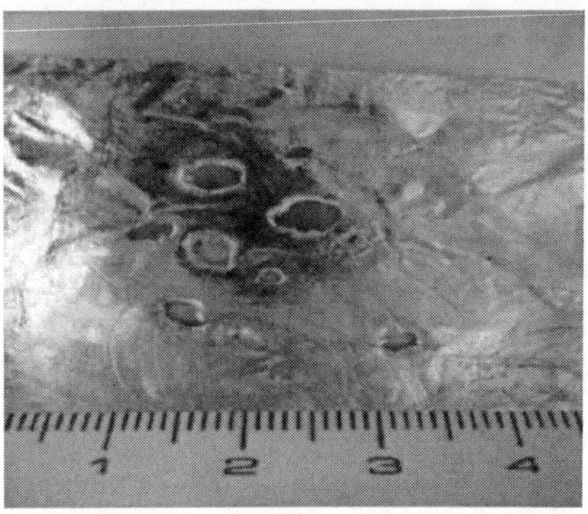

(a)

Figure 3. (Continued).

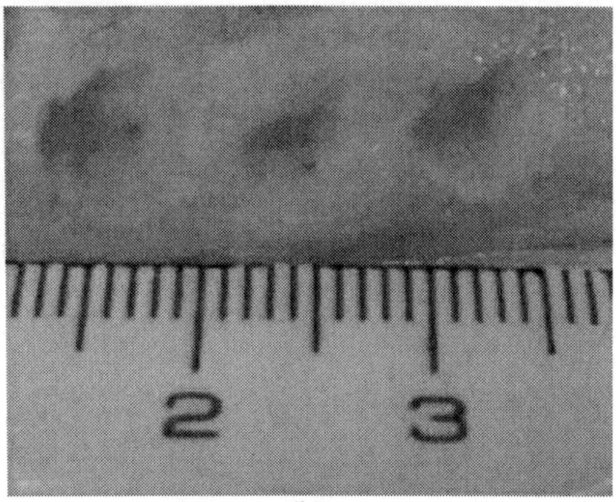

(b)

Figure 3. Photographs of cavitational holes in Al-foil (a) and cavitational lesions (b) in the porcine samples, induced by HIFU transducers at single and dynamical focusing excitation. Frequency excitation protocols: dynamical focusing - f1, f3 = 700 kHz, f2, f4 = 699 kHz (big holes and lesions), single focusing - f1, f2, f3, f4 = 700 kHz (small holes and lesion). The exposure time is 10 s, duty cycle is 1/10, burst length is 100 cycles, the acoustic intensity ISAL is 750 W/cm2.

Obviously, the cavitational tissue lesion generated by dynamical focusing method during one sonication is larger than at single frequency excitation under the same exposure conditions. Enhancement of the cavitational lesions and tissue ablations obvious from Figure 3 was reached due to dynamical focusing method (cyclic generation of different focal patterns) providing cavitation enhancement, inhomogeneous ultrasound beam patterns with higher pressure gradients and shear deformations, as well as nonlinear interactions of harmonics frequencies.

Additional increase in lesions activity is provided by optimal repetition frequencies and bursts lengths that correspond to specific resonance/relaxation times of cavitation "clouds" and/or tissues as well as to natural tissues reaction times. Possible applications of dynamical focusing method are ultrasonic surgery, hyperthermia, therapy and body aesthetics.

3.2. Multi-Frequency Harmonics Method

The multi frequency harmonics method consists in simultaneous or alternative (at optimal repetition frequency) excitation of single element HIFU transducer at frequencies corresponding to odd natural harmonics of piezoceramic element at ultrasound energy levels sufficient for producing cavitational, thermal or mechanical damage of fat cells at each of aforementioned frequencies.

Calculation and FEM modeling of transducer vibrations and acoustic field patterns for different frequencies sets were performed. Acoustic pressure in focal plane was measured in water using calibrated hydrophone and 3D acoustic scanning system. The results of *in-vitro* experiments on different tissues and phantoms confirming the advantages of multi-frequency harmonic method are presented.

3.2.1. Natural Harmonics of Piezoelement Vibrations

Any piezoelement can be excited at the odd harmonics of the main resonant frequency [59], but keff.n will be lesser than for the main mode. Effective electromechanical coupling factor keff.n for n-th harmonics of a piezoelement can be found as keff.n = 8 kt2/[(2n+1)π]2. However at proper electrical matching and excitation power a piezoelement can work on odd harmonics very efficiently. Odd harmonic frequencies can be calculated as fa.n = (2n + 1)fa.1, where fa.n is the anti-resonance frequency of n-th harmonics, fa.1 is the anti-resonance frequency of main mode, n is the integer. Anti-resonance frequencies must be used because of electromicanical contribution to the sound velocity of piezoelectric body. Corresponding resonance frequencies can be calculated from anti-resonance ones using standard formulas [59].

Impedance spectra illustrating odd harmonics of thickness vibrations of focusing piezoceramic element are shown in Figure 4. Frequency dependencies of impedance and phase of piezocomposite elements were determined by impedance spectroscopy method using 4294A Agilent Impedance Analyzer. Measurements were made for focusing piezoceramic element with the aperture of 85 mm and the radius of curvature of 55 mm having resonance frequencies of 205 kHz (1-st harmonic) and 670 kHz (3-rd harmonic), respectively.

FEM modeling was performed using ANSYS software package. Calculations were made for the same focusing piezoceramic element with harmonics frequencies of 205 kHz and 670 kHz, respectively. The results were compared with the focusing piezoceramic element with the first harmonic resonance frequency 670 kHz (Figure 5).

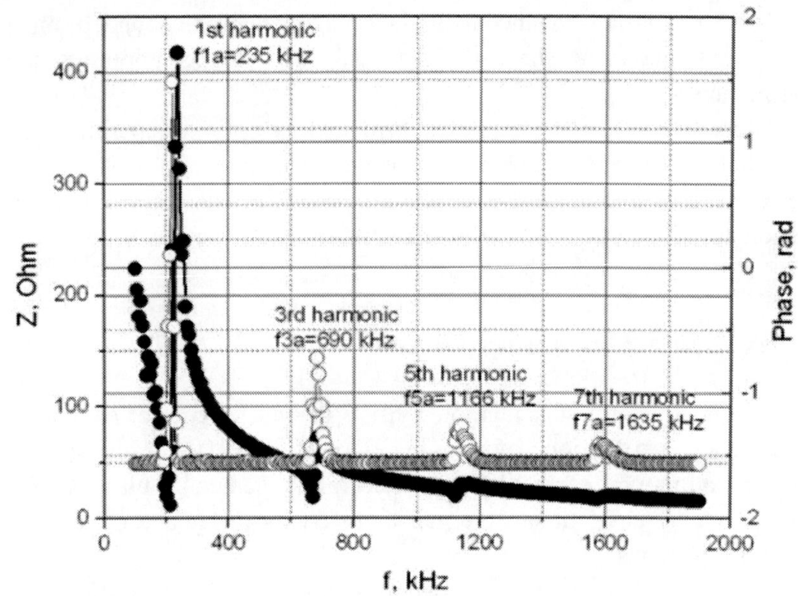

Figure 4. Impedance spectra (frequency dependencies of impedance and phase) of spherical focusing piezoelement with odd harmonics of thickness vibrations.

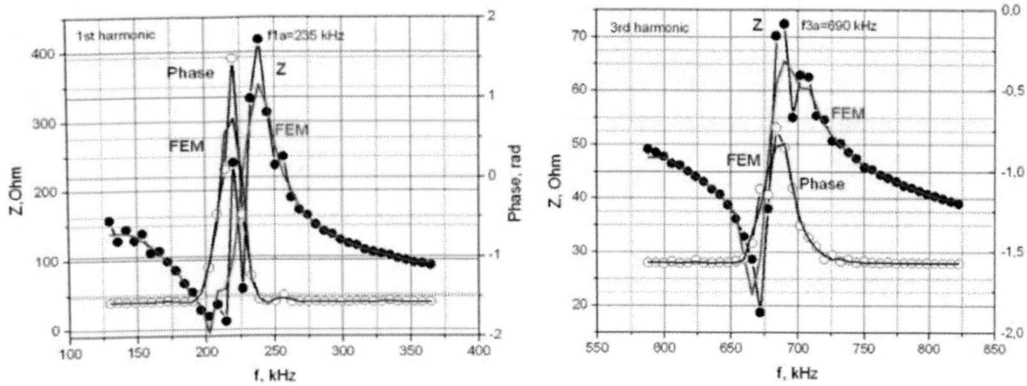

(a) (b)

Figure 5. Measured and FEM modeled impedance spectra for first (a) and third (b) harmonics of the focusing piezoceramic element.

FEM calculations of acoustic pressure and efficiency at the same input electric power show that the piezoceramic element working at the 1-st harmonic resonance frequency (670 kHz) produces 1.8 times higher focal pressure compared to the piezoceramic element excited at the same frequency at the 3-rd harmonic.

3.2.2. Acoustic Field Calculation and Measurements

Calculations and modeling of acoustic pressure in the focal plane for the 1-st and the 3-rd harmonics excitation of the focusing piezoceramic element were made by solving standard wave equations at the Fresnel approximation using Matlab software. Calculations were made for water loading and different pressures on the surface of piezoelement.

Acoustic measurements of the focusing piezoceramic elements were performed in oil bath using calibrated hydrophone and acoustic intensity measurement system (AIMS). The electric power and hydrophone signal measurements were made using LeCroy WS 24Xs oscilloscope. Piezoceramic element was powered using function generator Agilent 33220A and power amplifier AG 1012 LF. The total acoustic output was measured by acoustic force balance AFB. The results of AFB measurements at 205 kHz and 690 kHz are listed in Table 2.

Table 2.

Frequency [kHz]	Power [W]	Force [g]	Efficiency [mg/W]
205	20	0.25	125.1
690	20	0.09	45.3

2D-scans of acoustic pressure distribution in the focal plane of the focusing piezoceramic element measured by the hydrophone using acoustic intensity measurement system (AIMS) at low power excitation (4W, burst length equals 20 cycles) are shown in Figure 6. The results of acoustic field measurements for the 1-st and the 3-rd harmonics are listed in Table 3.

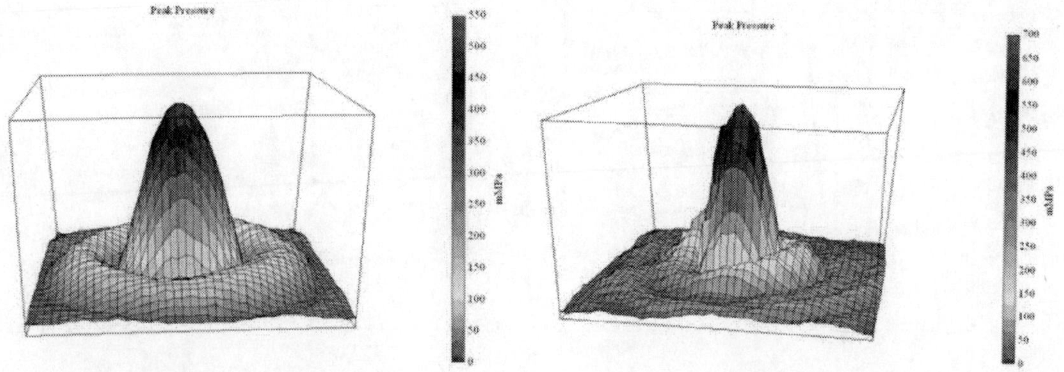

(a) (b)

Figure 6. AIMS 2D-scan of acoustic pressure in the focal plane for 1-st and 3-rd harmonics (fr = 205 KHz , f3 = 690 kHz).

Table 3.

First harmonic f_r = 205 kHz			
Focal Length [mm]	Focal Diameter [mm^2]	Peak Pressure [kPa]	Power @40dB [mW]
21.0	5.5×5.4	521	619
Third harmonic f_r = 690 kHz			
Focal Length [mm]	Focal Diameter [mm^2]	Peak Pressure [kPa]	Power @40dB [mW]
6.3	1.5×1.7	695	196

3.2.3. Ex-vivo Experiments on Tissues

Fresh porcine liver and muscle tissues were obtained from a butcher within 24 h of slaughter. A single element spherical PZT transducer (aperture of 85 mm, focal length of 55 mm, resonance frequency of 205 kHz) was used for the experiments. For comparison, spherical PZT transducer with the 1-st harmonic resonance frequency of 690 kHz was measured also. Piezoceramic element was powered simultaneously at the 1-st and the 3-rd harmonic frequencies of 205 kHz and 690 kHz using two function generators and power amplifiers. The acoustic intensity in the focal plane was kept as 250 and 500 W/cm^2 (I_{SAL}) for the 1-st and the 3-rd harmonics respectively. The acoustic intensity for PZT transducer with the main resonant frequency of 690 kHz was 750 W/cm^2 (I_{SAL}). The samples were placed in the oil bath and positioned right under the transducer such that focal point was placed inside the sample. The samples were irradiated by the harmonics frequency HIFU for 3 – 20 s at different duty cycles (from 1/2 to 1/100) and burst lengths (from 10 to 200 cycles) of the signals. After the exposure the samples were sectioned along the beam axis respectively to compare the dimensions of the lesions. Preliminary experiment on thin Al foil positioned at the focal plane of HIFU transducer in oil bath were also made for cavitation detection.

The photographs of cavitational holes in thin Al foil as well as thermal and cavitational lesions in the liver and porcine samples induced by HIFU transducer at simultaneous excitation on 1-st (205 kHz) and 3-rd (690 kHz) harmonics and by HIFU transducer working on 1-st harmonic (690 kHz) only are shown in Figure 7.

(a) (b) (c)

Figure 7. Cavitational holes in Al-foil (a), thermal (b) and cavitational (c) lesions in the liver and porcine samples induced by HIFU transducers at single- (small holes and lesions) and dual-frequency harmonics excitation. Treatment parameters: (a) exposure time equals 3 s, duty cycle equals 1/20, burst length equals 10 cycles; (b) exposure time equals 9 s, duty cycle equals 1/5, burst length equals 200 cycles; (c) exposure time equals 9 s, duty cycle equals 1/20, burst length equals 10.

Enhancement of the lesions and tissue ablations obvious from Figure 7 was reached due to simultaneous or sequential influence of harmonic frequencies providing cavitational enhancement of tissue heating, inhomogeneous ultrasound beam patterns with enhanced pressure gradients and shear deformations, as well as nonlinear interactions of harmonics frequencies. Additional increase in lesions activity is provided by optimal repetition frequencies, bursts lengths and sonification times that correspond to specific resonance/relaxation times of cavitation "clouds" and/or tissues as well as to natural tissues reaction times.

The multi-frequency harmonic technique brings a new approach to accelerate treatment of big volumes of adipose tissue or tumors ablation with HIFU and also provides a miniaturization of the conventional HIFU transducers. Possible applications of the multi-frequency harmonic technique are ultrasonic surgery, hyperthermia, and therapy, as well as body aesthetics.

3.3. Applications of HIFU for Hemostasis

Acoustic hemostasis may provide an effective method in surgery and prehospital settings for treating trauma and elective surgery patients. Application of HIFU therapy to hemostasis was primarily initiated in an attempt to control battlefield injuries on the spot. High-intensity ultrasound (ISA = 500 – 3,000 W/cm^2) is usually adopted for hemostasis. Many studies on animal models have been successful for both solid organ and vascular injuries [60].

The thermal effect has a major role in hemostasis. The proposed mechanisms of its action are as follows. Structural deformation of the parenchyma of a solid organ due to high temperature induces a collapse of small vessels and sinusoids or sinusoid-like structures. Heat also causes coagulation of the adventitia of vessels, and subsequently, fibrin-plug formation. The mechanical effect of acoustic cavitation also appears to play a minor role in hemostasis. Microstreaming induces very fine structural disruption of the parenchyma to form a tissue homogenate that acts as a seal and induces the release of coagulation factors [60]. No statistically significant hemolysis or changes in the number of white blood cells and platelets have been observed when blood is exposed to HIFU with intensities up to 2000 W/cm^2 [61].

The main drawback of the hemostasis applications is low ultrasound absorption ability of blood and, as a result, low heating and coagulation rate at real blood flow. In this section, HIFU transducer design, nonlinear acoustic field calculations and *in-vivo* experiments on blood vessels confirming enhanced hemostasis are described.

3.3.1. HIFU Transducers Design

HIFU transducer comprised 1.6 – 2 MHz spherical element made from porous piezoceramics [11–13] with 80 mm aperture and 40 mm center hole having radius of curvature 54 mm. The piezoelement was sealed in custom-designed cylindrical housing filled with the mineral oil providing acoustic contact and cooling of the element. The housing had an acoustic window made of very thin (0.15 mm) PVC membrane. Center opening was reserved for ultrasonic imaging transducer (Figure 8).

3.3.2. Acoustic Field Calculations

Calculations of acoustic fields of HIFU transducers were made using the models and algorithms described in Section 2. Figure 9 shows two-dimensional distributions of acoustic intensity in HIFU transducer's acoustic axis plane. Intensity levels are represented in absolute values (kW/cm^2). Calculations were made for two frequencies 1.6 MHz (a) and 2 MHz (b), respectively.

In Figure 10 acoustic pressure signals in the focus calculated at different initial intensities for 1.6 MHz and 2 MHz are shown. It is obvious that even at initial intensity level of 5 W/cm^2 non-linear effects lead to pressure profile asymmetry that transforms to a shock front in focus at initial intensity of 20 W/cm^2 that give rise to extreme heating [33].

(a)

(b)

Figure 8. Focusing piezoelement (a) and assembled HIFU ultrasonic transducer (b).

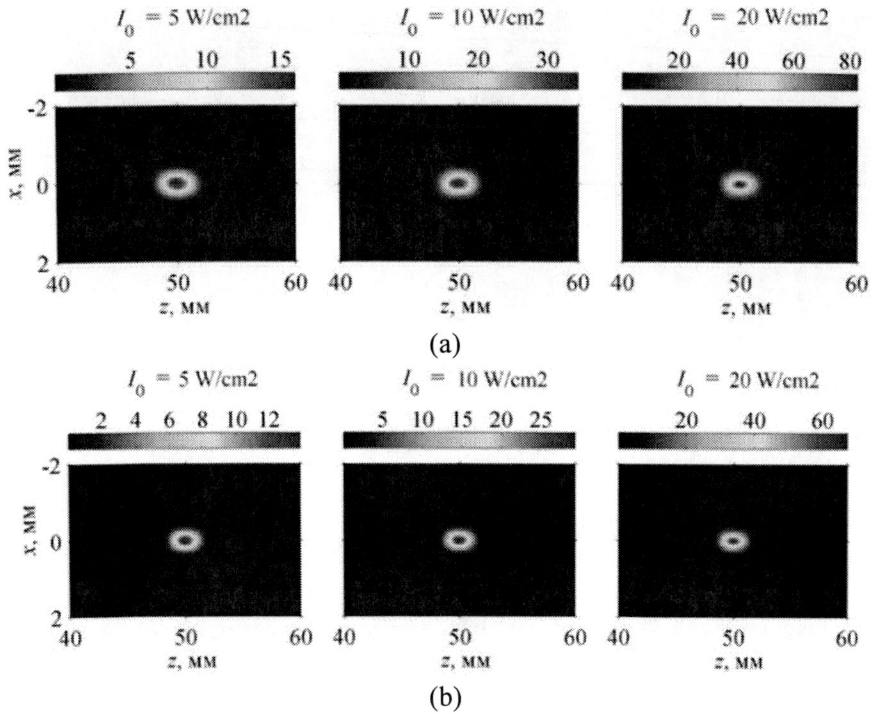

Figure 9. Two-dimensional distributions of acoustic intensity in acoustic axis plane of HIFU transducer for 1.6 MHz (a) and 2 MHz (b) frequencies.

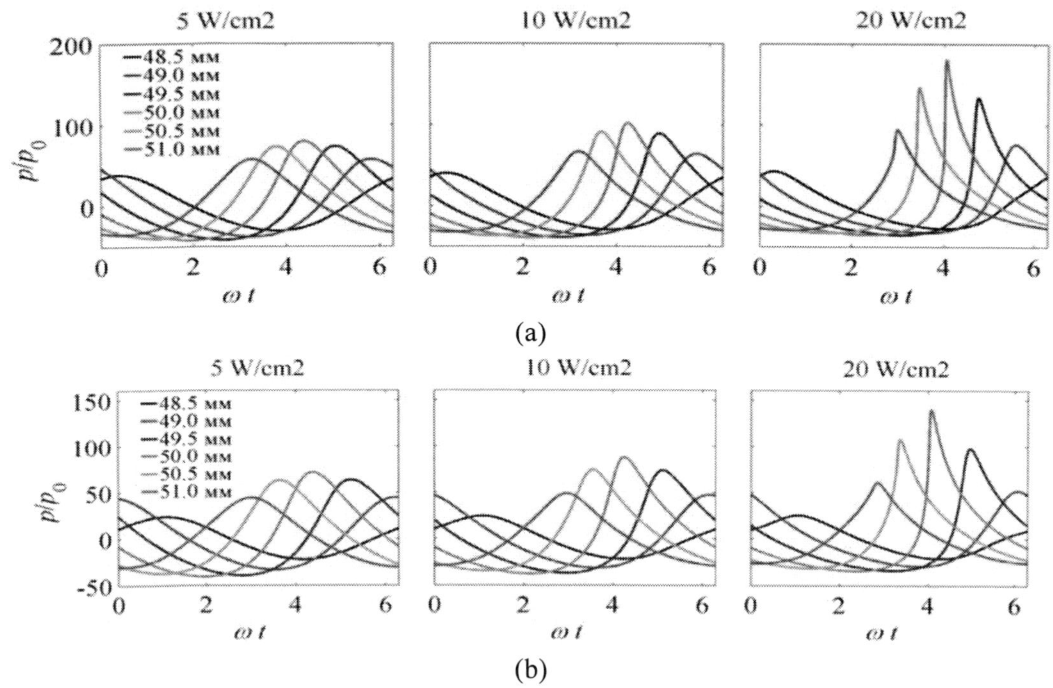

Figure 10. Acoustic pressure signals in the focus calculated at different initial intensities for 1.6 MHz (a) and 2 MHz (b) frequencies.

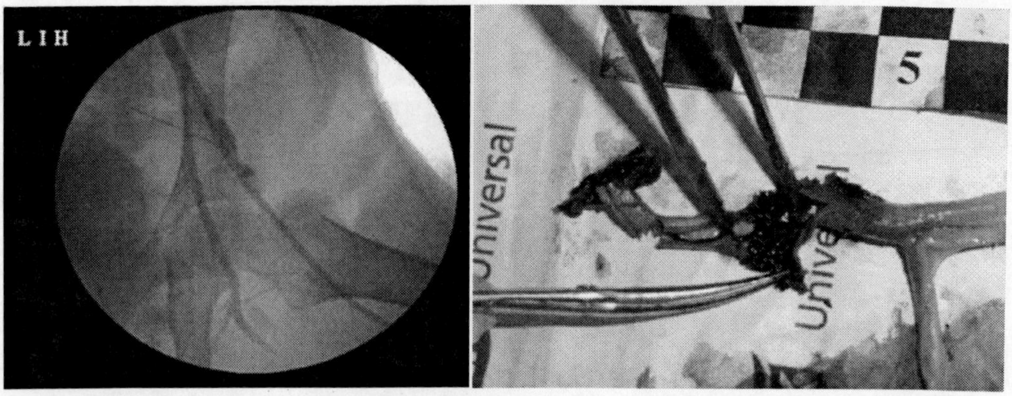

Figure 11. *Angiography image* of blood vessels showing ultrasound hemostasis and photograph of vessel thrombus in dissected femoral artery.

3.3.3. In Vivo Experiments on Blood Vessels

The experiments were made on lamb's femoral artery at a standard protocol. During ultrasound exposure arterial blood flow was temporarily stopped using intravascular balloon. Ultrasonic transducer with 2 MHz frequency described in previous sections was used for experiments. All acoustic measurements were performed in 3D Scanning System (UMS3) using a fiber-optic hydrophone (FOPH 2000) from Precision Acoustics Ltd. Waveforms from the hydrophones and the driving voltage were recorded using a digital oscilloscope LeCroy.

The transducer was driven by a function generator Agilent 33521B and a linear rf amplifier E&I model 2400L RF operating in a CW mode. The acoustic intensity in the focal plane measured in water tank at 5000 W/cm^2 (I_{SAL}) was kept for the object treatment.

After sonification procedure and angiography study the samples of femoral artery were extracted to confirm hemostasis and disclose vessel thrombus. The X-ray image of blood vessels obtained using contrast agents and photograph of dissected femoral artery are shown in Figure 11.

CONCLUSION

The results of theoretical modeling and experimental study of different HIFU transducers were presented. *Ex-vivo* experiments in tissues (fresh porcine adipose tissue, bovine liver) and *in-vivo* experiments in lamb's femoral artery were carried out using different protocols. The results of theoretical modeling and tissue experiments prove the efficacy, safety, and selectivity of the developed HIFU transducers and methods enhancing the tissue lysis and hemostasis and can be used for various therapeutic, surgical and cosmetic applications.

REFERENCES

[1] *Physical Principles of Medical Ultrasonics*; Second Edition. C. R. Hill, J. C. Bamber and G. R. ter Haar (Eds.), John Wiley & Sons Ltd. 2004.

[2] ter Haar, G. R. *Prog. Biophys. Mol. Biol.* 2007, *vol.* 93, 111–129.

[3] Wood, R. W; Loomis, A. L. *Phil. Mag.* 1927, *vol. 4(7)*, 417–36.
[4] ter Haar, G. R. In: *Physical Principles of Medical Ultrasound*, C. R. Hill (Ed.), Chichester, UK: Ellis Horwood, 1986; pp. 1-128.
[5] Kennedy, J. E.; ter Haar, G. R.; Cranston, D. *Br. J. Radiol.* 2003, *vol. 76*, 590–599.
[6] Crocker, M. J. *Encyclopedia of Acoustics.* 1st Ed. New York, John Wiley & Sons, 1997, p. 1727.
[7] *Biological Effects of Ultrasound: Mechanisms and Clinical Implications*, NCRP Report #74. National Council on Radiation Protection and Measurements (NCRP); 1983.
[8] Summer, W.; Patrick, M. K. *Ultrasonic Therapy - A Textbook for Physiotherapists.* London: Elsevier, 1964.
[9] Dalecki, D. *Annu. Rev. Biomed. Eng.* 2004, *vol. 6*, 229–248.
[10] Leslie, T. A.; Kennedy, J. E. *Ultrasound Q.* 2006, *vol. 22*, 263–272.
[11] Rybyanets, A. N. *IEEE Trans. UFFC*, 2011, *vol.58*, 1492-1507.
[12] Rybyanets, A. N. *Ferroelectrics*, 2011, *vol. 419*, 90-96.
[13] Rybyanets, A. N.; Rybyanets, A. A. *IEEE Trans*. UFFC 2011, vol. 58, 1757-1774.
[14] Rybyanets, A. N. In: *Piezoelectrics and Related Materials: Investigations and Applications*, Ivan A. Parinov (Ed.). Nova Science Publishers Inc. NY, 2012, Chapter 5, pp.143-187.
[15] Rybyanets, A. N. In *Piezoceramic Materials and Devices*, Ivan A. Parinov (Ed.). Nova Science Publishers Inc. NY, 2010, Chapter 3, pp. 113-174.
[16] Rybyanets, A. N.; Naumenko A. A.; Shvetsova. N. A. In: *Nano- and Piezoelectric Technologies, Materials and Devices*, Ivan A. Parinov (Ed.). Nova Science Publishers Inc. NY, 2013, Chapter 1, pp. 275-308.
[17] Rybjanets, A. N.; Nasedkin, A. V.; Turik A. V. *Integrated Ferroelectrics*, 2004, *vol. 63*, 179-182.
[18] Rybjanets, A. N.; Razumovskaja, O. N.; Reznitchenko, L. A.; Komarov, V. D.; Turik A. V. *Integrated Ferroelectrics*, 2004, *vol. 63*, 197-200.
[19] *Rybianets, A. N.; Nasedkin A. V.* Ferroelectrics, *2007,* vol.360*, 57-62.*
[20] Rybianets, A. N. *Ferroelectric* 2007, *vol. 360*, 84-89.
[21] Rybianets, A. N.; Tasker R. *Ferroelectrics* 2007, *vol. 360*, 90-95.
[22] Harris, G. *Proc. IEEE Ultrason. Symp.* 1999, *vol.* 2, 1341-1352.
[23] Bacon, D. *Ultrasound. Med. Biol.* 1984, *vol.*10, 189-195.
[24] Khokhlova, V.; Ponomarev, verkiou, M.; Crum, L. *Acoust. Phys.* 2006, *vol.* 52, 481-489.
[25] Averkiou, M.; Hamilton, M. *J. Acoust. Soc. Am,* 1997, *vol.* 102, 2539-2548.
[26] Filonenko, E.; Khokhlova, V. *Acoust. Phys.* 2001, *vol.* 47, 541-549.
[27] Khokhlova, V.; Bailey, M.; Reed, J.; Cunitz, B.; Kaczkowski, P.; Crum, L. *J. Acoust. Soc. Am.* 2006, *vol.* 119, 1834-1848.
[28] *O'Neil H.T.* J. *Acoust. Soc. Am.* 1949, *vol. 21*, 516–526.
[29] Strohm, E.M.; Kolios, M.C. *Proc. IEEE International Ultrasonics Symp.* 2011, pp. 2368-2371.
[30] Varslot, T.; Taraldsen G. *IEEE Trans. UFFC* 2005, *vol. 52*, 1473-1482.
[31] Ginter, S.; Liebler, M.; Steiger, E.; Dreyer, T. ; Riedlinger, R. *J. Acoust. Soc. Am.,* 2002, *vol. 111*, 2049–2059.
[32] Kuznetsov V. P. *Sov. Phys. Acoust.* 1971, *vol. 16*, 467-470.

[33] Canney, M. S.; Bailey, M. R ; Crum L. A.; Khokhlova, V. J.; Sapozhnikov, O. A. *J. Acoust. Soc. Am.* 2008, *vol.124* , 2406-2420.

[34] Dogra, Vikram S.; Man, Zhang; Bhatt, Shweta. *Ultrasound Clinics,* 2009, *vol. 4*, 307-321.

[35] Hynynen, K. *Ultrasonics.* 2010, *vol. 50(2)*, 221-229.

[36] Sarvazyan, Armen; Ostrovsky, Lev A.; Rubyanets, Andrey *Acoustical Society of America. Proceedings on Meeting on Acoustics.* 2009, *vol.* 6.

[37] Rybyanets, Andrey N. *9th International Symposium on Therapeutic Ultrasound. American Institute of Physics Conference Proceedings.* 2010, *vol.1215*, 287-290.

[38] Rybyanets, Andrey N.; Lugovaya, Maria A.; Rybyanets, Anastasia A. *9th International Symposium on Therapeutic Ultrasound. American Institute of Physics Conference Proceedings.* 2010,*vol.1215*, 291-294.

[39] Rybyanets, Andrey; Nasedkin, Andrey; Domashenkina, Tamara, Rybyanets, Anastasia; Lugovaya, Maria . *Proc. 2009 IEEE Ultrason. Symp.* 2009, pp. 1699-1702.

[40] *Rybyanets, Andrey. Nasedkin, Andrey; Rybyanets, Anastasia; Lugovaya, Maria; Domashenkina, Tamara* Proc. 2009 IEEE Ultrason. Symp. *2009, pp. 1765-1768.*

[41] Nasedkin, A.; Rybjanets, A. Kushkuley, L.; Eshel Y. *Proc. 2005 IEEE Ultrason. Symp.* 2005, pp. 1648-1651.

[42] Rybianets, A.; Kushkuley, L.; Eshel, Y.; Nasedkin A. *Proc. 2006 IEEE Ultrason. Symp.* 2006, pp. 1533-1536.

[43] Rybianets, A.; Eshel, Y.; Kushkuley L. *Proc. 2006 IEEE Ultrason. Symp*. 2006, pp.1911-1914.

[44] Rybyanets, A. *Proc. 2007 IEEE Ultrason. Symp*. 2007, pp. 1909-1912.

[45] Goland, V.; Kushkuley, L.; Mirman, S.; Zadok, Y.; Ben-Ezra, S.; Shalgi, A.; Rybianets, A.. *Proc. 2007 IEEE Ultrason. Symp*. 2007, pp. 1305-1308.

[46] Nudelman, Igor; Rybyanets, Andrey. US Patent 7,709,997. *Multi-element Piezoelectric Transducers*. Filled March 13, 2008; published May 4, 2010.

[47] Rybyanets, Andrey; Nudelman, Igor. US Patents Applications No. 20090171251. *Ultrasound Treatment of Adipose Tissue with Vacuum Feature*. Filled February 7, 2008; published July 2, 2009.

[48] Rybyanets, Andrey. US Patents Applications No. 20090171250. *Ultrasound Treatment of Adipose Tissue with Fluid Injection*. Filled February 7, 2008; published July 2, 2009.

[49] Rybyanets, Andrey; Allan, Entis. US Patents Applications No. 20090171255. *Apparatus and Method for Ultrasound Treatment*. Filled December 27, 2007; published July 2, 2009.

[50] Rybyanets, Andrey. US Patents Applications No. 20090171249. *Ultrasound Treatment of Adipose Tissue with Vacuum Feature*. Filled December 27, 2007; published July 2, 2009.

[51] Rybyanets, Andrey. US Patents Applications No. 20090171248. *Ultrasound Treatment of Adipose Tissue with Fluid Injection*. Filled December 27, 2007; published July 2, 2009.

[52] Rybyanets, Andrey. US Patents Applications No. 20090099483. *Apparatus and Method for Ultrasound Treatment*. Filled October 11, 2007; published April 16, 2009.

[53] Rybyanets, Andrey. US Patents Applications No. 20090048544. *Apparatus and Method for Ultrasound Treatment*. Filled August 16, 2007; published February 19, 2009.

[54] Rybyanets, Andrey. US Patents Applications No. 61/187407. *Moving Standing Waves (Provisional)*. Filled June 16, 2009. PCT/IB2010/052624 International filing date 12 June 2010.

[55] Rybyanets, Andrey. US Patents Applications No. 61/330,449. *Supersonic Shear Waves (Provisional)*. Filled May 3, 2010.

[56] Rybyanets, Andrey. US Patents Applications No. 61/385,163. *Synchronously Pumped Ultrasonic Waves and Shear Wave Generation by Same (Provisional)*. Filled September 22, 2010.

[57] Rybyanets, A. N. Patent Russian Federation No. 2413492 RU. *Method and Apparatus for Complex Cosmetic Treatment of Superficial Tissues of Patient*. Filled 27.01.2009; Published 10.03.2011.

[58] Sarvazyan, A. P.; Fillingera, L.; Gavrilov, L. R. *Acoustical Physics*, 2009, *vol. 55*, 630–637.

[59] Kino, Gordon S. *Acoustic Waves*, Inglewood Clifs, New Jersey: Prentice-Hall Inc., 1987.

[60] Vaezy, S.; Martin, R.; Crum, L. *Echocardiography* 2001, *vol. 18*, 309–315.

[61] Poliachik, S. L.; Chandler, W. L.; Mourad, P. D.; Bailey, M. R.; Bloch, S.; Cleveland, R. O., et al. *Ultrasound. Med. Biol.* 1999, *vol. 25*, 991–998.

In: Advanced Materials - Studies and Applications
Editors: I.A. Parinov, S.-H. Chang, S.Theerakulpisut
ISBN: 978-1-63463-749-7

Chapter 24

Advanced Investigations of Energy Efficiency of Piezoelectric Generators

***V. A. Akopyan**[*,1], **I. A. Parinov**[1], **Yu. N. Zakharov**[2], **V. A. Chebanenko**[1] and **E. V. Rozhkov**[1]*

[1]Vorovich Mathematics, Mechanics and Computer Sciences Institute, Southern Federal University, Rostov-on-Don, Russia

[2]Research Institute of Physics, Southern Federal University, Rostov-on-Don, Russia

Abstract

The paper presents theoretical and experimental studies of the piezoelectric generator (PEG) with stack configuration in the regimes of quasi-static loading, low-frequency impulse excitation and harmonic mechanical loading. In this paper, we present experimental setups for the harmonic and impulse excitations, and also measurement method of output voltage and power of the PEG. The impulse loading of the PEG allowed one to reach a regime of non-distorted transfer of piezoelectric signal, at which maximal part of the mechanical energy, applied to it, transformed into electric one with minimal loss. In the case of impulse mechanical loading, we reached the value of PEG power density was equal to 27.565 mW/cm3. This value is higher on two orders of magnitude than power density of multilayer PEG under quasi-static loading. It was also shown that peak values of PEG output power at harmonic excitation on the frequency of 34 Hz were almost two times greater than the value at the frequency of 17 Hz, but only for load resistance around 10 kΩ. This result does not agree with results of the paper [6]. Moreover, we observed in the plots of dependence of the normalized output power on load resistance under harmonic excitation a specific feature consisting in extremum value of the PEG output power at definite values of frequency and load resistance. It was stated that this dependence for maximal output power corresponded to common law in the form of power function of 3rd exponent. In general form, the problem of optimal choice of parameters of the PEG with stack configuration could be solved by using the obtained 3-D plots of dependence of the output power on frequency of the harmonic excitation and load resistance.

---

* Email: akop@math.rsu.ru

1. INTRODUCTION

In the last years, many studies have been devoted to estimation of influence of different electrophysical and geometrical factors on characteristics and energy efficiency of piezoelectric generators (PEGs). The main part of these investigations has touched the cantilever-type PEGs. Review of these works is given in detail in number of monographs and papers (see, for example [1–4]). Considerably lesser attention has been devoted to the PEGs with stack configuration (SC), while an actuality of development of this type of PEGs is obvious. The PEGs with stack configuration can operate at significantly greater mechanical stresses than the cantilever-type PEGs. For many applications this circumstance has a prevailing mean. In particular, there are devices for damping dangerous vibrations of helicopter blades, created by using the PEGs with stack configuration [5]. Based on the stack-type PEGs, similar devices have been developed for industrial robots, for which a great output power of PEG was also necessary. Several investigations [1–3] have been devoted to development of the energy efficiency criteria. In the last years, new results of study of the stack-type PEGs have been published [6,7]. In [6] this problem was considered with the help of analytic approach and numerical modeling of piezoelectric energy harvesting caused by the harmonic and random vibrations of multilayer SC PEG under axial compressive loading. To solve this problem, Fourier series representation of the vibration excitation was applied and used also Euler-Maruyama scheme to solve the constitutive electromechanical stochastic differential equation. As a result, the authors obtained dependences of voltage-to-frequency for multilayer PEG with sizes of $5{\times}5{\times}18$ mm^3 from PZT-5H for number of resistors and dependences of power output versus load resistance. These dependences were constructed not only on the base of calculations but using experimental data, too.

Other paper [7] discussed the research results of output characteristics of the multilayer SC PEG in the cases of quasi-static and fast impact excitations. As a result, it has been stated that the stack-type PEG produced a greater output voltage at the quasi-static loading than at the impact excitation. This result contradicts to known data (see, for instance [1,4]) on significant dependence of the PEG output voltage on excitation frequency. It leads to conclusion that the results of [7] describe only one particular case of the PEG impact excitation.

Analysis of known studies on energy efficiency of multilayer SC PEG showed that in the various cases of oscillatory excitation, the PEG output characteristics and hence the energy efficiency differ significantly and depend on parameters influencing output power of PEG.

The problem of searching optimal parameters, allowing one to reach maximal energy efficiency of the PEGs, is considered in this chapter.

2. ELECTROMECHANICAL MODEL

In the last time, the attention of researchers became to displace from the cantilever-type PEGs to the stack configurations of PEGs. Because the SC PEGs are able to resists to sufficiently high compression stresses, namely this factor allows their integration in the transport auto- and rail-road infrastructure.

There is a necessity to develop mathematical model for prediction of output characteristics of different PEGs. An analytical model of multilayer PEG has been proposed in [6] for axial compression $p(t)$ (see Figure 1).

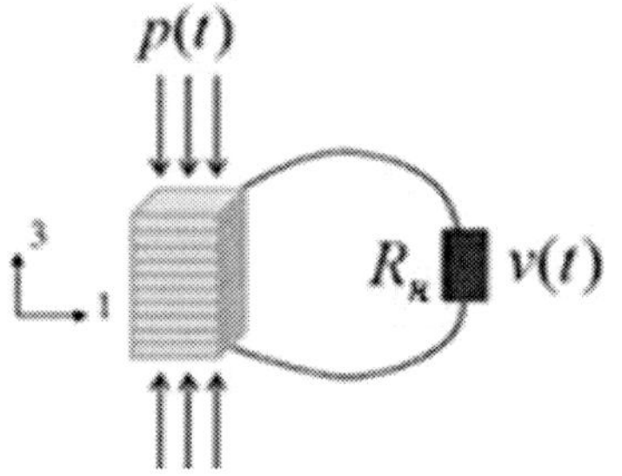

Figure 1. Electromechanical scheme of PEG, subjected to axial compression $p(t)$ [6].

It is assumed that the generator is excited on frequencies much lower than resonance eigen-frequency. In its turn, this allows one to describe dynamics of the system by using one-side electromechanical couple and ordinary differential equation of first order. If the PEG consists of N layers, polarized on thickness, and load resistance $R_н$ is switched on in parallel, then the constitutive equation describing this system has the form:

$$\sum_{i=1}^{N} \frac{d}{dt}\left(\int_{A_i} \mathbf{D} \cdot \mathbf{n} dA_i \right) = \frac{v(t)}{R_н}, \tag{1}$$

where $v(t)$ is the voltage on resistor, D is the vector of electric displacement, n is the vector of normal to surface of electrode, A_i is the square of electrode surface of i-th layer. Let us present the vector components of electric displacement in the form:

$$D_i = d_{ijk} T_{ij} + \varepsilon_{ij}^T E_j . \tag{2}$$

By submitting (2) into (1), we obtain the equation describing dynamics of the system, presented in Figure 1:

$$C_p^{eq} \dot{v}(t) + \frac{1}{R_н} v(t) = d_{33}^{eff} A \dot{p}(t), \tag{3}$$

where $C_p^{eq} = N\lambda\varepsilon_{33}^T A / h$ is the equivalent capacity, h is the thickness of one layer of the PEG, $d_{33}^{eff} = N\mu d_{33}$ is the effective longitudinal piezoelectric constant, λ and μ are empiric constants, introduced owing to difference between mono- and multilayer configurations of PEG, and also effects, caused by fabrication technology of the multilayer PEGs.

At the harmonic loading, the external exciting pressure can be written in the form of $p(t) = p_0 e^{i\omega t}$. Then a solution of Equation (3) is searched as $v(t) = Ve^{i\omega t}$, where ω is the circular frequency of external exciting pressure. Therefore, we can obtain an expression for amplitude value of output voltage in the form:

$$V = \frac{d_{33}^{eff} i\omega A p_0}{C_p^{eq} i\omega + \dfrac{1}{R_н}}. \tag{4}$$

Then, if we know voltage and load resistance, it can be found an expression for output power, generated by multilayer PEG at harmonic compression:

$$P = \frac{V^2}{R_н} = \frac{(d_{33}^{eff} A p_0 \omega)^2 R_н}{C_p^{eq2} \omega^2 R_н^{\ 2} + 1}. \tag{5}$$

By considering the expression (5) for output power as a function depending on $R_н$, and calculating extremum points, we define optimal load resistance, at which the power attains its maximum:

$$R_н^{opt} = \frac{1}{C_p^{eq} \omega}. \tag{6}$$

By substituting (6) into (5), we obtain maximum output power, which could be reached in PEG at the given mechanical loading:

$$P_{max} = \frac{\omega d_{33}^{eff\,2} A^2 p_0^2}{2 C_p^{eq}}, \tag{7}$$

where A is the square of cross-section of the PEG sensitive element, p_0 is the amplitude of mechanical stress, C_p^{eq} is the equivalent electric capacity; other parameters have been pointed above.

3. Experimental Validation

3.1. Experimental Setup for Harmonic Excitation

The structural scheme for low-frequency impulse loading of the PEG is depicted in Figure 2. The laboratory setup provides the low-frequency mechanical loading of the PEG at harmonic oscillation excitation in the regimes of program and manual control of the amplitude and frequency components of the forcing acting along the PEG axis with registration of input and output parameters of loading and response.

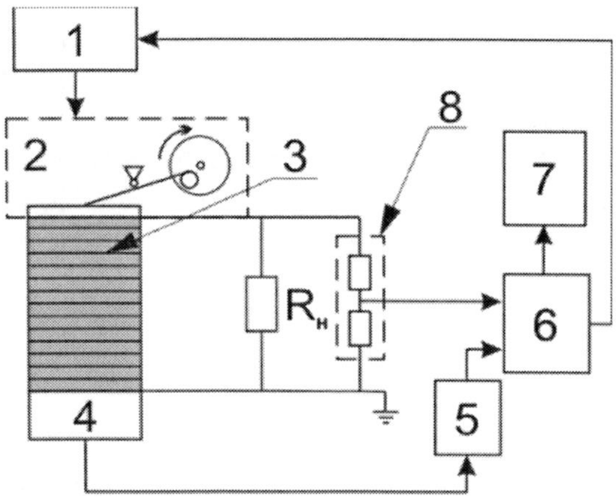

Figure 2. Structural scheme of test setup: 1 – frequency transducer ZVF11-H0004S2, giving the frequency of mechanical load; 2 – loading module; 3 – tested PEG; 4 – strain-gauge dynamometer; 5 – tension amplifier; 6 – ADT/DAT; 7 – PC with Power Graph software; 8 – voltage divider.

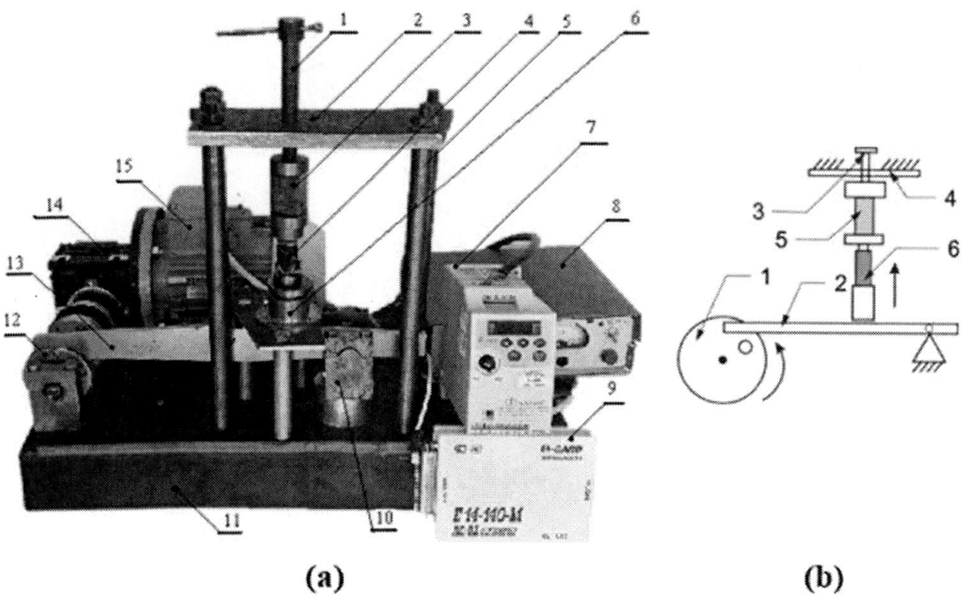

Figure 3. (a) Photo of laboratory setup for definition of stack-type PEG characteristics: 1 – screw, intended for disposition of PEG on thickness and for its initial preloading; 2 – immobile cross-arm; 3 – test PEG sample; 4 – strain-gauge dynamometer; 5 – force columns; 6 – directing cylinder with mobile cross-arm; 7 – frequency transducer, giving frequency of the engine rotation from 10 to 1,400 rpm; 8 – tension amplifier; 9 – transducer ADT/DAT; 10 – support-bracket; 11 – base;, 12 – eccentric disc with connecting-rod 13; 14 – reduction gear; 15 – engine; (b) kinematic scheme of the loading module: 1 – engine; reduction gear and eccentric disc; 2 – crank mechanism; 3 – preloading screw; 4 – cross-arm; 5 – PEG; 6 – strain-gauge dynamometer.

Mechanical loading of the test sample of PEG is carried out by loading module of the setup consisting of electric engine with reduction gear, eccentric oscillator and crank mechanism. Besides of the loading module, the setup includes strain-gauge dynamometer and tension amplifier, measuring voltage transducer ADT/DAT E14-440, electric load resistance, divider of output voltage, block device giving the frequency of impulse mechanical load, PC with Power Graph software. The test setup initiates the mechanical load on the studied PEG by using the eccentric oscillator. It is performed by electric engine with reduction gear and crank mechanism. The crank mechanism transforms a rotation of the reduction gear axis into translation of connecting-rod. The strain-gauge dynamometer is placed coaxially with the studied piezoelectric generator. In order to regulate the angle frequency of the engine rotation, we use frequency transducer ZVF11-H0004S2. Figure 3 presents photo of the laboratory setup. Operation of the loading module is explained by the kinematic scheme pointed in Figure 3b.

(a)

(b)

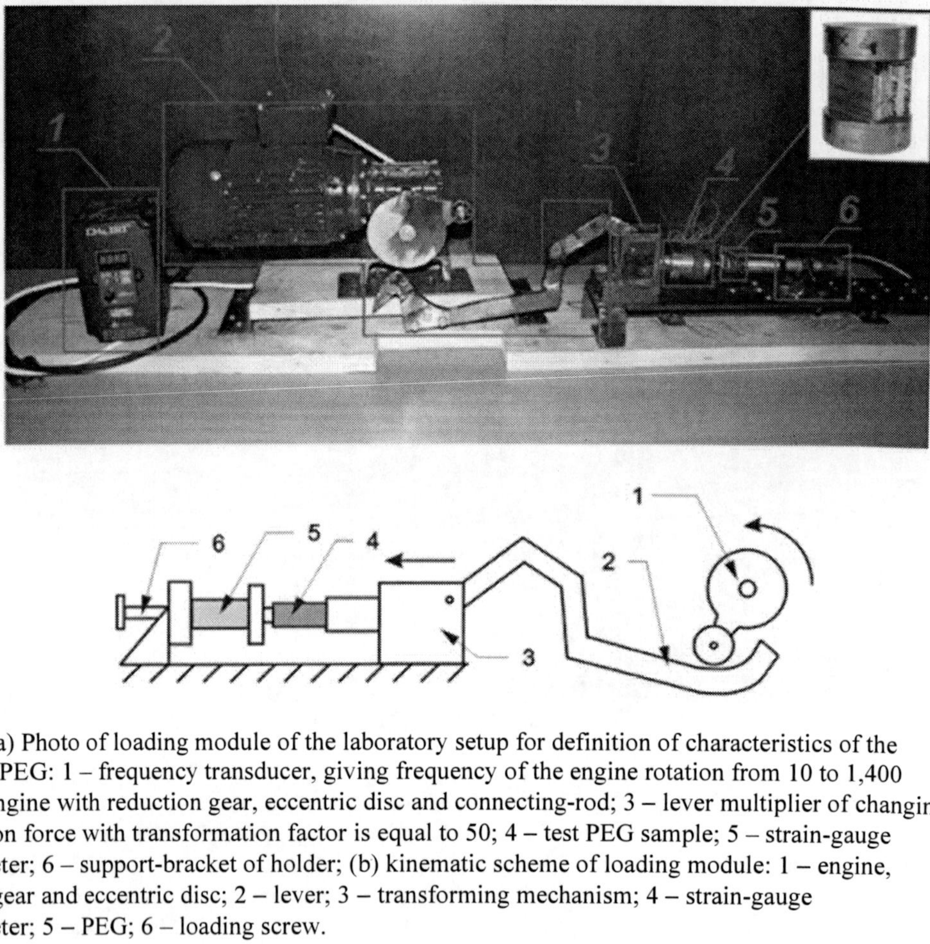

Figure 4. (a) Photo of loading module of the laboratory setup for definition of characteristics of the stack-type PEG: 1 – frequency transducer, giving frequency of the engine rotation from 10 to 1,400 rpm; 2 – engine with reduction gear, eccentric disc and connecting-rod; 3 – lever multiplier of changing compression force with transformation factor is equal to 50; 4 – test PEG sample; 5 – strain-gauge dynamometer; 6 – support-bracket of holder; (b) kinematic scheme of loading module: 1 – engine, reduction gear and eccentric disc; 2 – lever; 3 – transforming mechanism; 4 – strain-gauge dynamometer; 5 – PEG; 6 – loading screw.

In order to measure electric signals from the strain-gauge dynamometer and PEG and also for working out analogue and digital information, it is used measuring voltage transducer ADT/DAT E14-440 and Power Graph software. The voltage transducer provides measurement and scaling of voltage of the direct and alternating currents in single or several measurement channels (maximal number of the channels equals 16 or 32 in dependence on the connection diagram) by using 14-bit analogue-digital transducer (ADT) and multichannel commutation of input channels. Controlling operation of the voltage transducer E14-440 is performed through PC by means of standard interface of USB 2.0.

3.2. Experimental Setup for Impulse Excitation

Setup, providing impulse mechanical cyclic influence on the studied object, is similar in many details to the setup, described in Section 3.1. In new modification of the setup, the reduction gear disposes in vertical plane and eccentric disc is fixed on its axis. The crank mechanism transforms a rotation of the reduction gear axis into translation of connecting-rod. In order to regulate the angle frequency of the engine rotation, we use frequency transducer ZVF11-H0004S2. The amplitude of force impulse impact on PEG is restricted by the engine power, rigidity of dynamometer and support-bracket of holder. Figure 4 presents photo of loading module of the laboratory setup, but its structural scheme repeats completely the scheme presented in Figure 2.

The main difference of this setup from the setup for harmonic loading consists in the loading module. The principle of its operation is explained by kinematic scheme presented in Figure 4b.

3.3. Samples for Measurement of PEG output parameters

We studied multilayer SC PEGs of two configurations:

i) prismatic geometry with cross-section sizes of 24×16 mm^2 and a thickness of 21 mm (model 1);
ii) ring geometry with external and internal diameters of 18 and 8 mm, and heights of 26 and 38 mm, respectively (model 2).

The PEG of the model 1 consists of 40 piezoelectric elements with thickness of 0.5 mm of the PZT–19 piezoceramic (fabricated in the SDTB "Piezopribor" of the Southern Federal University, piezoelectric constant d_{33} = 530 pC/N). Each of 16 disc piezoelectric elements of the model 2 of the PZT–19 piezoceramic has a thickness of 2 mm and is polarized on thickness (piezoelectric constant d_{33} = 360 pC/N). Photos of the test PEG specimens are present in Figure 5.

Figure 5. Test PEG samples of two configurations: prismatic – model 1 (left) and ring – model 2 (right).

4. Measurement of Output Parameters

4.1. Estimation of Reliability of Experimental Results

The estimation of reliability of the measurement results of the PEG output voltage and compressing force, connected with it, was conducted taking into account the stated temporal equivalence of forms of the impulses of the compressing force to the PEG output voltage. Equivalence of these parameters is defined by criterion describing the non-distortion degree of translation of the electric signal (in our case, it is piezoelectric response of sensitive element). As is well-known [8], the regime of non-distortion translation of the piezoelectric impulse to input resistor $R_н$ for sensitive element of PEG with equivalent scheme, including in parallel electric capacity C_x of the sensitive element and active resistance R_x, is defined by the criterion:

$$R_н C_x \geq \tau_i, \quad (8)$$

where τ_i is the impulse duration of compressing force.

Taking into account this criterion, we performed experiments for PEGs differed by sizes and configuration. The sensitive element of first-type PEG (multilayer) consisted of 11 ring in parallel piezoelectric elements (PEs) with a thickness of 1 mm, external and internal diameters of 18 and 8 mm, respectively. Its common length was 11.3 mm and total electric capacity was equal to 20.22 nF. Each of PEs was polarized on thickness (piezoelectric constant $d_{33} = 360$ pC/N). Second-type PEG contained disc PE with a diameter of 10 mm and thickness of 3 mm; it was also polarized on thickness. Its electric capacity was 422 pF and piezoelectric constant d_{33} coincided with the first-type PEG.

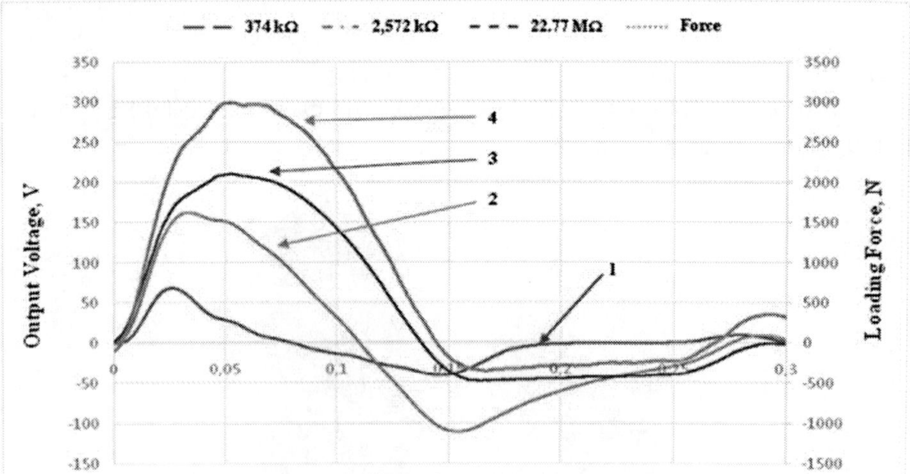

Figure 6. Forms of piezoelectric output voltage (curves 1 – 3) and impulses of compressing force (curve 4).

In accordance with above method, we conducted measurements, registered output characteristics of the PEGs and obtained temporal dependences. Figure 6 presents temporal dependences of output voltage of dynamometer corresponding to the force impulse of 3 kN (curve 4) and piezoelectric responses at the $R_н$ = 0.374, 2.572 and 22.77 MΩ (curves 1, 2 and 3) for the multilayer first-type PEG. Figure 6 demonstrates influence of load resistance on correspondence of forms of the force impulses and piezoelectric responses at increasing $R_н$ or transition from the regime of distorted translation of piezoelectric signal (with various degree of its differentiation) to the regime of non-distorted translation. As it follows from plots of dependences $u_{out}(t)$ in the cases of $R_н \geq 22.77$ MΩ, the amplitudes of compressing force and generated charge (and hence output voltage) attain maximal values simultaneously at $t = 0.05$ s. In its turn, this allows one to calculate effective value of output voltage of the PEG by integrating value of the response in respect on time of the impulse duration and define its actual magnitude and also its normalized value (ratio to piezoelectric constant d_{33}). Moreover, Figure 6 shows that greatest square covered by curve of output voltage in the range from 0 to 0.15 s, disposes under curve 3. Then it could be proposed that very great part of the mechanical energy, applied to the PEG sensitive element at the pointed value of $R_н$, transforms into electric energy with minimal loss satisfying the condition (8).

4.2. Measurement of PEG Output Voltage at Quasi-Static and Impulse Excitations

In accordance with measurement method of output parameters of the model 2 for stack-type PEG (Figure 5), we constructed dependences of output voltage and output power on loading frequency of the PEG by using the obtained values of output voltage at axial impulse loading with amplitude of σ_c = 17.2 MPa and different values of electric load resistance $R_н$ (see Figure 7). The plots of dependences for other geometry of the model 2 had similar forms.

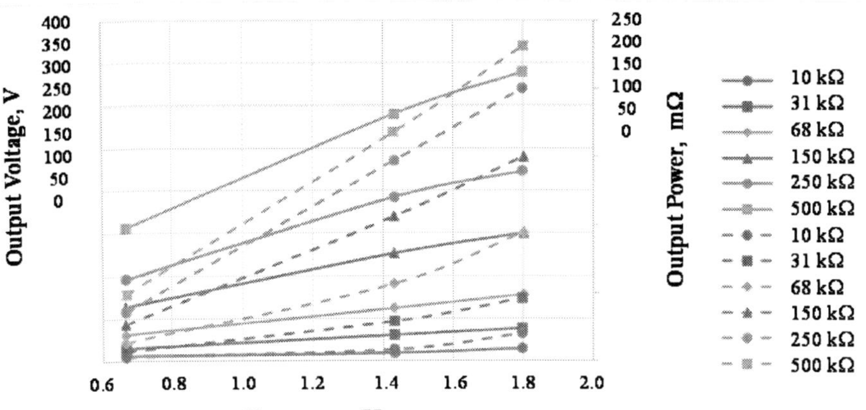

Figure 7. Output voltage (continuous lines) and output electric power (dashed lines) *vs* loading frequency at different values of electric load resistance for the ring type PEG with thickness of 38 mm at impulse mechanical load of 17.2 MPa.

As it follows from plots of the dependences, the PEG output voltage grows monotonously with increasing the frequency of its loading, demonstrating almost linear character at the electric load resistance $R_н$ in range from 68 up to 500 kΩ. Maximal peak value of output voltage is 327 V at $R_н$ = 500 kΩ. Then, based on the measured values of the output voltage and electric load resistance, peak values of output power were calculated by using formula (5), and constructed the plots of dependence of the output peak power on loading frequency (Figure 7).

The calculation results show that output power of 213.9 mW could be obtained at the impulse loading of 17.2 MPa and electric load resistance of $R_н$ = 500 kΩ. In this case, the ratio of power to volume of SE of the PEG model is equal to 27.565 mW/cm^3; this value is greater on two order of magnitude than corresponding power, obtained for the multilayer type PEG of the model 1 at the quasi-static loading (see Figure 8, PEG sample with a height of 36 mm). In this case, we take into account that the model 1 was subject to greater compression at the quasi-static loading than the model 2 at the impulse loading (40.3 MPa and 17.2 MPa, respectively).

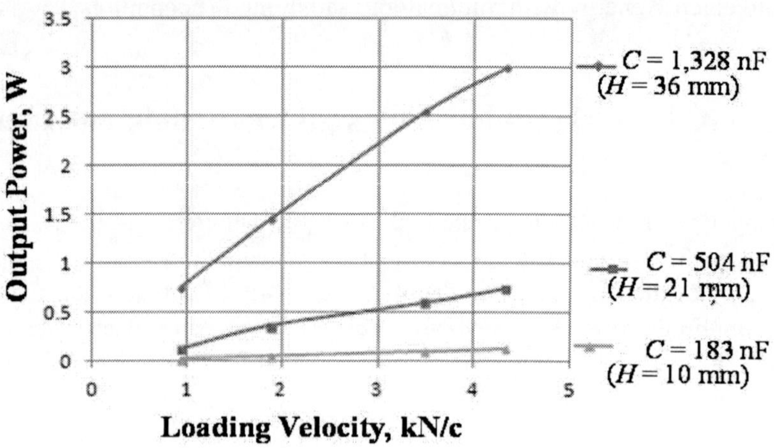

Figure 8. Peak value of output power *vs* quasistatic loading velocity of PEGs at different stack heights and PEG electric capacity.

Influence of compressing force on PEG power could be estimated by using new characteristic of its energy efficiency, namely ratio of PEG output power to maximal external mechanical stress P_{out}/σ_c [6]. This parameter characterizes energy efficiency factor of PEG. Then we use the dependence of increasing charge velocity on loading rate [13]:

$$\frac{dQ_i}{dt} = d_{33} S \frac{d\sigma_{33}}{dt}, \tag{9}$$

where Q_i is the generated charge, S is the cross-section square of PEG and σ_{33} is the applied stress.

Table 1. Comparative output characteristics of multilayer stack type PEGs of several configurations for three different kinds of mechanical loading

Loading Type	PEG Geometry	PEG Height (h), cm	Cross-section Square (S_c), cm^2	Maximum Compressing Stress (σ_c), MPa	Loading Rate (v_l), kN/s	Loading Frequency (f_l), Hz	Output Voltage (V_{out}), V	Peak Output Power (P_{out}), mW	Specific Output Power (P_{out}), mW/cm^3	Normalized Output Power (P_{out}/σ_c), mW/MPa	Output Power Factor ($\dot{P}_{out}/\sigma_c$), mW/($MPa \cdot cm^3$)
Quasi-static	Rectangular Cross-section, 24×16 mm^2	1.00	3.84	40.3	4.4	-	40	0.15	0.039	0.0037	2.4×10^{-5}
		2.03	3.84	40.3	4.4	-	130	0.75	0.096	0.0186	5.9×10^{-5}
		3.60	3.84	40.3	4.4	-	220	3.0	0.217	0.074	13×10^{-5}
	Square Cross-section, 48×48 mm^2 [11]	0.5	23.04	10	Loading time is 0.1 s	-	10	7.5	0.65	0.75	65×10^{-4}
Low-frequency Impulse	Ring Cross-section, 24×16 mm^2	2.6	2.538	19.9	-	1.8	106 (peak value)	54.6	10.284	2.74	4×10^{-2}
		3.8	2.538	19.5	-	1.8	327 (peak value)	213.9	27.565	10.96	12×10^{-2}
Harmonic	Square Cross-section, 5×5 mm^2 [6]	1.8	0.25	0.168	-	10	-	0.27	-	-	0.6
						20		0.017			
						30		0.007			

Taking into account (9), the values of the energy efficiency factor distinct more than two orders of magnitude for two models of the PEG at different external stresses and various kinds of loading (see, Table 1).

4.3. Measurement of Output Power by Harmonic Excitation of PEG

In Section 3.1, we presented description of the structural scheme and operation of setup at definition of PEG characteristics under harmonic excitation. By using setup, we conducted measurements of output voltage for the stack-type PEG of the model 1 (see Figure 6) with cross-section sizes of 24×16 mm^2 and a height of 21 mm at frequencies of 2.6, 3.3 and 4.0 Hz, electric load resistance varied from 6 to 800 kΩ. The loading frequency range was the same as in the known study [6]. However, the mechanical loading of the PEG was performed with magnitude of 0.36 MPa in difference of [6]. Based on the measurement results, it was obtained dependence of voltage output *vs* electric load resistance (Figure 9).

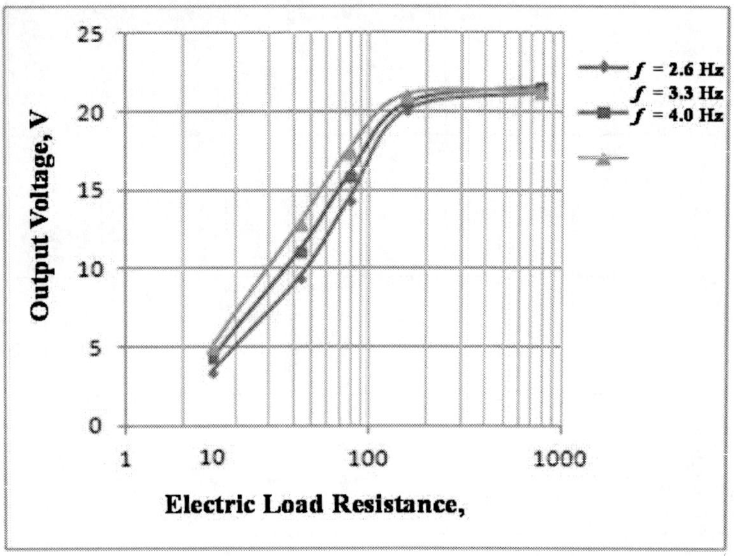

Figure 9. Voltage output *vs* load resistance under harmonic excitation up to 4 Hz at mechanical loading with magnitude of 8.8 MPa.

From analysis of these dependences, we concluded that if output voltage grows at increasing of electric load resistance up to 160 kΩ, then after this value the curve reaches the steady state and the growth of output voltage ends at the value of 22 V practically in a whole range of frequencies up to 4 Hz. Similar measurements of these dependences in the considered range of load resistance have been performed for two samples of the model 1 (see, Figure 5) with heights of 21 and 36 mm. Comparison of these data with calculation results is present in Figure 10; the comparison is sufficiently valid into limits of the measurement error.

To estimate influence of the loading frequency on output voltage of the PEG at the frequencies exceeding 4 Hz, we modified the setup (description of new setup did not present in this chapter) and performed measurements of the voltage at the frequencies of harmonic excitation were equal to 17 and 34 Hz at lower value of mechanical load (with magnitude of

0.36 MPa). The tests were conducted at lower amplitude of the mechanical load that was caused by the restrictions of technical characteristics of the modified setup.

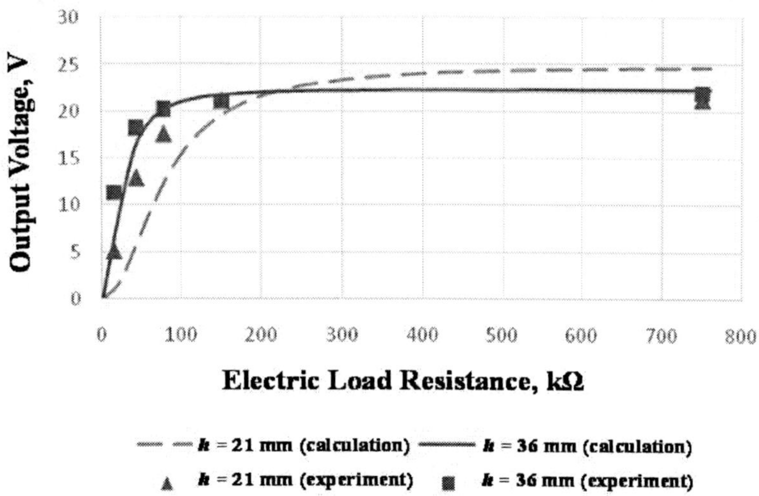

Figure 10. Comparison of calculations and experimental data for PEGs with heights of 21 and 36 mm at harmonic loading with the loading frequency of 4 Hz.

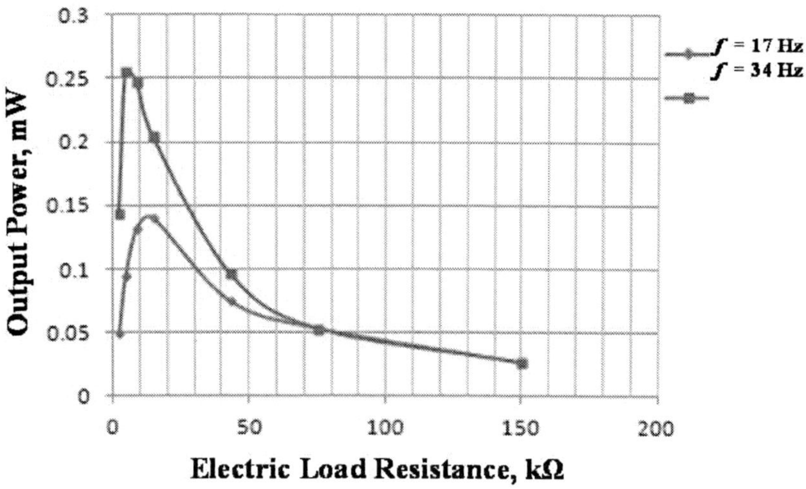

Figure 11. Output power *vs* electric load resistance in the case of harmonic excitation at frequencies of 17 and 34 Hz for model 1 (capacity of 504 nF) by loading with magnitude of 0.36 MPa.

We obtained that qualitative character of dependence of output voltage on the load resistance did not change compared to Figure 9 for the loading frequencies up to 4 Hz. However, in this case the output voltage reached the steady state only at the value of 2.2 V. Difference of these values of output voltage at the frequencies of 4 and 34 Hz is explained by significant difference of mechanical loads are equal to 8.8 and 0.36 MPa, respectively.

In order to analyze the output characteristics of PEG into range of load resistance up to 150 kΩ, we calculated values of its maximal output power by using formula (7) and then constructed dependence of output power on electric load resistance (see Figure 11).

Analysis of the dependences of output power on load resistance for frequencies of 17 and 34 Hz has showed that the peak value of the power at the frequency of 34 Hz was almost two times greater the power in the case of harmonic excitation at 17 Hz, but only at the load resistance around 10 kΩ. Ratio of the peak value of output power (0.25 mW) to the cross-section square of the PEG sensitive element gives the normalized value of peak power is equal to around 65 μW/cm^3, which differs significantly from the value of 17 μW/cm^3, obtained in [6] (but for PEG with other sizes). This difference is caused that the last result obtained at the mechanical load of 0.168 MPa, which is two times lower the applied load, considered by us.

To compare our results with data of [6], we calculated also normalized output power for the PEG model 1, normalized by amplitude of mechanical load. The calculation results are present in Figure 12.

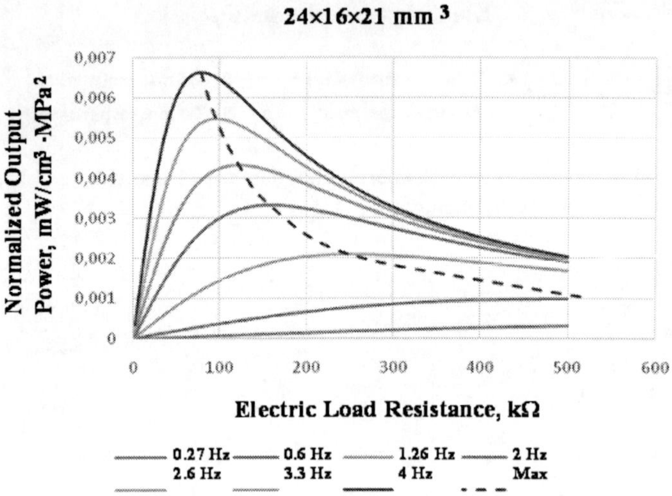

Figure12. Normalized output power *vs* load resistance under harmonic excitation at different frequencies from 0.27 up to 4 Hz.

Then to compare our results with known data [6], we recalculated values of the output power for PEG with sizes of 5×5×18 mm^3 and corresponding magnitudes of the load resistance at harmonic loading on frequency of 4 Hz. This recalculation (see Figure 13) was conducted in wider range of the load resistance in comparison with [6].

We stated common characteristic feature on both plots of the normalized output power (Figure12, 13), defined by shape of the curves, joining the points of maximal values of output power for given frequency and load resistance, which can be described by power functions.

Let us consider general form of power function $y = ax^{-n}$, where $a > 0$, n is the natural number. In our case, the relation of the normalized maximal outer power is defined as $P_n^{\max} = \dfrac{a}{R_н^3}$. We find a factor a for the both plots. In our case, $P_n^{\max} = 0.0033$ mW at $R_H = 158$ kΩ for f = 2 Hz, and we obtain a = 13,016 mW·kΩ^3. In the case of [6], $P_n^{\max} = 0.0384$ mW at $R_H = 72$ kΩ for f = 1.26 Hz, and we obtain a = 14,332 mW·kΩ^3.

The close values of the parameter *a*, obtained for the both cases, allow us to conclude that the dependence of the normalized output power on load resistance for the PEGs with different geometric parameters could be approximated by the same law in the form:

$$P_n^{\max} \propto \frac{a}{R_н^3}. \tag{10}$$

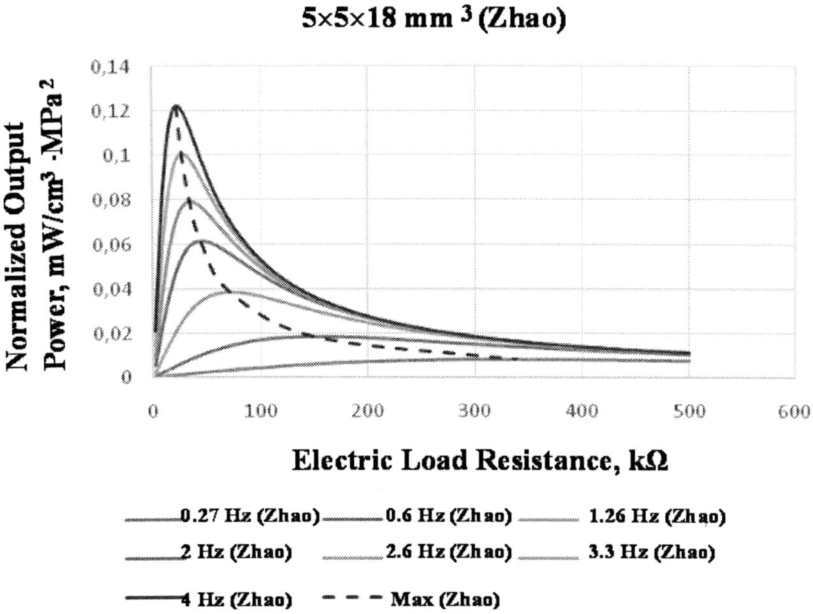

Figure 13. Normalized output power *vs* load resistance under harmonic excitation at different frequencies from 0.27 up to 4 Hz [6].

Analysis of relations (5), (6), (10) and test results, presented above in Section 3, shows that output power of the stack-type PEGs depends primary on the loading frequency and load resistance.

The obtained test values of output voltage for PEG of the model 1 with sizes of 24×16×21 mm^3 and corresponding magnitudes of load resistance and loading frequency were compared with calculations. As a result, we formulated a matrix of data for the treated type of PEGs. 3-D plots of output power versus load resistance in the frequency ranges up to 4 Hz (Figure 14) and up to 40 Hz (Figure 15) were constructed by using these matrix data in the Maple software. It follows from these plots that the maximal output power (at optimal load resistance) grows directly proportionally to frequency. This conclusion quietly agrees with relation (7) of the electric mechanical model, described in Section 2.

Analysis of the 3-D plots of the calculated dependence of the PEG output power on two parameters showed that with their help, we can choose optimal values frequency and load resistance, of which ranges the maximal power is reached, and hence maximal energy efficiency is defined based on specific values of the parameters.

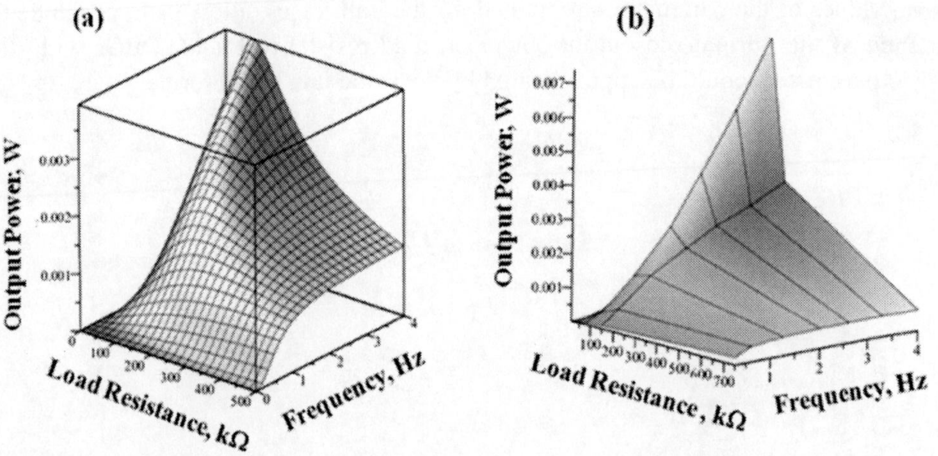

Figure 14. 3-D plots of output power of the PEG model 1 *vs* load resistance and frequency: (a) calculation, (b) experiment (in the range of harmonic oscillations up to 4 Hz).

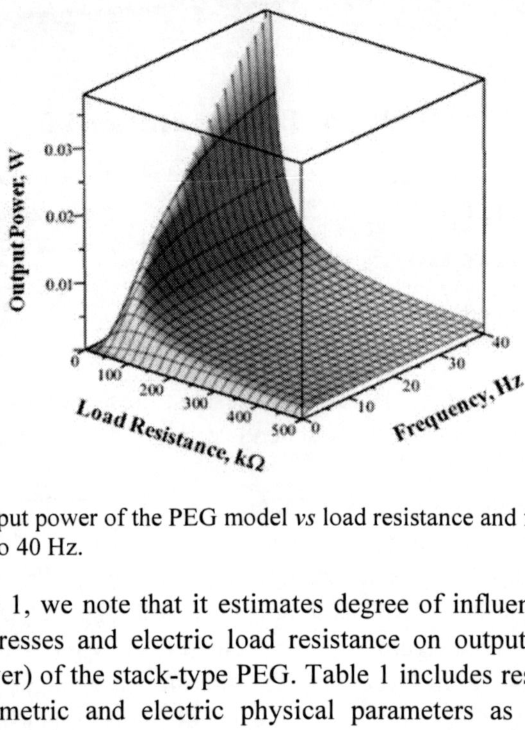

Figure 15. 3-D plot of output power of the PEG model *vs* load resistance and frequency in the range of harmonic oscillations up to 40 Hz.

Returning to Table 1, we note that it estimates degree of influence of the loading kind, level of mechanical stresses and electric load resistance on output characteristics (output voltage and output power) of the stack-type PEG. Table 1 includes research data obtained for samples close on geometric and electric physical parameters as in our study as other investigations [6,7,9–12].

So analysis of energy efficiency of the stack-type PEG under various loading regimes shows that in the case of impulse mechanical loading, we can reach a higher level (on one order of magnitude) of power density values compared to other types of loading at the equal other parameters of PEG (in particular, piezoelectric sensitivity of material of its sensitive element, geometric properties and processing technology of the generator). This conclusion defines favorability of application of the multilayer stack-type PEG under conditions of great dynamic loads, caused by external mechanical influences.

CONCLUSION

1) The test method and setup have been researched and developed for measurement and registration of output voltage of the stack-type PEG at impulse mechanical loading. The test method allows one to define mutually depended values of electric load resistance and amplitude of mechanical force, at which a regime of non-distorted translation of piezoelectric signal is reached. In this case, maximal part of the mechanical energy, applied to the PEG, transforms into electric energy with minimal losses.
2) Comparative values of output characteristics of the multilayer PEGs with stack configurations are obtained at their loading in impulse and quasistatic regimes. The calculation results showed that in the impulse mechanical loading the power density of PEG was 27.565 mW/cm^3. This value was greater than power density of multilayer PEG at quasi-static loading on two orders of magnitude. We compared our results with known data [7], defined at the peak impact loading of the PEG with stack configuration by force of 500 N. The value of 13.55 mW/cm^3 obtained in [7] is two times lesser than our result. It confirms that our method of impulse loading is more efficient.
3) The test method and setup have been researched and developed for measurement and registration of output voltage of the stack-type PEG at harmonic excitation of oscillations. It was stated that if the value of output voltage increases by increasing the load resistance up to 100 kΩ, then this dependence reaches steady state at the greater load resistance, and further growth of output voltage ends at the value of around 22 V in the frequency range up to 4 Hz. Similar character of this dependence was observed in [12], differing of our results by the steady-state value of output voltage was equal to 33 V (at frequency of 88 Hz) and 11 V (at 40 Hz) in contrast to 22 V in our case. This difference and scattering data are explained by use of piezoceramic PZT-5H in [12], demonstrating a higher piezoelectric sensitivity is equal to 700 pK/N in contrast to 530 pK/N for PZT-19, and different loading frequency in the compared cases. Therefore output voltage in [12] differs also from results obtained by us.
4) It was stated that the peak value of output power of the developed PEG with stack configuration was almost two times greater at harmonic excitation on frequency of 34 Hz, compared with the output power found at 17 Hz, but only at load resistance around 10 kΩ. In this case, the power density of PEG reached of 65 mW/cm^3 that did not agree with known value of [6] was equal to 17 mW/cm^3. The difference between these data was explained by that the last result has been obtained at the mechanical loading of 0.168 MPa. However this result is lesser on one order of magnitude, compared to load, applied to PEGs, studied in this chapter.
5) It was found a character feature in plots of the normalized output power density versus load resistance for PEGs under harmonic excitation, consisting in existence of extremum value of the PEG output power at various values of frequency and load resistance. This feature was stated in [6] for the stack-type PEG with geometry similar to studied by us. It was stated that the dependence of normalized output power on load resistance for PEG with other geometry is subject to common law in

the form of power function of third exponent for maximal output power. The relation for this function could be used in development of piezoelectric harvesters of energy with maximal value of output power at the condition of choice of optimal values of load resistance and loading frequency.

6) In general form, the problem of optimal choice of the stack-type PEG parameters with viewpoint of obtainment of the maximal output power in the case of harmonic oscillation regime with frequency up to 40 Hz could be solved by using 3-D plots of output power at varying frequency and load resistance. Similar results have been presented in [6], but for load resistance only up to 14.8 kΩ that restricts a possibility to analyze energy efficiency of the considered PEGs.

ACKNOWLEDGMENTS

The study was performed at the partial supporting of the Russian Foundation for Basic Research (grants Nos. 12-08-01137a, 13-08-00754-a) and the Ministry of Education and Science of Russian Federation (proposal No. 213.01-11/2014-25).

REFERENCES

[1] Erturk, A.; Inman, D. J. *Piezoelectric Energy Harvesting*, John Wiley & Sons, Chichester, UK, 2011 pp. 1 – 412.

[2] Shu, Y. C.; Lien, I. C. *Journal of Micromechanics and Microengineering*, 2006, *vol. 16*, 2429–2438.

[3] *Energy Harvesting Technologies*. Shashank Priya; Daniel J. Inman (Eds.), Springer, Berlin, 2011.

[4] Akopyan, V. A.; Parinov, I. A.; Rozhkov, E. V.; et al. In: In: *Advanced Materials. Physics, Mechanics and Applications*, S.-H. Chang, I. A. Parinov, V. Yu. Topolov (Eds.), Springer Proceedings in Physics, *vol. 152*, Chapter 29. Springer, Heidelberg, New York, Dordrecht, London, 2014, p. 355.

[5] Shevtsov, S.; Flek, M.; Zhilyaev, I. In: *Physics and Mechanics of New Materials and their Applications*. Chapter 21. Nova Science Publishers, New York, 2013, p.259-274.

[6] Zhao, S.; Erturk, A. Energy harvesting from harmonic and noise excitation of multilayer piezoelectric stacks: modeling and experiment. In: *Active and Passive Smart Structures and Integrated Systems*, Henry A. Sodano (Ed.). Proc. of SPIE, 2013, *vol. 8688*, p. 86881Q.

[7] Raunaq, Shah; Rahul, Khandelwal; Vishnukumar, A.; Sudha, R. *The International Journal of Engineering and Science (IJES)*, 2013, *vol. 2(5)*, 79-83.

[8] Morozov, A. G. Electro-techniques, Electronics and Pulse Techniques, High School, Moscow, 1987, pp. 1 – 448 (In Russian).

[9] Kazakov, V.; Klimashin, V., Nikiforov, V.; et al. *Components and Technologies*, 2007, *No 6*, 62-65 (In Russian).

[10] Afonin, S. M. Nano- and Microsystem Technique, 2011, *No. 3*, 30-38.

[11] Gritzenko, A.; Nikiforov, V.; Schegoleva, T. *Components and Technologies*, 2012, *No. 9*, 63-68 (In Russian).

[12] Wang, J.; Shi, Z.; Han, Z. Journal of Intelligent Materials Systems and Structures 2013, *vol. 24*, 1626-1636.

[13] Smazhevskaya, E. G.; Fel'dman, N. B. *Piezoelectric Ceramics*. Soviet Radio Press, Moscow, 1971 (In Russian).

In: Advanced Materials - Studies and Applications ISBN: 978-1-63463-749-7
Editors: I.A. Parinov, S.-H. Chang, S.Theerakulpisut

Chapter 25

NOVEL OPTIC DEVICES FOR MEASUREMENT OF DISPLACEMENTS BASED ON METHOD OF CONTROL OBJECT HIGHLIGHTING BY USING LASER INTERFEROMETER

***I. P. Miroshnichenko*[1,*], *I. A. Parinov*[2] *and Shun-Hsyung Chang*[3]**
[1]Don State Technical University, Rostov-on-Don, Russia
[2]Southern Federal University, Rostov-on-Don, Russia
[3]National Kaohsiung Marine University, Kaohsiung, Taiwan

ABSTRACT

Novel optic devices for displacement measurement are described based on the method of highlighting of the control object surfaces by laser interferometer. Moreover, the results of their ground are present using the conducted calculations and experiments, and also specific propositions are given on practical applications.

1. INTRODUCTION

In present time, research and development of mobile diagnostic complexes have very great actuality for non-accident operation and prevention of catastrophic loading of transport means and aircraft technique. These complexes allow us to conduct state diagnostics of construction materials of the main elements of cars and equipments by using non-destructive control methods directly at the sites of their operation (i.e., at the "field conditions"). Based on these studies, it could be taken a decision on possibility of their following exploitation and estimated a residual resource. One of the priority directions of equipment improvement of the complexes is development of the contactless measurement means with a high accuracy of displacements of the object control surfaces. These studies are directed to registration of

* Email: ipmir@rambler.ru

information on their state, for example by using acoustic active and passive methods of non-destructive control.

Russian patents [1–3] have presented the contactless means of displacement measurements of the object control surfaces based on application of the two-way laser interferometer with coupled branches [4,5] by using the modern laser technologies and novel methods of optical interferometry.

The devices [1–3] have been well grounded with help of the numerical calculations and experiments [6–11], approved and used at the solution of actual scientific and applied problems (see, for instance [12,13]).

1. THEORY

The optic interference means for measurement of displacements [1–3] could be present in general in the form of optically coupled and successively disposed a source of coherence electrical magnetic radiation, optic system, beam splitter, reflector and photo-receivers.

A possible optic scheme is present in Figure 1, where the beam splitter 1 with known amplitude transmission disposes in the plane *S*, with which Cartesian coordinate system $\{x, y, z\}$ is connected. The reflector 2 locates in the plane S_3, disposed under angle α at distance *h* from the beam splitter 1 being a sinusoidal diffraction lattice.

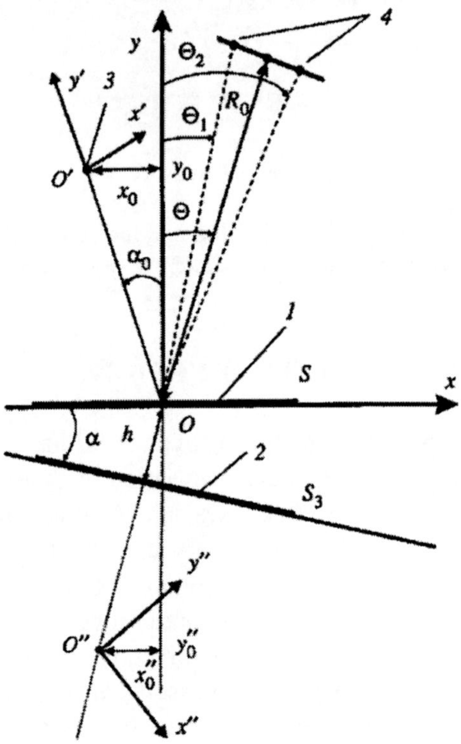

Figure 1. Optic scheme.

Coherent radiation from lighting point 3 falls on the beam splitter 1; an original location of the point is defined by coordinates $x_0, y_0, z_0 = 0$. The photo-receivers 4, with whose help we analyze interference picture, dispose in Fresnel zone.

The described device allows us by using the formed interference picture to define displacements of points of the control object surface in two cases:

(i) the lighting point 3, disposed on the control object surface, is obtained at reflection of focused laser radiation,
(ii) the reflector 2 states on the control object surface.

In the first case, it is necessary to define a changing coordinates $\Delta x_0, \Delta y_0$ by using the parameters of interference picture, and in the second case, is required to find values of α and h at constancy of x_0, y_0.

In order to define displacements of the control object surface, it is necessary to calculate an electrical magnetic field scattered by the considered devices in the Fresnel zone where the interference picture is observed. Moreover, we must define the field dependences on location of the lighting point x_0, y_0 and parameters characterizing location of the reflecting surface $\{\alpha, h\}$.

To solve this problem, we use the following Fresnel approximation [14]:

$$U(\theta,\psi) = U_0 \int_S U(x,z) e^{ik\frac{x^2+z^2}{2R_0}} e^{-ik\sin\theta(x\sin\psi + z\cos\psi)} ds, \tag{1}$$

where $U(\theta,\psi)$ is the complex amplitude of wave field; θ, ψ, R_0 are the spherical coordinates of observation points; x, z are the point coordinates of sources, which lies on the beam splitter surface.

The function $U(x,z)$ be into expression (1) and describes an optic field in the beam splitter plane. For its definition, we apply the image method [15], when a whole field on the beam splitter surface is represent by the wave superposition from the lighting point O' and image mirror point O'', formed by reflector. By this it is necessary to take into account the beam splitter transmission. For solution comfort, we introduce local Cartesian coordinate systems, which origins locate at the lighting point O' and image mirror point O'', respectively (see Figure 1).

The coordinate system $\{x', y', z'\}$ has been obtained by means of rotation about z-axis on angle α_0 in positive direction and subsequent parallel translation on value of $(-x_0)$ and y_0, respectively.

The coordinate system $\{x'', y'', z''\}$ is mirror in respect to $\{x', y', z'\}$, where y-axis direction changes to the contrary. It is obtained from $\{x, y, z\}$ by rotation on angle

$[-(\alpha_0 + 2\alpha)]$ and parallel shear along *x*-axis on value of $\left(-x_0^{"}\right)$ and *y*-axis on value of $\left(-y_0^{"}\right)$.

The relations between these coordinate systems and the base one have the next form:

$$x' = (x + x_0)\cos\alpha_0 + (y - y_0)\sin\alpha_0;\; y' = -(x + x_0)\sin\alpha_0 + (y - y_0)\cos\alpha_0;\; z' = z; \quad (2)$$

$$x'' = \left(x + x_0^{"}\right)\cos(\alpha_0 + 2\alpha) - \left(y + y_0^{"}\right)\sin(\alpha_0 + 2\alpha);$$
$$y'' = \left(x + x_0^{"}\right)\sin(\alpha_0 + 2\alpha) + \left(y + y_0^{"}\right)\cos(\alpha_0 + 2\alpha);\; z'' = z, \quad (3)$$

where

$$x_0^{"} = x_0 \cos 2\alpha + 2h\sin\alpha + y_0 \sin 2\alpha;\; y_0^{"} = -x_0 \sin 2\alpha + 2h\cos\alpha + y_0 \sin 2\alpha. \quad (4)$$

In expressions (2) – (4) values of the parameters x_0, y_0, α_0 and α, h define, respectively, a spatial disposition of the lighting point and reflecting surface (see Figure 1).

In accordance with the image method [15], the expression for complex amplitude of light flux on the beam splitter surface can be written as

$$U(x,z) = A_0 e^{ikr'} - T(x,z)e^{ikr"}, \quad (5)$$

where the Fresnel zone is defined by

$$r' = r_0 + \frac{(x + x_0)^2 - x_0^2 + z^2}{2r_0};\; r'' = r_0^{"} + \frac{\left(x + x_0^{"}\right)^2 - x_0^{"2} + z^2}{2r_0^{"}};$$

$$r_0 = \sqrt{x_0^2 + y_0^2};\, r_0^{"} = \sqrt{x_0^{"2} + y_0^{"2}}, \quad (6)$$

where A_0 is a constant, $T(x,z)$ is the beam splitter transmission.

By substituting (5) and (6) into (1), and also specifying $T(x,z)$, we obtain relation for calculation of the field in the observed points.

The intensity on which reacts photo-receiver is equal to square of absolute value of the expression (1).

Let us consider the case, when the field is independent on *z*-coordinate, and sinusoidal lattice is used as beam splitter. Diffraction on the lattice, as it is known, leads to formation of maxima only of first order $(m = \pm 1)$. Due to it is expedient to locate the photo-receivers in the areas corresponding to the given maxima and to treat an interference picture in these directions, which is formed by falling wave and waves diffracted on the lattice.

The optical field in the space $y > 0$ in accordance with (1) could be written as follows

$$U_1 = \int_{-d}^{d} \left(e^{ikr'} + e^{\pm ik\frac{\lambda}{\tau}x} e^{ikr''} \right) e^{ik\frac{x^2}{2R_0}} e^{-ikx\sin\theta} dx \cdot \tag{7}$$

Here we omitted an amplitude multiplier, τ is the lattice period, and a sign at the exponent is taken in accordance with considered main maximum, 2d is the reflector size.

By submitting values of r from (6) in this expression, we obtain relation for calculation of complex amplitude of the wave field in the form:

$$U_1 = e^{ikr_0} J_1 + e^{ikr_0''} J_2 , \tag{8}$$

where

$$J_1 = \int_{-d}^{d} e^{i\left[k\left(\frac{1}{2r_0} + \frac{1}{2R_0}\right)x^2 + k\left(\frac{x_0}{r_0} - \sin\theta\right)x\right]} dx ; \tag{9}$$

$$J_2 = \int_{-d}^{d} e^{i\left[k\left(\frac{1}{2r_0''} + \frac{1}{2R_0}\right)x^2 + k\left(\frac{x_0''}{r_0''} - \sin\theta \pm \frac{\lambda}{\tau}\right)x\right]} dx . \tag{10}$$

By using table data of [16], the integrals (9), (10) can be reduced to the Fresnel integral of the form:

$$J = \int_{-d}^{d} e^{i(ax^2 + bx)} dx = \frac{1}{\sqrt{a}} e^{i\nu} \int_{-\xi_1}^{\xi_2} \left(\cos\xi^2 + i\sin\xi^2\right) d\xi , \tag{11}$$

where

$$\nu = \frac{b^2}{4a} ; \xi_1 = -d\sqrt{a} + \sqrt{\nu} ; \xi_2 = d\sqrt{a} + \sqrt{\nu} \tag{12}$$

a is the coefficient before (ix^2) in the exponent of the expressions (9) and (10), respectively, but b is the factor before (ix) in these exponents.

The light intensity in accordance with (8) is calculated by using the formula:

$$I(\theta) = (e^{ikr_0} J_1(\theta) + e^{ikr_0''} J_2(\theta))(e^{ikr_0} J_1(\theta) + e^{ikr_0''} J_2(\theta))^* , \tag{13}$$

where sign (*) designs a complex-conjugate value.

If the beam splitter sizes are so great that diffraction on edges could be neglected and the integrating limits in (7) replaced to infinite ones, then integrals (9) and (10) could be calculated by using the following formulae [16]:

$$J_1 = e^{i\frac{\pi}{4}}\sqrt{\frac{2\pi R_0}{k}}\sqrt{\frac{r_0}{r_0 + R_0}}e^{ik\frac{r_0 R_0}{2(r_0+R_0)}\left(\frac{x_0}{r_0}-\sin\theta\right)^2}; \tag{14}$$

$$J_2 = e^{i\frac{\pi}{4}}\sqrt{\frac{2\pi R_0}{k}}\sqrt{\frac{r_0^{''}}{r_0^{''} + R_0}}e^{ik\frac{r_0^{''} R_0}{2(r_0+R_0)}\left(\frac{x_0^{''}}{r_0^{''}}-\sin\theta\pm\frac{\lambda}{\tau}\right)^2}. \tag{15}$$

In accordance with these expressions and with aim of simplification, taking $r_0 = r_0^{''} = R_0$ in the amplitude multipliers of interfered waves (14), (15), the intensity at the observed points can be written as

$$I = A_0(1-\cos\gamma), \tag{16}$$

where

$$\gamma = \frac{k}{2}\left[-r_0 + \frac{r_0 R_0}{r_0 + R_0}\left(\frac{x_0}{r_0} - \sin\theta\right)^2 + r_0^{''} - \frac{r_0^{''} R_0}{r_0^{''} + R_0}\left(\frac{x_0^{''}}{r_0^{''}} - \sin\theta \pm \frac{\lambda}{\tau}\right)^2\right]. \tag{17}$$

The dependence of spatial frequency on measured parameters at various points θ is defined as

$$\omega = \frac{\partial\gamma}{\partial\theta} = k\left[-\frac{r_0 R_0}{r_0 + R_0}\left(\frac{x_0}{r_0} - \sin\theta\right)\cos\theta + \frac{r_0^{''} R_0}{r_0^{''} + R_0}\left(\frac{x_0^{''}}{r_0^{''}} - \sin\theta \pm \frac{\lambda}{\tau}\right)\cos\theta\right]. \tag{18}$$

The reduced formulae define dependences of measured parameters of the interference picture along the directions of θ_1 and θ_2 (where photo-receivers locate) on coordinates of the lightened point of surface x_0, y_0, and also on disposition of reflector α, h.

So, if it is stated $\theta = \theta_1$ in (13), then we obtain in the right side an expression of intensity, depended on the measured parameters:

$$I_1(x_0, y_0, \alpha, h) = (e^{ikr_0}J_1(\theta_1) + e^{ikr_0^{''}}J_2(\theta_1))(e^{ikr_0}J_1(\theta_1) + e^{ikr_0^{''}}J_2(\theta_1))^*. \tag{19}$$

Similarly for the intensity measured by second photo-receiver disposed along direction of θ_2, we obtain

$$I_2(x_0, y_0, \alpha, h) = (e^{ikr_0}J_1(\theta_2) + e^{ikr_0^{''}}J_2(\theta_2))(e^{ikr_0}J_1(\theta_2) + e^{ikr_0^{''}}J_2(\theta_2))^*. \tag{20}$$

If the control object surface is lightened at constant parameters α, h of interferometer, then the expressions (19) and (20) present themselves a set of equations in respect to x_0, y_0. In the second case, when the reflector is the control object surface and the lighting point is immobile, we obtain a set of equations for definition of α, h.

Similar discussion could be also performed in the case of great sizes of beam splitter. So, by measuring, for example spatial circular frequency ω along the directions θ_1 and θ_2, we obtain a set of transcendental equations for determination x_0, y_0 or α, h, which can be solved by any known method.

To study possibilities of the developed method, mathematical modeling of measurement process has been conducted. The used diffraction lattice had a period of $\tau = 120\lambda$, but two photo-receivers were located at the distance of $R_0 = 0.2$ from the beam splitter along the directions of $\theta_1 = 2.4^o$ and $\theta_2 = 3.4^o$.

Let us consider first case, when it is required to define x_0, y_0. The location of the reflector is characterized by the parameters of $\alpha = 0.07^\circ$ and $h = 4.7 \cdot 10^{-4}$ m. The modeling results are present in Figure 2. The curves in Figure 2a show dependences of intensity at the point θ_1, calculated by using formula (19), on value of displacement of the lighting point on control object surface along *y*-axis at different values of the point displacement along *x*-axis.

Original site of the point of the control object surface is defined by coordinates of $x_0 = 0.1$ m and $y_0 = 0.2$ m.

Curve x_1 corresponds to initial site of the point on axis x_0, curve x_2 corresponds to displacement of $\Delta x = 3 \cdot 10^{-4}$ m and so on. Figure 2b presents similar curves for direction of $\theta = \theta_2$. As it follows from this figure, the intensity in the regions, corresponding to various main maxima at displacement of control object surface, changes differently, hence simultaneous measurement of the intensity along two directions gives a possibility to determine both coordinates x_0 and y_0.

Let us assume that measurements of the intensity along directions of θ_1 and θ_2 give the values of $I_1 = 0.79$ and 0.75, respectively. As it follows of Figure 2a, the intensity I_1 corresponds to a great number of values of Δx_0 and Δy_0, for instance $\Delta x_0 = 6 \cdot 10^{-4}$ m, $\Delta y_0 = 10^{-3}$ m; $\Delta x_0 = 1.2 \cdot 10^{-3}$ m, $\Delta y_0 = 1.45 \cdot 10^{-3}$ m and so on.

As it follows from the curve family of Figure 2b, the displacement can be also different at the $I_2 = 0.75$, namely $\Delta x_0 = 0$, $\Delta y_0 = 0.2 \cdot 10^{-3}$ m; $\Delta x_0 = 3 \cdot 10^{-4}$ m and $\Delta y_0 = 0.6 \cdot 10^{-3}$ m; $\Delta x_0 = 6 \cdot 10^{-4}$ m $\Delta y_0 = 10^{-3}$ m and so on. However, from a whole set of possible x_0 and y_0 only the value of $\Delta x_0 = 6 \cdot 10^{-4}$ m and $\Delta y_0 = 10^{-3}$ m occurs simultaneously in the both cases. Hence these values correspond to displacement of the control object surface.

We can also define coordinates x_0 and y_0, measuring spatial circular frequency ω. Figure 3 gives the calculation results on formula (18) of the circular frequency ω versus value of

Δy_0 at different values of Δx_0 for directions of θ_1 (Figure 3a) and θ_2 (Figure 3b), respectively.

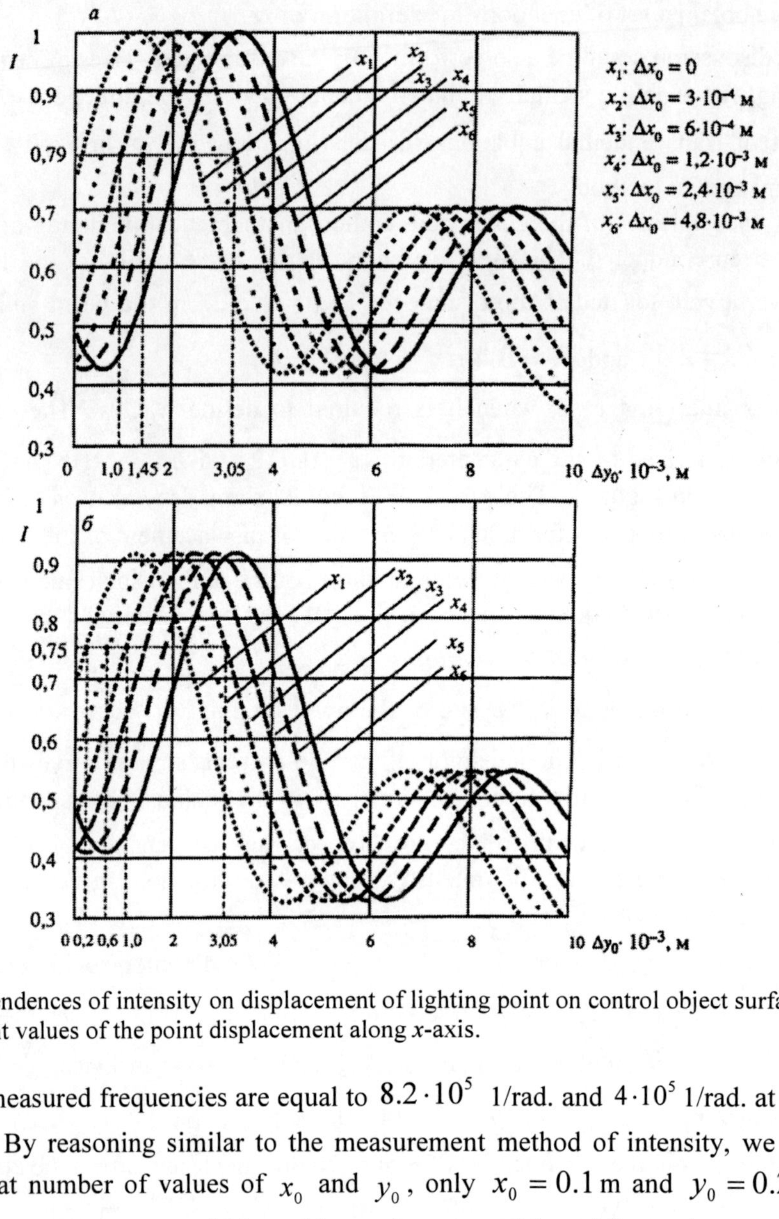

Figure 2. Dependences of intensity on displacement of lighting point on control object surface along y-axis at different values of the point displacement along x-axis.

Let the measured frequencies are equal to $8.2 \cdot 10^5$ 1/rad. and $4 \cdot 10^5$ 1/rad. at θ_1 and θ_2, respectively. By reasoning similar to the measurement method of intensity, we obtain that from the great number of values of x_0 and y_0, only $x_0 = 0.1\,\text{m}$ and $y_0 = 0.2\,\text{m}$ are the solutions of the equation set: $\omega(\theta_1) = f_1(x_0, y_0)$ and $\omega(\theta_2) = f_2(x_0, y_0)$.

By comparing the results of the displacement definition with help of the intensity measurement along two directions of θ_1 and θ_2, and by way of measurement of the spatial circular frequency along these directions, it should be noted that sensitivity and therefore accuracy is higher in the case of the intensity measurements compared to the measurement of frequency. However, the measurement ranges of displacements are significantly wider in the second case because in the first case non-uniqueness arises at lesser change of coordinates.

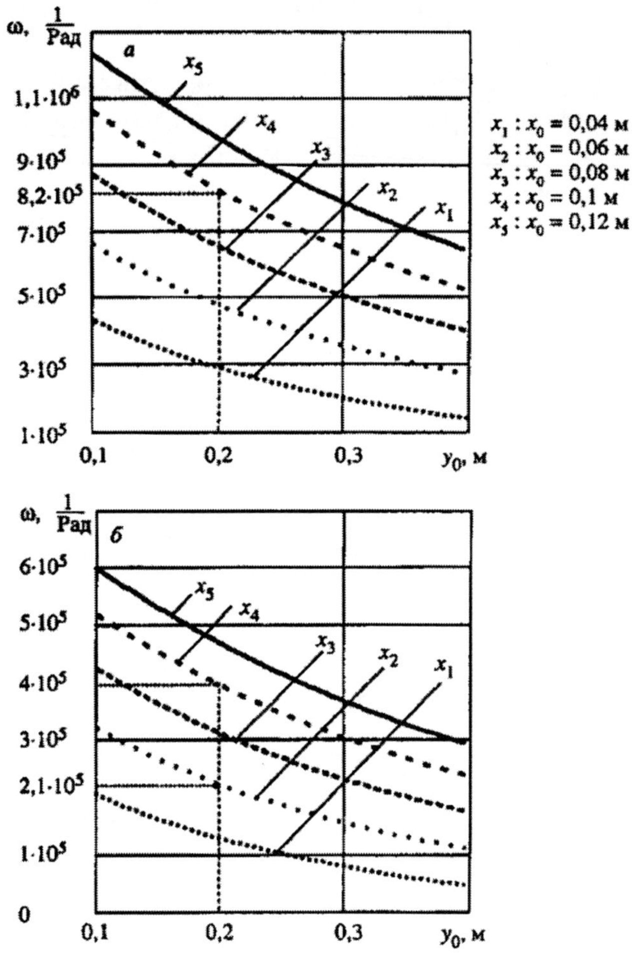

Figure 3. Calculation results of circular frequency ω *vs* value of Δy_0 at different values of Δx_0 for directions of θ_1 (a) and θ_2 (b).

It follows of Figure 2 that besides the values of $\Delta x_0 = 6 \cdot 10^{-4}$ m and $\Delta y_0 = 10^{-3}$, the values of $\Delta x_0 = 6 \cdot 10^{-4}$ m and $\Delta y_0 = 3.05 \cdot 10^{-3}$ m satisfy also the equations $I(\theta_1) = f_1(x_0, y_0)$ and $I(\theta_2) = f_2(x_0, y_0)$. At simultaneous measurement of ω and *I*, we obtain a possibility to expand the measurement range by keeping measurement accuracy.

Let us treat the second case, when the control object surface coincides with interferometer reflector. For example, we give the results, when occurs only linear displacement *h* without rotation ($\alpha = 0.0674^{\circ} = \text{const}$). These results are shown in Figure 4, where value of Δh is measured from 0 to λ along axis of abscissa and intensity in directions of θ_1 and θ_2 is measured along axis of ordinate.

As it follows from this figure, the measurement of intensity only in one direction (for instance θ_1) gives a possibility to define uniquely the difference Δh into limits of $\lambda/2$ (the range of Δh = (0.1 – 0.35) λ in Figure 4a). The measurement of intensity in the second direction θ_2 (Figure 4b) can be used to elucidate non-uniqueness, providing an increase of the measurement range.

Measurements along two directions are necessary also, when reflecting surface additionally to linear displacement also rotates on some angle $\Delta\alpha$. Algorithm of calculation of the parameters α and h is similar to described above the first case, when coordinates of the lighting point x_0 and y_0 are found.

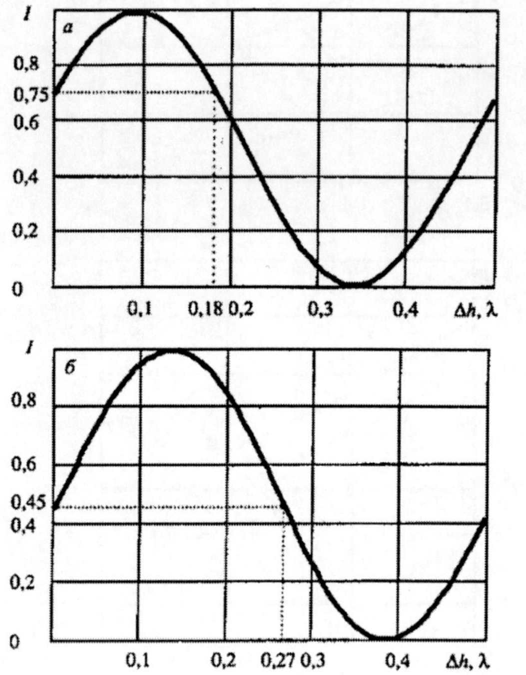

Figure 4. Results of intensity measurement, when occurs only linear displacement without rotation.

From results of the mathematical modeling, it follows that for measurement of small (part of the wave length of laser radiation) displacements it is expedient to use control object surface as reflector of interferometer. Application of the method of lighting the control object surface gives a lesser accuracy, but in this case a range of measured values of displacements expands. Moreover, in the last case a construction of measuring device simplifies significantly. Visible change of spatial frequency occurs at significant displacement of lighting point. Therefore, this method is expedient to use at great displacements of control object surface (order of several centimeters), and also to elucidate a non-uniqueness at application of other methods.

2. Novel Optic Interference Device

To extend functional possibilities of technical solutions [1 – 3] for account of ensuring change of measurement range of the displacements during measurements, we developed an optic interference device for the displacement measurements of object control surfaces [17].

The essence of the device [17] is explained by the scheme present in Figure 5. The device [17] consists of optically coupled and disposed in series source 1 of coherent optic radiation; optic system 2; control object 3 with surface 4; common base 5, on which there are rigidly fixed beam splitter 6, reflector 7 and photo-receiver 8, and also transducer 9. The base 5 has mechanism 10 for its displacement along optical axis between the object surface 4 of the control object 3 and external surface of the beam splitter 6.

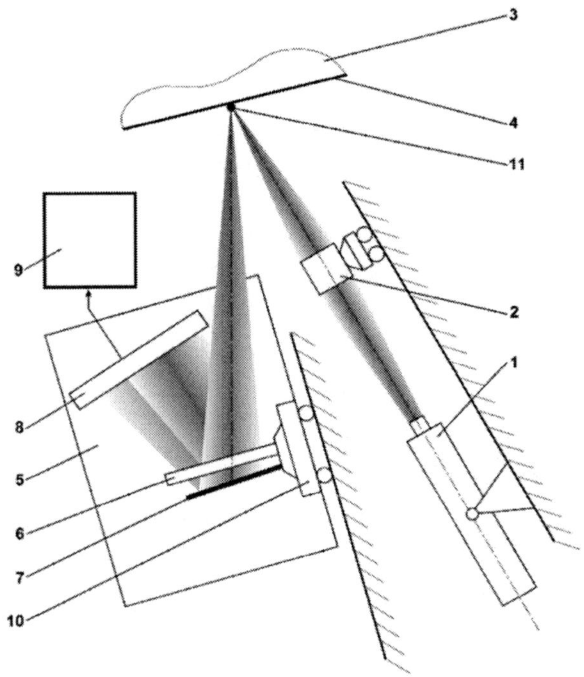

Figure 5. Scheme of optic interference device for displacement measurements of object control surfaces.

The optic scheme 2 focuses radiation to the surface 4 of the control object 3 in the form of lightening point 11. The beam splitter 6, reflector 7 and photo-receiver 8 are rigidly fixed to the base 5 at the fixed disposition one relatively of others. The transducer 9 is electrically coupled with the photo-receiver 8 and intended for definition of displacements of the surface 4 of the control object 3 based on the measurement results of optic field intensity into given region of interference picture by using the photo-receiver 8, and also to register the obtained results of displacement measurements.

The operation principle of the device [17] consists in the following. Before testing the control object 3, and respectively the displacement measurements of its surface 4, *a priori* we know a range of possible displacements arising during the tests. Therefore, the base 5 is disposed with fixed on it the beam splitter 6, reflector 7 and photo-receiver 8 by using the mechanism 10 at the distance R between the surface 4 of the control object 3 and external surface of the beam splitter 6, which corresponds to the given range of displacement change. This operation can be performed by using a dependence of sensitivity T on distance R between the surface 4 of the control object 3 and external surface of the beam splitter 6, defined in experiment during the metrological service (see, for instance [18–20]). We increase (decrease) the distance R at the increasing (decreasing) range of the measured displacements by means of corresponding change of the base 5 location.

The optic system 2 focuses radiation of the source 1 to the surface 4 of the control object 3 in the form of the lightening point 11. The lightening point 11 is the point source of radiation and disposes on surface of the control object 3. Then the radiation from the lightening point 11 falls on the beam splitter 6, which splits it into two beams, namely one beam reflects from surface of the beam splitter 6, and other reflects from surface of the reflector 7. The spatial coincidence of these beams takes place in the region of disposition of the photo-receiver 8, i.e., in the observation area of interference picture. The photo-receiver 8 registers intensity of optical field in the given region of interference picture, and the transducer 9, coupled with it, defines displacement of the surface 4 of the control object 3 by using measurement results of optic field intensity obtained by the photo-receiver 8, and also register the obtained results of the displacement measurement.

The device [17] realizes a novel measurement method of the control object surfaces, namely the method of lighting the control object surface by laser interferometer which has been proposed, described and well grounded in works [18–22].

It should be noted that the above-described device [17] is intended for using in laboratory conditions and could not be applied to solution of practical problems in structure of mobile diagnostic complexes due to a great labor expenditure of measurement preparation and conduction and decreasing of the result accuracy at fulfillment of the measurements in “field conditions”.

The great labor expenditure of measurement preparation and conduction is caused by that before the measurement performance we should carry out separately mounting and tuning each of optical elements to obtain interference picture in the information registration region. The decreasing of the result accuracy is caused by influence of external sources of optical radiation inserting distortions in the measurement results.

In order to elucidate these shortcomings preventing the device [17] application in the structure of mobile diagnostic complexes, we developed new technical solution, of which scheme is present in Figure 6.

The proposed device consists of base 1; device 2 for regulation and fixing the position, fixed to the base 1 and edge 3 of cylindrical frame 4. In internal hole 5 of the frame are located source 6 of coherent optic radiation with device 7 for regulation and fixing the source position and optic system 8 with device 9 for regulation and fixing the system position. Turning hinge 10 locates at edge 11 of the cylindrical frame 4 turned to surface 12 of control object 13. It joins the cylindrical frame 4 with supporting beam 14. Moreover the device consists of moving support 15 with device 16 for regulation and fixing the support position, which locates at surface 17 of the supporting beam 14. Supporting plate 18 fixes rigidly to the moving support 15. Moreover, there are opaque protective frame 19 and window 20. Beam splitter 22 and reflector 23, fixed rigidly one to other in fix location, are disposed into internal hole 21 of the frame together with device 24 for regulation and fixing their location, and also screen 25 with device 26 for the screen regulation and fixing.

Photo-receivers 27 dispose on the screen 25 in the given regions of interference picture. The supporting beam 14 is composite of the same type cylindrical elements 28, coupled one to other by connecter junction, for example thread junction 29. The cylindrical frame 4 and the supporting beam 14 are joined one to other by using device 30 to change and fix an angle location between them by using the turning hinge 10. Also, the proposed device consists of transducer, electrically connected with the photo-receivers 27 and intended for definition of displacements of the surface 12 of the control object 13 by using the measurement results of optical field intensity in the given regions of interference picture by the photo-receivers 27 and also for registration of the obtained results of measured displacements. The optic system 8 focuses radiation 31 of the source 6 on the surface 12 of the control object 13 in the form of lightening point 32. In its turn, this point is a point source of radiation, disposed on the surface 12 of the control object 13.

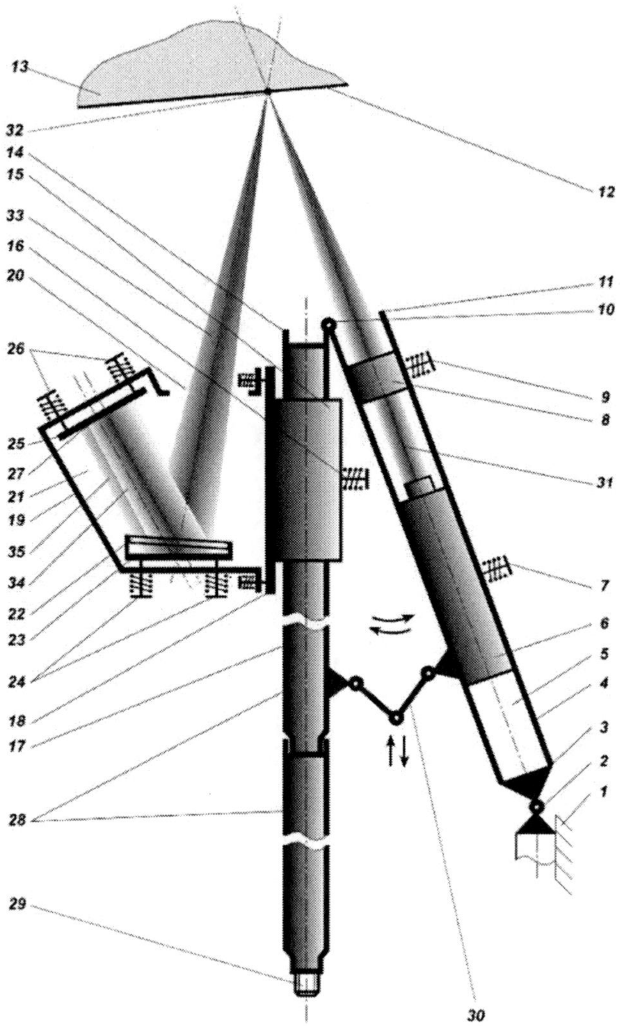

Figure 6. Scheme of device for application in structure of mobile diagnostic complexes.

The window 20 of the opaque protective frame 19 ensures optical contact of radiation 33 from the lightening point 32 with the beam splitter 22 and the reflector 23. Then the radiation 33 of the lightening point 32 falls on the beam splitter 22, which splits it into two beams, namely one beam 34 reflects from surface of the beam splitter 22 and other beam 35 reflects from the reflector 23. The spatial coincidence of these beams occurs in the area of the screen 25 disposition, i.e., in the region of observation of the interference picture.

Conclusion

The described device has been patented in Russia [23].

The results of experimental exploitation of the device model [23] have showed, that its constructive-composition scheme allowed one keep the device possibilities [1–3,17] and provided its application in the "field conditions" in the structure of mobile diagnostic complexes. Its construction, by keeping the possibilities of analog [17] ensures decreasing on 40 % of labor expenditure and time expanses on measurement preparation and conduction in the field conditions. It confirms expediency of its application in the structure of mobile diagnostic complexes.

The described technical solution can be successively applied during high-accurate measurements of linear and angle displacements of control object surface at fulfillment experimental studies of prospective constructions, buildings, etc., at estimation of their technical state and diagnostics in structure as stationary as mobile diagnostic complexes, at investigation of acoustic-emission processes in solids, damage processes in advanced construction materials, wave processes in layered constructions and constructions from anisotropic materials in machinery building, ship building, aircraft building, etc.

Acknowledgments

The work was supported in part by the Russian Foundation for Basic Research under grant No.13-08-00754.

References

[1] Miroshnichenko, I. P.; Parinov, I. A.; Rozhkov, E. V.; Serkin, A. G. Optic Device for Measurement of Displacements. Russian Patent No. 2373492, 11.20.2009.

[2] Miroshnichenko, I. P.; Serkin, A. G.; Sizov, V. P. Measurement Method of Linear and Angular Displacements. Russian Patent No. 2388994, 05.10.2010.

[3] Miroshnichenko, I. P.; Parinov, I. A.; Rozhkov, E. V.; Serkin, A. G. Optic Device for Measurement of Displacements. Russian Patent No. 2407988, 2010, 12.27.2010.

[4] Matveev, A. N. Optics; Vysshaya Shkola: Moscow, 1985; pp. 1 – 351 (In Russian).

[5] Batrakov, A. S.; Butusov, M. M; Grechka, G. P. Laser Measuring Systems; Radio i Svyaz: Moscow, 1981; pp. 1 – 456 (In Russian).

[6] Miroshnichenko, I. P.; Serkin, A. G. *Russian Journal of Nondestructive Testing*. 2007. *vol. 43(4)*, 234 – 239.

[7] Miroshnichenko, I. P.; Serkin, A. G. *Proc. Rostov State Transport University*. 2009. *No. 1(8)*. 85 – 90 (In Russian).

[8] Miroshnichenko, I. P. Control. Diagnostics. 2010. No. 1, 45 – 49 (In Russian).

[9] Alekhin, V. E.; Miroshnichenko, I. P.; Serkin, A. G. *Nanoindustry*. 2010, *No. 2*, 38-41 (In Russian).

[10] Miroshnichenko, I. P.; Parinov, I. A.; Rozhkov, E. V.; Serkin, A. G.; Sizov, V. P. In: Piezoelectrics and Related Materials: Investigations and Applications. Ivan A. Parinov (Ed.), Nova Science Publishers, New York, 2012, Chapter 7, pp. 238-282.

[11] Miroshnichenko, I. P.; Parinov, I. A.; Rozhkov, E. V.; Sizov, V. P.; Shevtsov, V. A. In: Physics and Mechanics of New Materials and their Applications. Ivan A. Parinov, Shun Hsyung Chang (Eds.), Nova Science Publishers, New York, Chapter 13, 2013; pp. 145-154.

[12] Miroshnichenko, I. P.; Parinov, I. A.; Rozhkov, E. V.; Serkin, A. G. *Metallurgist*. 2006. vol. 50(7 – 8), 408 – 410.

[13] Miroshnichenko, I. P.; Serkin, A. G. *Metallurgist*. 2010. vol. 54(3 – 4), 189 – 191.

[14] Vinogradova, M. B.; Rudenko, O. V.; Sukhorukov, A. P. Theory of Waves, Nauka, Moscow, 1979, pp. 1 – 383 (In Russian).

[15] Felsen, L.; Markuwits, N. Radiation and Scattering of Waves, Mir, Moscow, 1978, Vol. 2, pp. 1 – 152 (In Russian)..

[16] Yanke, E.; Emde, F.; Lesh, F. Special Functions, Nauka, Moscow, 1977, 1 – 342 (In Russian)..

[17] Alekhin, V. E.; Miroshnichenko, I. P.; Serkin, A. G.; Sizov, V. P. Optic Device for Measurement of Displacements of the Control Object Surfaces. Russian Patent No. 2343402, 10.01.2009.

[18] Alekhin, V. E.; Miroshnichenko, I. P.; Nesterov, V. A.; Sizov, V. P. *Russian Journal of Nondestructive Testing*, 2007, vol. 43(9), 592 – 596.

[19] Alekhin, V. E.; Miroshnichenko, I. P.; Serkin, A. G. *Measurement Techniques*, 2008, vol. 51(10), 1085.

[20] Miroshnichenko, I. P. *Control. Diagnostics*. 2010. No. 2, 34 – 37 (In Russian).

[21] Alekhin, V. E.; Miroshnichenko, I. P.; Sizov, V. P. *Russian Journal of Nondestructive Testing*, 2007, vol. 43(2), 113 – 118.

[22] Miroshnichenko, I. P.; Serkin, A. G.; Sizov, V. P. *Nanotechniques*, 2008. No. 4, 56 – 59 (In Russian).

[23] Miroshnichenko, I. P.; Parinov, I. A.; Rozhkov, E. V. Optic Interference Device for Measurement of Displacements of the Control Object Surfaces. Russian Patent No. 2512697, 04.10.2014.

In: Advanced Materials - Studies and Applications
Editors: I.A. Parinov, S.-H. Chang, S.Theerakulpisut
ISBN: 978-1-63463-749-7

Chapter 26

COLORFUL FLASHING LED NIGHT PEARLS FOR MARINE APPLICATION

Chyi-Da Yang[1,*], Fong-Jheng Lin[1], Chia-Hsiang Chou[1], Yu-Cheng Kung[1], Cheng-Liang Huang[1], Min-Yen Yeh[1], Jiing-Kae Wu[2], Huoo-Yuan Jenq[3], Jenq-Der Chen[4], Chih-Yu Lee[1], Chiung-Hsing Chen[5] and Shun-Hsyung Chang[1]

[1]Department of Microelectronics Engineering, National Kaohsiung Marine University, Kaohsiung, Taiwan
[2]Department of Marine Engineering, National Kaohsiung Marine University, Kaohsiung, Taiwan
[3]Department of Fishery Production and Management, National Kaohsiung Marine University, Kaohsiung, Taiwan
[4]Department of Electronic Engineering, National Quemoy University, Kinmen, Taiwan
[5]Department of Electronic Communication Engineering, National Kaohsiung Marine University, Kaohsiung, Taiwan

ABSTRACT

Based on limited power supplies at sea and the phototaxis of various seasonal fish species, this study developed an energy-saving color modulating sequential flashing light-emitting diodes (LEDs) lighting devices suitable for offshore fishery for underwater environments. These equipments include a self-generated power supply using solar or wind power and a radio frequency transceiver module for setting the flashing duration of the super bright colorful LED for emergency flasher (aquatic LED night pearls) and full-color sequential flashing LED fish lamps (underwater LED night pearls) which can be used 40 meters underwater. The lamp was mainly composed of high-brightness and low-consumption LEDs of the three primary colors (red, green, and blue) with waterproof resistance to moisture and salt. A control panel on a floating platform was used for modulating light color and sequential flashing. This LED lamp capable of full-color

[*] Email: cdyang@webmail.nkmu.edu.tw

modulation can not only serve as a background light source for underwater cameras and as a nighttime warning light (usually red) for sea fishing grounds, but also can enhance the management of set, set-net and cage-aquaculture fisheries. By integrating offshore power, underwater noise monitoring, underwater cameras, and lighting, the fish catches in cage nets and set nets and the use of the nets can become observable. With the assistance of underwater monitoring devices, owners of these two types of coastal fisheries can attain suitable production management methods.

According to previous studies, fish reacts differently to lights of different wavelengths. Therefore, if the color of a light source can be adjusted to one that attracts a specific fish, that fish species can be attracted by the light source and approach it. To effectively apply this technology in coastal fisheries to improve the fish catches of set-net, this study comprehensively investigated and experimented on this full-color sequential flashing fishing lamp in terms of control system structure, underwater horizontal illumination, attenuation rate, and various fishes' reactions to different light colors. The basic experimental results confirmed that the features of this full-color sequential flashing fishing lamp, such as its energy efficiency, fishing capabilities, and environmental friendliness, can effectively enhance the production management and applications of coastal fisheries.

1. INTRODUCTION

The development of new fishing technologies is based on fundamental subjects such as opto-acoustics, marine optics, electronics, fish physiology, and fisheries science. The principles of fishing are based on assessing the individual behavioral reaction of fishes to various acoustic, optic, and electronic parameters, selecting appropriate parameters according to these behavioral characteristics for inducing or driving the individuals into groups, and then catching fish by using suitable fishing tackle.

The promising future of fisheries inheres in coastal fisheries that involve specialized personnel and technologies, institutionalized operation and management, exquisite fish marketing, and recreational fishing activities and production. Fisheries based on set-net and cage nets that integrate production, life, and ecology are the optimal fishing methods for managing resources, promoting fish farming, cultivating recreational fishing, and improving quality of life.

Figure 1. Aerial photographs of two set-nets (length of 500 m; width of 200 m) connected by guiding nets.

Set nets are a passive type of trap net that are long stationed fixedly in specific waters near the coast for guiding and catching migratory aquatic organisms, such as fish, that pass through the nets. Fisheries that use set-net tackle for catching fish are known as stationary set-net fisheries. Figure 1 presents an aerial photograph of set-nets.

The rationalization of business operations as well as the appropriate management and cultivation of resources are crucial for developing coastal and offshore fisheries. In addition, to facilitate operational efficiency, more automatic video surveillance, underwater sound source, and communication equipment must be introduced to set-net and cage-net fisheries for monitoring fishes and fish catches, and warning lights must be used for maintaining safe operations in fishing grounds and on vessels sailing surrounding the fishing areas.

Presently, the operation of a set-net fishery requires staff members to set sail to specific set-net fishing grounds to catch lured fishes daily or regularly (Figure 2). Because of the special environment at sea, the power required for the operation of fishing grounds is currently supplied by the dynamos of rafts or boats. For periods outside operation times, batteries are used as the major power supply. However, because battery capacity is limited, battery power is only used for maintaining fishing ground indicator lights. To prevent fishing ground indicator lights from failing because of battery exhaustion, staff members must replace batteries during fishing. Thus, this type of operation mode may result in the following production and management problems:

- Staff members must regularly sail to fishing grounds to replace batteries, increasing personnel costs, fuel costs, and therefore the entire operating costs of fishing grounds;
- when fishing ground environments cannot be effectively maintained because power supplies are interrupted or insufficient, the safety of boats passing by the fishing grounds may be at stake;
- the catches of set-net and surrounding vessels cannot be effectively controlled using remote monitoring methods.

Figure 2. Operation of current set-net fisheries.

To solve the problems mentioned above, attain rationalized management, and enhance the safety of sailing vessels and the environments of set-nets and cage nets, a highly efficient electronic lighting and telecommunication system is required for instant control of safety warnings and fishing conditions surrounding the fishing grounds.

In rationalized management, instant data from fishing grounds is provided for set-net and cage-net fishery businesses by using an effective automated system for controlling the fishing conditions in fishing grounds. To achieve efficient, scientific, and systematic operation of fishing grounds, decrease the operating costs (lowering the frequency of sailing) of coastal fisheries such as set-net and cage-net fisheries, and increase revenues, staff members can adopt underwater monitoring cameras to monitor whether any high-priced fish have entered the nets, shut the net exits, and sail to the fishing grounds to catch the fish. Additionally, underwater microphone systems can be adopted to monitor the movements of the vessels around the cage nets for deterring fish theft and decreasing catch loss.

1.1. Phototaxis of Fish

The phototaxis of fish generates behavioral reactions oriented toward light stimulation in fish. Positive phototaxis denotes orienting behavior towards light, whereas negative phototaxis indicates orienting behavior that deviates from light. Phototactic behavior is common among numerous fishes such as sauries, sardines, pacific round herrings, chub mackerels, Japanese horse mackerels, ocellaris clownfish, Old World silversides, and needlefish.

The phototactic process of fish can be roughly divided into two stages: in the first stage, fish are stimulated by light and swim towards the light; in the second stage, fish swim below the light.

However, after a period of time, phototactic fish swim away from the light because of adaptation to light, fatigue, or environmental variations. The phototaxis of the same type of fish varies according to growth stage, sex, food intake, and swimming bladder structure. For example, the phototaxis level of juvenile fish is greater than that of adult fish, fish with small food intake are relatively easier to lure with light, and several fishes such as sauries exhibit less phototaxis when bearing eggs. Additionally, fish with different swim bladder structures have different floating levels after being lured and grouped together.

The sensitivity curves fish eyes for light with different wavelengths. Demersal fish can only sense a narrow spectrum of light (approximately 410 – 650 mm), whereas pelagic fish can sense a wider spectrum of light (approximately 400 – 750 mm). Only a certain range of illumination is suitable for fish, and therefore fish gather in the aquatic layers with this illumination range.

The illumination suitable for fish to gather varies with changes in environmental conditions. For example, when irradiated from a light source above water, sauries gather and stay approximately 2 m below the light source for long time. When irradiated by an underwater light source, even if the quantity of light is smaller than that of an above water light source, fish swims around and towards the underwater light source but leaves the light source after staying there for a short time. Fishes with strong phototaxis such as sardines, chub mackerels, sauries, and sand lances exhibit strong shoaling and schooling tendencies.

Fish often stay under a light source for a long time for protecting themselves, eating bait, or finding companions, and therefore exhibit shoaling and schooling reactions.

The phototaxis of fish is related to water temperature and moonlight. Generally, phototaxis decreases with rising water temperature and increases with lowering water temperature. Strong moonlights result in low phototaxis and poor shoaling and schooling effects. Increasing the illumination of light sources strengthens shoaling and schooling effects.

1.2. Light Fisheries That Take Advantage of the Phototaxis of Fish to Lure and Catch Fish Schools

Fishes are distributed in various aquatic layers of the ocean. Fishes living in the bottom of the ocean are commonly called demersal fish, which include large and small yellow croakers, and silver pomfrets. Fishes living near the sea surface are called pelagic fish, and include mackerels, chub mackerels (commonly known as "oil barrel fish"), and pacific herring (commonly known as herring). Pelagic fish are relatively more difficult to catch because they generally have keen vision, developed lateral lines, and body shapes suitable for rapid swimming.

Light influences substantially the habits of fish, as well. The movements of fish in the morning and at night under natural light as well as their reactions to the irradiation of artificial light have been observed. The results indicated that the irradiated fish (particularly eels and mackerels) react differently with changes in light color.

Since people have found that light influences the habits of fish, in addition to the use of demersal fish resources, fisheries with light sources and seine nets have become a focus of development efforts worldwide. Light fishing is an advanced method that involves luring and gathering fish by light sources and catching them by seine nets.

If fishermen know the wavelengths of the light colors that fishes are attracted to, they can lure fishes to nets, increase catches, and attain highly efficient fishing outcomes on boat. Thus, unnecessary costs and idle periods can be reduced.

Currently, the waterproof lamps used by most fishery owners are combined with buoys, and power is supplied by batteries. However, because batteries must be replaced regularly, current waterproof lamps are not convenient, effort-saving, power-saving, or environmentally friendly. Furthermore, if the buoy light bulbs on nets are not maintained properly, malfunction or insufficient brightness of bulbs may affect the safety of the vessels sailing near the coast at night. In addition, cases in which small boats mistakenly enter set-net or cage-net areas have occurred from time to time. Such accidents have often led to disputes regarding damaged propeller blades, cables, or fishing tackle.

2. Motivation and Research Objectives

Previous studies have found that fish react differently to light sources of different wavelengths. Therefore, the color of light sources can be adjusted according to the colors that are favored by different fishes to lure those fishes.

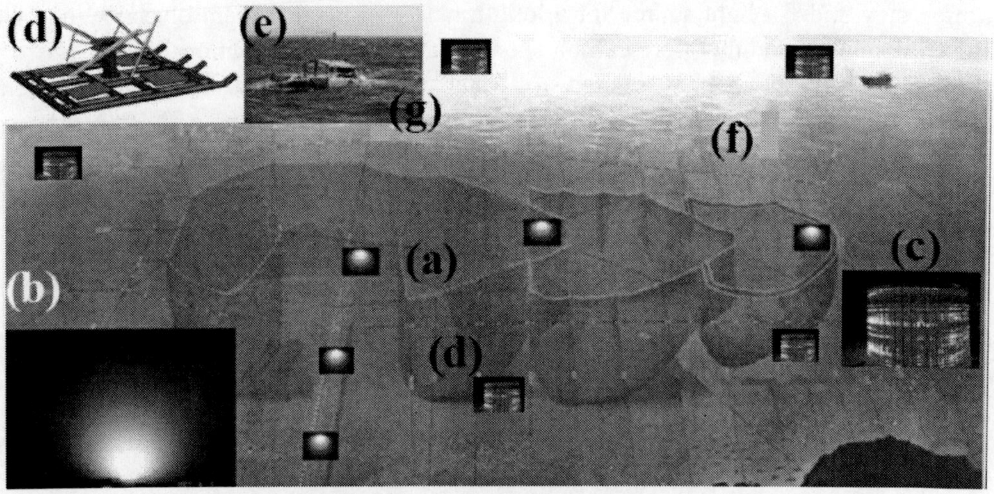

Figure 3. Illustration of the application of colorful flashing LED night pearls in set-net fishery: (a) set-net, (b) underwater LED night pearls, (c) aquatic LED night pearls (red), (d) wind power supply, (e) solar power supply, (f) underwater video camera, (g) hydrophones.

To achieve the goal mentioned above, a set of electronic lamps and signal transmission components that are suitable for use on sea surfaces and in underwater environments must be designed. Considering the limited power supply at sea, the light devices to be designed must consume low levels of power. In addition, multiple light-emitting diodes (LEDs) with high power and low power consumption have been developed because of advances in semiconductor fabrication process in recent years. LEDs have been applied widely both in scientific research fields and in daily life because they are small, power-saving, durable, and environmentally friendly. Current LED technology has enabled technicians to mix differently colored lights to acquire multiple colors. Therefore, this study aimed to design ideal electronic lamps based on studies on LEDs.

Because of the flourish of the LED industry, researchers can apply LED technologies to fisheries research. LEDs can be adopted to investigate the utility of color determination in fish and the effects of light intensity on fish injury, trapping, and dispersal. Consequently, exploring the phototaxis of fish by using LEDs will yield great benefits for fisheries. In this study, a set of electronic lamps and signal transmission components suitable for use both on sea surfaces and in underwater environments was designed. Moreover, a set of multicolor and full-color modulated LED lights based on pulse-width modulation (PWM) technology was developed. The designed lights can serve as background light sources for underwater cameras and as nighttime warning lights (generally red) for fishing grounds. Additionally, the lights can be used to develop underwater lamps for collecting or dispersing fish based on the light colors to which the various fishes in the waters of set-net fishing grounds are attracted (Figure 3).

Due to limited power supply in the sea, this chapter develops an energy-saving color modulating sequential flashing LED lighting devices for offshore fishery. These equipments includes a self-generated power supply using solar or wind power and a radio frequency transceiver module for setting the flashing duration of the super bright colorful LED for emergency flasher (aquatic LED night pearls) and full-color sequential flashing LED fish lamps (underwater LED night pearls) which can be used 40 meters underwater. This new fishing technology integrates modern oceanic photoelectric techniques with the phototaxis of the target species and navigation emergence flasher suitable for human vision to increase the underwater catch per unit effort (CPUE) in set-net fishery and provide emergence flasher on the seaplane for the fishing zone. Through sustainable incorporation with the colorful flashing LED night pearls brings the fishing community to be the wealthy and green fishing village.

3. Design Principles and Analysis

To achieve real-time monitoring, this study employed LEDs' diversity in color-matching and adopted a microcontroller as the command center of an underwater lighting system. With light source components that generate red, green, and blue colors, users can easily change LED colors through computers to instantly modulate the color of the underwater lighting system.

The full-color LEDs in this study are aimed to develop featured various characteristics such as low-power consumption, controllable light color, high brightness, and airtight packaging (waterproof and resistant to moisture and salt). Because various fishes have different reactions to different light colors, this study adopted these multicolor full-color modulated LEDs not only to provide background light sources for underwater cameras and serve as nighttime fishing ground warning lights (usually red), but also to conduct a series of investigations on fish species in the waters of set-net and cage-net fisheries based on how fishes are attracted to or dispersed by the LED lighting system.

The color mixing principles, LED color-mixing control, system design, waterproof structure design required by this study are elaborated in the following sections.

3.1. Color Mixing Principles

Generally, visible light is primarily composed of red (R), green (G), and blue (B) light. Therefore, light sources of different colors can be acquired by mixing the three light colors in different proportions (Figure. 4).

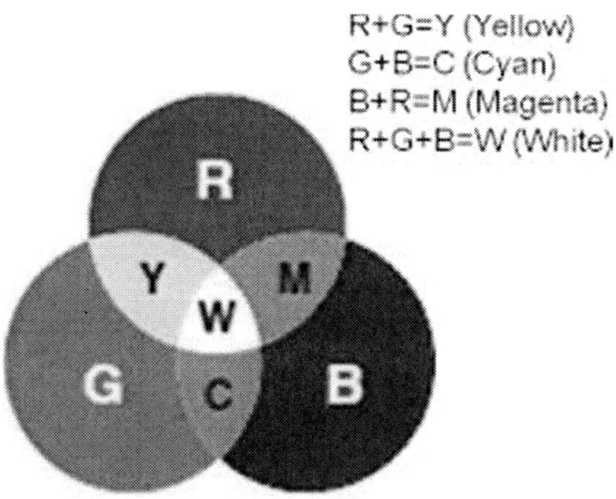

Figure 4. Basic hue diagram composed of RGB, the three primary colors of light.

Composed of three pure primary color LEDs tightly aligned on the same pixel level, a full-color LED was placed at a distance that produces a resolution smaller than the minimum angular resolution of the human eye for color mixing. When the resolution condition mentioned above cannot be fulfilled, LEDs colors are mixed in advance and packaged in transparent and semitransparent colloids.

3.2. LED Color-Mixing Control

3.2.1. Current Control

LEDs function under specific bias voltages and are driven by currents. Therefore, for LEDs to emit full-color lights, the three primary colors (RGB) must be mixed in different proportions to control individual LED brightness. In other words, the grayscale of each primary color is redistributed (grayscale refers to the number of the color levels of a single color). The methods of using currents to easily modulate individual LED brightness include: (i) fixing resistance and changing the methods for controlling voltage, and (ii) fixing voltage and changing resistance values. However, digital-to-analog converters are required for digital circuits such as microcontrollers to enable the application of LEDs to real-time monitoring. The hardware of analog circuits is relatively expensive, and resolution is limited in digital-to-analog converters.

3.2.2. Pulse-Width Modulation

PWM is a digital control mode that switches voltage output into a pulse mode to transform pulse width and number in order to acquire the required voltage and frequency.

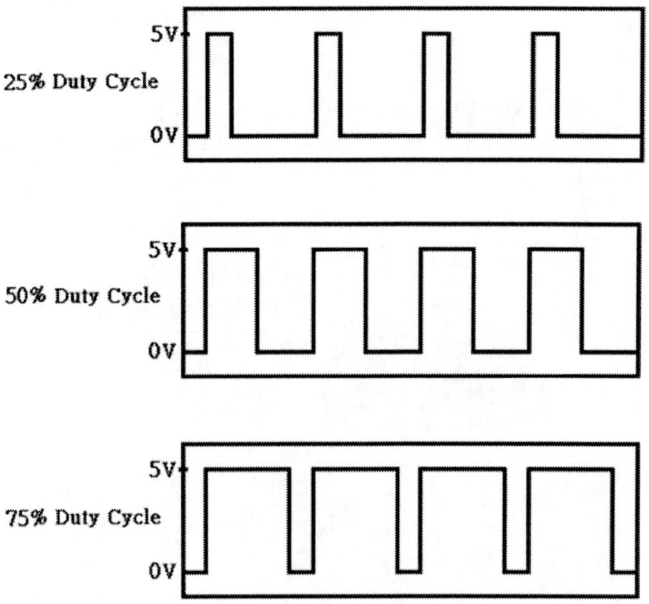

Figure 5. Using microcontroller to control PWM for generating different brightness levels.

Taking advantage of the visual inertia of human and fish eyes, this technique controls grayscale and periodically changes the optical pulse width (i.e., duty cycle) by modulating pulse width (Figure 5). As long as the repeating cycle of a light dot is short enough (i.e., a sufficiently high refresh rate, approximately 15 – 20 Hz), human and fish eyes perceive the vibration of the light-emitting pixel. Because PWM is relatively more suitable for digital control, this technique is adopted for grayscale control in almost every full-color LED nowadays as computers are widely used for displaying LED content.

3.3. System Design

The parameters for setting colors are composed of RGB light. For instance, each color has three statuses (full-dark, half-bright, and full-bright), and the number of colors that can be mixed depends on the number of color levels of a single color. Higher grayscales produce more refined images and richer colors, although they complicate the corresponding control system. Therefore, typical personal computers (PCs) were employed to control adjust colors by assigning the parameters required for various colors.

3.3.1. Basic End Control

To enable the rapid response of LEDs, a microcontroller was installed at the basic end of the circuit for enabling LEDs to modulate impulse width. Only a typical circuit was required for such a work, and the program could be revised repeatedly for fast and convenient adjustments, which met demands for repetitive verifications and debugging processes in an experiment. In addition, the microcontroller had a built-in memory for saving the parameters for setting colors. When an instruction was given from the exterior, export values could be changed through a table lookup.

3.3.2. Control Design

Initially, only the microcontroller and the basic export/import components were adopt for adjusting parameters. Until test results showed that full-color LEDs could be properly activated, a PC was then employed as a remote control end to transmit the required parameters through RS-232 (Figure 6).

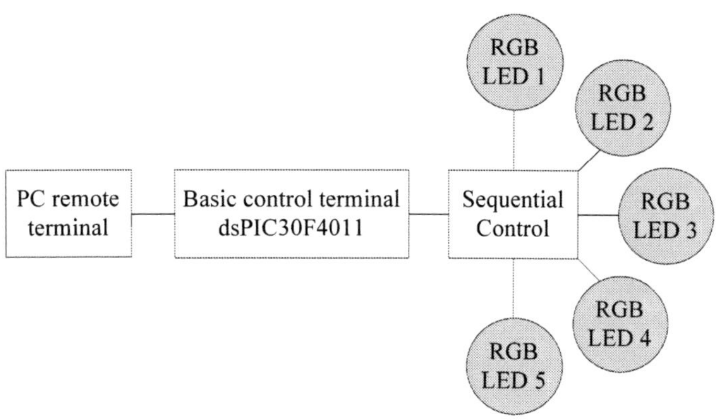

Figure 6. Hardware structure.

3.3.3. Serial Communication Control

The interface mentioned above was combined with a PWM program to modulate LED brightness (Figure. 7).

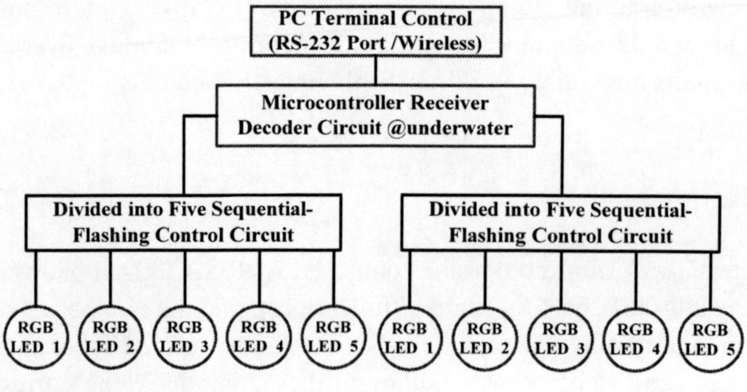

Figure 7. Hardware interface.

3.3.4. Labview Program Procedure

Labview was adopted as a controlling program at the PC end. The operating procedure and plan are illustrated in Figures 8 and 9.

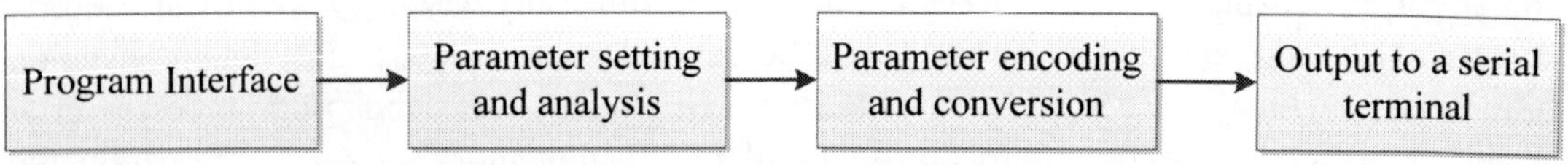

Figure 8. Labview program control procedure.

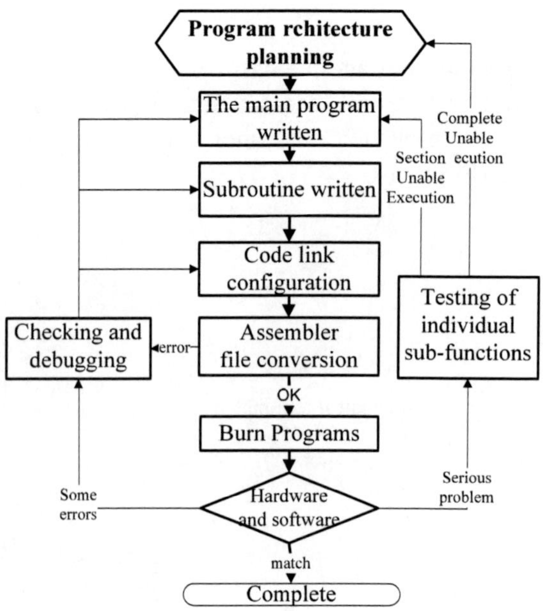

Figure 9. Program writing and procedure testing.

3.3.5. Program Procedure of the Microcontroller

Microchip dsPIC30F4011 was adopted as the core of the microcontroller and the C programming language was employed to control the basic circuit end. Figure 10 illustrates the program procedure. This study adopted the PWM function of the microcontroller to set the period of every color as 256 segments. Therefore, the total number of mixed colors was approximately 28 × 28 × 28 ≈ 1,670,000, which qualified the LEDs as full-color. The interface of LabVIEW was used for demonstrating different colors and the colorful flashing LED night pearls show the characteristics of color tunable as follows, respectively (Figure 11 and Figure 12).

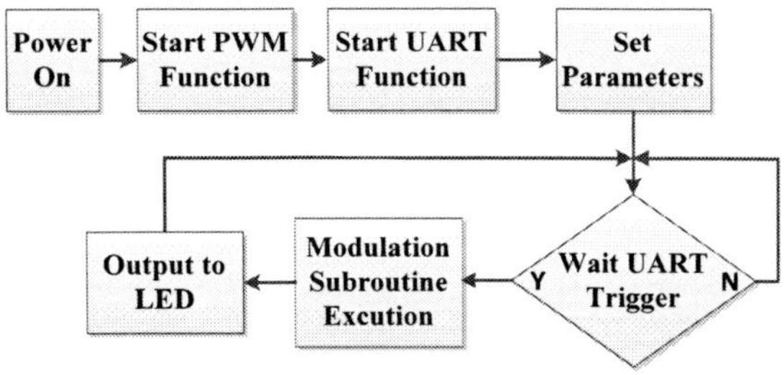

Figure 10. Program procedure of the microcontroller.

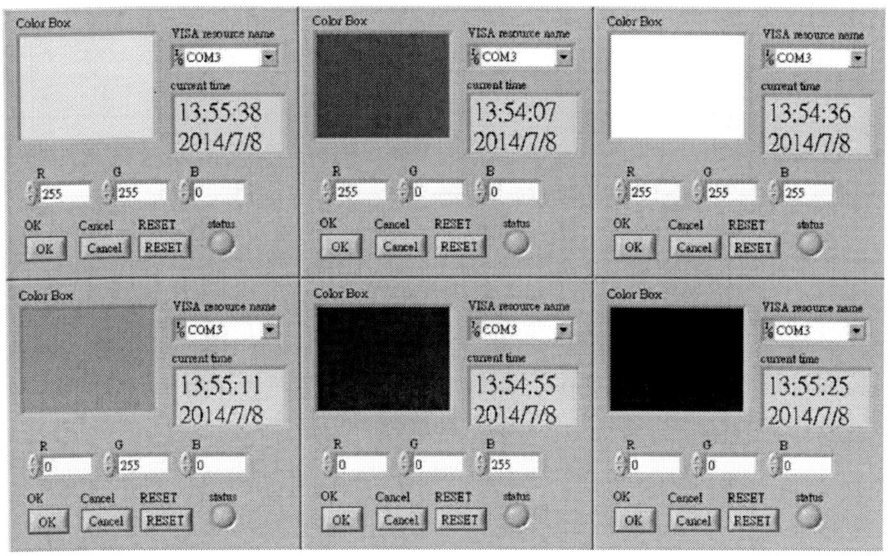

Figure 11. Setting various colors on the VB interface.

3.3.6. Design of Light-Emitting Diode Sequential Flashing

3.3.6.1. Sequential Design

Fishes have positive phototaxis and therefore can be lured to set fishing nets because of curiosity incurred by light stimulation; however, when fishes are constantly irradiated, the gradually decreasing stimulus effects of light on fish reduce curiosity among fish and affect fish catches.

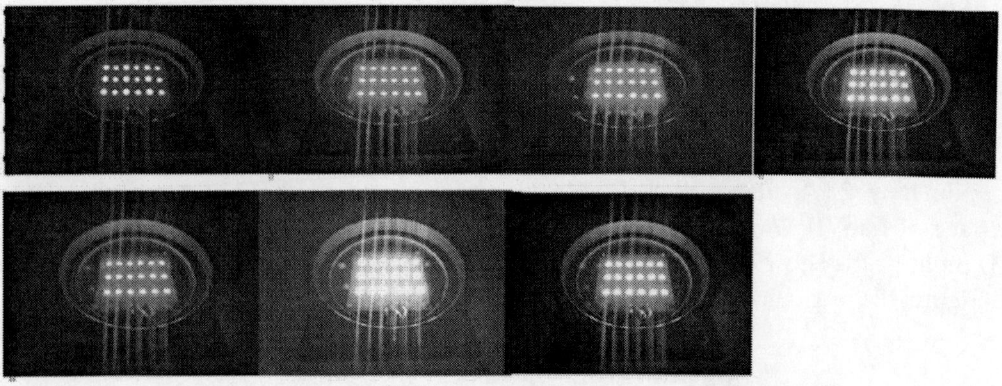

Figure 12. Colorful flashing LED night pearls with the characteristics of color tunable.

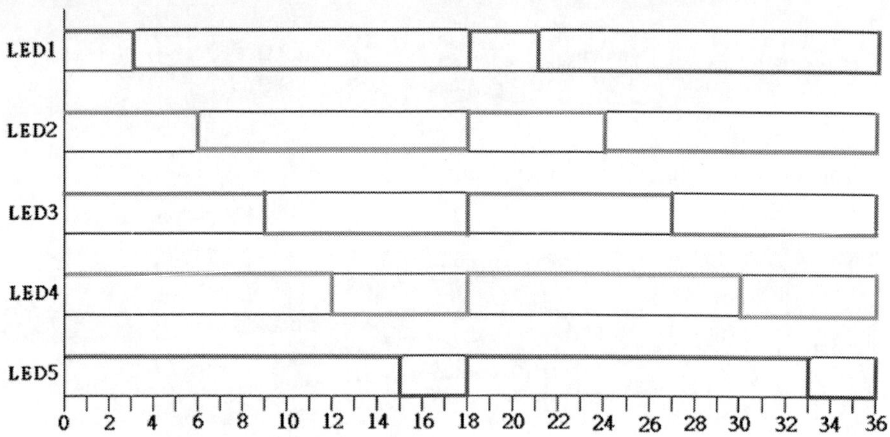

Figure 13. LED flashing sequence.

Therefore, a suite of sequential flashing LED lamps was designed in this study. Figure 13 illustrates the sequential design that enables LED lamps to flash like electronic scrolling texts to maintain the curiosity of fishes toward light sources, reduce power consumption caused by long-term illumination, and achieve power-saving effects.

3.3.6.2. Hardware Design

An AT89S51 microcontroller manufactured by Atmel Corporation was adopted to control the sequential function. Although the performance of the AT89S51 microcontroller does not surpass that of the Microchip dsPIC30F4011, the function of the AT89S51 microcontroller is sufficient for basic circuits.

A simple UART module built in an AT89S51 microcontroller was adopted to receive PC signals to start the sequential function. On the control panel for the sequential function, sequential signals from the AT89S51 and the PWM signals from dsPIC30F4011 were combined by an AND logic gate and exported to the full-color LEDs. Figures 14 and 15 present this procedure and Figure 16 shows the sequential characteristics of the .flashing LED night pearls.

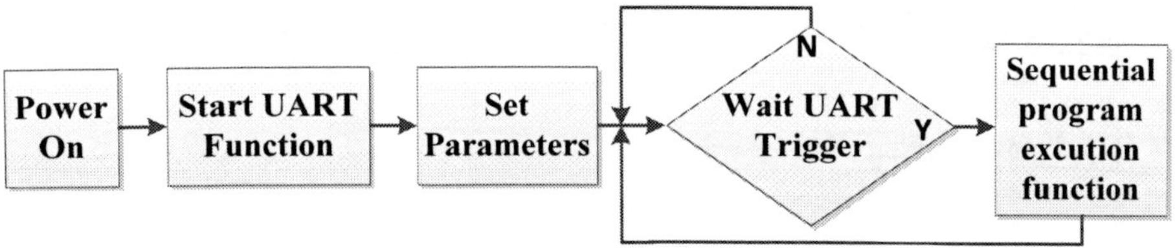

Figure 14. Microcontroller program procedure of the sequential circuit.

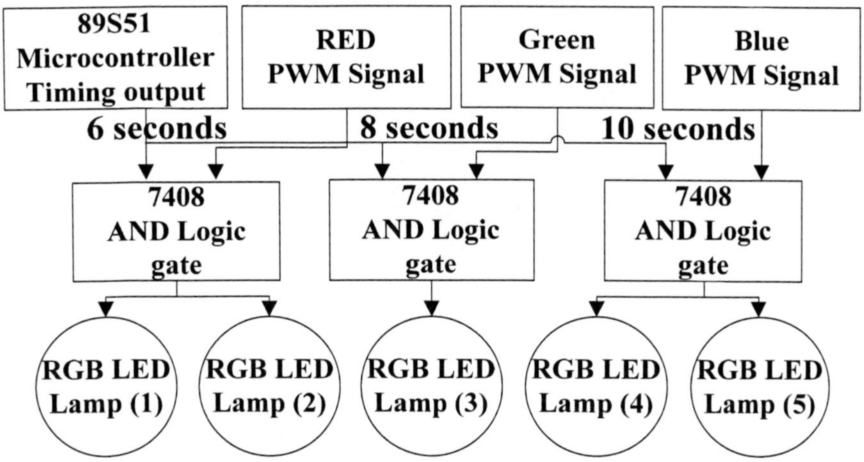

Figure 15. LED sequential flashing design.

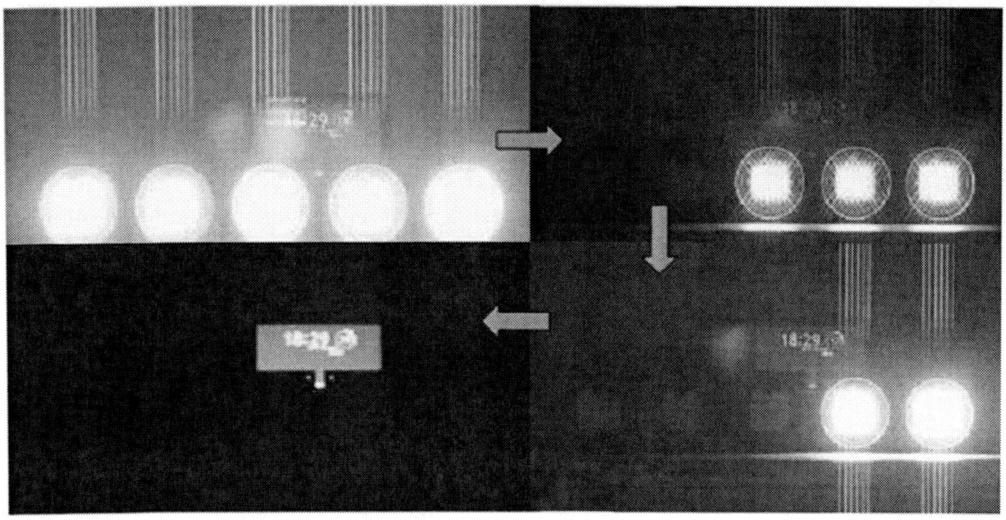

Figure 16. Colorful flashing LED night pearls with the characteristics of sequential flashing.

3.3.6.3. Waterproof Structure Design

When the PC end and hardware circuit were able to function normally, the circuit was packaged in a waterproof module that ensured the hardware circuit functioned properly inside.

Figure 17. Developed procedure for the waterproof structure.

Figure 18. High power RGB LED lamp with waterproof design.

To achieve this goal, a lamp case capable of working 40 m underwater that can carry all of the electronic components and ensuring their proper function in underwater environments must be designed. Consequently, experts of underwater packaging in relevant industries were consulted and invited to discuss and evaluate all the design requirements with the authors of this study. Subsequently, based on the specifications for the hardware components, a prototype model suitable for the required underwater components was designed for various tests and further development (Figure. 17).

To attain high underwater illumination efficiency, several high-power (3 W) LEDs capable of individually controlling the levels of RGB light were adopted and placed on a module substrate in a rectangular arrangement to achieve equidistant illumination and maximum illumination efficiency. Figure 18 presents the arrangement of the LEDs. The module must be sufficiently waterproof for working 40 m underwater. In addition, aluminum alloy must be adopted for fulfilling the cooling requirements of the module, and 8-mm tempered glass was used for LED irradiation faces to enable the entire module to resist water pressure and function properly.

The following requirements must be considered by designing underwater lamps:

- Equipped with a highly translucent panel
- Completely waterproof and dustproof
- Equipped with an O-Ring
- Resistant to sea water erosion
- Resistant to pressure at 40 m underwater
- Equipped with watertight connectors
- Enabling heat exchange between internal heat and external seawater
- Equipped with buckles for being affixed to set fishing nets
- Combined in a manner capable of expansion.

4. LED Lamp Testing and Discussions

4.1. Hardware Design for Measuring the Horizontal Illumination Attenuation of Underwater High-Power Full-Color LED Lamp Modules

When fish-luring lights transmit artificial light underwater, considerable light energy loss and light absorption and dispersion occur because of the high density of water and substantial suspended solids in seawater. For example, if the vertical illumination distribution of a 1500-w fish luring light produces 100% of brightness on water surface, only 50% of brightness is left 3 m under water, 33% is left 5 m under water, 25% is left 10 m under water, and 1% is left 35 m under water.

Horizontal illumination attenuates even faster than vertical illumination does. Illumination attenuation, either horizontal or vertical, varies according to an inverse-square law. Underwater illumination transmission of a light source can be expressed as $E = I\cos\theta / R^2$, where E is the illumination on a receiving surface, I is the luminous intensity of a light source, R is the distance between a light receiver and a source, and θ is the incident angle of a light source. Therefore, to ascertain the correct illumination of the underwater LED lamps, actual devices must be adopted to measure and calculate correct parameters to minimize the degree of error during application.

When the underwater high-power full-color LED illuminating module was completed, the characteristics and various parameters of the module were measured to improve the product in the future. To achieve the objectives mentioned above, a tank measuring environment was established. The parameters of the underwater full-color LED lamps to be measured and acquired are listed as follows:

(i) Horizontal illumination attenuation
(ii) Wavelength and color mixing spectral distribution.

Therefore, the established measuring model included the following devices:

(i) Electronic piezometer (for measuring water pressure)
(ii) Underwater illuminometer (for measuring illumination level)
(iii) Underwater spectrometer (for measuring light wavelength).

Figure 19 presents the measuring process, during which the measuring parameters were changed sequentially, the position of the LED module was then changed, and the measurements were recorded:

1) Prepare a hollow plastic cylinder with a length of 5 m and an inner diameter of approximately 40 cm and affix it firmly.
2) Cut a top lid from the cylinder for placing the LED module inside and adjusting its distance.
3) Paint the internal wall of the cylinder black to prevent light from reflecting through the wall.

4) Seal both ends of the cylinder with a waterproof baffle, and place an underwater illuminometer header on one end.
5) Pour tap water into the cylinder and close the top lid.

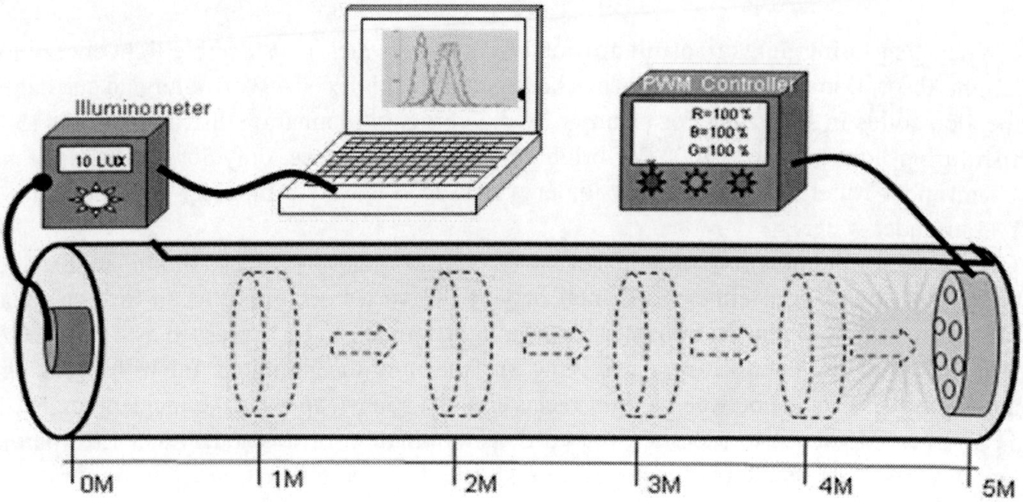

Figure 19. Underwater illumination attenuation measuring process.

4.2. Process of Measuring Optical Characteristics

The distribution of horizontal illumination attenuation levels in water and underwater wavelength and color mixing spectra at various distances were measured (Figure 20). To minimize external disturbances, the measurement was conducted at night to ensure that the accuracy of experimental data was not disturbed by sunlight or lamp lights.

Water impurities, sloshing, and receiver distance and position were crucial factors that affected the experimental results. Sloshing water resulted in various deflections in light of different colors, which led to incorrect luminous intensity measurements on the receiver and erroneous attenuation calculations.

Moreover, long-term use of the underwater high-power full-color LED electronic lamp produced heat that resulted in red-shift effects, causing wavelength and intensity shifts and inaccurate measurements. Therefore, the LED illuminating module was placed and left in the tank for 5 min until water waves completely stopped before the LED module was turned on and the measurements were obtained.

Light attenuation rates at various distances can be calculated by using the following formula: $\frac{A-B}{A}\times 100\%$, where A denotes the luminous intensity at the distance of 1 m, and B represents a luminous intensity at other distance. The detector in the DR-2000 illuminometer had different responsibilities to distinct wavelengths (Figure 21). Therefore, the wavelengths of various colors driven by an equal number of PWM segments had different candela (cd) readings on the illuminometer. The attenuation rates of the wavelengths of various colors can only be relatively compared (Table. 1 and 2), and arithmetic operations do not apply to these values.

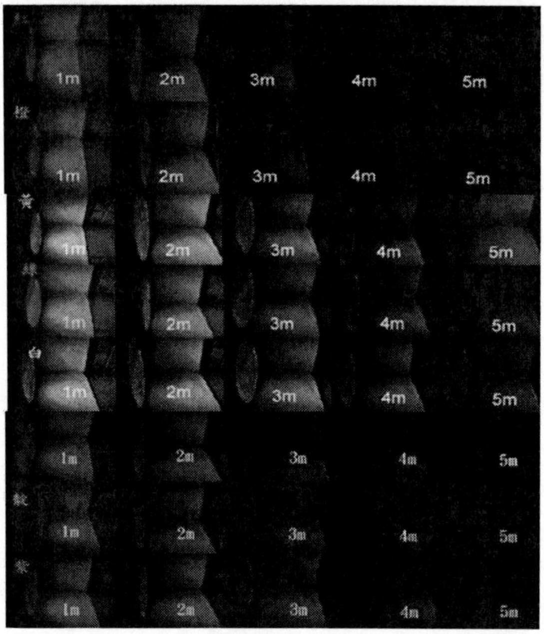

Figure 20. Brightness levels of the various modulated colors of the underwater high-power full-color LED electronic lamp at different distances in the water tank observed from a glass light hole.

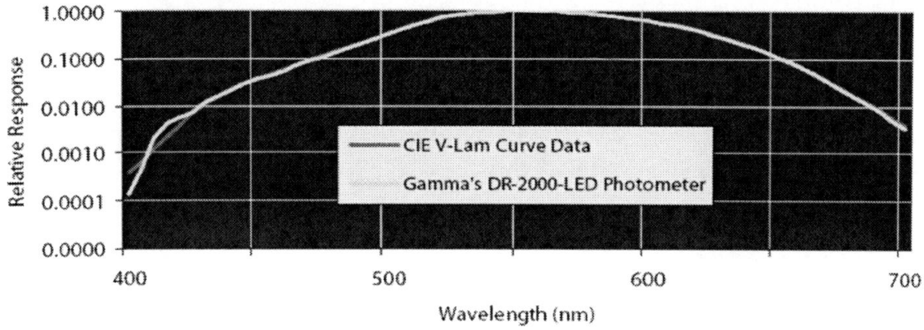

Figure 21. Relationship between wavelength and responsibility on DR-2000 illuminometer.

Table 1. Horizontal illumination attenuated results (reaction time: 1 s, unit: cd)

Distance / Color	1 m	2 m	3 m	4 m	5 m
Red	1.688	0.787	0.425	0.18	0.028
Green	7.953	5.3	4.093	2.966	1.88
Blue	1.955	1.306	0.942	0.648	0.339
White	10.838	7.147	5.378	3.84	2.35
Purple	3.68	2.33	1.575	0.901	0.504

Table 1. (Continued)

Color \ Distance	1 m	2 m	3 m	4 m	5 m
Yellow	9.685	6.54	4.784	2.97	2.038
Orange	2.4	1.348	0.892	0.483	0.253
Indigo	1.96	1.279	0.915	0.536	0.305

Table 2. Calculated results of horizontal illumination attenuation rates

Color \ Distance	2 m	3 m	4 m	5 m
Red	53.38	74.82	89.34	98.34
Green	33.36	48.54	62.71	76.36
Blue	33.2	51.82	66.85	82.66
White	34.06	50.38	64.57	78.32
Purple	36.68	57.2	75.52	86.3
Yellow	32.47	50.6	69.33	78.96
Orange	43.83	62.83	79.88	89.46
Indigo	34.74	53.32	72.65	84.44

4.3. Measurements and Discussion of Optical Characteristics

In addition to the three primary colors, yellow and white were mixed as the illumination was measured in the water tank. The duty cycle of the modulation was divided into eight segments, and therefore R8 denoted that the color red was at the highest part of its duty cycle and at full power. In the condition of identical PWM levels, the illumination values of the three primary colors were differently distributed (Figure 22), the results of which corresponded with the relationship between wavelength and responsibility on DR-2000 (Figure 21). The brightness of yellow and white increased because the two colors were generated by color mixing and overlapping.

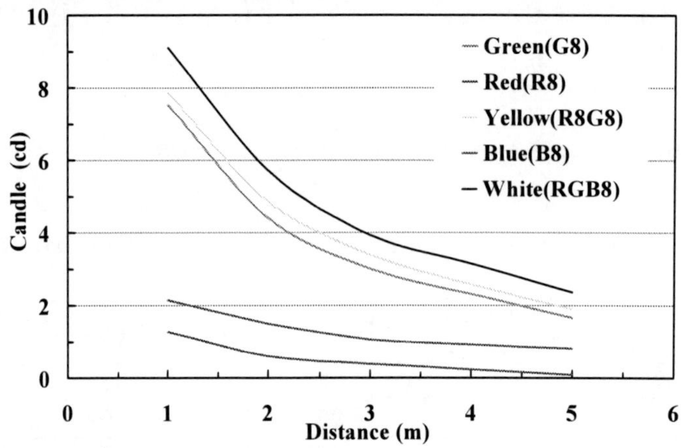

Figure 22. Relationship between illumination and distance of the various color lights produced by the underwater high-power full-color LED electronic lamp in the water tank.

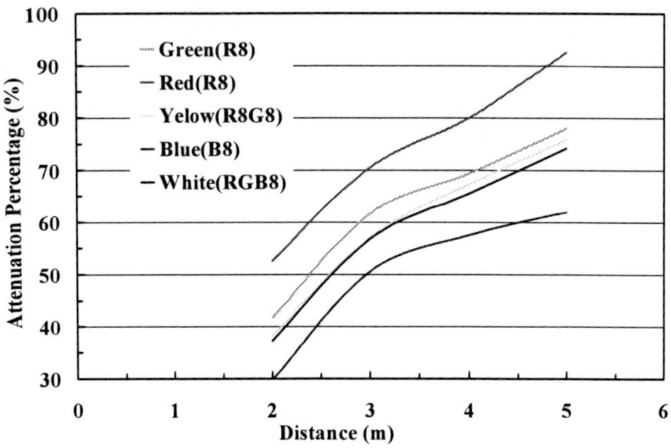

Figure 23. Illumination attenuation rates of the various colors of light emitted by the underwater high-power full-color LED electronic lamp calculated by using the distance of 1 m as a baseline.

With a distance of 1 m as a baseline, the attenuation rates of various modulated color lights in correspondence with various distances were calculated by using the attenuation formula. According to Figure 23, among the primary colors, red had the highest attenuation rates in correspondence with distance, followed by green and blue. The attenuation rates of the mixed colors were within the range of the attenuation rates of the primary colors. Based on the wavelength distribution of the five colors at the various distances shown in Figure 22, the attenuation rates of the relative intensities of the primary colors' dominant wavelengths corresponding to distance were calculated and are present in Figure 24.

Based on a comparison between the relative intensities of an individual dominant wavelength (Figure 24), the problem of varying responsibilities among various wavelengths on DR-2000 (Figure 21) was excluded. Therefore, after the attenuation rates were recalculated (Figure 25), the results showed that blue light had short wavelengths and high energy, and therefore could transmit for long distances. By contrast, red light had long wavelengths and low energy. Therefore, the energy of red light could be easily dispersed or absorbed by water impurities, which shortened transmission distances. Green light was situated within a range between blue light and red light.

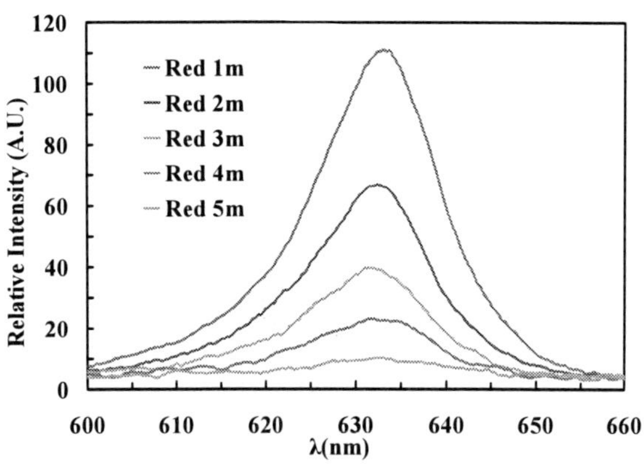

Figure 24. (Continued)

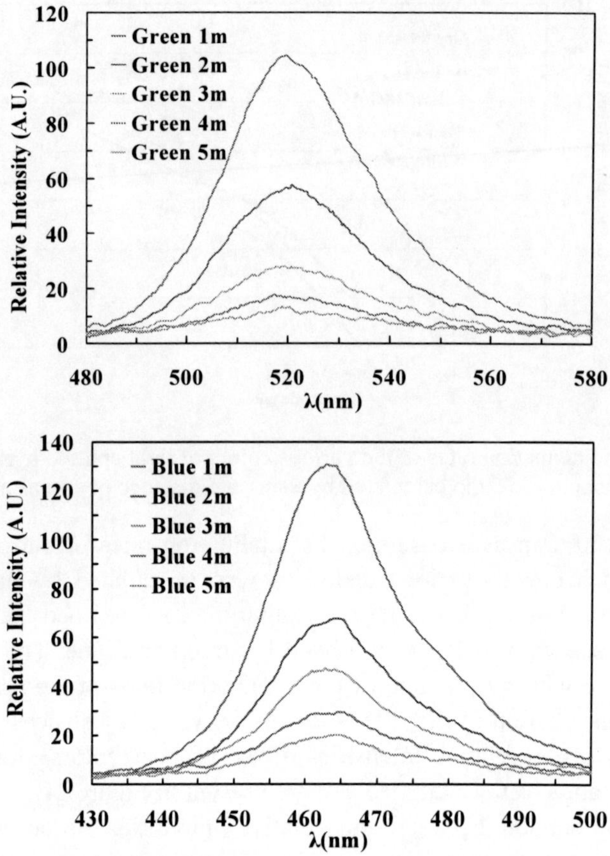

Figure 24. Responsive intensities of the dominant wavelengths of the three primary color lights emitted by the underwater high-power full-color LED electronic lamp at different distances in the water tank.

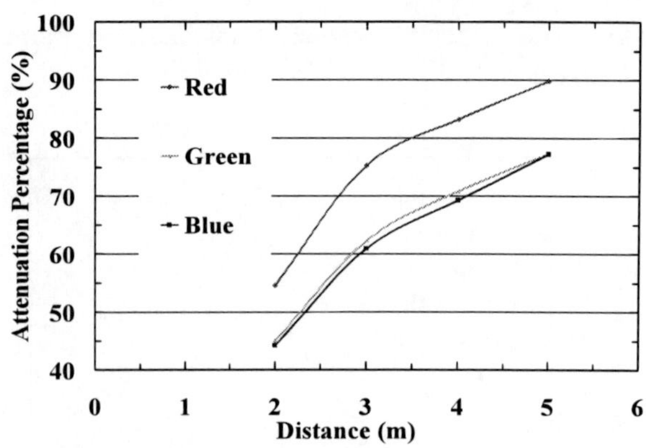

Figure 25. Attenuation rate curves of the relative intensities of the dominant wavelengths of the primary color lights emitted by the underwater high-power full-color LED electronic lamp in the water tank (1 m as the baseline).

Conclusion

Traditionally, mercury-vapor lamps of several tens of thousands of candelas are adopted for fish-luring lights. Although the wide spectra of such lamps can attract various types of fish, the excessive brightness damages the photoreceptor cells in the retina of the larval or juvenile fish that are exposed to such light for over 10 min. Blind larval or juvenile fish thus have no choice but to die. Although fish species essential to the economy can be caught, such a fishing method results in the unselective killing and sacrificing of larval and juvenile fish that could probably grow into fish essential to the economy.

Compared with traditional fishing lights, the underwater, high-power, full-color LED electronic lamp could lure and gather fish through phototaxis by using a small number of high-brightness and low-consumption LEDs that were arranged on a panel. Generally, during light transmission in water, illumination decreases with increasing travel distance, and the attenuation levels of different light colors vary as well. For example, blue light has short wavelengths and high energy, and therefore has relatively longer transmission distances. By contrast, red light has long wavelengths and low energy that can be easily absorbed. Consequently, red light has comparatively shorter transmission distances. Based on these optical characteristics, an underwater, high-power, full-color LED electronic lamp was developed in this study. This lamp could mix and generate colors of the specific light wavelengths and intensities to which certain economic fish species were particularly sensitive, and therefore protect fishery resources and meet fisheries regulations through selective fish luring and gathering.

- Product Features

1. Using advanced high-power LEDs to achieve power-saving and high-brightness luminous efficiency.
2. Using microcomputer control to vary light color for selective fish luring and gathering.
3. Using a human-computer interface to attain instant analysis and control of color light.

- Benefited Industries and Relevant Products to be Developed

In addition to being used in coastal fisheries such as light fisheries with seine nets, this product can be applied for selectively luring particular fish species in a specific place. The lamp enables light color modulation and can fulfill the specific requirements of products such as signal light fishing vessels and underwater lamps.

- Technical Features

By analyzing the phototactic behavior and spectral response range of fish, this study developed an LED lamp based on the VB human-computer interface and the integration of software and hardware. By using the serial transmission method to connect LEDs and the microcomputer controller, PWM technology was adopted to control the intensity and color mixing proportion of the LEDs and the interval of sequential flashing. By using this color-modulating device to lure and gather specific fish species and by heightening this effect through sequential flashing, fish catches can be increased and fish species can be selected during fishing for enhancing the competitiveness of offshore fisheries in Taiwan.

- Promotion and Application

1. Using advanced high-power LEDs to achieve power-saving and high-brightness luminous efficiency.
2. Using microcomputer control to vary light color for selective fish luring and gathering.
3. Using a human-computer interface to attain instant analysis and control of color light.

REFERENCES

[1] Jingcheng, Deng; Danjingpian, Zhumen yu Shizuo. *Single-Chip Introduction and Implementations*. Hong-Yu Books, Taipei, 1996.

[2] Yiming, Li; Danjingpian, Wechuli ji Yingyong. *Single-Chip Microprocessor: Principles and Applications*. Wenjing Publishing, Taipei, 1999.

[3] Mingbo, Lin; Shuwei, Xitong Sheji. *Digital System Design*. Chuan Hwa Book, Taipei, 1997.

[4] Guangzhong, Du; Chuanlie, Ziliao; Chuansong, Jishu Rumen. *Introduction to Serial Data Transmission Technology*. Chien Hong Publishing, Taipei, 1993.

[5] Shaolian, Mao; Yulei, Xingdong yu Buyujishu. *Fish Behavior and Fishing Techniques*. China Ocean Press, Beijing, 1985.

[6] Zhangying, Xia; Bulao, Xinjishu; Shengguandian, yu Buyu. *New Fishing Techniques: Sound, Light, Electricity, and Fishing*. China Ocean Press, Beijing, 1991.

In: Advanced Materials - Studies and Applications
Editors: I.A. Parinov, S.-H. Chang, S.Theerakulpisut
ISBN: 978-1-63463-749-7

Chapter 27

High-Q Biquadratic Bandpass Filter Synthesis Using Nodal Admittance Matrix Expansion

***H. D. Tran*[1], *H. Y. Wang**[1], *Q. M. Nguyen*[1], *L. T. Yin*[2] *and T. Y. Chen*[2]**

[1]Department of Electronic Engineering,
National Kaohsiung University of Applied Sciences, Kaohsiung, Taiwan, R.O.C.
[2]Department of Optometry,
Chung Hwa University of Medical Technology, Tainan, Taiwan, R.O.C.

Abstract

A systematic method for synthesizing high-Q biquadratic bandpass filters is proposed. The proposed approach is based on the prescribed transfer functions of synthesized circuits by using the nodal admittance matrix expansion method. Both cascadable voltage-mode and current-mode bandpass biquads can be synthesized. The Q factor of the derived filters can be adjusted without changing the natural frequency, so the high-Q bandpass filters are attainable. The obtained filters can be extended further to derive sinusoidal oscillators. The workability of the synthesized filters and oscillators are verified by PSPICE simulations.

1. Introduction

Filters and sinusoidal oscillators play essential roles in most electronic systems. They find numerous applications in instrumentation, measurement, control and communication systems. In the field of analog signal processing, the design of high-Q bandpass biquadreceived much attention since it could pass frequencies within a small frequency band

* Email: hywang@kuas.edu.tw

and attenuate frequencies outside the band [1–3]. The biquads can be cascaded to realize high-order filters.

Recently, the systematic methods adopting admittance matrix expansion with the motivation to synthesize active circuits from a mathematical viewpoint without any detailed prior knowledge of the circuit form were proposed in [4–7]. The matrix expansion process begins by introducing blank rows and columns, representing new internal nodes in the admittance matrix. Then nullators and norators are used to move the resulting admittance matrix elements to their final locations, describing either floating or grounded passive elements properly.

Thus, the final nodal admittance matrix (NAM) is obtained including finite elements representing passive circuit components and unbounded elements, so called infinity-variables [4], representing nullators and norators. Based on the above synthesizing method of active network, several filters and sinusoidal oscillators have been proposed recently [8–11]. In this chapter, a novel method for deriving cascadable voltage-mode and current-mode biquadratic high-Q bandpass filters is presented, by using the prescribed transfer functions of synthesized circuits for NAM expansion.

The Q factor of the derived filters can be adjusted without changing the natural frequency, thus the high-Q bandpass filters are attainable. The cascadable characteristic of a filter makes it easy to construct high-order filters without requiring any interconnecting network. By the slight modification of the derived filters, linear oscillators can be easily obtained. The oscillators possess the independent adjustable capability for their oscillation condition and oscillation frequency. The realization procedure is given and the derived circuits are verified with PSPICE simulator. The simulation results demonstrate the feasibility of the proposed approach.

2. SYNTHESIS METHOD OF NAM EXPANSION

The synthesis of high-Q bandpass filter using the symbolic framework of linear active circuits based on admittance matrix expansion was presented in [4 – 7]. To obtain an $n \times n$ admittance matrix from a starting 3×3 or 4×4 port matrix, one should select a suitable admittance matrix type in Table 1, where N and D respectively represent the numerator and denominator of the transfer function and Q represents an arbitrary admittance function parameter. The linked infinity-variable ∞_i represents the connected nullor.

In general, the Type I matrices in Table 1 is chosen for circuit with positive transfer function and the terms in the numerator appear also in the denominator. Type II matrices will be effective when the transfer function is positive and the terms in the denominator appear also in the numerator.

Type III matrices will be effective when the transfer function is negative and the terms in the numerator and denominator are independent and Type IV matrices will be effective where there is a combination of positive and negative terms in the numerator and/or denominator [5]. To carry out an $n \times n$ admittance matrix for circuit synthesis, several following techniques must be considered [5].

Table 1. Admittance matrices for circuits realizing [5]

a) Prescribed voltage transfer functions

Type I	Type II
$\left[\begin{array}{cc|c} 0 & 0 & 0 \\ 0 & \infty_1 & -\infty_1 \\ \hline -N & 0 & D \end{array}\right]$	$\left[\begin{array}{cc|c} 0 & 0 & 0 \\ \infty_1 & 0 & -\infty_1 \\ \hline 0 & -D & N \end{array}\right]$
$A_V = \dfrac{N}{D}$	$A_V = \dfrac{N}{D}$
Type III	**Type IV**
$\left[\begin{array}{cc|c} 0 & 0 & 0 \\ 0 & 0 & \infty_1 \\ \hline -N & -D & Q \end{array}\right]$	$\left[\begin{array}{cc|cc} 0 & 0 & 0 & 0 \\ 0 & 0 & \infty_1 & -\infty_1 \\ \hline -N_1 & -D_1 & P_1 & 0 \\ -N_2 & -D_2 & 0 & P_2 \end{array}\right]$
$A_V = -\dfrac{N}{D}$	$A_V = \dfrac{N_2P_1 - N_1P_2}{D_1P_2 - D_2P_1}$

a)

b) Prescribed current transfer functions

Type I	Type II
$\left[\begin{array}{cc|c} \infty_1 & 0 & 0 \\ 0 & 0 & -N \\ \hline -\infty_1 & 0 & D \end{array}\right]$	$\left[\begin{array}{cc|c} 0 & 0 & -D \\ \infty_1 & 0 & 0 \\ \hline -\infty_1 & 0 & N \end{array}\right]$
$A_I = \dfrac{N}{D}$	$A_I = \dfrac{N}{D}$
Type III	**Type IV**
$\left[\begin{array}{cc|c} 0 & 0 & -D \\ 0 & 0 & -N \\ \hline \infty_1 & 0 & Q \end{array}\right]$	$\left[\begin{array}{cc|cc} 0 & 0 & -D_1 & -D_2 \\ 0 & 0 & -N_1 & -N_2 \\ \hline \infty_1 & 0 & P_1 & 0 \\ -\infty_1 & 0 & 0 & P_2 \end{array}\right]$
$A_I = -\dfrac{N}{D}$	$A_I = \dfrac{N_2P_1 - N_1P_2}{D_1P_2 - D_2P_1}$

b)

(1) For a single case of a ∞-variable, arbitrary finite elements may be added to any position in the row and column where the ∞-variable is in.

(2) For the case of two ∞-variables in the same row or same column, arbitrary finite elements may be inserted anywhere in the row or column that the elements are in, and elements may be moved between the columns or rows.

(3) A row of zeros and column of zeros can be always introduced at the bottom and to the right of an admittance matrix; this corresponds to introduction of an isolated node.

(4) We can swap non-port rows and non-port columns without affecting the port behavior. It follows that we can add a row and column of zeros anywhere in the non-port part of an admittance matrix, moving existing columns to the right and existing rows downwards.

(5) We can link a node which corresponds to a column of zeros to any other node by a nullator, and a node which corresponds to a row of zeros to any other node by a norator.

(6) Relaxation of the zero inputs admittance condition for VCVS transfer functions and the zero output admittance condition for CCCS transfer function by moving certain matrix elements.

The NAM expansion method is applied to the synthesis of high-Q bandpass filters. We use two practical examples to demonstrate the procedures. They are addressed in Section 3.

3. SYNTHESIS OF THE HIGH-*Q* BANDPASS FILTERS

We first consider the transfer function of a voltage-mode bandpass biquad given in Equation (1), where ω_o and Q denote the center frequency and quality factor of the filter, respectively, and a_1Q/ω_o is the center-frequency gain. To synthesize a bandpass filter, we restrict ourselves to the use of less passive elements and the prescribed transfer function of the filter can be arranged as Equation (2), where each of a, b, c, d and e is a single admittance. In addition, we also consider the easy realization of high Q value and the independence between the terms in the denominator for designing bandpass filter. So a and c are capacitors and b, d and e are resistors in Equation (2).

$$T(s) = -\frac{a_1 s}{s^2 + \frac{\omega_o}{Q}s + \omega_o^2}, \tag{1}$$

$$T(s) = -\frac{ab}{d(a+e) + a(c-b)}. \tag{2}$$

To find the circuit implementation of the voltage transfer function in Equation (2), the Type III of the starting matrix of VCVS in Table 1(a) is chosen according to the NAM expansion method in Section 2. We can write the Type III matrix in Table 1(a) as

$$\begin{bmatrix} 0 & 0 & 0 \\ 0 & 0 & \infty_1 \\ -N & -D & Q \end{bmatrix} \rightarrow \begin{bmatrix} 0 & 0 & 0 \\ 0 & 0 & \infty_1 \\ -\frac{ab}{Q_1} & -\frac{d(a+e)+a(c-b)}{Q_1} & Q \end{bmatrix} \tag{3}$$

In (3), the function Q_1 is a first order function introduced to ensure that all of the elements have the dimensions of admittance. Let us take d as the first pivot, then $Q_1 = a + e$. Carry out pivotal expansion to (3), the result of the expansion matrix is shown as (4).

$$\begin{bmatrix} 0 & 0 & 0 & 0 \\ 0 & 0 & \infty_1 & 0 \\ 0 & -d & Q & -a \\ -b & b-c & 0 & a+e \end{bmatrix}. \tag{4}$$

The Q = a + d is chosen in order to simplify the function for synthesis [5]. By the virtue of ∞_1, we can add terms ±d and ±a to Equation (4) to satisfy the form of floating and grounding elements, as shown in Equation (5).

$$\begin{bmatrix} 0 & 0 & 0 & 0 \\ 0 & d & -d+\infty_1 & 0 \\ 0 & -d & d+a & -a \\ -b & b-c & -a & a+e \end{bmatrix}. \tag{5}$$

In order to make all the terms in Equation (5) appear in the form of admittance elements, the introduction respectively of two rows and columns of zeros to the bottom and right of (5) are needed. The two nullors are introduced as shown in (6).

$$\begin{bmatrix} 0 & 0 & 0 & 0 & 0 & 0 \\ 0 & d & -d+\infty_1 & 0 & 0 & 0 \\ 0 & -d & d+a & -a & 0 & 0 \\ -b & b-c+\infty_2 & -a & a+e & -\infty_2 & -\infty_3 \\ 0 & -\infty_2 & 0 & 0 & \infty_2 & 0 \\ 0 & 0 & 0 & 0 & 0 & \infty_3 \end{bmatrix}. \tag{6}$$

Following the method in Section 2, by the existence of the infinity variable ∞_2, terms ±b and –c in Equation (6) can be moved to row 5 and 6, respectively. When the infinity variable exists in the matrix, terms in Equation (6) are added to the corresponding positions. Relaxing the zero input admittance condition, we can move term ±b from row 0 to row 1 [5]. Finally, the obtained nodal admittance matrix of a high-Q bandpass filter after the process of expansion is expressed by (7).

$$\begin{bmatrix} b & 0 & 0 & 0 & -b & 0 \\ 0 & d+c & -d+\infty_1 & 0 & 0 & -c \\ 0 & -d & d+a & -a & 0 & 0 \\ 0 & \infty_2 & -a & a+e & -\infty_2 & -\infty_3 \\ -b & -\infty_2 & 0 & 0 & b+\infty_2 & 0 \\ 0 & -c & 0 & 0 & 0 & c+\infty_3 \end{bmatrix}. \tag{7}$$

The matrix in (7) describes a high-Q bandpass filter with three nullors, three resistors and two capacitors. The circuit topology of the obtained filter is shown in Figure 1. The nullators and norators are paired for feasible realization, and b, d, e, c and a are selected as passive components G_1, G_2, G_3, C_1 and C_2, respectively. If the nullors in Figure 1 are implemented by operational amplifier and CCII-s, the bandpass filter in Figure 2 is obtained. The obtained filter possesses output cascadable property and its transfer function and parameters can be expressed as below.

$$T(s) = -\frac{sC_2G_1}{s^2C_1C_2 + sC_2(G_2 - G_1) + G_2G_3} \tag{8}$$

$$\omega_0^2 = \frac{G_2G_3}{C_1C_2} = \frac{1}{R_2R_3C_1C_2}, \tag{9}$$

$$\frac{\omega_0}{Q} = \frac{1}{C_1}\left(\frac{1}{R_2} - \frac{1}{R_1}\right) \Rightarrow Q = \frac{R_1}{R_1 - R_2}\cdot\sqrt{\frac{R_2C_1}{R_3C_2}} \tag{10}$$

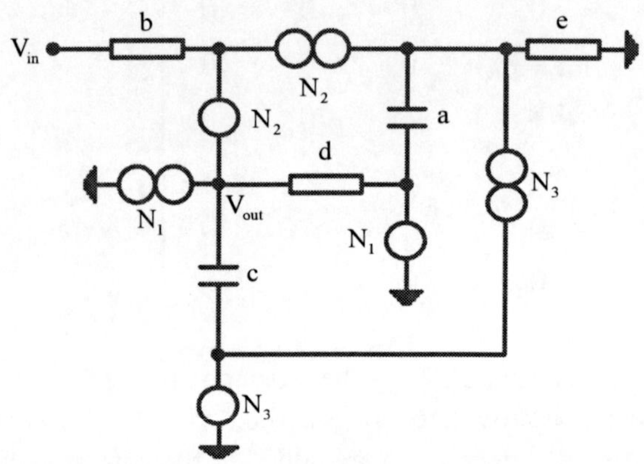

Figure 1. Synthesized nullor equivalent by the transfer function in Equation (2).

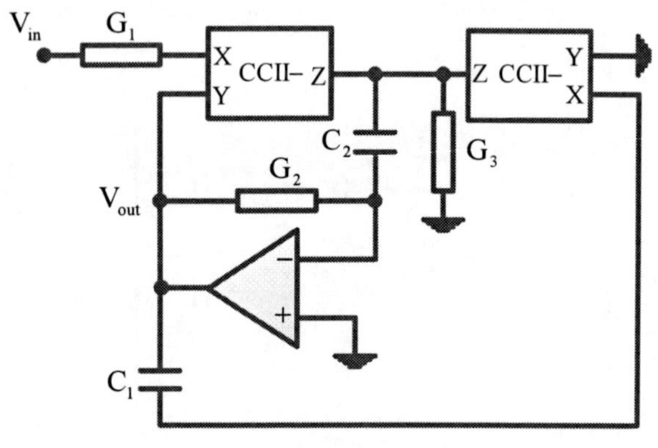

Figure 2. Second-order voltage-mode high-Q bandpass filter.

A possible design procedure is developed with the selection of equal capacitance values so that $C_1 = C_2 = C$ (components matching and tuning can be achieved in monolithic integration [12]); the R_1 can be used to decide Q, and C is used to decide ω_o; then Q and ω_0 become independently adjustable. Thus the high-Q bandpass filter is realizable. In addition, we can realize a sinusoidal oscillator by setting $V_{in} = 0$, as shown in Figure 3. By observing (9) and (10), it is found that the oscillation condition and oscillation frequency can be independently controlled by R_1 and R_3, respectively.

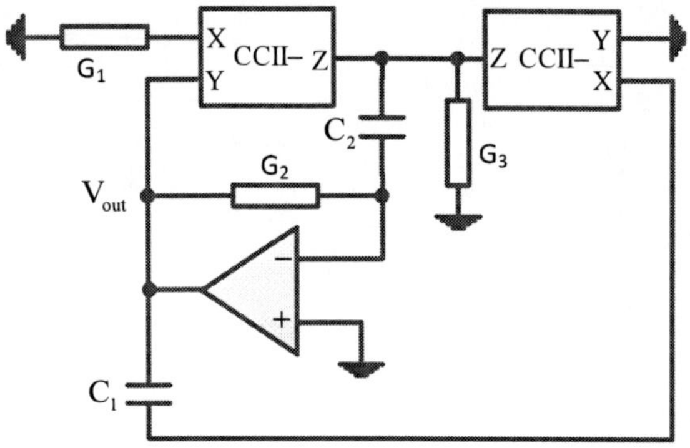

Figure 3. Sinusoidal oscillator obtained from the bandpass filter.

The second example is the synthesis of a current-mode high-Q bandpass biquad with its general transfer function shown in (11). The prescribed transfer function of the filter can be arranged as Equation (12) with the consideration of high Q value and tunability, where each of a, b, c, d, e and m is a single admittance. So a and c are selected as capacitors and b, d, e and m are selected as resistors in Equation (12).

$$T(s)=\frac{a_1 s}{s^2+\frac{\omega_o}{Q}s+\omega_o^2}, \tag{11}$$

$$T(s)=\frac{ab}{a(c+d-m)+de}. \tag{12}$$

To find the circuit implementation of the current-mode transfer function in Equation (12), the Type I of the starting matrix of CCCS in Table 1(b) is chosen according to the NAM expansion method [5]. We can write the Type I matrix in Table 1(b) as

$$\begin{bmatrix} \infty_1 & 0 & 0 \\ 0 & 0 & -N \\ -\infty_1 & 0 & D \end{bmatrix} \rightarrow \begin{bmatrix} \infty_1 & 0 & 0 \\ 0 & 0 & -\frac{ab}{Q_1} \\ -\infty_1 & 0 & \frac{a\left(c+d-m\right)+de}{Q_1} \end{bmatrix}. \tag{13}$$

The function Q_1 in (13) is a first order function introduced to ensure that all of the elements have the dimensions of admittance. Let us take d as the first pivot, then $Q_1 = -e$. Carry out pivotal expansion to (13), the result of the expansion matrix is shown as (14).

$$\begin{bmatrix} \infty_1 & 0 & 0 & 0 \\ 0 & 0 & 0 & -b \\ 0 & 0 & -d & c+d-m \\ -\infty_1 & 0 & -a & -e \end{bmatrix}. \quad (14)$$

Element c in set of elements of (14) describes circuit elements but it must be on the main diagonal. Elements –d and –e are not suitable for main diagonal terms, which should consist of a sum of positive element admittances. Now we use the technique 4 in Section 2 to create two columns of zeros (columns 5 and 6), and two rows of zeros (rows 3 and 6), moving existing rows 3, 4 to the bottom, and we introduce two nullors represented by ∞_2 and ∞_3 as shown in (15).

$$\begin{bmatrix} \infty_1 & 0 & 0 & 0 & 0 & 0 \\ 0 & 0 & 0 & -b & 0 & 0 \\ 0 & 0 & 0 & 0 & \infty_2 & 0 \\ -\infty_1 & 0 & -d & c+d-m & 0 & 0 \\ 0 & 0 & -a & -e & 0 & 0 \\ 0 & 0 & 0 & -\infty_3 & 0 & \infty_3 \end{bmatrix}. \quad (15)$$

By virtue of $\pm\infty_1$ and $\pm\infty_3$, we can bring terms –b, –m, –e from column 4 to column 6, respectively; by virtue of ∞_1, ∞_2 and ∞_3 we can add further elements to complete the symmetrical element sets for terms a, e, d, m. Finally, by virtue of ∞_3 and using technique 6 in Section 2, we complete the required form of admittance elements for term b. Then the matrix becomes (16).

$$\begin{bmatrix} m+\infty_1 & 0 & 0 & 0 & 0 & -m \\ 0 & b & 0 & 0 & 0 & -b \\ 0 & 0 & a+d & -d & -a+\infty_2 & 0 \\ -\infty_1 & 0 & -d & c+d & 0 & 0 \\ 0 & 0 & -a & 0 & a+e & -e \\ -m & -b & 0 & -\infty_3 & -e & e+b+m+\infty_3 \end{bmatrix}. \quad (16)$$

The matrix in (16) describes a high-Q bandpass filter with three nullors, four resistors and two capacitors. The nullor equivalent circuit of the derived filter is shown in Figure 4. Similarly, after selecting the elements b, d, e, m, a and c as passive components G_1, G_2, G_3, G_4, C_1 and C_2, respectively, and realizing the nullors in Figure 4 by operational amplifier and CCII-s, the bandpass filter in Figure 5 is obtained. The obtained circuit in Figure 5 possesses input cascadable property and its transfer function and parameters can be expressed as below.

$$T(s)=\frac{sC_1G_1}{s^2C_1C_2+sC_1(G_2-G_4)+G_2G_3}; \quad (17)$$

$$\omega_0^2=\frac{G_2G_3}{C_1C_2}=\frac{1}{R_2R_3C_1C_2}; \quad (18)$$

$$\frac{\omega_0}{Q}=\frac{1}{C_1}\left(\frac{1}{R_2}-\frac{1}{R_4}\right)\Rightarrow Q=\frac{R_4}{R_4-R_2}.\sqrt{\frac{R_2C_2}{R_3C_1}}. \quad (19)$$

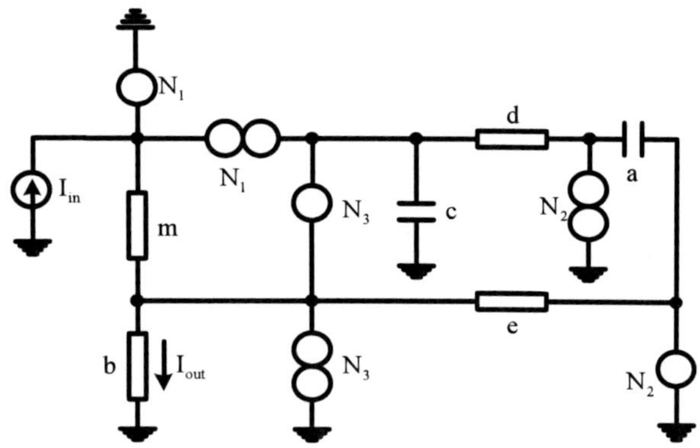

Figure 4. Synthesized nullor equivalent by Equation (12).

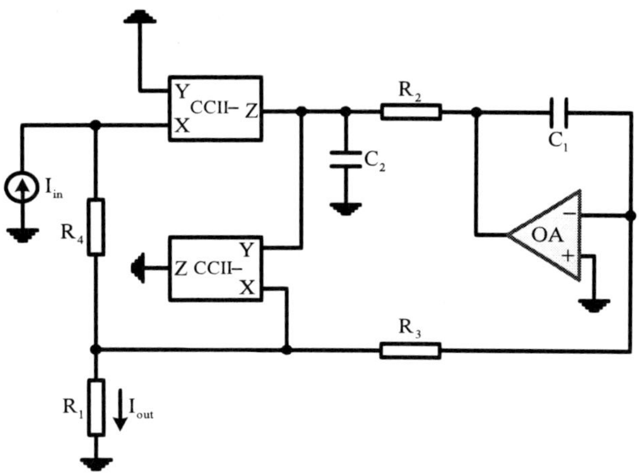

Figure 5. Second-order current-mode high-Q bandpass filter.

The center-frequency gain can be expressed as $R_2R_4/R_1(R_4 - R_2.)$. A possible design procedure is developed with the selection of equal fixed capacitance values so that $C_1 = C_2 = C$. The center-frequency gain can be independently adjusted by R_1 without affecting ω_0 and Q. Without disturbing each other, the ω_o and Q can be independently adjusted by C and R_4, respectively. So the high-Q bandpass filter can be realized. In addition, due to the identical characteristic equation, we can realize a sinusoidal oscillator with $I_{in} = 0$, as shown in Figure

6. It is easy to find that the oscillation condition and oscillation frequency can be independently controlled by R_4 and R_3, respectively.

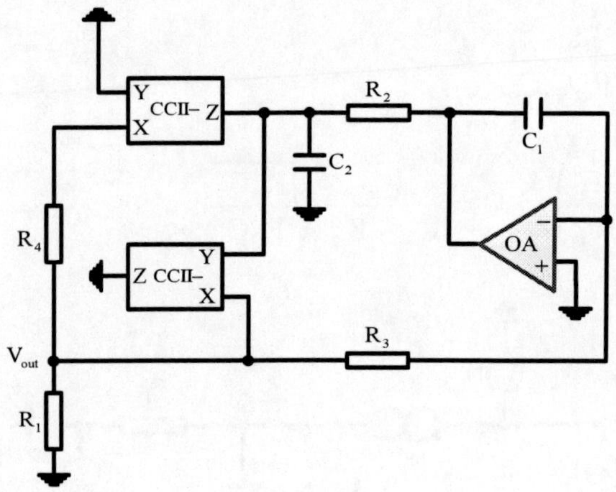

Figure 6. Sinusoidal oscillator obtained from the bandpass filter.

4. Simulation Results

To verify the workability of the proposed method, PSPICE simulations were performed for the synthesized bandpass filters in Figures 2 and 5 and the oscillators in Figures 3 and 6. The Op-Amp and CCIIs- were realized by uA741 and AD844 ICs, respectively. The macromodels of AD844 and uA741 were employed for the simulations with a DC supply of ±10 V. The passive component values of the circuit in Figure 2 are $R_2 = R_3 = 15.92$ kΩ and $C_1 = C_2 = 10$ nF.

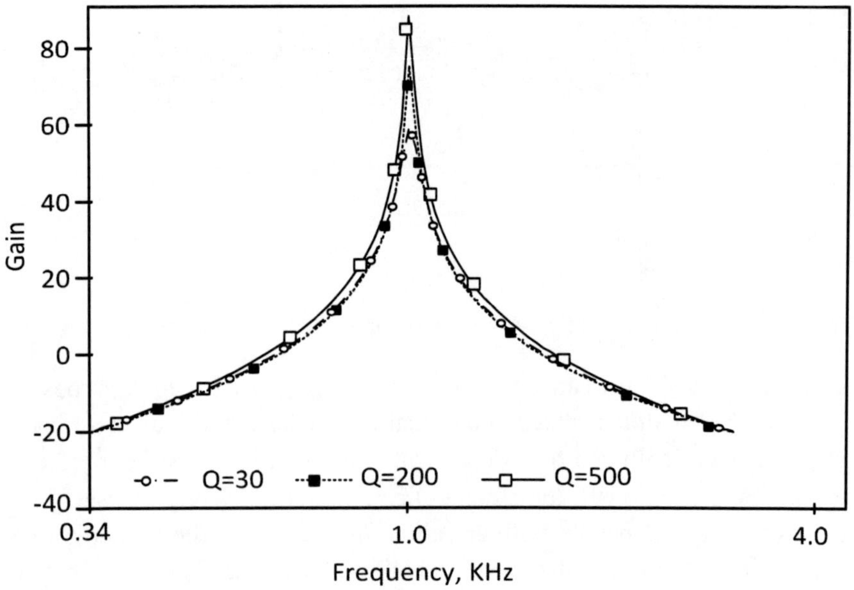

Figure 7. Frequency response for the bandpass filter in Figure 2 with tunable Q factor.

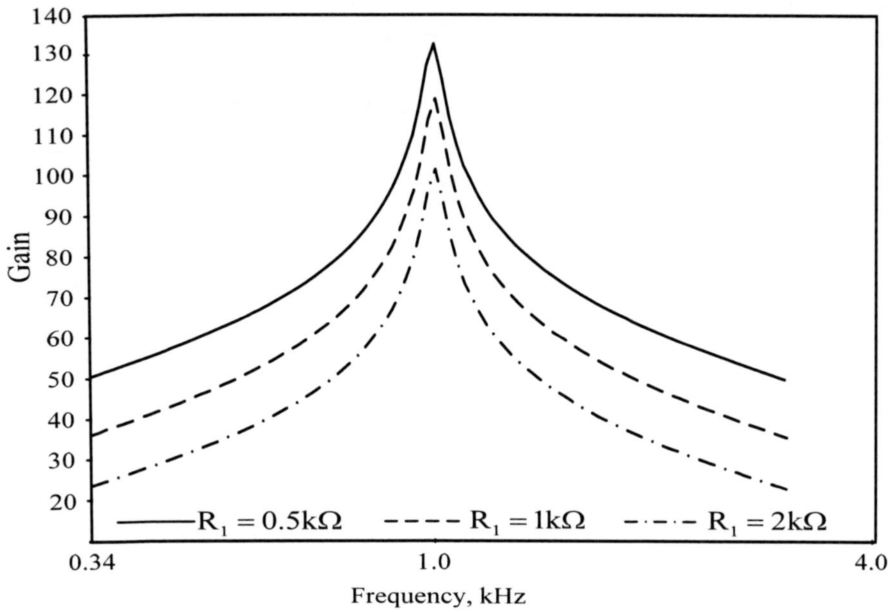

Figure 8. Independent tuning of the center-frequency gain with different R_1 values.

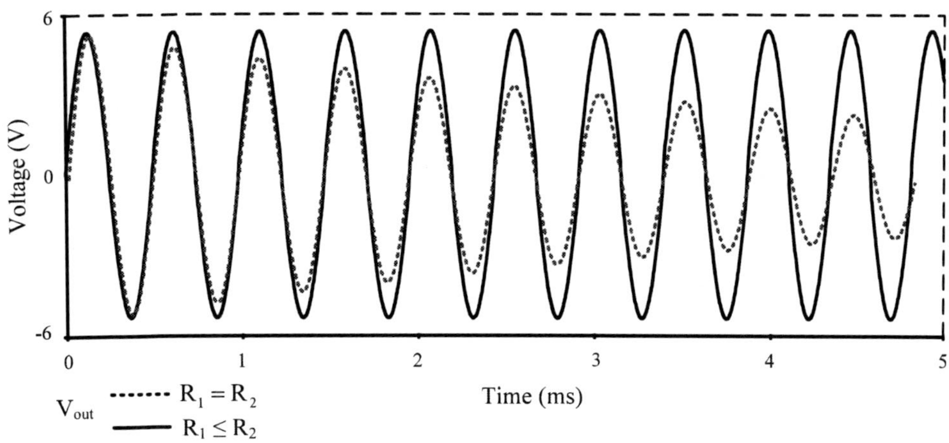

Figure 9. Oscillation waveforms for different oscillation conditions for the circuit in Figure 3.

Figure 7 shows the tuning of the Q with different R_1 values. The simulated results is consistent with our prediction. For the filter in Figure 5, to verify the center-frequency gain can be independently adjusted by R_1 without affecting ω_0 and Q, the passive component values $R_2 = R_3 = 15.92$ kΩ, $C_1 = C_2 = 10$ nF and $R_4 = 16.47$ kΩ are used for simulation. The tuning of the center-frequency gain with different R_1 values is shown in Figure 8. In addition, the workability of the oscillator circuits in Figures 3 and 6 were also verified successfully by PSPICE simulations. Figure 9 shows the output waveform of the circuit in Figure 3 with $R_1 = R_2$ and $R_1 \leq R_2$ [13]. The result is consistent with our theoretical prediction.

CONCLUSION

A systematic method for synthesizing high-Q biquadratic bandpass filters is proposed in this chapter. The proposed approach is based on the prescribed transfer function of synthesized circuits and takes advantage of the nodal admittance matrix expansion method. The prescribed transfer function of synthesized filter is arranged and nodal admittance matrix expansion has been used to derive the final topology of synthesized circuits. The derived high-Q filters with parameter tunable characteristic are suitable for practical applications. The filters can be extended to obtain sinusoidal oscillators with parameter tunable characteristic. The workability of the proposed method is demonstrated by circuit examples.

REFERENCES

[1] Toker, A; Ozoguz, S; Cicekoglu, O.; Acar, C. *IEEE Transactions on Circuits and Systems II*, 2000, *vol. 47*, 949 – 954.

[2] Comer, D. T.; Comer, D. J.; Gonzalez, J. R. *IEEE Transactions on Circuits and Systems II*, 1997, *vol. 44*, 856 – 861

[3] Fabre, A.; Saaid, O.; Wiest, F.; Boucheron, C. *Electronic Letters*, 1995, *vol. 31(20)*, 1727 – 1728.

[4] Haigh, D. G.; Clarke, T. J. W.; Radmore, P. M. *IEEE Transactions on Circuits and Systems I*, Regular Papers, 2006, *vol. 53(9)*, 2011 – 2024.

[5] Haigh, D. G. *IEEE Transactions on Circuits and Systems I*, Regular Papers, 2006, *vol. 53(12)*, 2715 – 2728.

[6] Haigh D. G.; Tan, F. Q; Papavassiliou, C. *Analog Integrated Circuits and Signal Processing*, 2005, *vol. 43(3)*, 297 – 315.

[7] Haigh, D. G.; Radmore, P. M. *IEEE Transactions on Circuits and Systems I,* Regular Papers, 2006, *vol. 53(10)*, 2214 – 2223.

[8] Soliman, A. M. *Analog Integrated Circuits and Signal Processing*, 2010, *vol. 65*, 43 – 59.

[9] Soliman, A. N. *International Journal of Electronics and Communications (AEU)*, 2010, *vol. 64*, 971 – 977.

[10] Tan, L.; Bai, Y.; Teng, J.; Liu, K.; Meng, W. *Circuits, Systems, and Signal Processing*, 2013, *vol. 32*, 1467 – 1476.

[11] Saad, R. A.; Soliman, A. M. *IEEE Transactions on Circuits and Systems I,* 2008, *vol. 55*, 2726 – 2735.

[12] Durham, A.; Redman-White, W. *IEEE Journal of Solid-State Circuits*, 1993, *vol. 28*, 835-839.

[13] Sedra, A. S.; Smith, K. C. *Microelectronic Circuits, Fifth edition,* Oxford University, New York, 2004.

In: Advanced Materials - Studies and Applications
Editors: I.A. Parinov, S.-H. Chang, S.Theerakulpisut
ISBN: 978-1-63463-749-7

Chapter 28

HILBERT-HUANG TRANSFORM BASED FEATURES FOR UNDERWATER VOICE (II) TRANSMISSION

C. F. Lin*[*1], *T. K. Chan*[1], *C. C. Wen*[2], *S. H. Chang*[3], *I. A. Parinov*[4] and *S. N. Shevtsov*[5]

[1]Department of Electrical Engineering,
National Taiwan Ocean University, Kaohsiung, Taiwan, ROC
[2]Department of Shipping Technology,
National Kaohsiung Marine University, Kaohsiung, Taiwan, ROC
[3]Department of Microelectronics Engineering,
National Kaohsiung Marine University, Kaohsiung, Taiwan, ROC
[4]Vorovich Mathematics, Mechanics and Computer Sciences Institute,
Southern Federal University, Rostov-on-Don, Russia
[5]South Scientific Center of Russian Academy of Sciences,
Rostov-on-Don, Russia

ABSTRACT

This chapter examines Hilbert-Huang transform (HHT)-based features for space-time block code (STBC) orthogonal variable spreading factor (OVSF)/orthogonal frequency division modulation (OFDM) underwater voice (II) transmission. The voice (II) samples were recordings of three male voices for 5 s duration. The voice (II) sample was inputted into a Matlab-based STBC OVSF/OFDM underwater multimedia communication platform.

Hilbert-Huang time-frequency analysis on the intrinsic mode functions (IMFs) features of the received voice (II) signal resulted in transmission bit error rates of 0, 10^{-3}, and 10^{-1}. According to the analysis results, the ratios of the energy of IMF3 to its refereed total energy for samples 5a, 5b, and 5c were 31.99%, 32.10%, and 15.19%, respectively, for a transmission bit error rate of 0. The ratios of the energy of IMF3 to its refereed total energy for samples 5a, 5b, and 5c were 31.42%, 31.21%, and 15.29%, respectively, for the transmission bit error rate of 10^{-3}.

[*] Email: lcf1024@mail.ntou.edu.tw

The ratios of the energy of the IMF3 to its refereed total energy for samples 5a, 5b, and 5c were 27.81%, 27.40%, and 18.02%, respectively, for the transmission bit error rate of 10^{-1}. Novel and substantial enhancements were observed for applying the HHT feature to voice (II) in underwater digital speech communication.

1. INTRODUCTION

Underwater digital speech communication transmission is a compelling research topic. Sari et al. [1] considered a new design methodology that involved implementation of speech-signal processing algorithms and transmission and receipt of speech in digital formats. Wei et al. [2] discussed the performance of an underwater digital speech communication system involving a low-density parity-check code. Luo et al. [3] designed an underwater speech communication codec system equipped with the AMBE-3000 multi-rate speech codec chip. Seyyed et al. [4] presented an orthogonal frequency division multiplexing (OFDM) based system, that was designed and implemented to transmit voice signals, via underwater acoustic channels. Junyeong et al. [5] proposed an acoustic multimedia data transmission system involving multiple-input multiple-output (MIMO) OFDM in an acoustic communications system. Lin et al. [6] proposed an MIMO-OFDM based data, voice, and image acoustic multimedia communications system.

The Hilbert-Huang Transform (HHT) [7 – 10] is a notable signal analysis method. Compared with the Fourier transform and wavelet transform, the HHT exhibits a higher analysis resolution. Additional physical phenomena and features of signals can be observed using the HHT analysis scheme. Lin et al. [11] used the HHT analysis method for examining the instantaneous frequency features of underwater voice (I) transmission. This study, investigated the intrinsic mode functions (IMFs) features of underwater voice (II) transmission.

2. SIMULATION AND HHT ANALYSIS MODEL

Figure 1 illustrates the simulation and HHT analysis model. The voice (II) samples were recordings of three male voices for 5 s duration, shown in Figure 2. The voice (II) samples were input into G.729 encoder, and the G.729 voice (II) bit streams were extracted as outputs. The G.729 voice (II) bit streams were input into 2 × 2 MATLAB-based underwater MIMO-OFDM simulator [6], and the received G.729 voice (II) bit streams at a bit error rate of 10^{-3} were extracted as outputs.

The received G.729 voice (II) bit streams at a bit error rate of 10^{-3} were input into a G.729 decoder, and the decoded voice (II) samples at a bit error rate of 10^{-3} were extracted as outputs. The decoded voice (II) samples at a bit error rate of 10^{-3} were analyzed in order to be applied to the HHT scheme.

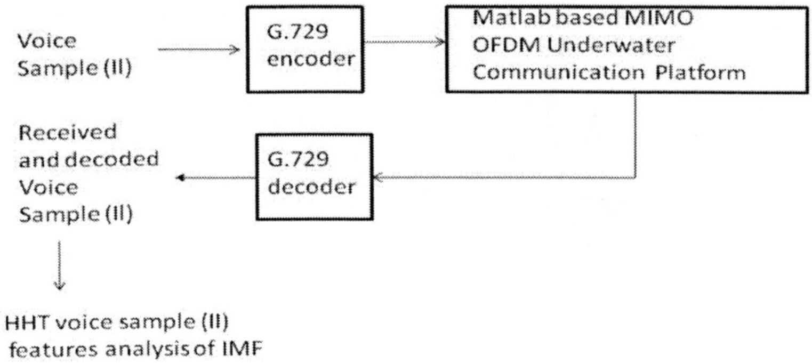

Figure 1. Simulation and HHT analysis model.

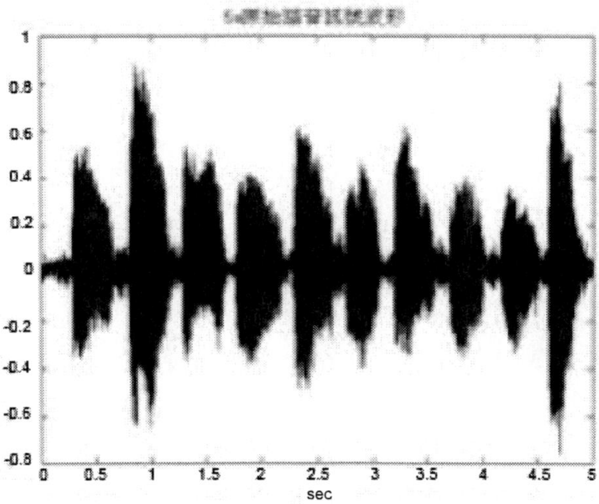

Figure 2. (a) voice (IIa) with 5 sec duration.

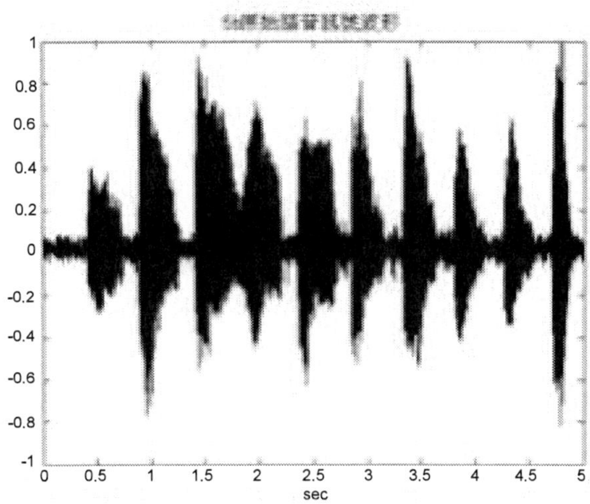

Figure 2. (b) voice (IIb) with 5 sec duration.

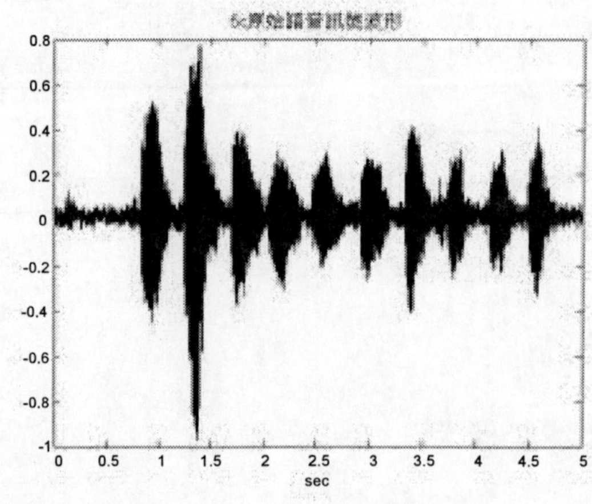

Figure 2. (c) voice (IIc) with 5 sec duration.

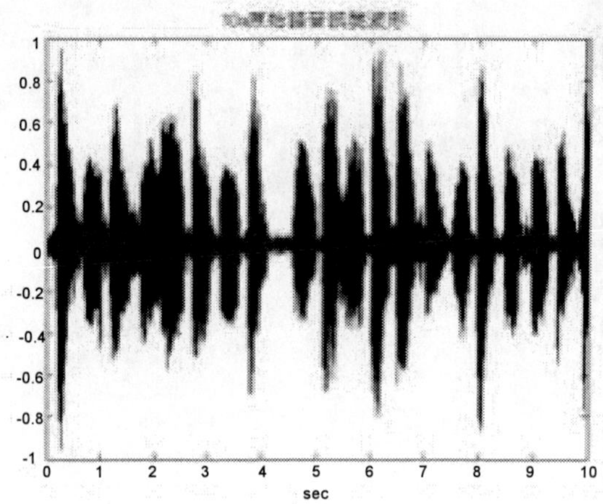

Figure 2. (d) voice (IId) with 10 sec duration.

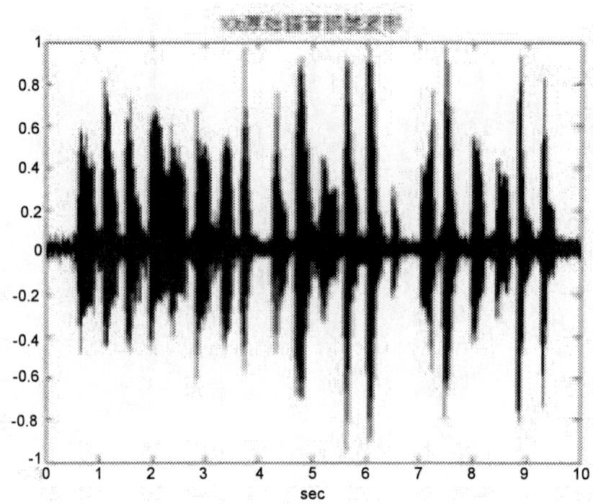

Figure 2. (e) voice (IIe) with 10 sec duration.

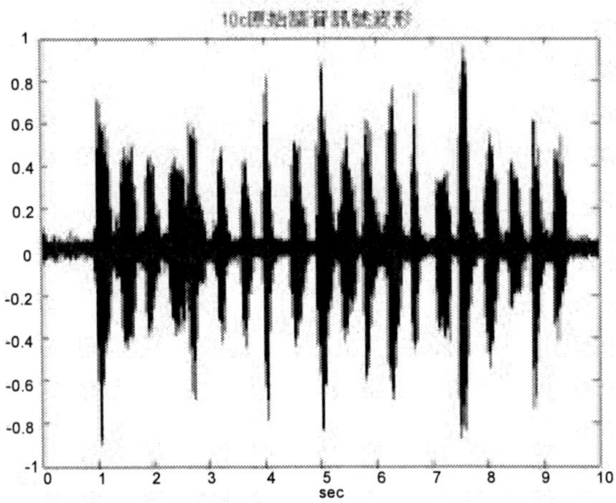

Figure 2. (f) voice (IIf) with 10 sec duration.

3. HHT Voice Sample II Feature of IMF

The empirical mode decomposition (EMD) method was used for resolving the received and decoded voice (IIa) sample at a bit error rate of 10^{-3}. Figure 3 shows the IMFs of the voice (IIa) sample at a bit error rate of 10^{-3}. As shown in the figure, 10 IMFs and a residual function were decomposed for the received and decoded voice (IIa) sample at a bit error rate of 10^{-3}. Moreover, 10 IMFs and a residual function were decomposed for the received and decoded voice (IIb), (IIc), (IId), (IIe), and (IIf), samples.

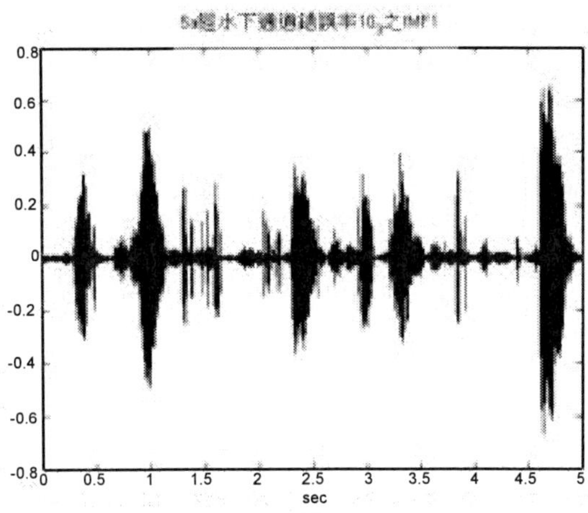

Figure 3. (a) IMF1 of the voice (IIa) sample with bit error rate of 10^{-3}.

Table 1 lists the statistical characteristics of the IMFs for the received and decoded voice (II) samples at a bit error rate of 0, 10^{-3}, and 10^{-1}. The refereed total energy of HHT-based signal is defined as

$$E_r(t) = \sum_{i=1}^{L} IMF_i^2(t) + RF^2(t)\,, \tag{1}$$

where $IMF_i(t)$ is the *ith* IMF, RF(t) is the residual functions.

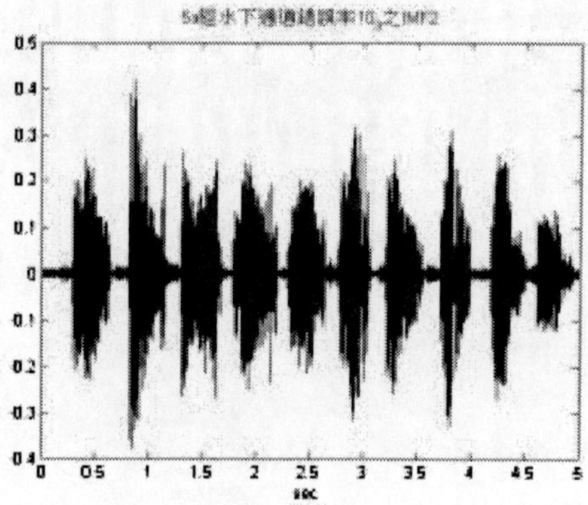

Figure 3. (b) IMF2 of the voice (IIa) sample with bit error rate of 10^{-3}.

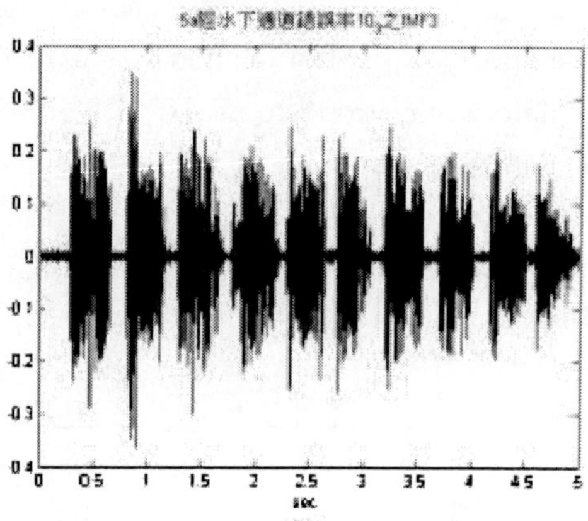

Figure 3. (c) IMF3 of the voice (IIa) sample with bit error rate of 10^{-3}.

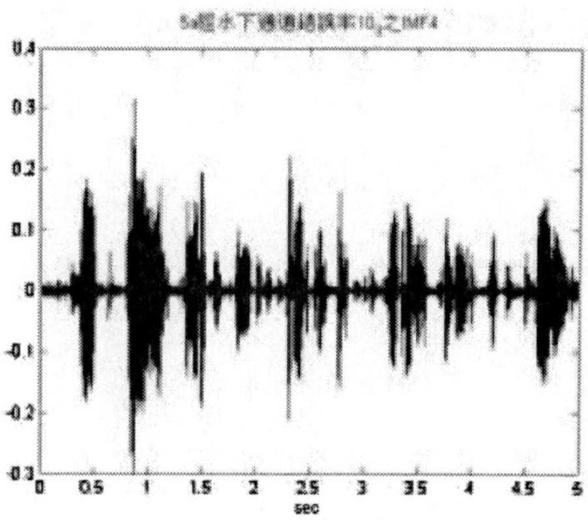

Figure 3. (d) IMF4 of the voice (IIa) sample with bit error rate of 10^{-3}.

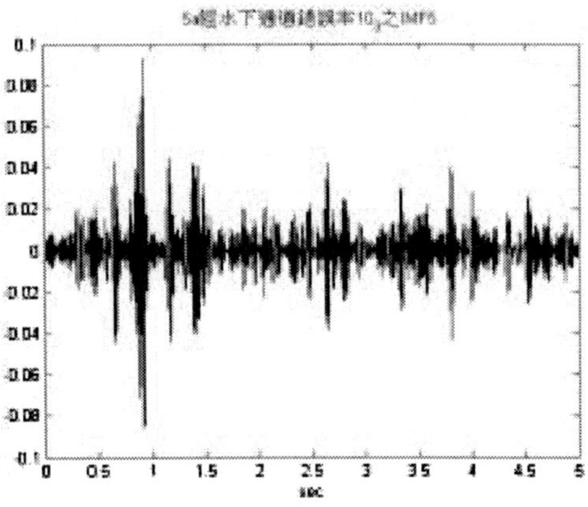

Figure 3. (e) IMF5 of the voice (IIa) sample with bit error rate of 10^{-3}.

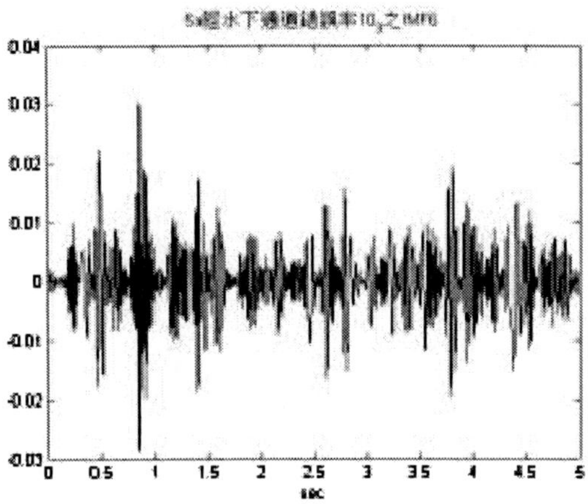

Figure 3. (f) IMF6 of the voice (IIa) sample with bit error rate of 10^{-3}.

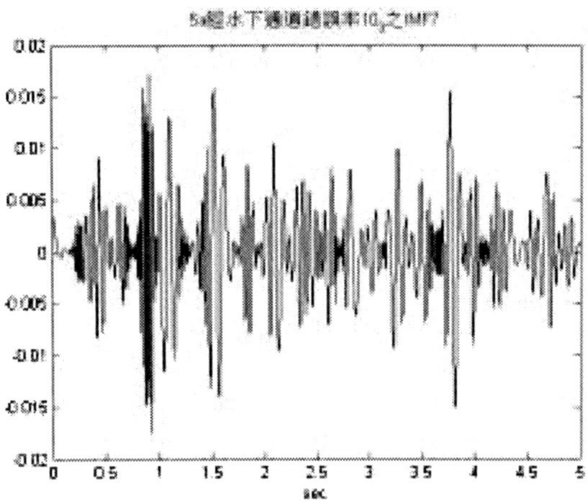

Figure 3. (g) IMF7 of the voice (IIa) sample with bit error rate of 10^{-3}.

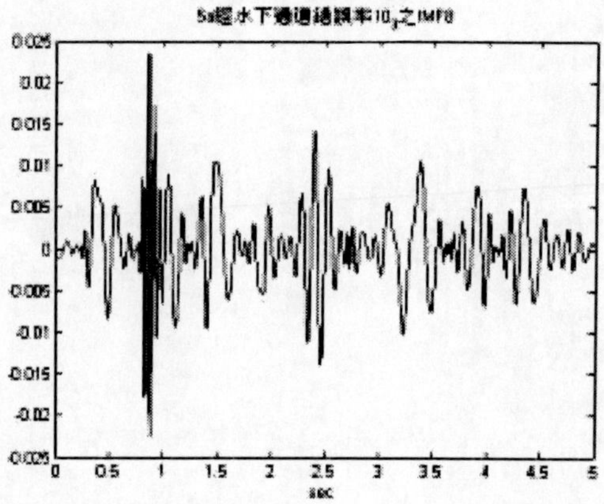

Figure 3. (h) IMF8 of the voice (IIa) sample with bit error rate of 10^{-3}.

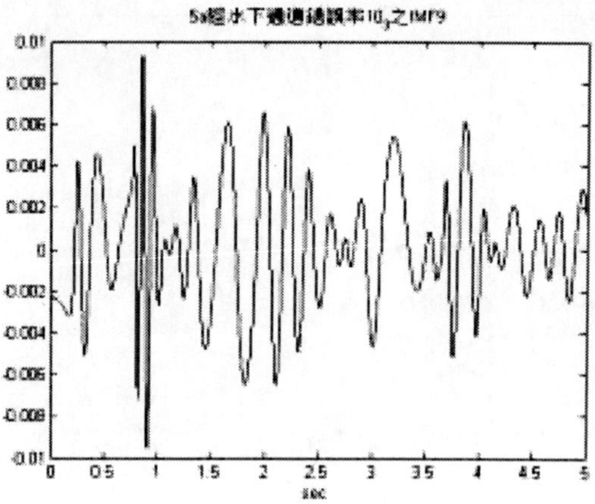

Figure 3. (i) IMF9 of the voice (IIa) sample with bit error rate of 10^{-3}.

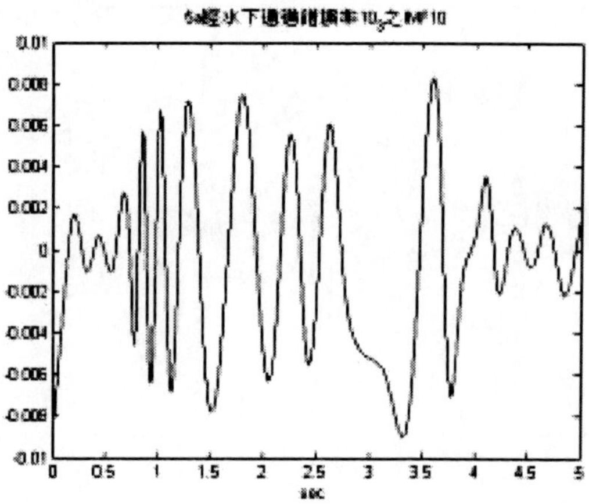

Figure 3. (k) IMF10 of the voice (IIa) sample with bit error rate of 10^{-3}.

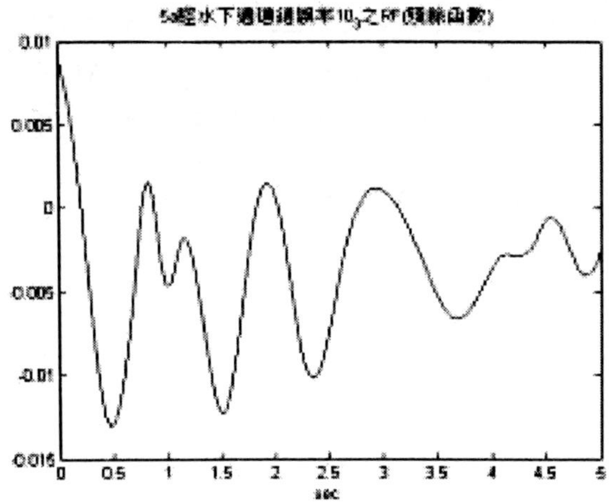

Figure 3. (l) Residual function of the voice (IIa) sample with bit error rate of 10^{-3}.

The ratios of the wave energy to its referred total energy for the *i*-th IMF is as follows:

$$E_{imf} = \frac{IMF_i^2(t)}{E_r(t)} \times 100\% . \tag{2}$$

The analysis results revealed that the ratios of the energy of the voice (IIa) sample at a bit error rate of 0 to its total energy for IMF1, IMF2, IMF3, and IMF4 were 24.16%, 30.96%, 31.99%, and 11.42%, respectively. The ratios of the energy of the voice (IIb) sample at a bit error rate of 0 to its total energy for IMF1, IMF2, IMF3, and IMF4 were 12.60%, 23.96%, 32.10%, and 27.47%, respectively. The ratios of the energy of the voice (IIc) sample at a bit error rate of 0 to its total energy for IMF1, IMF2, IMF3, and IMF4 were 21.10%, 51.42%, 15.19%, and 10.63%, respectively. The ratios of the energy of the voice (IIa) sample at a bit error rate of 10^{-3} to its total energy for IMF1, IMF2, IMF3, and IMF4 were 23.70%, 30.43%, 31.42%, and 13.06%, respectively. The ratios of the energy of the voice (IIa) sample at a bit error rate of 10^{-1} to its total energy for IMF1, IMF2, IMF3, and IMF4 were 21.63%, 34.60%, 27.81%, and 14.49%, respectively. Based on the analysis result, the transmission bit error rate was large, and the ratios of the energy of the voice (II) samples to its total energy for IMF1, IMF2, IMF3, and IMF4 were large.

Conclusion

Based on the results of the analysis, the ratios of the energy of IMF3s to their total energy were the highest for the voice (IIa) and (IIb) samples, respectively. The ratios of the energy of IMF2 to their total energy were the highest for the voice (IIc) sample. The difference among the ratios of the energy of the voice (IIa) sample at a bit error rate of 0 and those of the voice (IIa) sample at a bit error rate of 10^{-3} to its total energy for IMF1, IMF2, IMF3, and IMF4 were 0.46%, 0.53%, 0.57%, and −1.64%, respectively. The difference among the ratios of the energy of the voice (IIa) sample at a bit error rate of 0 and those of the voice (IIa) sample at a bit error rate of 10^{-1} to its total energy for IMF1, IMF2, IMF3, and IMF4 were 2.53%, –3.64%, 4.18%, and –3.15%, respectively. By applying the HHT method to the data analysis, critical new insights into the voice (II) samples as they relate to underwater digital speech communication were revealed. Table 1 shows the statistical characteristics of the IMFs for the received and decoded voice (II) samples with bit error rate of 0, 10^{-3}, and 10^{-1}, respectively.

Table 1. Statistical characteristics of IMFs

sample	BER	IMF	eng	E_{imf}(%)
IIa	0	1	189.4365	24.1649
IIa	0	2	242.7515	30.9659
IIa	0	3	250.8500	31.9990
IIa	0	4	89.5325	11.4210
IIb	0	1	101.1473	12.6072
IIb	0	2	192.2816	23.9664
IIb	0	3	257.544	32.1008
IIb	0	4	220.4659	27.4793
IIc	0	1	62.0716	21.1066
IIc	0	2	151.2406	51.4274
IIc	0	3	44.6843	15.1943
IIc	0	4	31.2735	10.6341
IIa	10^{-3}	1	173.9607	23.7092
IIa	10^{-3}	2	223.2837	30.4315
IIa	10^{-3}	3	230.5796	31.4259
IIa	10^{-3}	4	95.8925	13.0693
IIb	10^{-3}	1	94.2025	12.6395
IIb	10^{-3}	2	193.9520	26.0233
IIb	10^{-3}	3	232.6328	31.2133
IIb	10^{-3}	4	191.8462	25.7408
IIc	10^{-3}	1	69.8316	22.2774
IIc	10^{-3}	2	159.7581	50.9655
IIc	10^{-3}	3	47.9583	15.2995
IIc	10^{-3}	4	31.7828	10.1392
IIa	10^{-1}	1	105.6118	21.6310
IIa	10^{-1}	2	168.9707	34.6079
IIa	10^{-1}	3	135.8276	27.8197
IIa	10^{-1}	4	70.6494	14.4701
IIb	10^{-1}	1	61.3800	14.7206
IIb	10^{-1}	2	129.1052	30.9630
IIb	10^{-1}	3	114.2680	27.4047
IIb	10^{-1}	4	95.4527	22.8922
IIc	10^{-1}	1	47.0066	30.5603
IIc	10^{-1}	2	58.3883	37.9599
IIc	10^{-1}	3	27.7227	18.0233
IIc	10^{-1}	4	17.1750	11.1660

Acknowledgments

The Authors Acknowledge the Support of The National Science Council of Taiwan NSC 101-2221-E-022 -005, and the valuable comments of the reviewrs.

References

[1] Sari, H.; Woodward, B., In: *Underwater Acoustic Digital Signal Processing and Communication Systems*, R. S. H. Stepanian, M. Stojanovic (Eds.), 127 – 165, Kluwer Academic Publishers, 2002.

[2] Han, W.; Huang, J.; Jiang, M., *Proceeding of ICIEA*, 2009, pp. 567-570.

[3] Luo, J.; Hong, W.; Li, Z. Y.; Lu, M., *Proceeding of International Conference on Mechatronic Science, Electric Engineering and Compute*r, 2011, 2471-2475.

[4] Mahmood, S.; Sadeghi, J.; Derakhtian, M.; et al., *Proceeding of IEEE Symposium on Computers and Communications*, 2012, pp.49 – 52.

[5] Bok, J.; Ryu, H. G., In: *ICHIT 2011*, G. Lee, D. Howard, D. Ślęzak (Eds.), *LNCS 6935*, pp. 123–130, Springer-Verlag, Berlin, Heidelberg, 2011.

[6] Lin, C. F.; Chang, S. H.; Lee, C. C.; et al., *Wireless Personal Communications*, 2013, *vol. 71*, 1231 – 1245.

[7] Huang, N. E.; Hen, S .S. P., *Hilbert-Huang Transform and its Applications.* World Scientific Publishing Co., Singapore, 2005, pp. 1 – 307.

[8] Huang, N. E.; Shen, Z.; Long, S. R.; et al., *Proceedings of the Royal Society of London Series A – Mathematical Physical and Engineering Sciences*, 1998, *vol. 454*, 903 – 995.

[9] Wu, M. C.; Huang, N. E., In: *Advanced Biosignal Processing*, A. Nait-Ali (Ed.), Springer Publishers, Berlin, Heidelberg, 2009, pp. 335 – 350.

[10] Yan, R.; Gao, R. T., *IEEE Instrumentation & Measurement Magazine*, 2007, *vol. 10*, 11-15.

[11] Lin, C. F.; Hsiao, K. J.; Wen, C. C.; et al., In: *Advanced Materials. Physics, Mechanics and Applications*, S.-H. Chang, I. A. Parinov, V. Yu. Topolov (Eds.), Springer Proceedings in Physics, *vol. 152*, Chapter 24. Springer, Heidelberg, New York, Dordrecht, London, 2014, pp. 305-319.

In: Advanced Materials - Studies and Applications
Editors: I.A. Parinov, S.-H. Chang, S.Theerakulpisut
ISBN: 978-1-63463-749-7

Chapter 29

PROTECTION OF EQUIPMENT FROM EXCESSIVE PRESSURE

A. S. Yudin* and N. V. Belikov†
Institute of Mathematics, Mechanics and Computer Science,
Southern Federal University, Rostov-on-Don, Russia

ABSTRACT

The chapter considers modeling and manufacture technologies of flapping safety membranes with high accuracy of actuation, applied in devices, protecting industrial equipment from pressure overloads.

1. INTRODUCTION

The convex dome shells are used as elements of safety system devices, which protect industrial equipment and reservoirs from destruction by excessive pressure. These shells can lose stability and then fracture at a certain level of the pressure, applied to their convex parts, thus keeping integrity of the protected equipment. These shells are named by the flapping safety membranes (FSMs) because of characteristic sound of flap, accompanying the process of stability loss. Other names are the buckling safety membranes (BSMs) and bursting discs (BDs). These shell-like structures were developed as elements of the membrane safety devices to protect equipment in chemical and petroleum-refining industry; afterwards they were successfully employed in thermal and atomic engineering. The implementation of experimental and theoretical R and D of the FSMs and corresponding manufacture technologies in the Southern Federal University (Rostov-on-Don, Russia) was closely connected with scientific activity of the Soviet academician I. I. Vorovich.

* A. S. Yudin e-mail: yudin@math.sfedu.ru.
† N. V. Belikov e-mail: bel@math.sfedu.ru.

He was not only the head of the Mechanics and Applied Mathematics Research Institute, but also developed actively nonlinear theory of the shell stability [1–7].

Importance of study of the FSMs is determined by a wide variety of protective device constructions, complexity of the membrane response mechanisms, and their high sensitivity to small stochastic deviations from idealized design schemes.

2. EXPERIMENTS AND MANUFACTURE TECHNOLOGIES

The flapping membranes are ones of the most effective fractured elements of the safety devices. There are also other types of membranes, triggered directly by a critical hydrostatic pressure: bursting membranes, rupturing membranes, shearing membranes, tearing membranes, combined membranes and membranes, fabricated for specific purposes. Further classification could be also based on the construction features and methods of determining the subcritical and post-critical mechanical behaviors, namely: smooth membranes; membranes, containing notches or weakened zones; membranes with complicated geometric shapes (corrugated or with variable curvature of a meridian); composite, multi-layer structurally-anisotropic membranes; twisted fractured elements of shell type; embittered membranes and so on. Membranes could be also distinguished by materials, protective coatings and types of restraints, for example rigid flat or conical clamps, elastic supports and free supports [8].

In manufacture technologies of the FSMs, many factors exist that significantly affect difference between actual critical loads of the shells and loads, given by computational models of perfectly spherical elastic domes with clamped edges. These factors are called initial imperfections and include geometrical imperfections, caused by small bending and features of supporting in the inflection zone near the FSM dome. Moreover, the performed investigations have found other deviations from design models that affect value of the critical load:

1 It is usually assumed that physical and mechanical properties are averaged in respect to the material of construction and can be represented by simple continuum models with constant values of elasticity moduli, strength etc. However, these parameters are insufficient for study of thin-walled domes, fabricated of flat sheets. There are significant differences of the physical and mechanical properties of the flat sheets along and across the direction of rolling.

2 Sheet materials are manufactured with considerable variation in thickness of the good. To reduce the influence of this factor, it is often necessary to define sections with acceptable thickness variance by choosing work pieces.

3 Actual nature of the membrane support differs from the idealized model support. Indeed, edges are clamped to ensure a required level of tightness. It cannot usually be identified with employed models of "rigid clamping", "plastic hinge" or "free support". Unfortunately, attempts to evaluate the actual conditions for the membrane supports have been unsuccessful. Various spacers, plastic or collapsible peripheral rings, stress concentrators, etc. were introduced into the model to idealize clamping surfaces. However, it is still impossible to link the estimated results of micro-

deformations in the clamping rings of the holders with the observed changes in the pressure response of the actual FSM.

4 Micro-structure analysis of the near-edge zones of membranes, which can be fabricated by plastic molding, shows that the thickness of the membrane material is non-uniform due to presence of micro-structural grains, local zones of tightening and aeration in the inflection regions.

5 There is also a set of factors, related to the characteristics of media, which are in contact with surface of the safety membranes during technological process. Apart from the well-known factors, associated with corrosion of the material, the analysis of operation capacity of the safety membranes has demonstrated a phenomenon of mass transfer from the material of holder rings to the dome of membrane. This effect can be explained by dissolving the carbon in near-surface layers with its subsequent absorption in the near-edge zones of the membrane dome, fabricated of mild austenitic steel. This process causes a change in the physical and mechanical properties of the membrane.

Experimental methods are the most reliable in predicting the response of the FSMs are subject to pressure loading. Conventional methods for the determination of a critical pressure value for flapping membranes include the controlled fracture of a statistically sufficient number of samples. The accuracy of definition of the critical pressure for a single membrane is relatively low (± 5%), so the confidence range of pressure is established for a whole setting of the tested samples on the base of data dispersion.

Note, that the destructive tests have been employed in the early stages of information selection on behavior of the membranes. Numerous experiments have been performed for smooth shells, fabricated of different materials, by using conventional technology of free plastic drawing. The geometric characteristics (thickness of original sample, diameter of clamping ring and height of dome), loading path (multiple loading and unloading) and clamping force of membrane in the ring varied in these experiments.

The critical heights of drawing out were found, including the failure moment. The influence of the membrane characteristics on the critical loads was analyzed, as well as the spread of the pressure response in the set of nominally the same membranes. The methods for prediction of the critical loads were thoroughly tested. The obtained results were analyzed with the help of statistics; dispersions and confidence ranges of critical pressures were estimated and the forms of stability loss classified.

Considerable experience was obtained during these investigations, including study of the membrane characteristics in the subcritical stage of deformation. It allowed one to develop methods and fabricate the hardware support systems for nondestructive testing, patented in the Soviet Union. The design of membranes was improved and serial samples were obtained, demonstrating high accuracy [9–19]. The problem of processing the FSMs with high accuracy response was resolved by development of equipment for manufacturing the membranes and test devices for prediction of the pressure response by using non-destructive methods.

The developed methods of non-destructive testing the pressure response of the membranes are based on physically independent principles of diagnostics. Extrapolation of the critical load, based on the nonlinear diagram "pressure – displacement" is used in the first group of the methods [9, 10]. The direct acoustic emission (AE) method or employment of acoustically active coatings is applied in the second group [11, 12], and optical methods present the third group of these methods [13]. By analyzing subcritical behavior, membranes were divided into two classes, namely thin-walled and thick-walled membranes [8, 20–22].

The acoustic emission activity for membranes of the first type (thin-walled) appears just before a "flap" in the form of a very short pulse spike; at the same time, AE response grows asymptotically for membranes of the second type (thick-walled). Therefore, in the first case, subcritical elastic deformations occur everywhere, while in the second case, elastic-plastic deformations appear in extending plastic strain zones, only.

These two types of membranes have different forms of compression marks generated at the initial stage of the stability loss. For thin membranes, these compression marks locate, mainly, in the near-edge zone. Rarely, the compression marks locate in the zone close to the center of the dome. In contrast, the thick-walled membranes have compression marks of a crescent (conchoidal) type near the edge, only.

It should be noted that above classification differences are clearly observed for membranes, which have a lesser than 1–2% variation in critical pressure. Moreover, for membranes with high-accuracy actuation, buckling modes are surprisingly similar. Otherwise, the compression marks of an ill-defined geometry can be observed If the analysis takes into account the influence of initial imperfections, then, obviously, membranes with stable values of critical loads have the same type of deviations from the ideal spherical shell as well as similar main parameters. The scatter of the pressure values is associated with variations in the initial imperfections. Usually, it depends on properties of the flat sheets, dome shell building material, and also on variations in the thickness, non-uniformity and anisotropy of physical and mechanical properties.

These observations led to the idea to control shell properties with the help of artificial initial imperfections, which ensure the stable buckling modes and critical load. Moreover, they have great importance for an analysis of real membrane materials, which age under real operation conditions.

One of these procedures, which have been implemented in the construction of membranes and their manufacture technologies [14-19], was aimed to provide initial axisymmetric modes of buckling and initialize clipping, directed from the top of the dome. The procedure improved the dynamics of stability loss and effectively "opened" a cross-section of the membrane during cutting the membrane on legs. The developed methods of artificiality (i.e., fabrication of artificial or man-made membrane goods) can be divided into two groups. So-called constructive methods have been developed for testing the stiffness parameters, and geometric methods studied the form of the membrane middle surface.

The research method of constructive artificial imperfections was first proposed in [14]. A membrane has stepwise-varying thickness with two sections *A* and *B*. Zone *A* has annular grooves which increase from periphery to depth center. At the same time, zone *B* is uniform and has the smallest thickness in the vicinity of the dome top. An optimal diameter of the central zone is about 0.4 times of a membrane diameter. The buckling occurs in the weakened zone in the vicinity of the apex, while axisymmetric deformation occurs in the circumferential direction of the section *A*, where stiffness, caused by structural anisotropy, is higher. The fabrication process of shells with these features includes several stages: coverage of a plate work piece with a photo-layer, exposure to a pattern of concentric rings from a negative on this work piece, film processing and fixation, drawing of a dome-shape shell, increasing the resistance of the photo-layer, and chemical etching of grooves. For this purpose, only suitable reactive materials, such as aluminum, should be employed.

The structural anisotropy for chemically stable materials can be created with the help of corrugation. An appropriate way to perform this operation is described in [15]. The annular corrugation is stamped on a flat blank; an adhesion-resistant sealant with auto-solidification properties is used. First, this flat smooth plate is placed on the sealant, then the resulting "sandwich" is clamped in the rings and, finally, the dome is drawn out. Afterwards, a shell with a new corrugated surface separates from the supporting membrane and frees from the sealant.

Constructively, these artificially unperfected shells may also include membranes with rigid over-layers on top of the dome [16, 17]. The over-layers replace outer legs and ensure effective destruction of the membrane at the critical load. Figure 1 shows several types of the membranes with the artificially constructed imperfections.

Further investigations and applications of the different types of membranes have shown that smooth thin-walled membranes have the best characteristics, while smooth thick-walled membranes are the second best. It was observed that membranes with notches, grooves and other stress concentrators ensure the effective opening but have lower operation performance. It can be explained by micro-plastic damage accumulation, occurring due to repeated loads and a decrease in the critical pressure response from cycle to cycle.

Therefore further studies have been concentrated on the geometric artificial imperfections [23, 24], and aimed to fabricate the smooth membranes with a zone of decreased curvature in the vicinity of tip. This fabrication is possible with pressure in addition to that what is used for molding. To implement this approach, three methods have been proposed. The first of them assumes to use the device as proposed in [18] and includes not only a basic molding chamber but also a back pressure chamber with a flexible plate. During the final stage of the drawing out process, the dome-shape shell is in contact with the flexible plate, which creates resistance in the apex zone. As a result, this zone has smaller curvature, compared to the curvature, obtained in the free drawing out. It should be noted that employed sensors test the curvature value of the shell as well as the back pressure. In the second method, the dome is in contact with a material layer of experimentally chosen stiffness at the final stage of molding. The third method uses a concentrated force, applied to the top of the shell. This method is most efficient and currently used as the primary method.

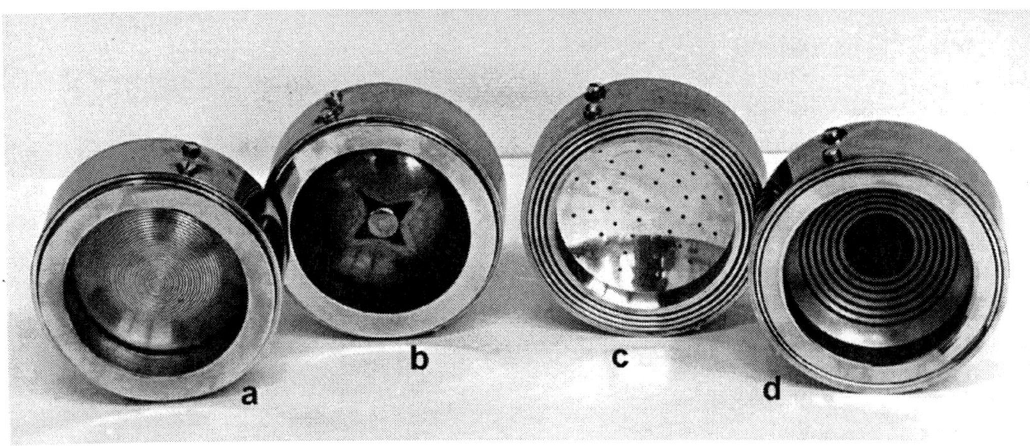

Figure 1. Structurally anisotropic membranes.

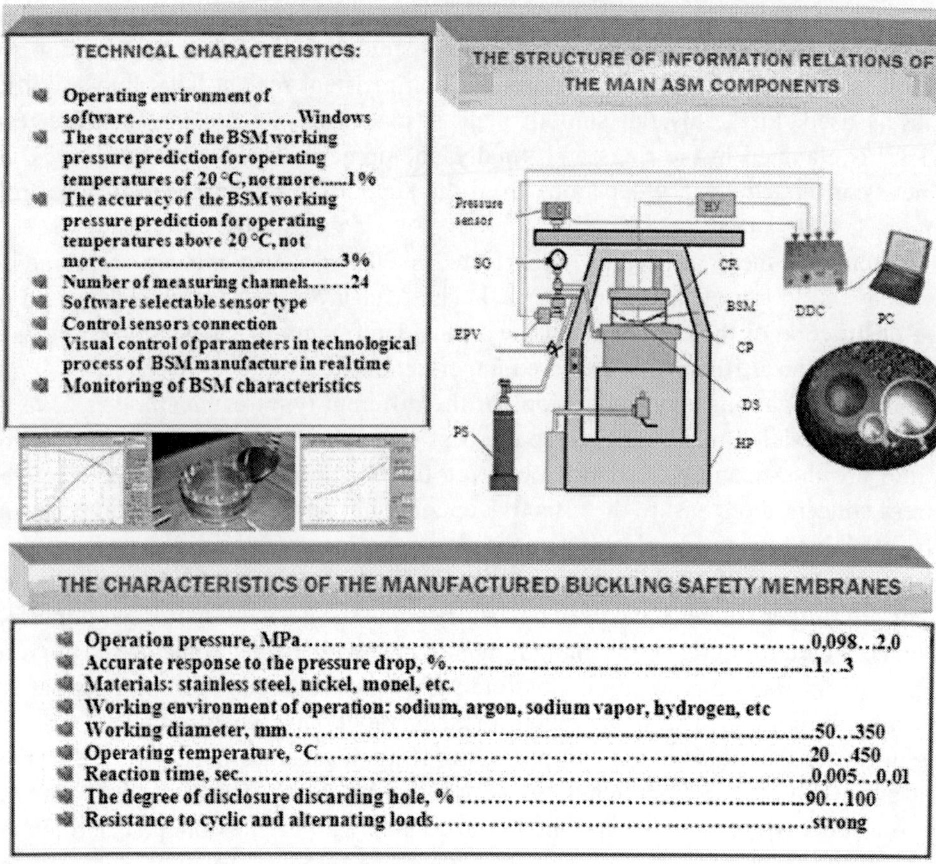

Figure 2. Setup for manufacturing and non-destructive testing of membranes.

Modern implementation of these manufacture procedures and methods for the behaviour prediction is based on computer testing and mathematical processing of data with the help of a software/hardware setup [25–27] (see Figure 2; The accuracy of the critical pressure prediction is of ~ 1 – 3% for operation temperatures in the range of 20-450 °C).

Because of the sensitivity of shells to the initial random imperfections, the artificial imperfections overlap the effect of the random imperfections, stabilize the critical load value and ensure favourable shell top buckling modes.

Moreover, since the critical load is estimated individually for each membrane, artificial imperfections eliminate inconsistencies, related to the edge zone, and diminish the effect of other random imperfections, caused by initial sheet material quality.

Moreover, the buckling modes ensure the shell motion has high dynamics after the stability loss, the effective destruction of the shell and an almost 100%-guaranteed opening of the cross-section by breakaway along the contour by the ring knives [19].

3. Mathematical Modeling

In the early stages of the research, the FSM has been considered as a segment of an elastic spherical shell of constant thickness in mathematical models, based on the solution of boundary-value problems. In recent years, the problem of nonlinear stability of such shells, loaded by uniform external pressure, has attracted significant attention.

This model has stimulated the development of the concept of stability loss, the bifurcation theory of equilibrium, the geometric theory of shell stability, the sensitivity theory of critical loads to initial imperfections, an application of asymptotic methods to stability analysis of shells, development and implementation of direct numerical methods and algorithms for the solution of nonlinear boundary-value problems. The effects of the complicated mechanical properties of the material were analyzed; the supported and layered shells were studied with solution of corresponding dynamic problems [28-43].

The compared theoretical and experimental results have demonstrated consistency in several precise experiments for which idealized conditions, close to those, assumed in theoretical investigations, were ensured.

For example, the author in work [35] has tested a very thin and highly elastic membrane, manufactured by the deposition of copper onto a spherical steel substrate in vacuum conditions.

In work [44], a very flat spherical segment was fabricated from a thick plate work piece. A ring, through which pinching is realized, was attached to this work piece. Undoubtedly, these successful results have reduced the gap between theory and experiments.

The vast majority of test results have demonstrated that, in fact, the experimental values of critical loads for elastic shells are much lower than the theoretical ones. As noted before, the reason for this is in the random initial imperfections. That is why semi-empirical formulae [8, 20–22] are usually used to select preliminary parameters in the manufacture of membranes by the free drawing out process.

To match the mathematical models with the behaviour of real shells, it is necessary to develop a theory, which not only reproduces real experimental conditions, but considers more comprehensive shell geometric features and mechanical properties.

Initially, modeling techniques of geometric imperfections have been developed on the base of the elastic flat shell theory. An initial deflection was often chosen to be close to the expected buckling value. Moreover, boundary conditions have been varied due to an assumption on the possible flexibility of the shell's contour restraints. Another factor, which can improve modelling, is the consideration of work, done by the shell material beyond the elastic limit [33, 42]. The role of plasticity may also increase significantly, when initial imperfections are taken into account. However, collaborative consideration of all these factors has proven too challenging for a comprehensive analysis.

In work [45], the author has investigated geometric aspects of a dome shape, which are related to the technology of plastic molding for flapping membranes by free drawn pressure. In this procedure, a work piece, given by a flat circular thin plate, is clamped by annular flanges so that its edges do not slip out during the formation. The flanges are hermetically attached to the container, in which a medium (e.g., air or oil) is injected under pressure. The transmitted pressure deforms plastically the work piece and transforms it to a dome. Free drawing process is stable and allows one to fabricate a shell with the required height. In this method of molding, there is an inflection zone between the flat zone of the flange and the dome. This zone has a negative Gaussian curvature and could be modeled with a circular torus element, which is smoothly connected to the dome part. Thus, the main dome represents an ellipsoid of revolution. Comparative analysis of these shells and a spherical shell of the same height has indicated significant differences in curvatures.

Moreover, the authors have considered non-uniform thickness and approximated it by simple functions, based on experimental results [46]. All these factors were taken into account in the computational modelling [47]. For flapping membranes that lose stability at non-symmetric shapes, it was shown that the calculated bifurcation critical loads tend to converge to the experimental values.

In fact, the critical loads decreased by 2 – 3 times, compared with the idealized model. The presence of the torus edge zone (arm) has changed significantly the critical shape of the membrane. The number of circumferential waves for the idealized models was between 6 and 18, while for the shells with the arm this number was equal to two. This was consistent qualitatively with the buckling shapes which were often experimentally observed when strain developed from the edge zone in the form of the crescent-dents.

Significant progress, made in works [48–54], has helped to reach agreement between theory and experiment. The main reason was due to solving the problems by using a two-stage approach. The first stage simulates the manufacture of the artificially unperfected membrane. The second stage analyzes the stability of the fabricated membrane. For the molding stage, mathematical models of the physically nonlinear shells of revolution with large displacements and rotation angles which account for changes in metrics were developed and successfully utilized.

Constitutive relations of the Davis–Nadai type, which take into account the material thickness heterogeneity, induced by deformations, have been derived. Semi-inverse methods have been employed to obtain approximate analytical solutions of the molding problem for spherical and ellipsoidal domes, fabricated from plates. Moreover, it has allowed us to gain ground theoretically on the concept of using the artificial concentrated force. Comparison of theory and experiments showed a good agreement [53].

CONCLUSION

Flapping membranes, fabricated with the help of new technologies, have been tested under operation conditions in the State Scientific Center of Russian Federation "Leipunsky Physical and Energetic Institute". The results of the tests have confirmed a compliance of the membranes to strict technical standards of durability, including influence of sodium coolant, accuracy and stability of critical pressure.

ACKNOWLEDGMENTS

The chapter was prepared under theme No. 213.01-11/2014-29 as a basic part of the government research, financed by the Russian Department of Education and Science.

REFERENCES

[1] Vorovich, I. I., Zipalova, V. F., *Engineering Journal Mechanics of Solids*, 1966, No. 2, 150 (In Russian).

[2] Vorovich, I. I., In: *Actual Problems of Science*, Rostov University Press, Rostov-on-Don, 1967, p. 111 (In Russian).

[3] Vorovich, I. I., Minakova, N. I., *Journal of Applied Mathematics and Mechanics*, 1968, vol. 32(2), 332 (In Russian).

[4] Vorovich, I. I., Minakova, N. I., *Izvestiya of Academy of Sciences of the USSR, Mechanics of Solids*, 1969, No. 2, 121 (In Russian).

[5] Vorovich, I. I., Minakova, N. I., In: *Results of Science and Technology. Mechanics of Deformable Solids*, 1973, vol. 7 (In Russian).

[6] Vorovich, I. I., Minakova, N. I., *Journal of Applied Mathematics and Mechanics*, 1973, vol. 37(5), 934 (In Russian).

[7] Vorovich, I. I., *Mathematical Problems of Nonlinear Theory of Shallow Shells*, Nauka, Moscow, 1989.

[8] Pyankov, B. G., Kakurin, A. M., Yudin, A. S., *Proceedings of 4th International Conference on Modern Problems of Mechanics of Solid Media,* vol. 2, North-Caucasus Scientific Center of High School Press, Rostov-on-Don, 1998, p. 134 (In Russian).

[9] Pyankov, B. G., Tischenko, S. G., Tsaryuk, L. B., *Method for Determination of Flapping Membranes Actuation Pressure*, USSR Inventor Certificate No. 470724, 1975 (In Russian).

[10] Bocharov, Yu. N., Pyankov, B. G., Tischenko, S. G., Tsaryuk, L. B., In: *Increase of Operation Safety of Refineries and Petrochemical Plants with Aid of Protective Membranes*, Neftehim Press, Moscow, 1979, vol. 19, p. 118 (In Russian).

[11] Zatsarinny, V. P., Pyankov, B. G., Rozhkov, E. V., *Method of Control of Loaded Thin-Wall Structures*, USSR Inventor Certificate No. 725019, 1980 (In Russian).

[12] Pyankov, B. G., *Acoustic-Emission Method of Structure Control*, USSR Inventor Certificate No. 1089512, 1984 (In Russian).

[13] Pyankov, B. G., Kakurin, A. M., *Device for Diagnostics of Stress-Strain State*, USSR Inventor Certificate No. 1610265, 1990 (In Russian).

[14] Yudin, A. S., Fedorov, V. V., Pyankov, B. G., *Flapping Membrane*, USSR Inventor Certificate No. 388160, 1973 (In Russian).

[15] Zatsarinny, V. P., Pyankov, B. G., Yudin, A. S., *Method of **Fabrication of Flapping** Membranes,* USSR Inventor Certificate No. 514144, 1976 (In Russian).

[16] Pyankov, B. G., Tsaryuk, L. B., Yudin, A. S., *Protection Valve*, USSR Inventor Certificate No. 399674, 1973 (In Russian).

[17] Kakurin, A. M., Pyankov, B. G., Tsaryuk, L. B., *Protection Valve*, USSR Inventor Certificate No. 815398, 1981 (In Russian).

[18] Olhovsky, N. E., Pyankov, B. G., Tsaryuk, L. B., *Device for Punching Pressure of Working Environment*, USSR Inventor Certificate No. 576143, 1977 (In Russian).

[19] Pyankov, B. G., Kakurin, A. M., Chomutov, A. S., Olhovsky, N. E., *Safety Device*, USSR Inventor Certificate No. 945562, 1982 (In Russian).

[20] Pyankov, B. G., In: *Increase of Operation Safety of Refineries and Petrochemical Plants with Aid of Protective Membranes*, Neftehim Press, Moscow, 1979, vol. 19, p. 39 (In Russian).

[21] Pyankov, B. G., *Proceedings of 12th All-Union Conference on Theory of Shells and Plates,* vol. 3, ESU Press, Erevan, 1980, p. 174 (In Russian).

[22] Pyankov, B. G., *J. Fire-Explosion-Safety*, 2001, No. 1, 12 (In Russian).

[23] Pyankov, B. G., Kakurin, A. M., Yudin, A. S., *Proceedings of 10th International Conference on Modern Problems of Mechanics of Solid Medium,* North-Caucasus Scientific Center of High School Press, Rostov-on-Don, 1998, vol. 2, p. 129 (In Russian).

[24] Pyankov, B. G., Kakurin, A. M., Yudin, A. S., Izvestiya of North-Caucasus Scientific Center of High School. *Natural Sciences*, 1999, No. 3, 22 (In Russian).

[25] Belikov, N. V., Zanimonets, Yu. M., Yudin, A. S., *Proceedings of 33rd School-Seminar on Mathematical Modeling in Problem of Rational Nature Management*, North-Caucasus Scientific Center of High School Press, Rostov-on-Don, 2005, p. 133 (In Russian).

[26] Belikov, N. V., Zanimonets, Yu. M., Yudin, A. S., *Proceedings of 9th International Scientific-Practical Conference on Modern Problems of Mechanics of Solid Media,* vol. 1, CVVR Press, Rostov-on-Don, 2005, p. 36 (In Russian).

[27] Belikov, N. V., Zanimonets, Yu. M., Pavlov, V. V., *Proceedings of 5th International Conference on Research, Development and Application of High Technologies in Industry*, St. Petersburg, 2008, vol. 12, p. 402, (In Russian).

[28] Srubstchik, L. S., *Buckling and Post-buckling Behavior of Shells*, Rostov University Press, Rostov-on-Don, 1981 (In Russian).

[29] Yudin, A. S., *Izvestiya of Academy of Sciences of the USSR, Mechanics of Solids*, 1974, No. 3, 122 (In Russian).

[30] Yudin, A. S., In: *Physical Mathematical Investigations*, Rostov University Press, Rostov-on-Don, 1972, p. 42, (In Russian).

[31] Yudin, A. S., In: *Mathematical Analysis and its Applications*, Rostov University Press, Rostov-on-Don, 1974, vol. 6, p. 148, (In Russian).

[32] Zipalova, V. F., Yudin, A. S., *Proceedings of 10th All-Union Conference on Theory of Shells and Plates,* Tbilisi, 1975, vol. 1, p. 610 (In Russian).

[33] Yudin, A. S., In: *Calculation of Shells and Plates*, RISI Press, Rostov-on-Don, 1978, p. 170 (In Russian).

[34] Kokorina, G. N., Yudin, A. S., Sizova, T. P., In: *Increase of Operation Safety of Refineries and Petrochemical Plants with Aid of Protective Membranes*, Neftehim Press, Moscow, 1979, vol. 19, p. 27 (In Russian).

[35] Pogorelov, A. V., *Geometric Theory of Shells Stability*, Nauka, Moscow, 1966 (In Russian).

[36] Hutchinson, J. W., Koiter, W. T., *Applied Mechanics Reviews*, 1970, No. 12, 1353.

[37] Akkas, N., Bauld, N. R. (Jr), *J. Eng. Mech. Div., Proc. Amer. Soc. Civ. Eng.*, 1971, vol. 97(3), 727.

[38] Petrenko, M. P., In: *Theory of Plates and Shells*, Nauka, Moscow, 1971, p. 220 (In Russian).

[39] Timashev, S. A., *Stability of Reinforced Shells*, Stroiizdat, Moscow, 1974 (In Russian).

[40] Karmishin, A. V., Lyaskovets, V. A., Myachenkov, V. I., Frolov, A. N., *Statics and Dynamics of Thin-Walled Shell Structures*, Mashinostroenie, Moscow, 1975 (In Russian).
[41] Valishvili, N. V., *Computing Methods for Shells of Revolution*, Mashinostroenie, Moscow, 1975 (In Russian).
[42] Grigolyuk, E. I., Kabanov, V. V., *Stability of Shells*, Nauka, Moscow, 1978 (In Russian).
[43] Grachev, O. A., *Applied Mechanics* (Ukraine), 1985, vol. 21(1), 53.
[44] Krenske, M. A., Kiernan, T. J., *AIAA Journal*, 1963, vol. 1(12), 2855.
[45] Yudin, A. S., In: *Modern Problems of Mechanics of Solid Medium*, Kniga, Rostov-on-Don, 1995, p. 210 (In Russian).
[46] Kamishev, V. V., Sizova, T. P., Surkin, R. G., Chernovisov, G. N. In: *Increase of Operation Safety of Refineries and Petrochemical Plants with Aid of Protective Membranes*, Neftehim Press, Moscow, 1979, vol. 19, p. 82 (In Russian).
[47] Yudin, A. S., Shepeleva, V. G., *Proceedings of 2nd International Conference on Modern Problems of Mechanics of Solid Medium,* Kniga, Rostov-on-Don, 2006, vol. 3, p. 156 (In Russian).
[48] Yudin, A. S., Izvestiya of North-Caucasus Scientific Center of High School. *Natural. Sciences*, 1978, No. 3, p. 34 (In Russian).
[49] Yudin, S. A., Yudin, A. S., *Proceedings of International Conference on Models and Algorithms for Simulation of Physical-Chemical Processes*, TSPI Press, Taganrog, 2006, p. 212 (In Russian).
[50] Yudin, S. A., *Proceedings of Post-graduate Students and Competitors of Rostov State University*, TerraPrint Press, Rostov-on-Don, 2006, vol. 12, p. 38 (In Russian).
[51] Yudin, A. S., Yudin, S. A., *Proceedings of 10th International Conference on Modern Problems of Mechanics of Solid Medium,* CVVR Press, Rostov-on-Don, 2006, vol. 1, p. 290 (In Russian).
[52] Yudin, A. S., Yudin, S. A., *Proceedings of 10th International Conference on Modern Problems of Mechanics of Solid Medium,* CVVR Press, Rostov-on-Don, 2006, vol. 1, p. 295, (In Russian).
[53] Yudin, A. S., Yudin, S. A., *Proceedings of 11th International Conference on Modern Problems of Mechanics of Solid Medium,* CVVR Press, Rostov-on-Don, 2007, vol. 1, p. 255, (In Russian).
[54] Yudin, A. S., Yudin, S. A., *Proceedings of International Conference on Models and Algorithms for Simulation of Physical-Chemical Processes*, TSPI Press, Taganrog, 2008, p. 86 (In Russian).

In: Advanced Materials - Studies and Applications
Editors: I.A. Parinov, S.-H. Chang, S.Theerakulpisut
ISBN: 978-1-63463-749-7

Chapter 30

RIGIDITY OF NOVIKOV GEARING TEETH

***V. I. Korotkin**[*,1], **D. A. Gazzaev**[1] **and E. V. Kovalenko**[2]*

[1]Vorovich Mathematics, Mechanics and Computer Sciences Institute,
Southern Federal University, Rostov-on-Don, Russia
[2]Financial Institute under Government of Russian Federation, Moscow, Russia

ABSTRACT

This chapter is devoted to modeling of 3D stress-strain state of solid with a complex shape on an example of teeth of the Novikov gearing with a basic rack profile, determined on base of the Russian Standard “GOST 30224-96”. This modeling is performed by using the finite-element software ANSYS. The derived engineering formulae define displacements in a contact area of the teeth, which are subject to a loading. Moreover, we define the teeth rigidity, required to study a multipoint Novikov engagement.

1. INTRODUCTION

Usually, axial overlapping factor of the designed Novikov gearings with double-point contact engagement is greater or equal to one. In this case, a minimum number of contact points (actual contact areas), participating in any time moment in the engagement, reaches as minimum two; a number of the points in various phases of the engagement exceeds the minimum number. By increasing the axial overlapping factor, a number of contact areas increases [1], therefore the Novikov gearing be a multipoint engagement, in principle. It is quite natural, that distributions of the loads and stresses are non-uniform in the contact areas. Different positions of the contact areas on tooth in respect to gear faces in various phases of the engagement explains these distributions; obviously, a distance from the gear faces causes different rigidity of teeth. Moreover, the effects of non-uniformity arise due to the technological errors of manufacture and assembly of gear wheel, etc.

[*] korotkin@math.rsu.ru.

In some methods [2, 3], to evaluate stress state of the teeth, first uniform distributions of loads and stresses are considered between minimum quantity of contact points (centers of contact areas). Then, the authors introduce, so-called, factor of non-uniform distribution of load along length of the contact lines. Russian Standard [4] defines this factor, intended for calculation of the involute gearing with theoretically linear contact.

Of course, it is difficult to agree with this approach, because one ignores completely the specific features of the Novikov gearing, operating only with axial overlapping of teeth at their theoretical point but no linear contact.

There are publications, which consider rigidity of the Novikov gearing teeth with different viewpoints. For example, paper [5] gives some theoretical background and [6] presents results of model experiments on determination of the rigidity, but there are no estimates of influence of gear faces, which can be very significant. An attempt to consider this effect [1] on base of restricted model experiment [7] gives an approximate estimate.

This chapter is devoted to definition elastic displacements during processes of the teeth contact and bending-shear, and determines rigidity of the Novikov gear teeth in different phases of the engagement, taking into account necessity to investigate stress-strain state within a multipoint engagement. We use method, based on finite-element modeling, for solution of 3D mixed problem with preliminary unknown shape and sizes of the contact area.

2. MODEL RESULTS

As the studied model, we consider tooth of an equivalent spur wheel, fabricated on a base of rack profile in accordance with the Russian Standard "GOST 30224-96" [8]. The base of the rack profile has restraint [9], which replaces a dropped part of the wheel rim. Elastic indenter (stamp) acts on the tooth addendum under certain calculated angle in respect to normal force F_n, replacing the tooth dedendum of the pair wheel. Such a replacement simplifies the model and, at the same time, does not decrease an accuracy of solution, since the tooth dedendum together with indenter are subject to only contact deformations, and bending strains are absent practically due to the extremely small arm of the force F_n, influencing on the tooth dedendum.

The working surface of elastic indenter (Figure 1) consists of points of the hyperbolic type, having at the contact point a profile radius ρ_f of curvature (concave part) and reduced main longitudinal radius ρ_β of curvature (convex part). Figure 2 depicts contact of indenter 1 with investigated tooth 2, and Figure 3 demonstrates a position of the indenter in the different phases of engagement, characterized by movement of contact point K from one gear face to other on distance $\pm S$, calculated from middle ($S = 0$) of the gear rim.

Simulation was performed with help of the finite-element software ANSYS. Computer programs on the macro-language APDL for the software ANSYS of Version 11.0 (and above) allowed us to solve above problems with introduction of parametric input data. In the simulation, we chose local zones of a regular partition in near surface layers of both contacting bodies. These local zones in the form of corresponding equidistant surfaces deepened into bodies on some distances. To construct the tooth surface, we used a spline-interpolation of second order. The finite-element model used quadratic elements SOLID95 and software for contacting pair CONTA174 - TARGE170. Boundary conditions for tooth 2 were similar to those in the problem on action of concentrated force on the gearing tooth [9].

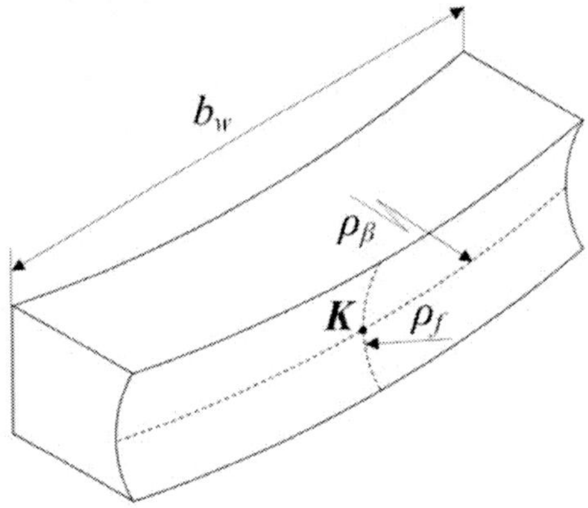

Figure 1. Common look of elastic indenter; point *K* is one of the theoretical contact points on the surface of hyperbolic type.

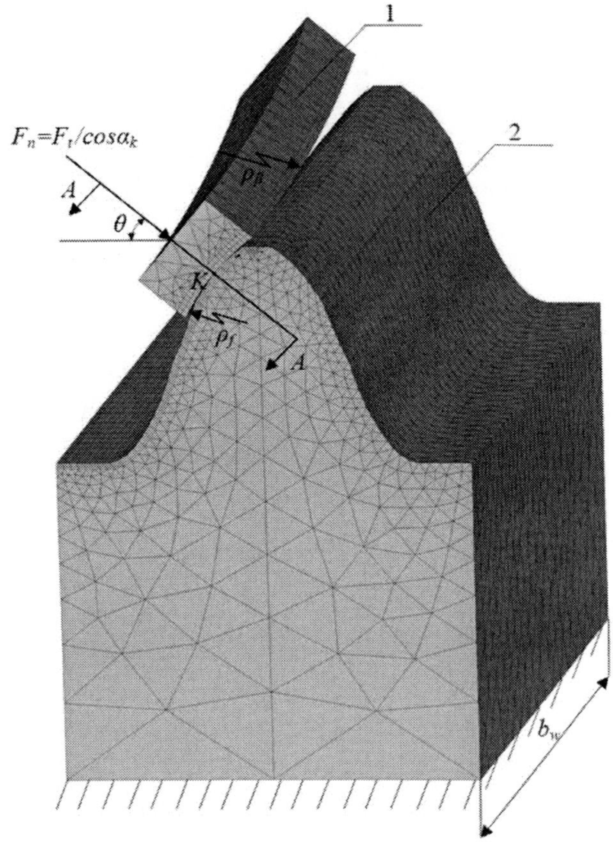

Figure 2. Scheme of indenter 1, contacting with investigated tooth 2.

We stated the following boundary conditions for indenter 1: concentrated force acted on upper surface of the indenter, which moved in its parallel direction strictly along the normal force, i.e., in perpendicular direction to the tooth surface 2 at point K.

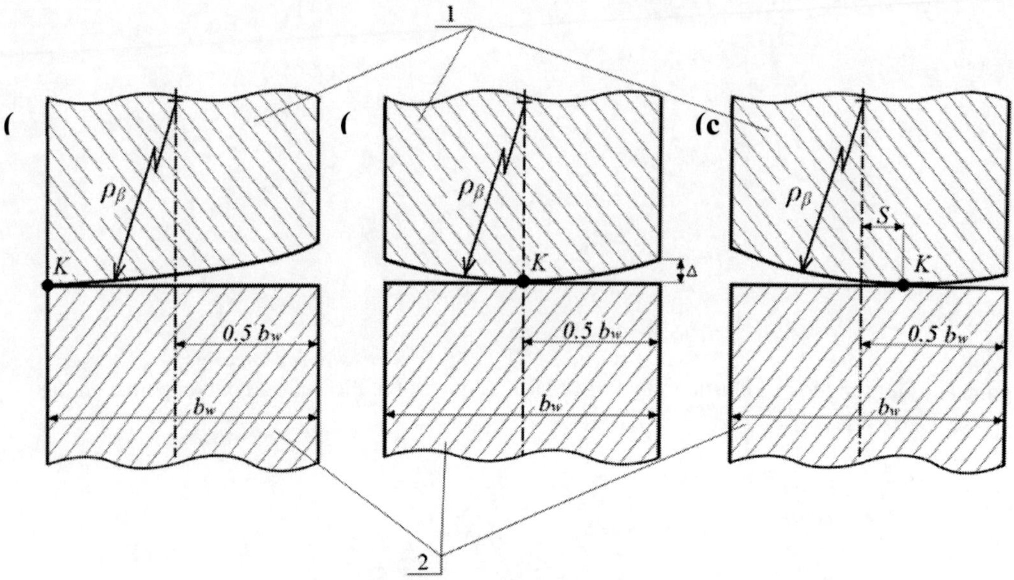

Figure 3. Position of indenter 1 in different phases of engagement (in respect to the cross-section along direction AA, see Figure 2): (a) contact point *K* at the gear face ($S = 0.5b_w$); (b) contact (base) point *K* in middle of the tooth length ($S = 0$); (c) common case of location of contact point *K* ($0 < S < 0.5b_w$).

Note the following circumstance. The long-term tests have proved [1], that teeth of the Novikov gearing (including high-hardened version) run-in sufficiently quickly, and difference $\Delta\rho = \rho_f - \rho_a$ between the profile radii of curvature of the tooth dedendum (ρ_f) and the tooth addendum (ρ_a) for pair teeth significantly decreased. Height of contact area at its characterization by value of b_H, being small semi-axis of a conditional ellipse, becomes equal to the value:

$$b_H = 0.5l = 0.5Z_l K_l l_k, \quad (1)$$

where l_k is the chordal height of active tooth addendum of the basic rack profile, K_l is the factor, taking into account decreasing a height of the active tooth addendum in teeth cutting, Z_l is the run-in factor that depends on hardness of the teeth surfaces [1]. Namely, by taking into account the run-in factor, we should determine the unknown elastic displacements of teeth, which we in the simulation. We obtained $l_k = 0.776m$ [1] (where *m* is the engagement modulus) for the considered below gearing, fabricated on the base of the basic rack profile [8]. Then we calculated dimensionless averaged value of $b_H^* = 0.33$ in accordance with (1), which we used in the computer program. (Hereinafter the parameters, marked with an asterisk, relate to the engagement modulus). All calculations were performed by using the values of elasticity moduli for the teeth material (Young's modulus $E = 2 \cdot 10^5$ MPa and Poisson's ratio $\mu = 0.3$).

The longitudinal extent of the contact area along the tooth influences on elastic displacements at the fixed width b_H. Obviously, the contact area is slightly different from an ellipse, but its longitudinal extent, defined by simulation, confirms results of [5]. This longitudinal size is slightly longer than the value of a_H, being great semi-axis of the Hertzian contact ellipse that is connected with influence of a bending deformation of the tooth. However, the analysis shows that close correlation between half-length of a real contact area and calculated (Hertzian) value of a_H^* allows us to use the last parameter in numerical calculations.

We use the theory of elliptic integrals [10] to define initial data for modeling, taking into account the run-in of teeth at the fixed value of $b_H^* = 0.33$. With this purpose, we define the ellipticity factor $\overline{\beta} = b_H / a_H$, giving the ratio $\psi = a_H / b_w$, where b_w is the tooth length. Then we use the full elliptic integrals K, F of the modulus ν, where $\nu = \arccos \overline{\beta}$, and determine the ratio $C_{\alpha\beta} = \rho_\alpha / \rho_\beta$ of reduced main curvature radius ρ_α of a profile for contacting teeth surfaces to the longitudinal radius ρ_β of curvature [11], and also itself radius ρ_β from the following relations:

$$C_{\alpha\beta} = (K - F)(F\overline{\beta}^{-2} - K), \tag{2}$$

$$a_H = n_e (F_n \rho_\beta / E)^{1/3}, \tag{3}$$

where $n_e = 1.2023\left[(K - F)/(1 - \overline{\beta}^2)\right]^{1/3}$.

Then we define the indenter radius $\rho_f = \rho_a + \Delta\rho$ of curvature, required in the modeling, where $\Delta\rho = (\rho_a)^2 / (\rho_\alpha - \rho_a)$, $\rho_\alpha = C_{\alpha\beta}\rho_\beta$.

A whole displacement W_n of the tooth along normal direction to the contact surfaces (i.e., in the direction of the force F_n action) consists of two components: (i) displacement as a result of deformation at the contact point (compression strain) and (ii) displacement due to the tooth deflection (bending-shear deformation). We defined the displacement W_n in simulation by using changes in coordinates of fixed points.

Table 1 shows the initial parameters, described above, and the values of W_n, obtained in the simulation at disposition of center of the contact area in middle of the tooth length, that is, in absence of influence of the gear faces. These values correspond to magnitude of $S = 0$ and hereinafter we call them the base values. We choose the following parameters of gearing: $F_t = F_n \cos\alpha_k$ is the tangential force, α_k is the theoretical pressure angle on the basic rack profile, $z_v = z / \cos\beta$ is the reduced tooth number of an equivalent spur gear, z is the number of teeth of helical gear, β is the helix angle of tooth, x is the shift of the basic rack profile at cutting the teeth. Then we compare the values of W_n with the values of W_n^0, calculated by using formula (4) (see below) and presented in Table 1.

Table 1. Initial parameters and base displacements (at $S^* = 0$) of teeth, obtained by calculation (W_n^0) and simulation (W_n)

$\psi = a_H^*/b_w^*$	a_H^*	$\bar{\beta}$	ρ_β^*	ρ_α^*	$\Delta\rho^*$	ρ_f^*	W_n^0, μm	W_n, μm
0.05	1	0.33	58.9	10.947	0.1923	1.550	14.191	14.296
0.10	2	0.165	338.6	21.031	0.0937	1.452	10.529	10.446
0.15	3	0.11	973.5	31.198	0.0618	1.420	8.093	8.157
0.20	4	0.0825	2,084	41.402	0.0461	1.404	6.567	6.640
0.25	5	0.066	3,784	51.622	0.0367	1.395	5.545	5.586
$F_t = 1$ kN; $m = 1$ mm; $z_v = 18$; $x^* = 0$; $b_H^* = 0.33$; $b_w^* = 20$								

The performed studies have shown that value of W_n depended not only on parameter ψ, but on values of $z_v, \Delta\rho, x, \rho_\beta, F_t, m$. In order to define these effects, we conducted simulation in a wide range of variation of these parameters. The results of the modeling are present below in tables and plots.

Table 2 contains data, which show that the displacement W_n decreases slightly at increasing the number z_v of the wheel teeth, and Table 3 shows that change of the curvature radii difference $\Delta\rho$ does not influence practically on the values of W_n.

Table 2. Simulation estimates of influence of the teeth number on wheel displacement

z_v	9	12	18	40	100
W_n, μm	13.160	12.122	11.251	10.457	10.101
$F_t = 1$ kN; $\rho_\beta^* = 250$; $m = 1$ mm; $x^* = 0$; $\Delta\rho^* = 0.2$; $b_w^* = 20$; $S = 0$					

Table 3. Simulation estimates of influence of the curvature radii difference $\Delta\rho^*$ on the displacement W_n

z_v	ρ_β^*	F_t, N	$\Delta\rho^*$	W_n, μm
9	250	400	0.2	6.266
			0.0565	6.071
18	1,000	2,000	0.2	13.584
			0.0963	13.453
40	1,000	500	0.2	4.568
			0.0398	4.440
100	250	1,000	0.2	10.101
			0.1094	9.964
$m = 1$ mm; $x^* = 0$; $S = 0$; $b_w^* = 20$				

Figure 4 shows decreasing the displacement W_n in dependence on increasing the modulus m, and Figure 5 presents features of growth of the displacement W_n with increasing the force F_t at different values of the longitudinal radius ρ_β of curvature. It is seen that the growth of

W_n occurs slower than increase of F_t, since the longitudinal size a_H of the contact area increases simultaneously, constraining the growth of W_n. Then an increase of the longitudinal radius ρ_β of curvature leads to a decrease of W_n at other equal conditions. At the same time, influence of shift x is very insignificant and can be accounted with the help of a special correction (see below a comment to formula (4)).

The processed results of simulation allowed one to propose the following dependence for calculation of the total base displacement W_n^0:

$$W_n^0 = 0.359 F_t^{0.77} m^{-0.585} (\rho_\beta^* + 100)^{-0.294} (0.111 z_v)^\gamma , \tag{4}$$

where $\gamma = -0.294(0.111 z_v)^{-0.423}$.

Comment: The result for W_n^0 must be multiplied by 0.96 at $x^* > 0$; at the same time, it must be multiplied by 1.07 at $x^* < 0$.

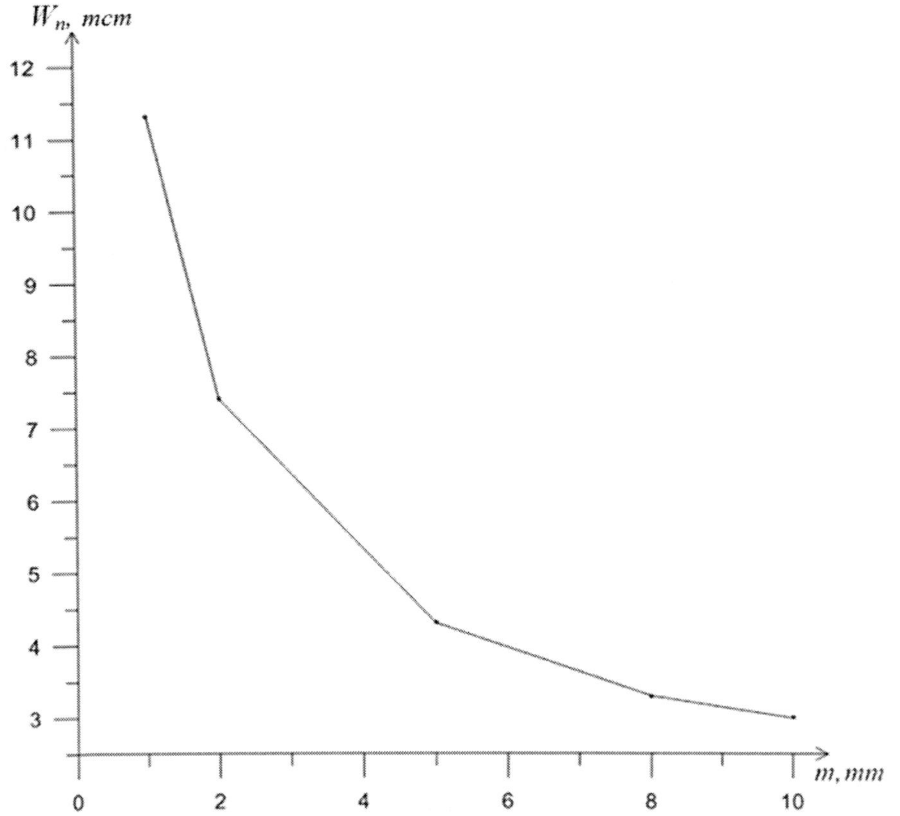

Figure 4. Dependence of base displacement W_n on modulus m for gearing with the parameters: $z_v = 18; x^* = 0; \rho_\beta^* = 250; F_t = 1\,\text{kN}; \Delta\rho^* = 0.2$.

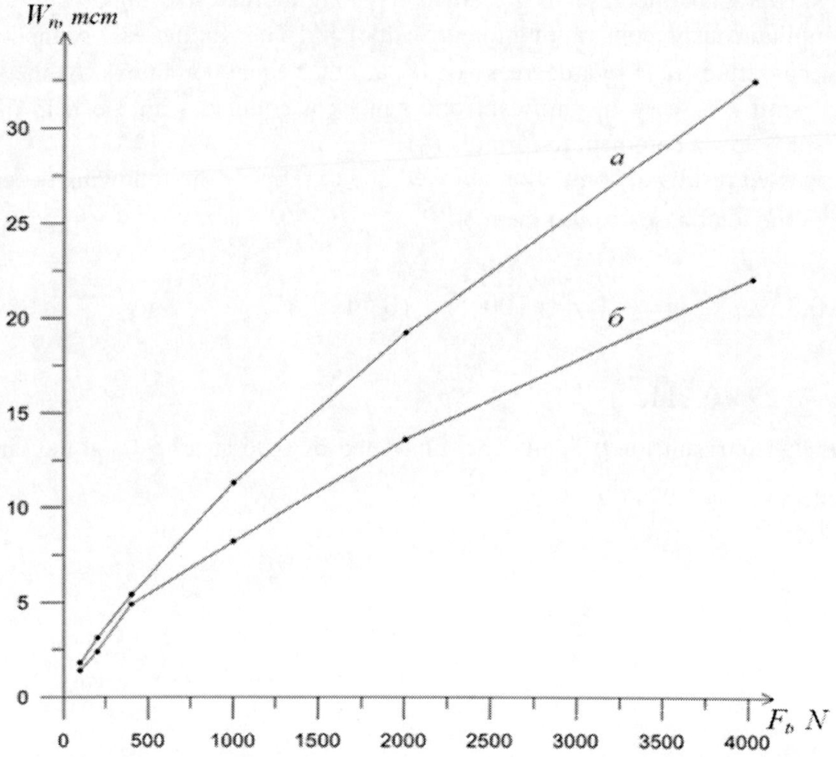

Figure 5. Dependences of base displacement W_n on force F_t for gearing with the parameters: $m = 1\,\text{mm}; z = 18; \bar{x}^* = 0; \Delta\rho^* = 0.2$ for (a) $\rho_\beta^* = 250$ and (b) $\rho_\beta^* = 1{,}000$.

The total base displacement W_t^0 along the tangential direction (in the face plane) is equal to

$$W_t^0 = W_n^0 (\cos\alpha_k \cos\beta)^{-1}. \tag{5}$$

We define the base rigidity along the tangential direction as

$$C_t^0 = F_t / W_t^0 = 2.786 F_t^{0.23} m^{0.585} (\rho_\beta^* + 100)^{0.294} (0.111 z_v)^{\gamma'} \cos\alpha_k \cos\beta, \tag{6}$$

where $\gamma' = -\gamma = 0.294(0.111 z_v)^{-0.423}$.

Comment: The result for C_t^0 must be multiplied by 1.04 at $x^* > 0$; at the same time, it must be multiplied by 0.93 at $x^* < 0$.

As we can see, the tooth rigidity increases nonlinearly with increasing the load F_t that coincides with result of [6].

Investigation of a stress state in various phases of engagement, i.e., at a displacement of the contact area along the tooth, shows that the stress state depends not only on position of the contact point (area) S along the tooth length in respect to the middle (base) point, but also on the ratio $\psi = a_H / b_w$.

Simulation in various stages of engagement was performed in the range of $S^* = 0 - 0.5b_w^*$, i.e., taking into account an effect of face engagement, and we added index "s" in designations of displacements (which were called by "phase"), determined at $S \neq 0$.

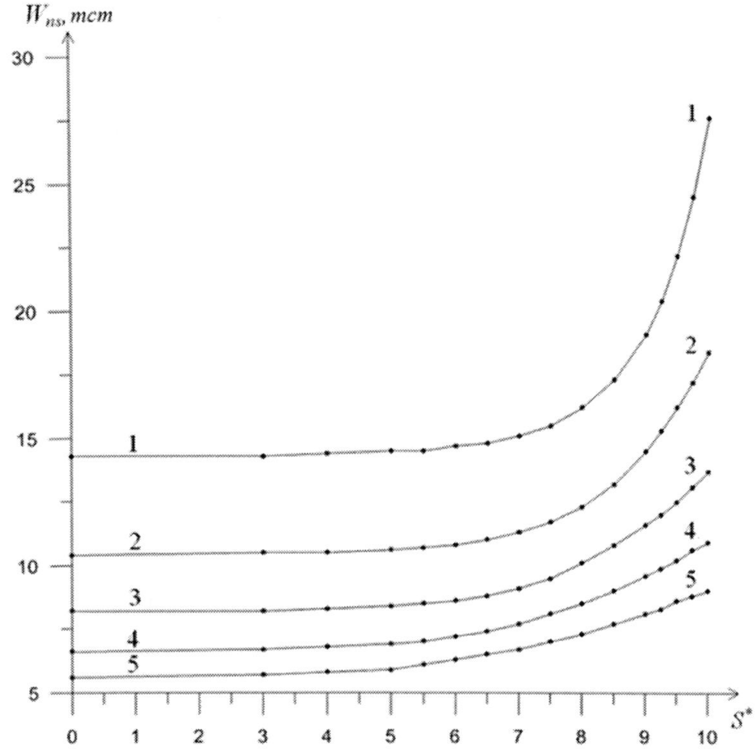

Figure 6. Dependences of stage displacement W_{ns} on position S^* of the contact area for gearing parameters: $F_t = 1$ kN; $m = 1$ mm; $z_v = 18$; $x^* = 0$; $b_w^* = 20$; 1 – $\psi = 0.05$; 2 – $\psi = 0.1$; 3 – $\psi = 0.15$; 4 – $\psi = 0.20$; 5 – $\psi = 0.25$.

Figure 6 shows plots of dependence of the phase normal displacement W_{ns} on position S of the contact area center for various values of ψ, and Table 4 presents the calculated ratios $K_W = W_{ns} / W_n$, where the position of the contact area is determined by the relative value of $t_W = S / b_w$.

The total phase displacement (for $S^* \neq 0$) along the tangential direction is defined as

$$W_{ts} = W_t^0 K_W, \tag{7}$$

and the phase rigidity in the tangential direction is equal to

$$C_{ts} = C_t^0 K_W^{-1}. \tag{8}$$

Table 4. Values of ratio $K_W = W_{ns} / W_n$ in various phases of engagement in dependence on ψ and t_W

	ψ= 0.05	ψ= 0.10	ψ= 0.15	ψ= 0.20	ψ= 0.25
t_W	K_W	K_W	K_W	K_W	K_W
0.500	1.93	1.76	1.68	1.64	1.62
0.488	1.72	1.65	1.61	1.59	1.57
0.475	1.55	1.55	1.54	1.53	1.53
0.463	1.43	1.46	1.48	1.49	1.49
0.450	1.33	1.39	1.42	1.44	1.45
0.425	1.21	1.27	1.32	1.35	1.38
0.400	1.13	1.18	1.23	1.28	1.32
0.375	1.09	1.12	1.17	1.22	1.26
0.350	1.06	1.08	1.12	1.16	1.21
0.325	1.04	1.05	1.08	1.12	1.16
0.300	1.03	1.04	1.06	1.09	1.12
0.275	1.02	1.03	1.04	1.06	1.09
0.250	1.01	1.02	1.03	1.04	1.07
0.200	1.01	1.01	1.01	1.02	1.03
0.150	1.00	1.00	1.00	1.,01	1.01
0	1.00	1.00	1.00	1.00	1.00

The obtained formulae are the ground for investigations of multipoint Novikov gearing.

ACKNOWLEDGMENTS

This work was partially supported by grant No. 13-08-00386 of the Russian Foundation for Basic Research.

REFERENCES

[1] Korotkin, V. I.; Onishkov, N. P.; Kharitonov, Yu. D. *Novikov Gearing: Achievements and Development.* Nova Science Publishers, New York, 2011, pp. 1 – 249.

[2] *Novikov Gearing with Tooth Surface Hardness HB≥350. Strength Calculation. Methodical Recommendations MP 221-86.* VNIINMASH, Moscow, 1987, pp. 1 – 86 (In Russian).

[3] Yakovlev, A. S. *Vestnik Mashinostroeniya*, 1984. *No. 6*, 18-20 (In Russian).

[4] State Standard GOST 21354-87. *Cylindrical Involute Gearing. Strength Calculation.* Standards Press, Moscow, 1988, pp. 1 – 125 (In Russian).

[5] Yakovlev, A. S.; Muzalevskaya, G. A. *Izvestiya Vuzov. Mashinostroenie*, 1985, *No. 2.* 30 – 34 (In Russian).

[6] Yakovlev, A. S.; Muzalevskaya, G. A. *Izvestiya Vuzov. Mashinostroenie*, 1985, *No. 4*, 23-27 (In Russian).

[7] Zablonsky, K. I. *Gearing. Load Distribution in Engagement.* Tekhnika, Kiev, 1977, pp. 1 – 208 (In Russian).

[8] Russian Standard "GOST 30224-96". *Novikov Cylindrical Gearing with Tooth Surface Hardness no Lesser than 35 $HRC_Э$. Basic Rack Profile.* Interstate Council on Standardization, Metrology and Certification. Minsk, 1996, pp. 1 – 5 (In Russian).

[9] Korotkin, V. I.; Gazzaev, D. A. *Vestnik Mashinostroeniya*, 2014, *No. 1*, 14 - 19 (In Russian).

[10] Korn, G.; Korn, T. *Handbook of Mathematics.* Nauka, Moscow, 1970, pp. 1 – 720 (In Russian).

[11] Kovalsky, B. S. *Calculation of Parts for Local Compression.* KhVKIU, Kharkov, 1967, pp. 1 – 233 (in Russian).

INDEX

#

A

B

C

D

E

F

G

H

I

J

K

L

M

N

O

P

Q

R

S

T

U